AF411443

Relative Sea-Level Changes and their Impact on Coastal Zones

Relative Sea-Level Changes and their Impact on Coastal Zones

Editors

Pietro Aucelli
Giuseppe Mastronuzzi
Gaia Mattei

MDPI • Basel • Beijing • Wuhan • Barcelona • Belgrade • Manchester • Tokyo • Cluj • Tianjin

Editors

Pietro Aucelli
Department of Science and
Technology
Parthenope University
of Naples
Naples
Italy

Giuseppe Mastronuzzi
Department of Earth and
Environmental Sciences
University of Bari
Bari
Italy

Gaia Mattei
Department of Science and
Technology
Parthenope University
of Naples
Naples
Italy

Editorial Office
MDPI
St. Alban-Anlage 66
4052 Basel, Switzerland

This is a reprint of articles from the Special Issue published online in the open access journal *Water* (ISSN 2073-4441) (available at: www.mdpi.com/journal/water/special_issues/Sea_Level_Changes).

For citation purposes, cite each article independently as indicated on the article page online and as indicated below:

LastName, A.A.; LastName, B.B.; LastName, C.C. Article Title. *Journal Name* **Year**, *Volume Number*, Page Range.

ISBN 978-3-0365-4634-6 (Hbk)
ISBN 978-3-0365-4633-9 (PDF)

Cover image courtesy of Gaia Mattei

Contents

Editorial

Relative Sea-Level Changes and Their Impact on Coastal Zones: Past and Future Scenarios from Cases Studies around the World

Pietro P. C. Aucelli [1], Giuseppe Mastronuzzi [2] and Gaia Mattei [1,*]

1   Dipartimento di Scienze e Tecnologie, Università degli Studi Parthenope di Napoli, 80133 Naples, Italy; pietro.aucelli@uniparthenope.it
2   Dipartimento di Scienze della Terra e Geoambientali, Università degli Studi di Bari Aldo Moro, 70121 Bari, Italy; giuseppeantonio.mastronuzzi@uniba.it
*   Correspondence: gaia.mattei@uniparthenope.it; Tel.: +39-081-547-66-35

Citation: Aucelli, P.P.C.; Mastronuzzi, G.; Mattei, G. Relative Sea-Level Changes and Their Impact on Coastal Zones: Past and Future Scenarios from Cases Studies around the World. *Water* 2022, 14, 1822. https://doi.org/10.3390/w14111822

Received: 16 December 2021
Accepted: 3 June 2022
Published: 6 June 2022

Publisher's Note: MDPI stays neutral with regard to jurisdictional claims in published maps and institutional affiliations.

In recent decades, the study of sea-level changes as recorded along entire coastlines around the world has been a primary scientific focus in climate change studies; it allows the exploration of past landscape evolution, geomorphological processes, human impact, and system response to develop future perspectives.

The relative local sea level is the sum of global, regional, and local processes, i.e., subsidence, isostasy, tectonics, and human activities; all these processes are spatially and temporally variable and cause complex sea-level changes at both regional and local scales, at long-/short-term scale or during a proximal time such as tsunami and storm surge. A multidisciplinary approach addressing paleo-sea level reconstructions at different scales is essential to understand the role of the underlying natural and anthropogenic forces in landscape evolution, as well as to discover past human adaptions to natural modifications of the landscape to develop future ones.

The classical approach to this type of study has thus far been based on the geological–geomorphological study of a coastal area aimed at the detection and measuring of erosive/depositional sea-level markers, interpreted as imprints of past sea levels. Furthermore, in recent decades, traditional surveys and analysis were progressively integrated by (1) digital high-resolution field survey methods, (2) new methods of post-processing analysis, and (3) improvement in geochemical and geoarchaeological methods of age attribution. Geoacoustic and remote sensing methods—including UAV- and UTV-mounted systems—can provide the high-resolution mapping of wide coastal areas and seabed morphologies. On the other hand, AMS dating techniques are increasingly precise owing to the need for small samples, together with daily advanced sedimentological, palynological, and paleontological analysis techniques, allowing the interface of their results with typical archaeological findings to highlight every change in coastal environments in the relation to sea-level history. Moreover, these methods are more precise techniques to evaluate the local vertical ground movement (VGM) that are a primary force in sea-level oscillations in the medium/short term. To be sure, multidisciplinary studies can be considered the most appropriate modern approach to paleo-landscape and paleo-sea level reconstructions.

These advances reinforce the increasingly accurate techniques for the survey of coastal sectors by acquiring a large volume of four-dimensional—i.e., 3D points and time—data that, in the specific case of geoarchaeological sites, have also a documental value in terms of cultural and natural heritage.

Detailed multidisciplinary survey and consequent analysis provide multiscale/temporal datasets about landscape reconstruction, its evolution, and dynamics over the last millennia. At the same time, investigating these regional and local patterns is urgently needed to reconstruct possible scenarios of the relative impact of a rise in sea levels and to prepare the adaptation of coastal communities threatened by future implications of climate change, defining hazard and vulnerability as components of the risk of coastal inundation and submersion.

This Special Issue aims to collect contributions addressed to discuss methodological and multidisciplinary approaches to studying relative changes in sea level and the related coastal modification, aiming to reconstruct possible future trends and evaluate the impact of such changes on coastal zones and their amount [1–12].

This collection deals with a broad range of contributions, in which different methodologies were applied to deepen this knowledge. Even if the Special Issue includes study cases from extra-Mediterranean sectors, such as North America, Baltic, and Iberian Atlantic coast, special consideration was given to the Mediterranean area, with cases from Croatian, Italian, and Iberian coasts, considering that this semi-enclosed basin is relevant in the framework of the ongoing climate change studies. In fact, it represents a transitional zone between mid-latitude and subtropical climate zones and, consequently, is affected by present and future global warming trends. Additionally, according to recent IPCC reports, changes in land use, increasing pollution, and declining biodiversity are exacerbating this warming process, making this area a climate hotspot.

The manuscripts amassed in this Issue focused on the main effects related to past and future trends related to climate change, both in terms of coastal changes and environmental risks.

They cover three fundamental topics, highlighting different aspects through an original angle of investigation: (1) late quaternary coastal changes in coastal plain and rocky sectors due to RSL variations, often exacerbated by vertical ground movements of tectonic and volcano-tectonic origin; (2) recent coastal changes due to short-term variations in sea levels, also related to extreme events, and their effects in terms of human adaptations; (3) future coastal changes and flooding scenarios due to ongoing climate change and subsequent accelerated trends in a rise in sea levels.

The first group of manuscripts is focused on four significant Italian case studies of paleo-landscape and morpho-evolutive reconstructions of coastal plains located in different morpho-dynamic contexts by using direct and indirect methods. In particular, Corrado et al. [1] analyzed a robust dataset of stratigraphic records to evaluate the tectonic subsidence rate and the morpho-evolutionary path of the Volturno plain before and after the Campania Ignimbrite super-eruption, reconstructing the interactions among eustatic changes, tectonics, and sedimentary input variations. In comparison, D'Orefice et al. [2] applied the same methodology to the study of the environmental and morphological evolution of the Burano paleo-lagoon during the last 8000 years to reconstruct a site-specific sea-level curve.

As regards the indirect methods applied to long-term paleoenvironmental reconstructions, De Santis et al. [3] analyzed a high-precision seismic dataset inferred from submerged boreholes to evaluate the Holocene morpho-evolution of the coastal barrier systems of Ofanto and Carapelle-Cervaro Valleys concerning the post-glacial sea-level rise and paleoclimatic effects. Finally, a multidisciplinary approach comprising photogrammetric and direct surveys was applied by Aucelli et al. [4] to outline the main coastal changes due to ground movements of volcano-tectonic origin and consequent human interventions during the Roman period of a well-known military and commercial hub (Portus Julius) built along the Gulf of Pozzuoli coasts, located in Campi Flegrei Caldera.

Moving from past to recent coastal changes, the second set of articles document advances in the measurement of coastal effects due to storm or atmospheric extreme events. In particular, Ferrando et al. [5] integrated several remote sensing techniques, such as photogrammetry from unmanned aerial vehicles (UAV), mobile laser scanners (MLS), and multibeam echo-sounders (MBES), to evaluate the strong coastal effects at Portosole Marina (Sanremo, Western Liguria) consequent to the storm Vaia that occurred in October 2018, well-known for catastrophically hitting the Tyrrhenian coasts. Scicchitano et al. [6] used the same terrestrial methods to reconstruct immersive virtual scenarios to geometrically analyze boulder displacements recorded in the first video account of coastal boulder displacements recorded during the impact of Medicane Zorbas on southeastern Sicily. Buonocore et al. [7] considered the combined effects of the global increase in sea levels and regional atmospheric perturbations that tripled the sea-level variations in the

Gulf of Naples over the last 20 years, by analyzing a tide-gauge dataset collected on the Island of Ischia. Special focus was placed on the Aqua Alta episodes whose frequency has increased in recent decades, with dramatic effect on coastal areas where intense social and economic activity occurs.

The third group of articles—contributions by Antonioli et al. [8], Scardino et al. [9], Abadie et al. [10], White et al. [11], and Vecchio et al. [12]—mainly addressed future coastal scenarios related to the present acceleration in sea-level rise in different morphological contexts around the world. This topical issue was faced at the Mediterranean scale by analyzing LiDAR and Copernicus Earth Observation data to provide estimates of potential marine submersion in 2100 for 16 small-sized coastal plains located in the Italian Peninsula and four Mediterranean countries (i.e., France, Spain, Tunisia, and, Cyprus), all characterized by different geological, tectonic and morphological features, as revealed by Antonioli et al. [8]. However, a high-precision multitemporal mathematical model of coastal submersion was implemented by Scardino et al. [9], to evaluate the sea-level rise and changes in shorelines at a larger scale in the Gulf of Taranto.

Finally, scenarios regarding sea-level variability were reconstructed in the Mediterranean scale by Vecchio et al. [12], and extra-Mediterranean scale, i.e., in the North American context presented by White et al. [11], taking into account the IPCC predictions and regional tectonic behaviors. Peculiar is the case study presented by Abadie et al. [10], in which the potential economic damage in 62 Iberian coastal cities from 2020 to 2100 was estimated using relative regional data on sea-level rises under three representative concentration pathways (RCP 8.5, RCP 4.5, and RCP 2.6).

The papers collected in this Special Issue underline, from different angles, the need for direct and indirect measurements of past and recent relative sea-level oscillations to also enhance our knowledge about their effects on future coasts. These different contributions perfectly fit into the international scenario of sea-level studies (cited in the Introduction), demonstrating the importance of collecting historical records of coastal evolution as key to predicting future modifications of littoral areas in a context of global climate change, rise in sea levels, and an increasing number of extreme events.

On the other hand, they emphasized the significance of high-resolution reconstruction, at both regional and local scales. Special emphasis was placed on Holocene coastal modifications, demonstrating the major influence of climatic modifications on changes in coastal landscapes, both in terms of varied rates of sea-level rise and increased intensity of extreme events during this period.

Anthropogenic adaptations to such climatic variability have also been a central focus of this Special Issue, both in terms of the risk of flooding in coastal areas and the costs that such flooding may produce for coastal communities.

Author Contributions: Conceptualization, P.P.C.A., G.M. (Giuseppe Mastronuzzi) and GM. (Gaia Mattei); methodology, P.P.C.A., G.M. (Giuseppe Mastronuzzi) and GM. (Gaia Mattei); validation, P.P.C.A., G.M. (Giuseppe Mastronuzzi) and GM. (Gaia Mattei); formal analysis, P.P.C.A., G.M. (Giuseppe Mastronuzzi) and GM. (Gaia Mattei); resources, P.P.C.A., G.M. (Giuseppe Mastronuzzi) and GM. (Gaia Mattei); data curation, P.P.C.A., G.M. (Giuseppe Mastronuzzi) and GM. (Gaia Mattei); writing—original draft preparation, P.P.C.A., G.M. (Giuseppe Mastronuzzi) and GM. (Gaia Mattei); writing—review and editing, P.P.C.A., G.M. (Giuseppe Mastronuzzi) and GM. (Gaia Mattei); supervision, P.P.C.A., G.M. (Giuseppe Mastronuzzi) and GM. (Gaia Mattei); project administration, P.P.C.A., G.M. (Giuseppe Mastronuzzi) and GM (Gaia Mattei). All authors have read and agreed to the published version of the manuscript.

Funding: This research received no external funding.

Institutional Review Board Statement: Not applicable.

Informed Consent Statement: Not applicable.

Conflicts of Interest: The authors declare no conflict of interest.

References

1. Corrado, G.; Amodio, S.; Aucelli, P.P.; Pappone, G.; Schiattarella, M. The subsurface geology and landscape evolution of the Volturno coastal plain, Italy: Interplay between tectonics and sea-level changes during the Quaternary. *Water* **2020**, *12*, 3386. [CrossRef]
2. D'Orefice, M.; Bellotti, P.; Bertini, A.; Calderoni, G.; Censi Neri, P.; Di Bella, L.; Fiorenza, D.; Foresi, L.M.; Louvari, M.A.; Rainone, L.; et al. Holocene Evolution of the Burano Paleo-Lagoon (Southern Tuscany, Italy). *Water* **2020**, *12*, 1007. [CrossRef]
3. De Santis, V.; Caldara, M.; Pennetta, L. "Continuous" Backstepping of Holocene Coastal Barrier Systems into Incised Valleys: Insights from the Ofanto and Carapelle-Cervaro Valleys. *Water* **2020**, *12*, 1799. [CrossRef]
4. Aucelli, P.P.C.; Mattei, G.; Caporizzo, C.; Cinque, A.; Troisi, S.; Peluso, F.; Stefanile, M.; Pappone, G. Ancient coastal changes due to ground movements and human interventions in the Roman Portus Julius (Pozzuoli Gulf, Italy): Results from photogrammetric and direct surveys. *Water* **2020**, *12*, 658. [CrossRef]
5. Ferrando, I.; Brandolini, P.; Federici, B.; Lucarelli, A.; Sguerso, D.; Morelli, D.; Corradi, N. Coastal Modification in Relation to Sea Storm Effects: Application of 3D Remote Sensing Survey in Sanremo Marina (Liguria, NW Italy). *Water* **2021**, *13*, 1040. [CrossRef]
6. Scicchitano, G.; Scardino, G.; Tarascio, S.; Monaco, C.; Barracane, G.; Locuratolo, G.; Milella, M.; Piscitelli, A.; Mazza, G.; Mastronuzzi, G. The First Video Witness of Coastal Boulder Displacements Recorded during the Impact of Medicane "Zorbas" on Southeastern Sicily. *Water* **2020**, *12*, 1497. [CrossRef]
7. Buonocore, B.; Cotroneo, Y.; Capozzi, V.; Aulicino, G.; Zambardino, G.; Budillon, G. Sea-Level Variability in the Gulf of Naples and the "Acqua Alta" Episodes in Ischia from Tide-Gauge Observations in the Period 2002–2019. *Water* **2020**, *12*, 2466. [CrossRef]
8. Antonioli, F.; De Falco, G.; Lo Presti, V.; Moretti, L.; Scardino, G.; Anzidei, M.; Bonaldo, D.; Carniel, S.; Leoni, G.; Furlani, S.; et al. Relative Sea-Level Rise and Potential Submersion Risk for 2100 on 16 Coastal Plains of the Mediterranean Sea. *Water* **2020**, *12*, 2173. [CrossRef]
9. Scardino, G.; Sabatier, F.; Scicchitano, G.; Piscitelli, A.; Milella, M.; Vecchio, A.; Anzidei, M.; Mastronuzzi, G. Sea-level rise and shoreline changes along an open sandy coast: Case study of gulf of taranto, Italy. *Water* **2020**, *12*, 1414. [CrossRef]
10. Abadie, L.M.; Sainz de Murieta, E.; Galarraga, I. The Costs of Sea-Level Rise: Coastal Adaptation Investments vs. Inaction in Iberian Coastal Cities. *Water* **2020**, *12*, 1220. [CrossRef]
11. White, E.D.; Meselhe, E.; Reed, D.; Renfro, A.; Snider, N.P.; Wang, Y. Mitigating the effects of sea-level rise on estuaries of the Mississippi Delta Plain using river diversions. *Water* **2019**, *11*, 2028. [CrossRef]
12. Vecchio, A.; Anzidei, M.; Serpelloni, E.; Florindo, F. Natural Variability and Vertical Land Motion Contributions in the Mediterranean Sea-Level Records over the Last Two Centuries and Projections for 2100. *Water* **2019**, *11*, 1480. [CrossRef]

Article

The Subsurface Geology and Landscape Evolution of the Volturno Coastal Plain, Italy: Interplay between Tectonics and Sea-Level Changes during the Quaternary

Giuseppe Corrado [1,*], Sabrina Amodio [2], Pietro P. C. Aucelli [2], Gerardo Pappone [2] and Marcello Schiattarella [1]

1   Dipartimento delle Culture Europee e del Mediterraneo (DiCEM), Università della Basilicata, 75100 Matera, Italy; marcello.schiattarella@unibas.it

2   Dipartimento di Scienze e Tecnologie, Università degli Studi di Napoli Parthenope, 80121 Napoli, Italy; sabrina.amodio@uniparthenope.it (S.A.); pietro.aucelli@uniparthenope.it (P.P.C.A.); gerardo.pappone@uniparthenope.it (G.P.)

*   Correspondence: giuseppe.corrado@unibas.it

Received: 30 October 2020; Accepted: 28 November 2020; Published: 2 December 2020

Abstract: The Volturno alluvial-coastal plain is a relevant feature of the Tyrrhenian side of southern Italy. Its plan-view squared shape is due to Pliocene-Quaternary block-faulting of the western flank of the south-Apennines chain. On the basis of the stratigraphic analysis of almost 700 borehole logs and new geomorphological survey, an accurate paleoenvironmental reconstruction before and after the Campania Ignimbrite (CI; about 40 ky) eruption is here presented. Tectonics and eustatic forcing have been both taken into account to completely picture the evolution of the coastal plain during Late Quaternary times. The upper Pleistocene-Holocene infill of the Volturno plain has been here re-organized in a new stratigraphic framework, which includes seven depositional units. Structural analysis showed that two sets of faults displaced the CI, so accounting for recent tectonic activity. Yet Late Quaternary tectonics is rather mild, as evidenced by the decametric vertical separations operated by those faults. The average slip rate, which would represent the tectonic subsidence rate of the plain, is about 0.5 mm/year. A grid of cross sections shows the stratigraphic architecture which resulted from interactions among eustatic changes, tectonics and sedimentary input variations. On the basis of boreholes analysis, the trend of the CI roof was reconstructed. An asymmetrical shape of its ancient morphology—with a steeper slope toward the north-west border—and the lack of coincidence between the present course of the Volturno River and the main buried bedrock incision, are significant achievements of this study. Finally, the morpho-evolutionary path of the Volturno plain has been discussed.

Keywords: well log correlation; late Quaternary environments; sea-level changes; Volturno plain; southern Italy

1. Introduction

The Tyrrhenian coast of Italy is edged by several coastal plains separated by more or less pronounced promontories, especially in its southern segment. In northern Campania, the region of Naples, one of the major alluvial-coastal plains is that of the Volturno River (Figure 1), characterized by a regular (i.e., roughly quadrangular) shape due to the regional-scale Pliocene-Quaternary block-faulting of the western flank of the southern Apennines, accompanying the back-arc basin opening.

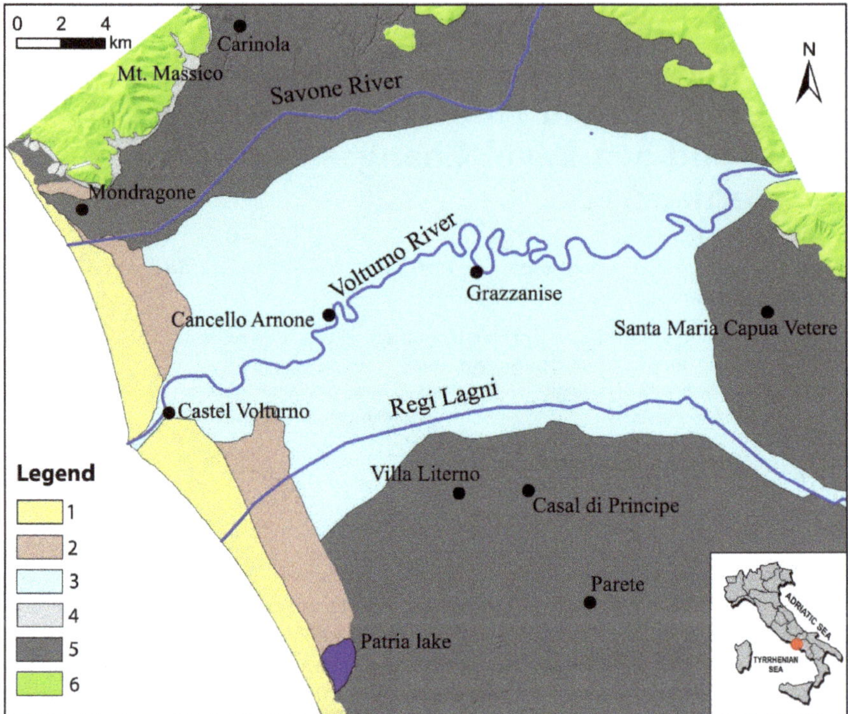

Figure 1. Simplified geological map of the Volturno Plain area; Legend: (1) dunal ridge deposits (Holocene); (2) marsh deposits (Holocene); (3) alluvial deposits (Late Pleistocene-Holocene); (4) slope waste deposits (Late Pleistocene-Holocene); (5) lavas and pyroclastic rocks of Roccamonfina Volcano, Phlegraean Fields, and Mt. Somma-Vesuvius (Middle Pleistocene-Holocene); (6) Pre-Quaternary bedrock.

The Volturno plain is bordered by the structural highs of Mt. Massico and Caserta Mts. to the north-west and north-east, respectively, and by the Ischia Island-Phlegraean Fields volcanic areas to the south-east. This coastal plain has been widely studied in the past, mostly from a stratigraphic viewpoint, defining the basin architecture by means of borehole analysis [1–8]. The aim of our work is to improve the number of observations of archives of boreholes (as yet partly unpublished), investigating the stratigraphy of the Quaternary infill of the plain on the basis of almost 700 well logs, so better reconstructing some trends in subsurface markers. In addition, the geomorphological analysis of the Late Quaternary landscape (mainly coastal features and foothill–plain connecting landforms) has been performed to complete the framework of the recent evolution of the plain. The final goal of this study is in fact to define a detailed paleoenvironmental and stratigraphic reconstruction after the well-known Campanian Ignimbrite (hereafter CI) eruption. In particular, we try to reveal how and when the tectonic subsidence was active during the last sea level rise in the study area, also characterized by relevant variation of sediment supply. It is worthy to note that kilometers-thick successions of marine, transitional, and alluvial sediments, besides great numbers of volcanic rocks, filled the accommodation space tectonically created during the Quaternary [9,10].

The Quaternary infill of the coastal sectors of the plain is not laterally isotropic and homogeneous because of the presence of layers of clastic sediments with different degrees of compaction. Dunal and beach sands, back-ridge (lagoonal and palustrine) silty clays, paleosols and thick peaty layers are the main facies of this sedimentary prism. The back-ridge depressions hosted palustrine and marshy environments that were artificially drained in the last centuries.

2. Large-scale Geological Setting and Morpho-Evolution

The Volturno River plain is hosted in a graben-like structure (Figure 1), bordered by high-angle faults with offsets of thousands of meters that run at the foot of carbonate slopes [3,9,11,12]. The absence of Pliocene marine deposits in the deepest boreholes [13] suggests that the plain was above sea level during the late Tertiary [14]. During the Quaternary, the tectonic subsidence favored the accommodation of more than 3000 m of both sedimentary deposits and volcanic products [7,9,10]. During the first part of the Late Pleistocene, in fact, the filling of the plain was helped by the onset of volcanic activity from different sources. At about 40 ky, the huge CI eruption occurred [15] and uniformly covered the entire plain with a tens of meters-thick pyroclastic flow deposit.

In the upper part of the Late Pleistocene, the whole plain emerged due to the subsidence rate reduction and the contemporary last glacial regression. During the Holocene, the plain underwent diffuse flooding in concomitance with the peak of the post-glacial transgression. The subsequent decrease of the sea level rise favored coastal progradation and development of lagoon and swamp systems several kilometers inland from the present coastline [2–5].

3. Previous Works

The first detailed study of the upper Quaternary (upper Pleistocene-Holocene) sedimentary infill of the Volturno plain, based on both outcrops and boreholes, was provided by Romano et al. [1]. Six stratigraphic units from the lowermost marine sediments (Cancello terrace at 126 ky, marine stage 5.5) to the upper (beach dune and lagoonal/swamp sediments, Holocene in age) have been set up. The first-stage geomorphological reconstruction depicts a wide paleo-gulf bordered by calcareous mountains of the Apennine chain during early Late Pleistocene; after the "pre-CI" pyroclastic and volcanoclastic deposition, the paleo-gulf was substantially restricted to the central and the northern sectors of the plain (50–39 ky), due to differential uplift from the south (more) to the north (less) of the plain. In the second half of Late Pleistocene, the complete emersion of the plain was induced by glacio-eustatic sea-level lowering and enhanced by huge pyroclastic aggradation from CI eruption and by tectonic uplift of the north and eastern edges of the plain, at least until historical times. During the early Holocene, the rising of the sea level produced the progressive retrogradation of the plain, accelerated by the deposition of younger Phlegrean pyroclastic eruptions (12–23 ky). Deltaic and lagoonal environments characterized the lower plain until 5000 years ago, after substituting for swamp and marsh deposition during the Late Holocene progradation. This latter was induced by slowing down of the sea-level rise.

Barra et al. [2] further detailed the Holocene evolution of the southern sector of the Volturno coastal plain by new three boreholes. The beach-ridges and back barrier deposits lie on subaerial erosional surface superimposed on CI tuff. The ^{14}C dating on back barrier deposits suggested a 1 mm/year sedimentation rate during the last sea level rise until 6 ky BP (before present). The sedimentation rate increased up to more than 3 mm/year until 5000 years ago (the authors considered such a rate valid also for more recent times). The maximum ingression of marine layers in the area to the south of Regi Lagni is 1.5 km from the present-day coastal position. Chronological data showed that at least 5000 years, the progradation of the coastal ridges had not yet started to grow. The reconstruction of sea-level changes displayed no appreciable subsidence occurring in that area during Late Holocene.

On the basis of field and borehole data, Putignano et al. [3] focused on the Late Quaternary evolution of the north-eastern margin of the Volturno plain, confirming the stratigraphic subdivision by Romano et al. [1], stressing the reactivation of the main NO-SE and E-W border faults. Displacement of about 50 m toward the southwest of the MIS5 marine deposits and the pre-CI pyroclastic unit, in fact, indicates a Late Pleistocene activity of the Apennine-orientated fault (CI would appear to seal this movement). The offset of about 10 m of the CI top along an E-W-directed fault shows a more recent faulting episode in the very Late Pleistocene-Holocene time span.

Santangelo et al. [4] individuated two distinct levels of lagoonal sediments in an 80 m-thick core drilled 28 km away from the present-day coastline. Such levels, buried at −40 m and −18 m, were dated at MIS7 (>160 ky) and MIS5 (105–130 ky), respectively. The sedimentation appears to be controlled by glacio-eustatism and tectonic subsidence. Between these deposits, a continental stage including the emplacement of pyroclastic products (140–160 ky Taurano Tuff) was clearly identified, as well as the continental stage between the second marine event and the CI unit, marked by a volcaniclastic unit (39–105 ky; Durazzano Ignimbrite [1,3]. The authors suggested a different tectonic history of the plain, not substantially stable, but with increasing subsidence mainly after the volcanic events.

Amorosi et al. [5] furnished a reconstruction of the CI top with a terraced and incised paleo-morphology created during the last sea-level falling. Moreover, the author identified upper Pleistocene, laterally discontinuous, alluvial bodies (14C-dated at 37 ky BP) between the CI and the post-CI volcaniclastic units. After the CI, a deepening-shallowing upward stratigraphic sequence reflects the backstepping of the wave-dominated delta system (latest Pleistocene-early Holocene, from 14C dating of lagoonal deposits at 8–12 ky BP), followed by the fore-stepping of pro-delta/strand-plain/alluvial facies, which was induced by deceleration of the last sea-level rise (post 5 ky). Modern alluvial and coastal plain deposits closed this sequence. The maximum ingression (about 7 ky BP) of the shoreline was about 6 km NE of its present position.

A detailed study of the Lake Patria coastal lagoon, remnant of a larger lagoonal-marsh area developed in the southern part of the Volturno delta system, was carried out by Sacchi et al. [6]. Sedimentological and stratigraphic analyses of 12 cores allowed the dating of the inception of the back-barrier area around 4500–4800 yeas ago, caused by decreasing sea-level rise. About 3000 years ago, the lagoonal area reached its maximum extension keeping, connection to the open sea; after that, a fresh water input was established in the brackish lagoon. From Roman times, a vast marsh area was present until the Bourbons' reclamation that gradually reduced Lake Patria to its present size by constructing the Regi Lagni channel through the diversion of the Clanio River directly into the sea.

Santangelo et al. [7] illustrated the Quaternary geomorphological evolution of the Volturno plain, furnishing a review of all data previously published. They focus on the tectonic and glacio-eustatic controls influencing the coastline variations. According to the authors, a significant subsidence occurred in the Middle-Late Pleistocene, but with different rates along the plain. Tectonic subsidence was more homogeneous in the last 130 ky, with increasing trends mainly after the volcanic events. 40 ky-old CI eruption was responsible for the strong and rapid aggradation of the plain, which totally emerged and was therefore affected by fluvial erosion.

A land-sea profile, interpreted in terms of sequence stratigraphy and based on an extended set of on-land core data and offshore seismic reflection profile, has been carried out by Ruberti et al. [16] with particular reference to the Late Quaternary. In the same year, Matano et al. [17] illustrated the subsidence trends of the plain in the 1992–2010 interval, as inferred by SAR interferometry data. Significant subsidence (up to −420 mm) along the channel of the Volturno River and in its river mouth area occurs. Low subsidence or stability is recorded along the dune ridges as well as in a 10 km-large strip across the Volturno River. According to the authors, such a scenario appears linked to the lithological nature of subsurface deposits.

Finally, the offshore deposits of the last glacio-eustatic cycle in front of the plain have been investigated by Iorio et al. and Misuraca et al. [18–20]. Seismic profiles and cores allowed detailing of the depositional sequence located in the outer shelf off the Volturno plain. The oldest pro-grading units emplaced during the sea-level fall (−120 m) are limited at the upper boundary by a ravinement surface (regional unconformity) on which the youngest units developed [18,19]. The 3D modelling of the LGM (Last Glacial Maximum) erosional surface revealed a topographic high bounded by two opposite-facing sets of normal faults 11.5 km offshore of the Volturno River mouth [20].

4. Materials and Methods

In order to reconstruct the paleo-morphology of the buried roof of the Campania Ignimbrite, to determine the paleoenvironmental conditions of the area coinciding with the present-day Volturno plain, and to provide the stratigraphic synthesis of the Quaternary succession, a large amount of stratigraphic data has been acquired by the analysis of a collection of borehole logs from previous works and public institutions, such as the technical offices of municipalities and ISPRA (Italian Agency for Environmental Research and Protection). The subsurface lithostratigraphic dataset is based on 680 wells drilled in the plain and integrated by geological cartography and field survey. Sedimentological analysis and facies interpretation were acquired from the literature [1–5], whereas other unpublished stratigraphic logs furnished less detailed lithological data. Stratigraphic observations have been performed directly in the field where Quaternary sediments crop out. Age determinations for stratigraphic correlations are based on chronological constraints of volcanic and marine deposits (see references in Results).

All these datasets permitted the reconstruction of the Quaternary stratigraphic succession and its representation in a synoptic scheme (cf. Section 5.1). Such a task required an effort to homogenize data from different sources by using log charts and their precise positioning in a GIS. Subsurface and field data have been used to construct many geological cross-sections constituting an orthogonal grid (with a set parallel to the coastline). Vertical exaggeration of cross-sections is about 30x, due to the need of restitution of some stratigraphic details. In such a way, units' geometry and their stratigraphic relationships are better identified and readable. Of course, this involved an increase in apparent inclination of geological bodies and slopes.

In order to obtain useful elements to define the Late Quaternary evolution of the area as well as the tectonic relationships between Mount Massico and the surrounding plains, a geomorphological and structural survey were carried out to better define the elements of the recent brittle deformation (i.e., faults).

Finally, a contour map of the CI upper boundary has been drafted by GIS interpolation techniques.

5. Results

5.1. Stratigraphic Synopsis of the Late Quaternary Succession from the Volturno Plain

The stratigraphic analysis of the wells has allowed us to individuate 7 lithostratigraphic units based on homogeneous sedimentological and paleontological criteria. Table 1 summarizes the main lithological characteristics for each unit, including their stratigraphic contacts with the over- and underlying units. Numbering follows from the younger to the older deposits. For each unit, environmental interpretation is also provided. Stratigraphic positions and radiometric constraints enable us to propose a viable chronology, attributing an age interval for each unit.

Table 1. Summarized description of the main lithostratigraphic units recognized by core data and environmental interpretation (ages from literature).

Unit	Description	References	Chronology (Age BP)	Depositional Environment
U1p	Grey silty clays and silt with local sandy intercalations. Small mollusc shells (euryhaline fauna) are common. Thickness range: few metres—6 m. Transitional boundary with U1d. This unit is only located to coastal sector.	[5]	Holocene (0–12 ky)	Pro-delta
U1d	Well sorted, medium to coarse grey or yellowish sands, with abundant marine shells fragments, plant and rare pumice fragments. A general coarsening-upward trend occurs. Thickness range: few meters–16 m. Basal erosional contact with U2 and U4. Transitional boundaries with U1m landward and U1p seaward.		Holocene (0–12 ky)	Shoreface-beach-dune-ridge complex

Table 1. *Cont.*

Unit	Description	References	Chronology (Age BP)	Depositional Environment
U1m	Grey to dark grey clays with abundant organic matter and freshwater gastropods. Peat intercalations and sandy layers with mollusc and pumice fragments occur at places (Radiometric age from lower part of swamp/lagoonal deposits: 8–12 ky). Thickness range: few meters—30 m. Lower erosive boundary with U2, U3 and U4. Transitional boundaries with U1a landward and U1d seaward.	[5]	Holocene (0–12 ky)	Back-dune ridge complex Lagoon/estuary, coastal plain/swamp
U1a	Rooted and pedogenized, brown to grey clays and silty clays with abundant whitish pumices and plant fragments. Gravel with sand lens locally occurs. Thickness range: few meters—35 m. Lower erosive boundary with U2 and U3 and transitional contact with U1m seaward.	[5]	Holocene (0–12 ky)	Alluvial plain
U2	Two distinct events form this unit: Giugliano Ignimbrite (GI) and Neapolitan Yellow Tuff (NYT). The GI is pyroclastic flow deposit (23–18 ky), separated by paleo-sols from Campania Ignimbrite (CI) below and the incoherent NYT fall deposits above. The NYT consists of alternations of pumice and ash beds (Pozzolana), zeolitized yellow tuff, volcanic sands, cinerite deposits, sometimes reworked by fluvial erosion. Thickness range: few meters—30 m. Basal erosional contact with U3 and U4 and upper erosive boundary with U1.	[15,21]	Late Pleist. Holocene (12–23 ky)	Continental (volcaniclastics)
U3	Gravels and sands deposited as lens-shape bodies and characterized by upward transition to silty clays or silts with local peat intercalations (radiometric age 37 ky BP). White pumices and lapilli as well as pedogenized horizons locally occur. Thickness range: few metres—20 m. Lower erosional contact with U4 and U5; cover is represented by U1 and U2.	[5]	Late Pleist. (23?–37 ky)	Alluvial plain
U4	CI is a laterally continuous body formed by a pyroclastic flow, consisting of coherent (breccia Piperno) to totally incoherent (Cinerazzo) reddish or grey tuffs, with sandy scoriae, pumices and lithic elements. Thickness range: few metres—40 m. CI rests on coastal marine (U5) seaward and on the pre-CI volcaniclastic (U6) unit landward. The U1 and U2 units cover it by erosional contact.	[22]	Late Pleist. (about 40 ky)	Continental (volcaniclastic)
U5	Yellowish and grey sands with mollusc shells, clays with local peat lens. Intercalations of gravels, pyroclastic and travertine deposits also occur. Thickness range: 20 to >50 m. Upper (U4 and U3) and lower (U6) erosional contacts are present.		Late Pleist. (40?–50 ky)	Coastal-marine, transitional with continental episodes
U6	Pre-CI pyroclastic deposits produced by different eruptions from Roccamonfina (youngest events) and Phlegraean Fields (oldest events), locally separated by thin paleo-sols. Tuffs, pozzolana, lapilli and cinerite; rare lavas and clay lens. Thickness range: 5 to >60 m. The U6 rests on U7 by erosional contact and is covered by the U4 and U5. Radiometric age at the base 105 ka.	[4]	Late Pleist. (50–105 ky)	Continental (volcaniclastic)
U7	Fossiliferous clays and silts interbedded with yellow sands or fine gravels (Radiometric age 126 ky;). Thickness range: 30 to >50 m. The U7 could be in contact at the base with the carbonate substrate and it is covered by the U6.	[1]	Late Pleist. (105–130 ky)	Coastal-marine, transitional

Unit 1—Holocene deposits (0–12 ky). This unit represents a complex and articulated sedimentary body formed by four sub-units: pro-delta (U1p), beach-dune ridge (U1d), lagoonal-swamp (U1m) and alluvial (U1a) deposits. The U1 bottom is normally the U4 piroclastic unit (CI) which also represent its lateral contact towards north and east. The Pleistocene alluvial (U3) and pyroclastic (U2)

deposits represent a very discontinuous substrate for this unit, which rests on these only in some areas. By integrating subsurface and literature [1,2,5,6] data, the U1 can be considered as the Holocene depositional system having a complex transgressive-regressive architecture. It is articulated by a laterally adjacent facies belt, parallel to the coast, that progressively moving inshore shows: pro-delta, beach-dune ridge and back barrier (U1p, U1d and U1m) depositional units, until arriving at the more internal alluvial plain (U1a).

Unit 2—Pyroclastic flow and fall deposits (12–23 ky). This volcaniclastic unit is composed of the Giugliano Ignimbrite (GI, age 18–23 ky [15,21]) below, and the Neapolitan Yellow Tuff (NYT, ~12 ky in age [21]) above, both deposited in subaerial conditions. The U2 mostly occurs in the southern sector of the Volturno plain (Giugliano, Qualiano, Melito di Napoli, Casavatore, Lake Patria) and shows a lower erosive contact with the U4 (CI) marked by a paleosol and, where it does not crop out, is covered by the U1 unit. The U2 is rarely documented on the right side of the Volturno River and has been mostly removed by fluvial erosion [12].

Unit 3—Pleistocene alluvial plain deposits (23?–37 ky). This highly discontinuous and subsurface unit is represented by channeled fills of paleo-valleys incised in the U4 or fluvial terraces created by pyroclastic material reworking of the underlying U4 unit. Amorosi et al. [5] dated this unit at about 37 ky BP. In places, the incision of the paleo-river appears to cross the entire U4, reaching the underlying coastal-marine unit (U5), as testified by erosive contacts of some alluvial lens directly with the U5, described below.

Unit 4—Campanian Ignimbrite (CI, about 40 ky). This unit is attributed to the highly explosive eruption of the CI, which entirely covered the Campania Plain at ~40 ky [22]. It borders the southern and western flanks of the Massico and Caserta Mountains, respectively, and represents a laterally continuous horizon with an articulate paleo-topography clearly traced in the subsurface. CI upper boundary appears at a progressively higher depth (35/40 m b.s.l.) in the central sector of the Volturno Plain [1,4,5].

Unit 5—Coastal-marine and transitional deposits with alluvial episodes (40?–50 ky). This subsurface unit is often the substrate of the Unit 4 at depths more than 40 m b.s.l. The upper limit is very irregular and is placed between 20 and 10 m a.s.l. in the Mondragone and Sparanise area, but between 10 and 40 m b.s.l. moving from the coastal towards the inner sector of the plain. Based on lithological characteristics and stratigraphic position, a correspondence with the unit 4 of Romano et al. [1] can be established.

Unit 6—pre-Campanian Ignimbrite volcaniclastic deposits (50–105 ky). This subsurface unit appears more discontinuous in the northern (top depth between 10 m b.s.l. and 30 m a.s.l.) than in central and southern (20/60 m b.s.l. top depth) sectors of the plain. U6 is situated stratigraphically under the U5 in the southern sector but directly under the U4 in the Villa Literno—Casal di Principe area. U6 is also horizontally adjacent to U5 under the Regi Lagni area suggesting possible dislocations of these units. Some correspondence with the Unit IV [4] and Unit 5 [1] is based on chronostratigraphic (~50–105 ky) and paleo-environmental constraints.

Unit 7—Coastal-marine and transitional deposits (105–130 ky). This subsurface unit appears to be continuous and the deepest, the upper boundary ranging between 40 and 70 m b.s.l. and, between 0 and 30 m b.s.l., on the left and right sides of the Volturno River, respectively. This differential depths could suggest a tectonic dislocation. The minimum thickness estimated is about 50 m even if the base of this unit has never been reached in the studied boreholes. These deposits cover lava bodies in the southern sector of the plain ascribed to the buried Parete volcano [23]. U7 corresponds to marine sediments of unit 6 from Romano et al. [1] (U/Th dating on *Cladocora*, about 126 ky) and unit III from [4] (age > 105 ky).

5.2. Geological Cross Sections

Mapping and correlation of outcrop and subsurface stratigraphic data in relation to the present sea level allowed us to outline the presence/absence, elevation, thickness and stratigraphic relationships of the seven stratigraphic units above illustrated. Six geological sections have been constructed (Figure 2); three of them are parallel to the major river axis, starting from the coast to the inland (A–A′, B–B′ and C–C′ from the north to the south, Figure 3), whereas other three are orthogonal to the first ones (i.e., NW-SE oriented) and labelled D–D′, E–E′ and F–F′ (Figure 4). All these sections show the late Pleistocene-Holocene succession of the marine-to-continental infill, resulting from interactions among tectonics, eustatic changes, and sediment supply variations, in turn largely affected by neo-tectonic faults.

The A-A′ section shows, in its deepest portion near the north-east boundary of the plain, the coastal-marine and transitional deposits (U7) covered by the pre-CI continental volcanoclastic unit (U6). These two units are attributed respectively to 105–130 ky and 50–105 ky ages [4]. Further, in this section the U7 directly rests on the Mesozoic carbonate bedrock and the CI (U4) directly covers the U6. The staircase trajectory of their upper boundaries buried at the base of the carbonate slope shows the downthrown structure of the north-eastern border of the plain, due to normal faulting (A–A′ section, Figure 3).

U4 crops out along the border ridges of the plain, showing thickness up to 50 m in different points. It corresponds to the volcanic products of the 40 ky-old major eruption of CI, forming a laterally continuous body cropping out along the carbonate slopes of border mountains and drilled by many cores in the plain. The geological sections illustrate a very irregular profile shaped by erosion due to the long subaerial exposure and articulated by exposed and buried faults. An erosive boundary (mountainward) and U5 coastal-marine and transitional deposits (seaward) clearly separate in subsurface this volcanic unit from the pyroclastic/volcanoclastic lower unit (U6) in the A–A′, C–C′, and D–D′ sections (Figure 3). Moreover, the U5 upper limit is progressively dislocated by faults at greater depths toward the central part of the plain, as one can observe in the D–D′ and F–F′ sections (Figure 4). It follows that a couple of major faults are parallel to the southern slope of Mt. Massico (i.e., NE-SW-directed), the carbonate ridge that closes the plains toward NW. An antithetic fault is placed in the central-southern sector of the plain, to the south of Volturno River, constrained by a lot of boreholes used for the construction of D-D′, E-E′ and F-F′ sections (Figure 4).

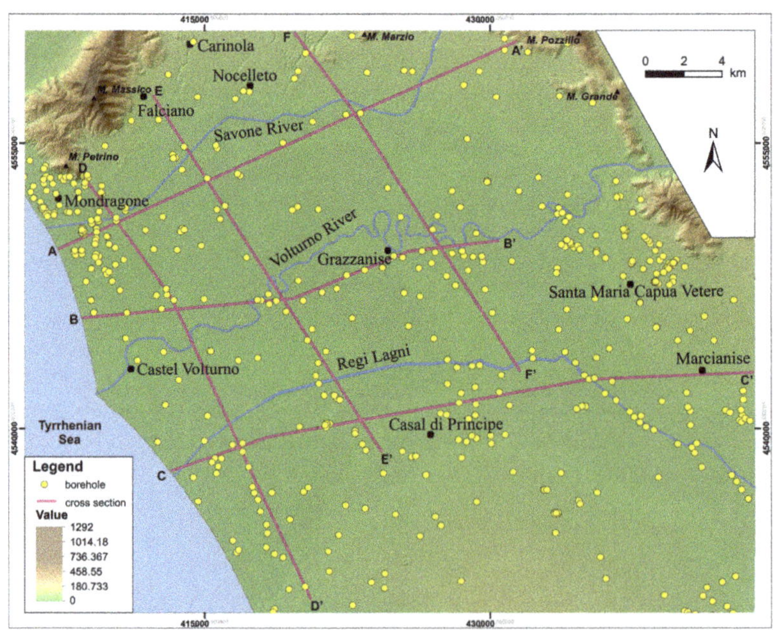

Figure 2. Locations of boreholes and geological cross-sections.

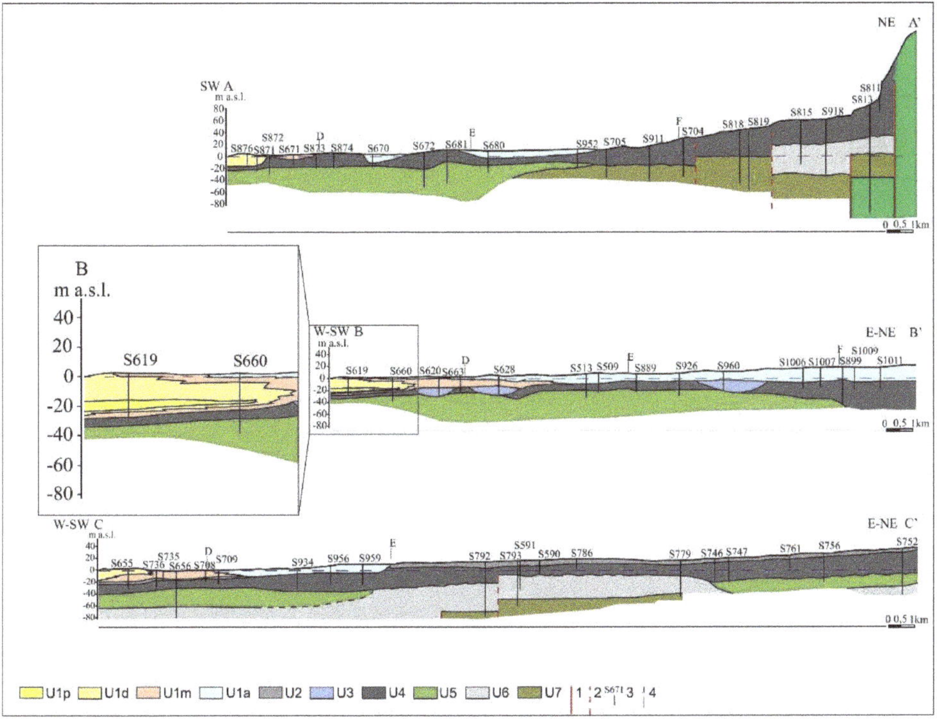

Figure 3. A–A′, B–B′, and C–C′ geological cross-sections. U1p: pro-delta deposits; U1d: beach-dune ridge deposits; U1m: lagoonal-swamp deposits; U1a alluvial deposits; U2: pyroclastic flow and fall deposits (12–23 ky); U3: Pleistocene alluvial deposits (23?–37 ky); U4: Campanian Ignimbrite deposits (CI, about 40 ky); U5: coastal-marine and transitional deposits with alluvial deposits (40?–50 ky); U6: pre-Campanian Ignimbrite volaniclastic deposits (50–105 ky); U7: coastal-marine and transitional deposits (105–130 ky); 1: fault; 2: uncertain fault; 3: borehole; 4: interception with cross sections.

The U4 thickness wears thinner in the plain where the fluvial erosion has been active until now, so that the maximum thickness of the overhead Holocene alluvial deposits (U1a, about 40 m) is in correspondence with the minimum of the U4 (about 10 m in the E-E′ section and 15 m in the F-F′ section, Figure 4). Where the past fluvial incision reached the underlying coastal-marine U5, upper Pleistocene alluvial deposits belonging to the U3 are preserved (see B-B′ and E-E′ sections, Figures 3 and 4); their lens-shaped geometry depicts the paleo-channels fillings. The fluvial erosion appears partly sets up along a tectonic lineament: buried channelled fill lies to the north of the *Regi Lagni*, where the antithetic fault above mentioned is tentatively traced (see E-E′ section, Figure 4).

The post-IC volcanic unit U2, originally covering the entire plain, was mostly removed by fluvial processes and occurs only south of Volturno River. It crops out as a quasi-continuous and thin level at the north of the Phlegraean Fields and mainly on morphological highs in the southern portion of the plain [5]. This horizon runs into subsurface reaching the coast at the south of the *Regi Lagni* (C–C′ section, Figure 3). This unit (12–23 ky BP, [15,24]) produced a shoreline progradation due to both volcanoclastic aggradation and volcano-tectonics [2] in spite of sea level rise starting after the Last Glacial Maximum.

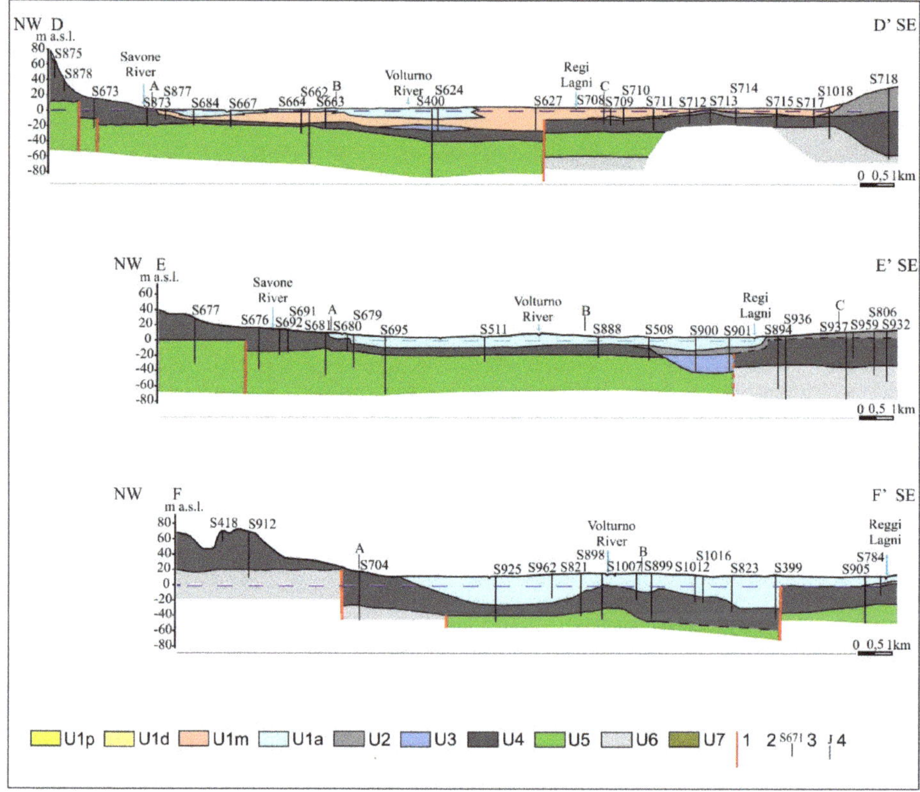

Figure 4. D–D′, E–E′, and F–F′ geological cross sections. U1p: pro-delta deposits; U1d: beach-dune ridge deposits; U1m: lagoonal-swamp deposits; U1a alluvial deposits; U2: pyroclastic flow and fall deposits (12–23 ky); U3: Pleistocene alluvial deposits (23?–37 ky); U4: Campanian Ignimbrite deposits (CI, about 40 ky); U5: coastal-marine and transitional deposits with alluvial deposits (40?–50 ky); U6: pre-Campanian Ignimbrite volcaniclastic deposits (50–105 ky); U7: coastal-marine and transitional deposits (105–130 ky); 1: fault; 2: uncertain fault; 3: borehole; 4: interception with cross sections.

The youngest unit (U1) mostly represents a sedimentary coastal wedge, Holocene in age, split into U1p, U1d and U1m equivalent to pro-delta, beach-dune, back-dune systems, respectively; the above mentioned U1a covers the inner sectors of the plain. The U1 rests on the U2, U3 and U4 units, separated by an erosional surface. The coastal wedge has retro-gradational facies architecture in the lower portion of the U1, clearly illustrated in the B-B′ and C-C′ sections (Figure 3). The Holocene sea-level rise may account for this facies arrangement [5]. A variable landward shifting of the marine-coastal wedge can be traced by the borehole data plotted along the orthogonal sections to the coast. The beach-dune system (U1d), up to 25 m thick, migrates landward from a minimum of about 1.5 km immediately south of Mondragone (A–A′ section) to a maximum of 3.5 km in the Castel Volturno area (B–B′ section). Only in this coastal sector are the pro-delta deposits (U1p), a few meters thick, individuated in the cores, above the transgressive barrier sands. The back-dune system (U1m) including lagoonal-estuarine, swamp and coastal plain environments, appears to reach the maximum landward position, about 9 km along the Volturno axis, but up to about 3 and 6 km at the north and south of the Volturno River, respectively. The maximum thickness of the U1m occurs in the coastal sector enclosed between the Volturno and *Regi Lagni* channels (see D–D′ section, Figure 4). Here, these deposits appear to suture the antithetic north-dipping fault roughly parallel to the counter-Apennine Mt. Massico fault system.

Upwards, the U1 facies shows a coastal prograding pattern produced after the maximum flooding surface aged about 7 ky [5]. The seaward shift of U1m and U1a, as well as of the sand beach-dune ridge and delta systems up to where they reach the present coast configuration, is delineated by the core stratigraphic succession more or less aligned to the depocenter (B–B' section, Figure 3). The most recent stage leads to a progressive continental aggradation, with the swamp and marshy areas disappearance in the last 2 ky [5,6].

The Late Quaternary Volturno valley infill architecture, above delineated, allowed us to create a more complete scheme of the main stratigraphic relationships among the seven units recognized in this work (Figure 5).

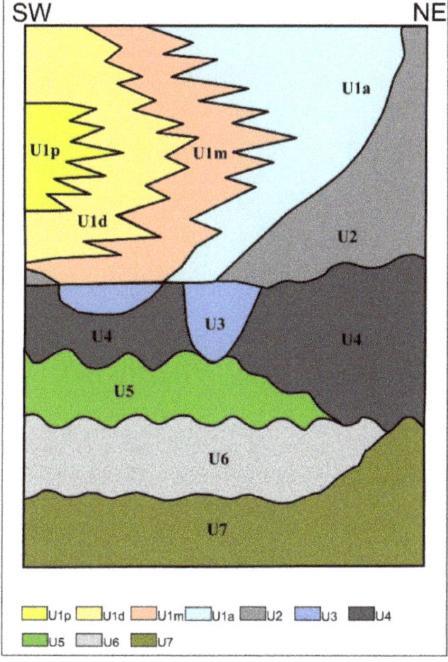

Figure 5. Scheme of stratigraphic relationships between the seven lithostratigraphic units of the study area. U1p: pro-delta deposits; U1d: beach-dune ridge deposits; U1m: lagoonal-swamp deposits; U1a alluvial deposits; U2: pyroclastic flow and fall deposits (12–23 ky); U3: Pleistocene alluvial deposits (23?–37 ky); U4: Campanian Ignimbrite deposits (CI, 40 ky); U5: coastal-marine and transitional deposits with alluvial deposits (40?–50 ky); U6: pre-Campanian Ignimbrite volcaniclastic deposits (50–105 ky); U7: coastal-marine and transitional deposits (105–130 ky). SW: south-west; NE: nord-est.

5.3. Tectonics

The genesis of the graben-like structure that hosts the present-day Volturno coastal plain and its tectonic evolution (including tectonic subsidence of the plain) can be traced back to polyphasic brittle deformation, exerted by a system of faults mainly constituted by two orthogonal sets (Figure 6). The most ancient deformational stage, which occurred between the Pliocene and early Pleistocene, created the embryonal low with a half-graben geometry, like other structural depressions of the Tyrrhenian side of the Apennines orogenic chain [11], and set the conditions for their evolution and acquisition of accommodation space for marine sediments. At the passage between the Early and the Middle Pleistocene, the extension direction switched from NW-SE to NE-SW, causing the formation or re-activation of Apennine-directed and/or E-W-oriented high-angle faults as normal faults (Figure 6). The same extensional axis characterizes the late Pleistocene-Holocene and present-day stress field, accompanied by relevant volcanic activity.

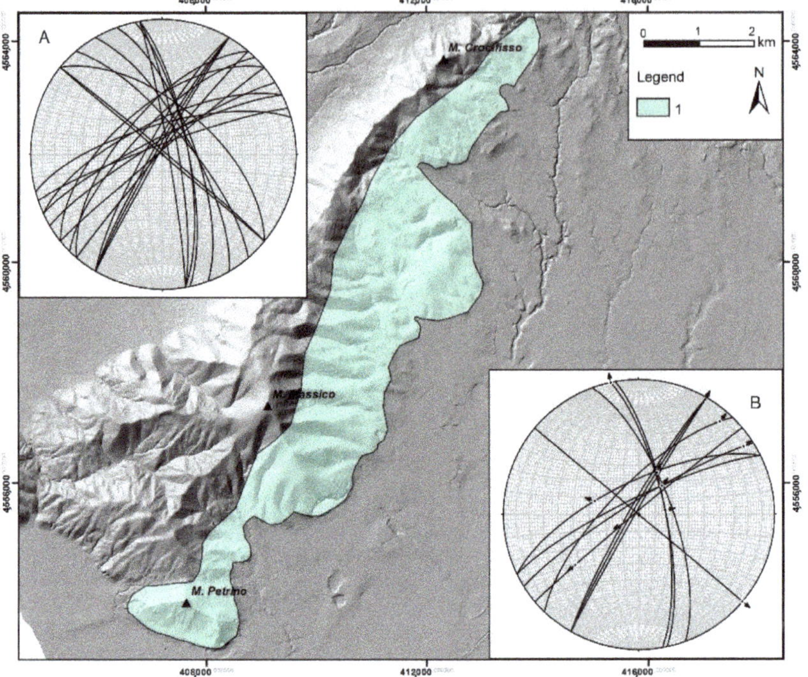

Figure 6. Measurement stations (box 1 in the legend) and related stereoplots of brittle structures from the southern slope of Mt. Massico. (**A**): stereoplot of faults; (**B**): stereoplot of faults with kinematics.

Geophysical data and stratigraphic constraints from deep boreholes suggest that the Apennine-trending faults dip towards the south-west and are the most superficial evidence of a complex system of listric faults that gave rise to the asymmetrical extensional structures strongly controlling the general physiography of the region [11].

Two sets of faults with clues of recent activity have been identified in the study area (Figures 6 and 7). Both displaced the CI, so showing post-40 ky offsets. Late Quaternary tectonics is evidenced by the decametric vertical separations operated by these faults. The interval of possible slip rates, calculated on the basis of chronostratigraphic constraints and considering offsets from 10 to 30 m, is equal to 0.25–0.77 mm/year. Their average slip rate, that would represent the tectonic subsidence rate of the plain, is about 0.5 mm/year.

The structural analysis performed along the southern slope of Mt. Massico permitted us to identify different kinematic indicators on the fault planes. Orientation data from three measurement stations showed that NW-SE-striking faults are featured by normal, transcurrent, and transtensive markers. In this case, the strike-slip component is always left-lateral. Regarding the NE-SW-directed planes (i.e., counter-Apennine set), both right-lateral and left-lateral components of slip are present, whereas dip-slip indicators are absent. Therefore, the counter-Apennine faults surveyed at the north-western border morphostructure of the plain can be interpreted as second-stage (i.e., re-activated) transfer faults that accommodated the youngest extensional deformation with a NE-SW-trending tensile axis. Large-scale evidence (volcanism, hydrothermal phenomena, morphological anomalies, sinkhole alignments) suggest that the counter-Apennine lineaments are expression of very recent tectonic activity [1,25,26].

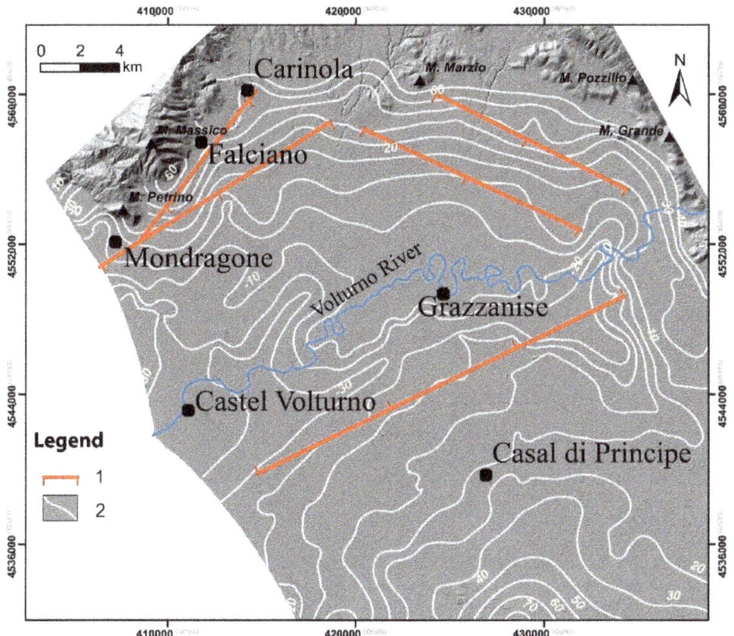

Figure 7. Map showing the CI unit upper morphological surface and the Quaternary main faults. Legend: (1) fault; (2) contour lines of the CI roof. In blue, the present Volturno River.

5.4. Geomorphology

Close to the present coastline, a composite sandy ridge testifies to the evolution of the Holocene barrier-lagoon system. It disappears inland under a flat, muddy, back-dune depression. These sandy ridges constitute the same dune system with a mean height of about 3 m a.s.l., interrupted by rivers and artificial drainage channels. The back-ridge depressions, spread over a large area of the plain with a mean height of about 0.50–1.5 m b.s.l., hosted palustrine and marshy environments, artificially drained in modern times.

At the base of the slopes there are alluvial fans and waste deposits, Late Quaternary in age. Carbonate slopes facing to the plain generally presents regularized profiles and rectilinear basal envelopes inherited by mechanisms of recession occurred under cold climate conditions. They are often dissected by minor streams—related to a warmer-humid later period—that in some cases have incised transversal valleys, then were partially filled by alluvial sediments.

On the basis of boreholes analysis, the trend of the CI roof was reconstructed. Such a paleomorphology shows an articulated buried surface, significantly eroded and sharply incised by streams. In particular, a paleovalley can be clearly recognized about two kilometres to the south of the present-day course of the Volturno River (Figure 7). Here, the paleoriver seems responsible for a fluvial incision of about 10 m, carved in the CI (the upper morphological surface of this unit is at −20 m at about 24 km from the coastline, whereas the same element is at −30 m at about 8 km from the present coast). This paleoriver can likely represent the ancient (i.e., post-CI and pre-Holocene) course of the same Volturno River.

6. Discussion and Conclusions

Geological and geomorphological data here presented allowed us to define a new and more detailed picture of the morphological, tectonic and sedimentary evolution of the Volturno River alluvial-coastal plain during the terminal part of the Quaternary (Figure 8).

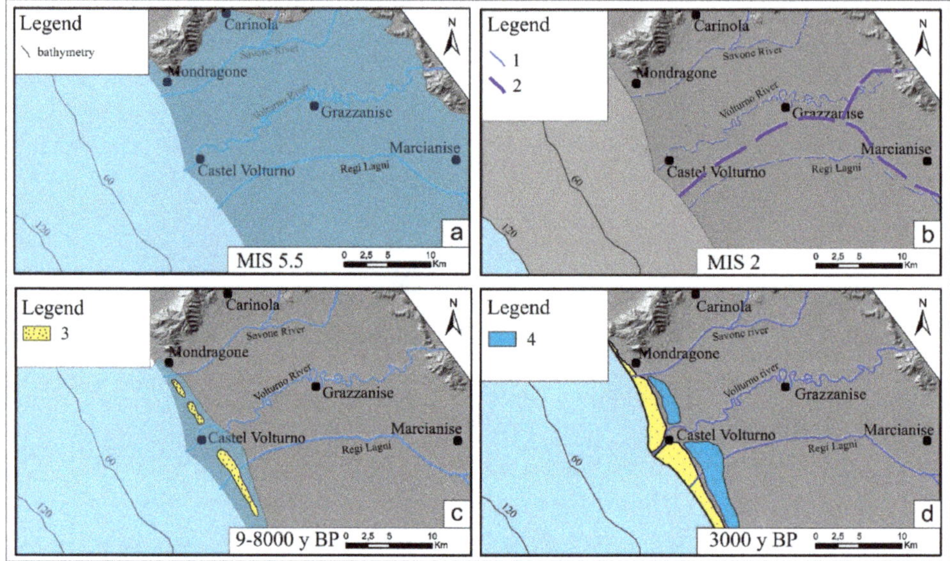

Figure 8. Morpho-evolutive stages (see text for explanation). Legends: (1) Current rivers; (2) Paleo-Volturno course; (3) dunal ridge; (4) swamp.

During the Tyrrhenian stage (MIS 5.5), a marine sedimentary environment (represented by deposits of U7) took place in the paleogulf previously generated and bordered to the north-west and north-east by faulted ridges mainly made of Mesozoic carbonates. A stage of pyroclastic aggradation (testified by U6 deposits), which perhaps occurred in the MIS 3, provoked a shift of the coastline towards the sea, especially in the southern part of the study area due to the greater thickness of such deposits observed in cross-sections C–C', D–D', E–E' and F–F' (Figures 3 and 4). Marine and transitional facies, with alluvial bodies, was deposited in this restricted paleogulf. They mirror a highstand phase, whose effect reaches the massive Mesozoic limestone of the Mt. Massico foothill (i.e., the sandy sedimentation today buried at the north-western edge of the current plain, see D-D' and E-E' cross-sections, Figure 4) around 50,000 years ago [1] (MIS 3.3). During this entire period (early Late Pleistocene), the permanence of the marine environment was undoubtedly favoured by subsidence, as testified by the share of beach rocks dated by Romano et al. [1] at 126 ± 11 ky BP, found at −50 m a.s.l., 56 m lower than the expected level (Figure 8a).

The scenario above described changed completely during the eustatic lowering of the final part of the Late Pleistocene (from MIS 3.1 to MIS 2), probably also coupled to the reduced subsidence. The eruption that caused the placement of the Campanian Ignimbrite (40 years BP [22]) occurred in a subaerial environment (Figure 8b). The resulting increase in height of the topographical surface in coincidence with a low eustatic level of the sea set the conditions for the triggering of a new stage of fluvial erosion that reshaped the morphology of the roof of this unit [1,12]. The U3 alluvial sediments were deposited in this stage near the Volturno River and south of it.

The subsurface trend of the CI roof shows that tectonics is still active in the final part of the Late Pleistocene, with the uplift of U4 in the boundary sectors of the plain. Two sets of faults, in fact, displaced the CI with decametric offsets (Figure 7). A paleovalley south of the current Volturno River axis – which migrated northward after the MIS2—is also well recognizable, as confirmed by the presence of buried upper Pleistocene alluvial deposits (U3) associated to this morphological feature. Such a northward migration may be attributed to the activity of the NE-SW-striking Mt. Massico fault system that would have generated a NW-directed tilt of the whole area.

Seismic data from the outer shelf in front of the Volturno River mouth [20] showed that a buried structure (i.e., a volcano older than 0.7 Ma) bordered by NE-SW oriented normal faults may have controlled the southward shifting of the sedimentary depocentres related to Holocene transgressive deposits. This would confirm the mobility processes (i.e., lateral migration) of the Volturno River detected in this study, indicating an "inversion" of sense of shift (southward to northward) during the Late Quaternary. It means that the major NE-SW-directed faults could have been alternatively activated (before the south-Volturno fault and then the Mt. Massico fault).

A new cycle of eruptions from the volcanic district of Phlegraean Fields and the consequent emplacement of the Giugliano Ignimbrite and the Neapolitan Yellow Tuff (U2) between 23 and 12 ky BP mainly interest the southern part of the study area, maybe suturing the southernmost counter-Apennine fault.

Finally, the Holocene marine transgression affecting the plain starting from about 10,000 years BP established a lagoon environment in the coastal belt (cf. U1m deposits), that probably lasted up to about 5000 years BP, as suggested by dating of lagoon deposits from the nearest Garigliano River plain [27]. The maximum ingression of the shoreline in the Volturno plain was about 3–4 km inland Figure 8c). The successive seaward migration of the sandy shore induced the sedimentary filling of the lagoon areas and their replacement by marshy stretches of water (Figure 8d). During this last stage, the central area of the plain has retained substantial tectonic stability, as suggested by the geometry of Holocene sediments, which levelled the paleomorphology.

In conclusion, the significant achievements of this research are a greater knowledge of distribution and evolution of the coastal-marine to continental palaeoenvironments of the Volturno plain during the Late Quaternary, and the reconstruction of the roof of the Campanian Ignimbrite that allowed upgrading of the 3D shape of the Volturno plain paleomorphology before the deposition of late upper Pleistocene—Holocene sediments. It is now possible to infer an asymmetrical shape of the ancient morphology of the plain—with a steeper slope toward the north-west border (i.e., Mt. Massico morphostructural high)—and the lack of coincidence between the present course of the Volturno River and the main buried bedrock incision. Taking into account also the subsidence data from the Volturno plain [17,28], Late Quaternary tectonics should therefore be considered active and able to produce the asymmetry the CI roof as well as the lateral migration of the Volturno River.

Author Contributions: Conceptualization, S.A., P.P.C.A., G.P. and M.S.; methodology, S.A., P.P.C.A., G.C., G.P. and M.S. software, G.C.; validation, S.A., P.P.C.A., G.P. and M.S.; formal analysis, S.A., P.P.C.A., G.C., G.P. and M.S.; investigation, S.A., P.P.C.A., G.C. and M.S.; resources, P.P.C.A., G.P. and M.S.; data curation, S.A., P.P.C.A., G.C. and M.S. writing—original draft preparation, S.A., G.C. and M.S.; writing—review and editing, S.A., P.P.C.A., G.C., G.P. and M.S.; visualization, G.C.; supervision, P.P.C.A., G.P. and M.S.; funding acquisition, P.P.C.A., G.P. and M.S. All authors have read and agreed to the published version of the manuscript.

Funding: This research received no external funding.

Acknowledgments: This study was financially supported by Basilicata University RIL 2020 granted to M. Schiattarella and University of Naples Parthenope research found granted to P.P.C. Aucelli and G. Pappone.

Conflicts of Interest: The authors declare no conflict of interest.

References

1. Romano, P.; Santo, A.; Voltaggio, M. L'evoluzione geomorfologica della pianura del F. Volturno (Campania) durante il tardo Quaternario (Pleistocene medio-superiore—Olocene). *Il Quat.* **1994**, *7*, 41–56.
2. Barra, D.; Romano, P.; Santo, A.; Campaiola, L.; Roca, V.; Tuniz, C. The Versilian transgression in the Volturno river plain (Campania, Southern Italy): Palaeoenvironmental history and chronological data. *Il Quat.* **1996**, *9*, 445–458.
3. Putignano, M.L.; Ruberti, D.; Tescione, M.; Vigliotti, M. Evoluzione tardo quaternaria del margine casertano della Piana Campana (Italia meridionale). *Boll. Soc. Geol. It.* **2007**, *126*, 11–24.

4. Santangelo, N.; Ciampo, G.; Di Donato, V.; Esposito, P.; Petrosino, P.; Romano, P.; Russo Ermolli, E.; Santo, A.; Toscano, F.; Villa, I. Late Quaternary buried lagoons in the northern Campania plain (Southern Italy): Evolution of a coastal system under the influence of volcano-tectonics and eustatism. *Ital. J. Geosci.* **2010**, *129*, 156–175.

5. Amorosi, A.; Pacifico, A.; Rossi, V.; Ruberti, D. Late Quaternary incision and deposition in an active volcanic setting: The Volturno valley fill, southern Italy. *Sediment. Geol.* **2012**, *242*, 307–320. [CrossRef]

6. Sacchi, M.; Molisso, F.; Pacifico, A.; Vigliotti, M.; Sabbarese, C.; Ruberti, D. Late-Holocene to recent evolution of Lake Patria, South Italy: An example of a coastal lagoon within a Mediterranean delta system. *Glob Planet Chang.* **2014**, *117*, 9–27. [CrossRef]

7. Santangelo, N.; Romano, P.; Ascione, A.; Russo Ermolli, E. Quaternary evolution of the Southern Apennines coastal plains: A review. *Geol. Carpath.* **2017**, *68*, 43–56. [CrossRef]

8. Aiello, G.; Barra, D.; Collina, C.; Piperno, M.; Guidi, A.; Stanislao, C.; Saracino, M.; Donadio, C. Geomorphological and paleoenvironmental evolution in the prehistoric framework of the coastland of Mondragone, southern Italy. *Quat. Int.* **2018**, *493*, 70–85. [CrossRef]

9. Ippolito, F.; Ortolani, F.; Russo, M. Struttura marginale tirrenica dell'Appennino campano: Reinterpretazione di dati di antiche ricerche di idrocarburi. *Mem. Soc. Geol. It.* **1973**, *12*, 227–251.

10. Bernasconi, A.; Bruni, P.; Gorla, L.; Principe, C.; Sbrana, A. Risultati preliminari dell'esplorazione geotermica profonda nell'area vulcanica del Somma-Vesuvio. *Rend. Soc. Geol. It.* **1981**, *4*, 237–240.

11. Milia, A.; Torrente, M.M.; Russo, M.; Zuppetta, A. Tectonics and crustal structure of the Campania continental margin: Relationships with volcanism. *Mineral. Petrol.* **2003**, *79*, 33–47. [CrossRef]

12. Corrado, G.; Amodio, S.; Aucelli, P.P.C.; Incontri, P.; Pappone, G.; Schiattarella, M. Late quaternary geology and morphoevolution of the Volturno coastal plain, southern Italy. *Alp. Mediterr. Quat.* **2018**, *31*, 23–26.

13. Balducci, S.; Vaselli, M.; Verdini, G. Exploration well in Ottaviano permit, Italy, Trecase 1. In Proceedings of the European Geothermal Update 3rd Intern. Sem., Munich, Germany, 29 November–1 December 1983; pp. 407–418.

14. Brancaccio, L.; Cinque, A.; Romano, P.; Rosskopf, C.; Russo, F.; Santangelo, N.; Santo, A. Geomorphology and neotectonic evolution of a sector of the Tyrrhenian flank of the Southern Apennines (Region of Naples, Italy). *Zeit. Geomorph.* **1991**, *82*, 47–58.

15. Rolandi, G.; Bellucci, F.; Heizler, M.T.; Belkin, H.E.; De Vivo, B. Tectonic controls on the genesis of ignimbrites from the Campania volcanic zone, southern Italy. *Mineral. Petrol.* **2003**, *79*, 3–31. [CrossRef]

16. Ruberti, D.; Sacchi, M.; Pepe, F.; Vigliotti, M. LGM incised valley in a volcanic setting. The northern Campania Plain (Southern Italy). *Alp. Mediterr. Quat.* **2018**, *31*, 35–38.

17. Matano, F.; Sacchi, M.; Vigliotti, M.; Ruberti, D. Subsidence Trends of Volturno River Coastal Plain (Northern Campania, Southern Italy) Inferred by SAR Interferometry Data. *Geosciences* **2018**, *8*, 8. [CrossRef]

18. Iorio, M.; Capretto, G.; Petruccione, E.; Marsella, E.; Aiello, G.; Senatore, M.R. Multi-proxy analysis in defining sedimentary process in very recent prodelta deposits: The Northern Phlegraean offshore example (Eastern Tyrrhenian Margin). *Rend. Fis. Acc. Lincei* **2014**, *25*, 237–254. [CrossRef]

19. Aiello, G.; Insinga, D.; Iorio, M.; Meo, A.; Senatore, M.R. On the occurrence of the Neapolitan Yellow Tuff tephra in the Northern Phlegraean Fields offshore (Eastern Tyrrhenian margin; Italy). *Ital. J. Geosci.* **2017**, *136*, 263–274. [CrossRef]

20. Misuraca, M.; Budillon, F.; Tonielli, R.; Di Martino, G.; Innangi, S.; Ferraro, L. Coastal Evolution, Hydrothermal Migration Pathways and Soft Deformation along the Campania Continental Shelf (Southern Tyrrhenian Sea): Insights from High-Resolution Seismic Profiles. *Geosciences* **2018**, *8*, 121. [CrossRef]

21. De Vivo, B.; Rolandi, G.; Gans, P.B.; Calvert, A.; Bohrson, W.A.; Spera, F.J.; Belkin, H.E. New constraints on the pyroclastic eruptive history of the Campanian volcanic Plain (Italy). *Mineral. Petrol.* **2001**, *73*, 47–65. [CrossRef]

22. Giaccio, B.; Hajdas, I.; Isaia, R.; Deino, A.; Nomade, S. High-precision 14C and 40Ar/39Ar dating of the Campanian Ignimbrite (Y-5) reconciles the time-scales of climatic-cultural processes at 40 ka. *Sci. Rep.* **2017**, *7*, 45940. [CrossRef]

23. Ortolani, F.; Aprile, F. Principali caratteristiche stratigrafiche e strutturali dei depositi superficiali della Piana Campana. *Boll. Soc. Geol. It.* **1985**, *104*, 195–206.

24. Deino, A.L.; Orsi, G.; De Vita, S.; Piochi, M. The age of the Neapolitan Yellow caldera-forming eruption (Campi Flegrei caldera—Italy) assessed by ^{40}Ar/^{39}Ar dating method. *J. Volcanol. Geotherm. Res.* **2004**, *133*, 157–170. [CrossRef]

25. Giordano, G.; Naso, G.; Trigari, A. Evoluzione tettonica di un settore particolare del margine tirrenico: L'area al confine tra Lazio e Campania. Prime considerazioni. *Stud. Geol. Camerti* **1995**, *2*, 269–278.

26. Billi, A.; Bosi, V.; De Meo, A. Caratterizzazione strutturale del rilievo del M. Massico nell'ambito dell'evoluzione quaternaria delle depressioni costiere dei Fiumi Garigliano e Volturno (Campania settentrionale). *Il Quat.* **1997**, *10*, 15–26.

27. Di Lorenzo, H.; Corrado, G.; Aucelli, P.; De Iorio, M.; Schiattarella, M.; Russo Ermolli, E. Environmental evolution and anthropogenic forcing in the Garigliano coastal plain (Italy) during the Holocene. *Holocene* **2020**, in press.

28. Aucelli, C.P.P.; Di Paola, G.; Incontri, P.; Rizzo, A.; Vilardo, G.; Benassai, G.; Buonocore, B.; Pappone, G. Coastal inundation risk assessment due to subsidence and sea level rise in a Mediterranean alluvial plain (Volturno coastal plain, southern Italy). *Estuar. Coast. Shelf Sci.* **2017**, *198*, 597–609. [CrossRef]

Article

Holocene Evolution of the Burano Paleo-Lagoon (Southern Tuscany, Italy)

Maurizio D'Orefice [1], Piero Bellotti [2], Adele Bertini [3], Gilberto Calderoni [4], Paolo Censi Neri [1], Letizia Di Bella [5,*], Domenico Fiorenza [1], Luca Maria Foresi [6,7], Markella Asimina Louvari [8], Letizia Rainone [3], Cécile Vittori [9], Jean-Philippe Goiran [10], Laurent Schmitt [9], Pierre Carbonel [10], Frank Preusser [11], Christine Oberlin [12], Francesca Sangiorgi [13] and Lina Davoli [5]

[1] Italian Institute for Environmental Protection and Research, ISPRA, Department for the Geological Survey of Italy, 00144 Rome, Italy; maurizio.dorefice@isprambiente.it (M.D.); paolo.censineri@isprambiente.it (P.C.N.); domenico.fiorenza@isprambiente.it (D.F.)

[2] AIGeo, Italian Association of Physical Geography and Geomorphology, c/o Department of Earth Sciences, Sapienza, University of Rome, 00185 Rome, Italy; piero.bellotti@gmail.com

[3] Department of Earth Sciences, University of Florence, 50121 Florence, Italy; adele.bertini@unifi.it (A.B.); letizia.rainone@stud.unifi.it (L.R.)

[4] Institute of Environmental Geology and Geoengineering, CNR, c/o Department of Earth Sciences, Sapienza, University of Rome, 00185 Rome, Italy; gilberto.calderoni@uniroma1.it

[5] Department of Earth Sciences, Sapienza, University of Rome, 00185 Rome, Italy; lina.davoli@uniroma1.it

[6] Department of Physical sciences, Earth and environment University of Siena, 53100 Siena, Italy; luca.foresi@unisi.it

[7] Institute of Geosciences and Earth resources, CNR, c/o Research Area of Pisa, 1–56124 Pisa, Italy

[8] Faculty of Geology and Geoenvironment Department of Historical Geology and Paleontology, National and Kapodistrian University of Athens, 15784 Athens, Greece; melinaluvari@geol.uoa.gr

[9] Laboratoire Image Ville Environnement (LIVE), UMR 7362, University of Strasbourg, Faculté de Géographie et d'Aménagement, 67083 Strasbourg CEDEX, France; cecile.vittori@live-cnrs.unistra.fr (C.V.); laurent.schmitt@unistra.fr (L.S.)

[10] Archéorient, UMR 5133 Maison de l'Orient et de la Méditerranée, Centre National de la Recherche Scientifique/Université of Lyon 2, 7 rue Raulin, 69007 Lyon, France; jean-philippe.goiran@mom.fr (J.-P.G.); carbonel@free.fr (P.C.)

[11] Institute of Earth and Environmental Sciences, Albert-Ludwigs, University of Freiburg, 79104 Freiburg, Germany; frank.preusser@geologie.uni-freiburg.de

[12] CNRS, French National Centre for Scientific Research, UMR 5138 ArAr, Lumière University Lyon 2, Claude Bernard University Lyon 1, 69622 Villeurbanne, France; christine.oberlin@univ-lyon1.fr

[13] Department of Earth Sciences, Marine Palynology and Paleoceanography, Utrecht University, 3584 CB Utrecht, The Netherlands; F.Sangiorgi@uu.nl

* Correspondence: letizia.dibella@uniroma1.it

Received: 3 February 2020; Accepted: 26 March 2020; Published: 1 April 2020

Abstract: The study of Burano paleo-lagoon—Wetland of International Value, has allowed us to better define and extend the reconstruction of the Holocene paleoenvironmental evolution of the paleo-lagoons previously studied, located on the Tyrrhenian coast in central Italy. The investigated area is located in Southern Tuscany near the Burano Lake. The area was investigated by means of field surveys, historical maps, 16 coring, sedimentological, palynological and microfaunal analyses (foraminifera and ostracods), combined with robust geochronological control provided by 52 datings (^{14}C and OSL). The study allowed us to reconstruct the environmental and morphological evolution of the Burano paleo-lagoon during the last 8000 years and to hypothesize a Rise Sea Level (RSL) curve. In this context, 5 main evolutionary phases have been recognized. (1) before 7.5 ka BP in the southern-eastern part, an open lagoon developed; (2) ~6 ka BP a barrier-lagoon system develops throughout the entire area and the lagoon progressively changed from open to closed one; (3) ~5 ka

BP the width of the lagoon increases and a lacustrine facies appears along the entire axis of the coastal basin; (4) ~4 ka BP the lacustrine facies shows a discontinuous distribution respect to the previous phase; (5) during the last 4 ka the lacustrine facies disappear and the lagoon turns into a wetland area.

Keywords: Holocene; coastal lagoon; geochronology; sea level change; bio-indicators; Mediterranean Sea

1. Introduction

Although lagoons are environments that develop and become extinct in geologically short times, their sediments often constitute an important data archive relating to rapid events of considerable impact on the coastal area, which can also derive from variations on a global scale. The morphological and dynamic characteristics of the lagoons are extremely variable [1] according to the regional setting, tidal range and sediment availability. The development and evolution of the Holocene lagoons, mostly still active, were strongly influenced by the glacioeustatic sea level rise. During this period, in many cases the lagoons first developed within an incised valley [2] when the transgressive phase was still active. Later, they changed in shape, size and hydrodynamic characters when they reached almost sea level still stand [3–8]. Even shorter-term climatic variations [9] have often influenced the lagoon systems due to changes in coastal hydrodynamics and freshwater inflow [10,11].

The Tyrrhenian coast of the central Italian Peninsula (Figure 1), prone to a microtidal regime, is characterized by differently developed Holocene beach-dune systems bordering landward small-depressed areas. Until the nineteenth century, these areas were constituted of wetlands, lakes and lagoons. The land reclamations, between the end of the nineteenth to the first half of the twentieth century, transformed some coastal basins into cultivated or urbanized back dune plains. However, in some cases wide coastal lakes such as the ones located in the Southern and Northern Latium (M. Circeo promontory and Lido di Tarquinia respectively) have persisted until today. The Holocene history of some of these plains, that are locally close to river mouths (i.e., Tiber and Garigliano), has also been studied because of their geoarchaeological and historical importance [12–19]. The study of the Burano paleo-lagoon has allowed us to better define and extend the reconstruction of the Holocene paleoenvironmental evolution of the paleo-lagoons previously studied along the Latium coast from the Campania to the Tuscany regions. The study is very relevant because, unlike the extensive literature on meso and macrotidal estuarine systems, it highlights the Holocene evolution of a microtidal lagoon system. This system evolved into a tectonically stable area, not influenced by the proximity of important river mouths.

Similar studies were conducted in other areas of the Mediterranean Sea characterized by the presence of Holocene deltas and/or the ancient ports [20–29].

This paper focuses on the Burano paleo-lagoon that until 1830, extended for about 13 km along the Southern Tuscany coast between the Chiarone River mouth and the Ansedonia Promontory (Figure 1). It was possible to access and carry out the surveys only in part of the area where the portion of the paleo-lagoon stretching parallel to the coast for about 8 km, partly to the WNW and partly to ESE of the present-day Burano Lake. After the reclamation, the Burano Lake is what remains of the ancient lagoon. The lake, together with the neighbouring areas, is part of a Natural State Reserve. The area is also a Site of Community Importance (SCI IT51A0031), a Special Protection Zone (SPZ IT51A0033) and a Wetland of International Value (Ramsar Conventions on Wetlands). The purpose of the study is the Holocene reconstruction of the Burano paleo-lagoon evolution by means of facies analysis and stratigraphic relationships of the sedimentary bodies. The acquired data also allow us to hypothesize an RSL rise curve.

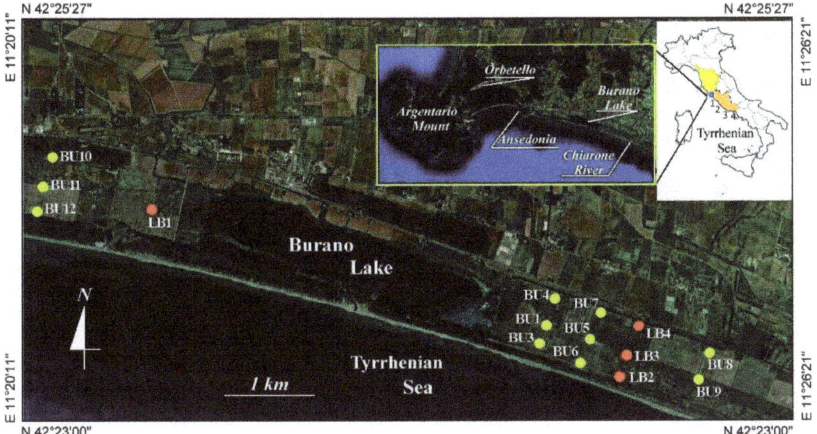

Figure 1. Location of the studied area. The position of the continuous (red dot) and percussion (yellow dot) drillings is indicated. The regions Tuscany (yellow) and Latium (orange) are indicated in the box at the top right. The numbers indicate the position of Lido di Tarquinia (1), Tiber river mouth (2), M. Circeo promontory (3) and Garigliano river mouth (4). From Google Earth image 2019. WGS84 Geographic Coordinate System refers to the vertices.

2. Regional Setting

From a geological point of view, the study area falls inside a narrow coastal belt, where Quaternary marine-coastal sediments outcrop, confined to the N and NW by hills made up of the pre-Quaternary substrate, which is progressively downthrown seaward by normal faults. The substrate is made up of lithotypes of the Tuscan metamorphic basement, on which intensely deformed Meso-Cenozoic sedimentary units tectonically overlap (Ligurian *l.s.* and the Falda Toscana Auct. units) [30,31]. Upper Miocene sediments follow upwards [30–32]; these sediments were deposited during the extensional phase related to the opening of the Tyrrhenian basin [33] and references therein Ref. [34]. The development of Pleistocene Vulsino volcanism, which is a few tens of kilometres from the study area, is linked to the same extensional phase [31].

3. Materials and Methods

In order to reconstruct the Holocene evolution of the Burano paleo-lagoon, at first a geomorphological analysis was carried out. Successively stratigraphical, sedimentological and geochronological analyses were conducted on sixteen cores drilled in the area (Figure 1). Moreover, palynological and microfaunal analyses were performed on selected cores in base of their chronology and sedimentary features.

3.1. Geomorphology

Historical maps relative to nineteenth century, bibliographic-cartographic literature, satellite images (Google Earth 2018) and aerial photos (1990) and field surveys (2016–2017) were utilized for the geomorphological analysis of the study area.

3.2. Lithology

Mechanical drillings and sampling—12 percussion and 4 mechanical rotary cores were carried out. The first ones (label BU1-12) were drilled by means of a coring system produced by AFgtc s.r.l. [35,36]. Both equipment and technique were suitable for the recovering of continuous cores, in which the sedimentary structures were preserved. The cores are 4–5 m long except for the core BU2, which is only 0.80 m long. Its position coincides with BU3 and is not shown in Figure 1. The second ones (label LB1-4) were drilled by the Geoambiente soc. Coop. a.r.l. company, which performed archaeological cores for the Special authority for the archaeological heritage of Rome and several dozen geomorphological cores in the Tiber delta (ANR-POLTEVERE, ERC Portus-Limen). The mechanical drill-coring machine is equipped with a casing to prevent the hole from collapsing during the drilling operations. Sediment cores were retrieved and emptied without water by using a hydraulic extruder in order to avoid any kind of contamination and/or disturbance. This method provides continuous and deep core reaching depths between 6 and 15 m from the ground level. The lithology was described for all cores, considering a core stretching/shortening error, and sediment samples were collected for the sedimentological analyses. The samples were numbered indicating the depth in centimetres from the core top. SEM and diffractometric analyses were performed.

3.3. Geochronological Analysis

^{14}C dating was carried on 22 samples from cores BU1, 4, 5, 7, 10, 11, and 12. In LB cores, 25 AMS ^{14}C and 5 datings by Optically Stimulated Luminescence (OSL) were carried out.

In BU cores, dating was performed on plant remains ranging from herbs to variably altered woody debris, that had been previously decontaminated to remove carbonates, as well as acid- and alkali-soluble organics. The residues were then burned, and the resulting CO_2 was used for the synthesis of benzene, the means for the ^{14}C analysis using Liquid Scintillation Counting (LSC). The conventional ages, corrected for the C isotope fractionation to $\delta^{13}C = -25\permil$ and calculated according to Stuiver & Polach [37], are reported in year BP (present time set at 1950). The conventional ages were calibrated according to Ref. [38]) and are given as calibrated years before the present (calibrated year BP). The uncertainties of both the conventional and calibrated ages are at the level of $\pm$ 1σ (i.e., 68% probability).

The radiocarbon datings relative to sediments in LB cores were mainly performed on *Posidonia* fibres, wood, plant remains and charcoal as well as occasionally on organic matter and shell. Charcoal and wood undergo an acid-base-acid pre-treatment (ABA). *Posidonia*, plant material and other organic matter are only treated with acid. Each acid or base wash is followed by several rinses with Merck Millipore MilliQ™ (Darmstadt, Germany) ultrapure deionized. After the final rinse, samples are dried prior the combustion. The standard pre-treatment method for shells involves surface cleaning by air abrasion with aluminium oxide powder to remove the outer surface and rinsing with ultrapure water. The sample is then dried. All glassware used is baked out at 500 °C prior to use for a minimum of 3 h to remove any organic contaminants. A few milligrams of the sample in solid form were placed in clean tin capsules and weighed. They were then combusted in an Thermo Finnigan Flash EA 1112 (San Jose, CA, United States) elemental NC analyser under a helium stream in the presence of copper oxide at 980 °C. The CO_2 produced in the combustion or by reacting in vacuo with phosphoric acid for the shell, was transferred to a glass ampoule, which was then sealed and sent to the Laboratory for Carbon 14 Measurement (LMC14) at Saclay (France). There, the CO_2 gas was reduced to solid graphite and measured by the Accelerator Mass Spectrometer or AMS (NEC tandem accelerator of 3MV), labelled "Artemis". The δ^{13} C values reflect the original isotopic composition in the sample only very roughly because the graphitisation process and the AMS spectrometer (unlike normal mass spectrometer) introduces significant isotopic fractionation.

Optical Stimulated Luminescence (OSL) dating of quartz extracts followed standard preparation techniques (chemical pretreatment, density separation, HF etching (cf. [39]). Measurements with the Single Aliquot Regenerative Dose (SAR) protocol [40] were done on a Freiberg Instruments Lexsyg Smart (Freiberg, Germany) device [41]. A preheat at 210 °C for 10 s was used and stimulation was for 50 s at 125 °C (detection at 380 nm). The distributions of the equivalent dose (D_e) expressed in grays (Gy), for most samples show a positive skewness, why the Minimum Age Model [42] was applied. For one sample (LB4), the D_e values spread from 50 Gy to more than 400 Gy without a distinct population and this sample is considered not datable. The concentration of dose-rate relevant elements was done by high-resolution gamma spectrometry (cf. [43]). Ages were calculated using ADELE-v2017 software (add-ideas.de).

3.4. Sedimentological Analysis

A total of 37 clastic sediment samples were taken from BU cores and analysed, except of BU2 core due to its short length.

A basic sedimentological description was carried out on all the cores, highlighting the main sedimentary structures. Sieving was carried out for the coarser fraction (0.074 mm), while the finest one was determined using the densiometric technique. Samples were classified based on gravel-sand-silt-clay ratio according to the Folk classification scheme [44]. Moreover, 136 non-organic sediment samples were also taken from LB cores. For their analysis, about 30 g of dry bulk sediment were deflocculated in 2.5% sodium hexametaphosphate solution for 24 h before being wet sieved (2 mm and 63 µm). The dried coarse (>2 mm) and sandy (2 mm to 63 µm) fractions were weighted and the fine fraction (<63 µm) calculated to discriminate textural groups (according to gravel-sand-mud triangular diagrams described by Folk [44]).

3.5. Microfaunal Analysis

Foraminiferal analyses were performed on BU1, 3, 5 and 7 cores, while Ostracoda were studied on samples from all LB cores.

3.5.1. Foraminifera

For this purpose, 21 samples from BU1, 14 samples from BU3, 32 samples from the BU5 and 30 from the BU7 cores were collected, with different sampling resolution, giving special emphasis on the lithological changes. For each sample, 50–100 g of sediment was soaked, treated with hydrogen peroxide, washed over a 63 µm sieve and then oven-dried at 50 °C. Qualitative analyses was conducted on all samples while quantitative and statistical analysis was very difficult to carry out for the scarcity of the foraminiferal content and the holigothipic feature characterizing the samples. In particular, quantitative analysis was performed when the fauna was sufficiently present and split in aliquots containing at least 200–300 benthic specimens. Species diversity was quantified considering the number of taxa occurred in the samples (S) [45] and, where the count was possible, Shannon Weaver (H) and Fisher α-index were calculated using the PAST (PAlaeontological STatistics) version 1.38 data analysis package [46]. In each sample, between 3 and 38 taxa were identified, most of them displaying very low frequencies (1–4 individuals/sample). For generic attributions of benthic foraminifera see Ref. [47]. Taxonomic identifications at species-level were based on [48–50] and complied with the latest name list of the online database of the World Register of Marine Species [51]. The distinction of different benthic foraminiferal biofacies throughout the cores relies chiefly on the ecological preferences of the most abundant species and microfaunal diversity.

3.5.2. Ostracoda

One hundred and thirty-six sediment samples from LB cores were analysed for ostracods (LB1:65, LB2:36, LB3:41 and LB4:21). Ostracods were recovered from 98 samples (LB1:45, LB2:10, LB3:36 and LB4:11) and 50 species were identified. The sandy fraction of the texture samples was dry sieved and the ostracods were extracted from fine to medium sand-sized sediment residues (500 to 125 µm). Up to 100 ostracods were picked using a stereomicroscope (Diaspore model from Nachet). The best-preserved ostracod shells were identified using Environmental Scanning Electron Microscopy (ESEM) images [52] with reference to [53–59] and the taxonomic attribution follows [51]. The ostracod density was expressed as the number of valves/g and the species richness (diversity) as the number of species in each sample. The ostracod taxa were divided in 5 ecological groups according to their salinity tolerance and their habitat: freshwater to low brackish, brackish, lagoonal coastal, phytal coastal and marine (Table 1).

Table 1. Ecological groups with the related taxa (in bold the common taxa > 5 valves).

Ecological Groups	Taxa (Binomial Nomenclature)
Freshwater to low brackish	*Candona* sp. **Candona angulatata** **Darwinula stevensoni** *Herpetocypris* sp. *Heterocypris salina* *Ilyocypris* sp. *Limnocythere inopinata* *Paralimnocythere* sp. *Pseudocandona* sp.
Brackish	**Cyprideis torosa** **Loxoconcha elliptica**
Lagoon/coastal	**Loxoconcha sp.** **Xestoleberis communis** **Xestoleberis dispar**
Phytal coastal	**Aurila sp.** **Aurila woodwardii** **Bairdia mediterranea** *Callistocythere pallida* **Cushmanidea turbida** **Cytheretta sp.** *Cytheridea neapolitana* **Hiltermannicythere rubra** *Leptocythere* sp. **Leptocythere bacescoi** *Leptocythere* cf. *ramosa* **Leptocythere fabaeformis** *Microcytherura angulosa* *Microcytherura fulva* **Neocytherideis fasciata** *Neocytherideis (Sahnia) subulata* *Paracytheridea depressa* *Procytherideis (Neocytherideis) subspiralis* *Sagmatocythere littoralis* *Semicytherura* sp. *Semicytherura* aff. *rara* *Semicytherura amorpha* *Semicytherura incongruens* *Semicytherura sulcata* *Urocythereis favosa*
Marine	*Carinocythereis whitei Costa punctatissima* *Cytheropteron latum Eucythere curta* *Eucytherura angulata* *Jugosocythereis* sp. **Paracytherois mediterranea** *Paradoxostoma bradyi*

3.6. Palynological Analysis

Forty-eight samples were taken between 0 and 430 cm from the BU1 core. Twenty-four samples were processed at the Palynological Laboratory of the University of Florence. An additional 24 samples were processed at laboratory of University of Utrecht as part of an Erasmus plus traineeship. Both laboratories used the exact same standard procedure for palynological processing. One *Lycopodium clavatum* tablet containing a known amount of spores was added to each exactly weighted dried sample to determine palynomorph concentrations. The samples, ~1.00 to 3.3 g in weight, were treated with 30% HCl, cold 38% HF, and briefly with KOH. Residues were then sieved in an ultrasonic bath using a 10 μm polymer mesh. Thirty-eight mobile slides were mounted using glycerol and analysed using optical microscopes. Slides are very rich in both pollen and Non-Pollen Palynomorphs (NPPs). In particular, land-derived palynomorphs include pollen, fungal remains (e.g., spores, hyphae, and fruiting bodies) and embryophyte spores. Freshwater palynomorphs include algal zygospores, especially Desmidiaceae (e.g., *Cosmarium*) and Zygnemataceae (e.g., *Mougeotia*), invertebrate mandibles, aquatic fungi, *Pediastrum* and *Botryococcus* algae. Marine palynomorphs include cyst of dinoflagellates, and foraminifer linings. The preservation of both pollen grains and NPPs is generally good; in fact, degraded, corroded or broken remains are very rare; moreover, there is no evidence for reworking of palynomorphs.

The results of the pollen analysis are summarized in a pollen percentage diagram including two main groups, i.e., Arboreal (AP) and Non Arboreal (NAP) taxa; here the main percentage sum is based on terrestrial pollen excluding NPPs. Moreover, on the right of the summary pollen diagram, the occurrence of selected NPPs is marked.

4. Results

4.1. Geomorphology

A depression parallel to the coast with WNW-ESE trend, limited landward by Pleistocene sandstone deposits and seaward by a coastal dune system, characterizes the studied area. Near the depression, the sandstone deposits rise to about 20 m a.s.l. Two ridges distant from each other, about 100 m and both approximately 70–80 m wide, constitute the dune system (primary and secondary dune). This system is confined between depression containing the paleo-lagoon and the present backshore. The maximum altitude of the outer dune system is close to 8 m a.s.l. and a dense shrub vegetation of Mediterranean maquis covers the secondary dune. The back-barrier is about 1 km wide. Before the nineteenth-century reclamation, the depression was a wetland periodically submerged; the Fosso Melone and the Fosso del Chiarone rivers flowed into the depression, in the central and eastern part, respectively (Figure 2). Currently, the depression is almost flat and at some decimeters below sea level. The depression is crossed by several reclamation canals and includes the Burano Lake, the only submerged area after the old lagoon reclamation. The lake is about 3.5 km long, 0.5 km wide and covers an area of about 1.5 km^2, with a maximum depth of about 1 meter. An inlet, located at the centre of the outer side, connects the lake to the sea. Washover fans are particularly evident in the north-western area of the lake.

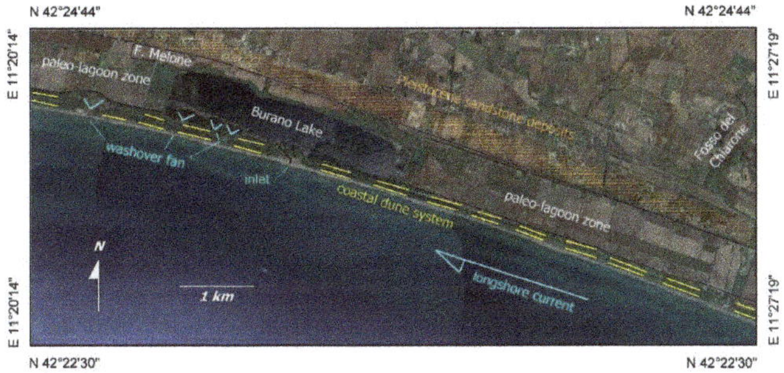

Figure 2. Geomorphological sketch map. The Fosso del Chiarone (river), which today flows into the sea, flowed into Lake Burano before the reclamation works. The washover fans indicate in the sketch are inactive today. From Google Earth image 2019. WGS84 Geographic Coordinate System are referred to the vertices.

4.2. Lithological Units

Five lithological units (Figure 3a,b) were recognized based on the cores.

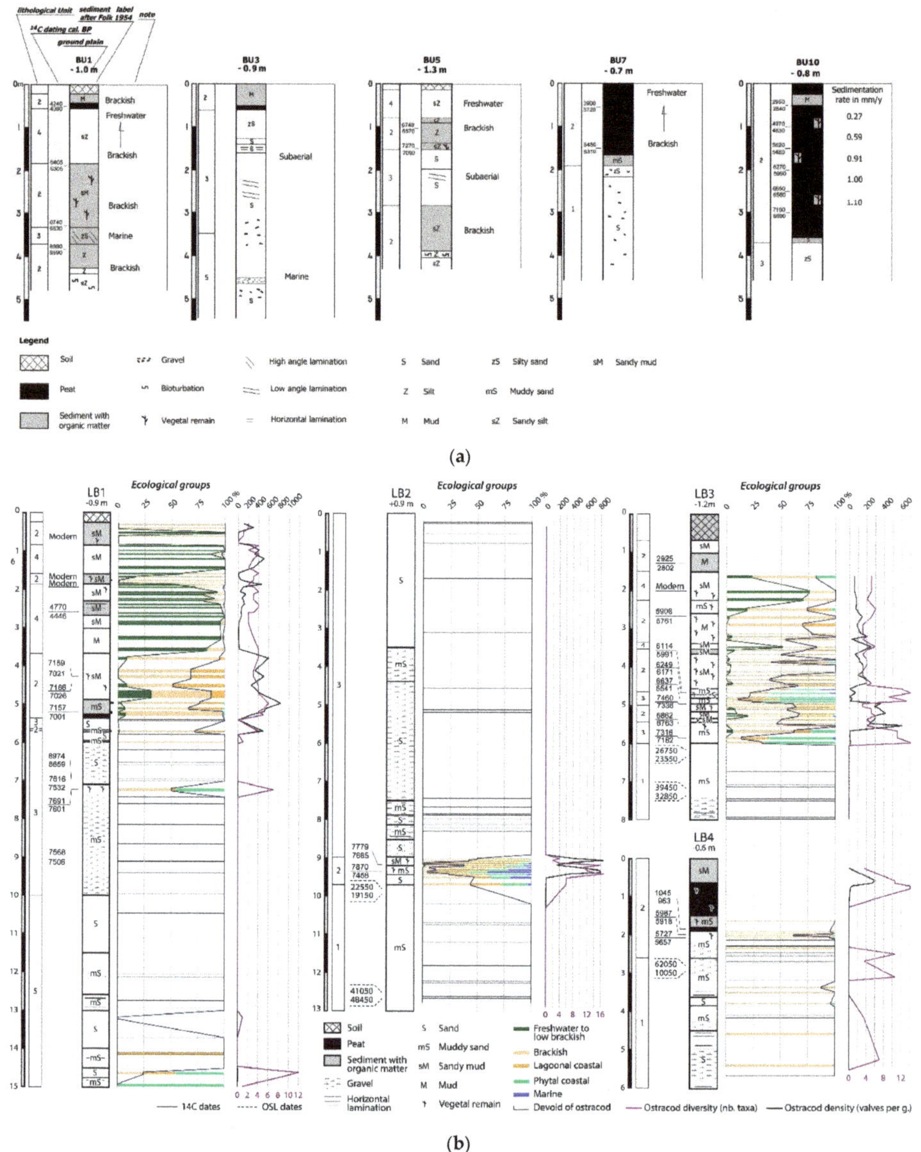

Figure 3. (**a**) Litho-stratigraphy and environmental evolution along BU1, 3, 5, 7, 10 percussion drillings. In BU10 core, the sedimentation rates in the peats are reported. (**b**) Litho-stratigraphy and ostracod ecological groups distribution along LB1, 2, 3, 4 continuous drillings. Age cal BP.

- Unit 1—Ocher and locally cemented quartz silty sands (zS) constitute this unit. Rare fine gravel with occasional centimetric pebbles and local plane-parallel lamination are present. The medium-sorted sediment shows uni or bimodal grain-size distribution. This unit outcrops along the inner edge of the depression and it was intercepted at the base of the BU4, 7 and 8, LB 2, 3 and 4 cores. The unit was crossed for about 3 m without ever reaching the bottom.
- Unit 2—It is a complex lithology unit mainly consisting of: soft peats, sandy mud (sM), mud (M), silt (Z) and muddy sand (mS) with abundant organic matter. Some thin levels with shell debris; are present more frequently in the lower part of the unit. They show a high silt content and are locally bioturbated. There are remains of *Posidonia* and undecomposed vegetal matter, occasional bivalve fragments and thin-shell gastropods. In the upper part, the unit is characterized by black-brown decametric thick soil outcropping into the depression. The unit was intercepted in the central and upper part of almost cores with a thickness ranging from 0.50 to over 3.0 m.
- Unit 3—It is made up of fine to medium-coarse greyish, more rarely yellowish, no diagenized feldspathic quartz sands (S). Sands are well sorted with a unimodal grain size distribution curve. Muddy sand (mS) and silty sand (zS) are more rarely present. Horizontal and medium-low angle plane-parallel laminations are present and centimetric to decimetric dark levels with femic minerals (in particular augite) were observed too. SEM and diffractometric analyses showed rare garnets, amphibole, pyrite, ilmenite, titanite and magnetite. There are locally present oblate or flattened pebbles, rare fine gravel levels and bivalves with complete shell or more frequently fragmented. This unit was intercepted mainly in the drills closest to the sea (where sometimes it represents the only body intercepted) with maximum thicknesses of about 7 m (LB2) and with more reduced thicknesses in the central ones; it is not present in the inner cores except BU10. Locally, this unit is interbedded to Unit 2 where it sometimes shows a high angle lamination.
- Unit 4—Whitish $CaCO_3$ enriched silt (Z), scarce fine sand (S) and a bioclastic fraction characterize this unit. SEM and diffractometric analyses showed a composition constituted mainly of euhedral calcite crystals and subordinately secondary of gypsum. Downwards rare dark horizontal laminations were observed. This unit was intercepted in the upper part of the cores drilled in the central sector of the depression. The thickness is greater northward where it reaches about 2 m (LB1). In most parts of the cores, the whitish silt is interbedded with thin blackish peaty levels and in any case is always embedded in the Unit 2.
- Unit 5—This unit consists of medium to fine incoherent silty sand (zS) and muddy sand (mS). Moreover, there are horizontal plane-parallel laminations, rare gravel constituted of flat grains rare oxidized foraminifera and shell debris. This unit was intercepted, below Unit 3, in LB1 core for a thickness of about 7 m without ever reaching the bottom, and in BU3 core.

For the geochronological data, see Tables 2 and 3.

Table 2. Radiocarbon datings (LB samples [52]).

Sample Identifier	Core # (Depth in Core, m)	Material	Conventional ^{14}C Age (Year BP)	Calibrated Age (Calibrated Year BP)	$\delta^{13}C^{(2)}$ (‰, *vs.* SMOW (Standard Mean Ocean Water))
Rome-2334	BU-1 (0.45–056)	sandy peat	3775 ± 40	4200–4080	−21.7
Rome-2335	BU-1 (1.82–1.89)	clay peat	5770 ± 40	6405–6305	−26.0
Rome-2336	BU-1 (3.10–3.19)	silt peat	5860 ± 40	6740–6630	−24.8
Rome-2337	BU-1 (3.59–3.64)	clay peat	7940 ± 50	8980–8690	−23.1
Rome-2338	BU-4 (0.50–0.60)	peat debris	1010 ± 40	960–850	−23.7
Rome-2339	BU-4 (1.0–1.1)	peat level	3570 ± 40	4090–3920	−25.0
Rome-2340	BU-4 (1.75–1.85)	clay peat	5940 ± 50	6860–6670	−25.8
Rome-2341	BU-5 (0.98–1.08)	peat debris	5860 ± 45	6740–6570	−23.4
Rome-2342	BU-5 (1.44–1.54)	silt clay	6280 ± 50	7270–7090	−24.9
Rome-2342*	BU-7 (0.50–0.60)	peat level	3550 ± 40	3900–3720	−22.5

Table 2. *Cont.*

Sample Identifier	Core # (Depth in Core, m)	Material	Conventional ^{14}C Age (Year BP)	Calibrated Age (Calibrated Year BP)	$\delta\,^{13}C^{(2)}$ (‰, *vs.* SMOW (Standard Mean Ocean Water))
Rome-2343	BU-7 (1.44–1.57)	clay peat	5620 ± 45	6450–6310	−23.2
Rome-2349	BU-10 (0.50)	peat level	2790 ± 40	2950–2840	−22.9
Rome-2360	BU-10 (1.00)	peat level	4320 ± 40	4970–4830	−24.1
Rome-2364	BU-10 (1.50)	peat level	4840 ± 40	5620–5480	−22.5
Rome-2361	BU-10 (2.00)	peat level	5310 ± 40	6270–5990	−24.1
Rome-2365	BU-10 (2.50)	peat level	5840 ± 40	6650–6560	−22.6
Rome-2350	BU-10 (3.00)	peat level	6115 ± 40	7160–6890	−25.0
Rome-2348	BU-11 (1.31–1.46)	clay peat	3830 ± 35	4290–4150	−23.4
Rome-2347	BU-11 (2.83–2.97)	silty peat	5270 ± 40	6170–5940	−24.3
Rome-2346	BU-11 (3.53–3.67)	clay peat	6445 ± 45	7430–7320	−25.1
Rome-2344	BU-12 (0.50)	peat level	3700 ± 40	4090–3930	−23.5
Rome-2345	BU-12 (1.00)	peat level	4560 ± 45	5320–5050	−24.2
Lyon-14228(sacA-49736)	LB1 (0.70)	Wood	Modern	Modern	Unavailable
Lyon-14223(sacA-49731)	LB1 (1.75–1.78)	Plant material	Modern	Modern	Unavailable
Lyon-15107(sacA-53026)	LB1 (1.93–1.96)	Plant material	Modern	Modern	Unavailable
Lyon-14224(sacA-49732)	LB1 (2.57)	Charcoal	4060 ± 30	4779–4446	Unavailable
Lyon-15109(sacA-53028)	LB1 (3.64–3.71)	Wood	6180 ± 30	7159–7021	Unavailable
Lyon-15110(SacA-53029)	LB1 (3.93–3.96)	Wood	6200 ± 30	7166–7026	Unavailable
Lyon-14226(sacA-49734)	LB1 (4.25–4.28)	Wood	6155 ± 35	7157–7001	Unavailable
Lyon-15117(SacA-53036)	LB1 (5.35–5.40)	Wood	7945 ± 35	8974–8659	Unavailable
Lyon-15108(sacA-53027)	LB1 (5.47)	*Posidonia*	7095 ± 40	7616–7532	Unavailable
Lyon-14227(sacA-49735)	LB1 (5.56–5.62)	*Posidonia*	7190 ± 40	7691–7601	−14.66
Lyon-15111(sacA-53030)	LB1 (7.13)	Wood	6640 ± 30	7568–7506	Unavailable
Lyon-15112(SacA-53031)	LB2 (7.67–7.70)	*Posidonia*	7275 ± 35	7779–7685	Unavailable
Lyon-14229(sacA-49737)	LB2 (7.78–7.81)	*Posidonia*	7350 ± 40	7670–7458	−14.51
Lyon-14231(sacA-49739)	LB3 (0.40–0.43)	Organic matter	2775 ± 30	2925–2802	−26.14
Lyon-14230(sacA-49738)	LB3 (0.72–0.75)	Plant material	Modern	Modern	Unavailable
Lyon-15113(SacA-53032)	LB3 (1.03–1.06)	Shell	5095 ± 30	5908-5761	Unavailable
Lyon-14232(sacA-49740)	LB3 (3.06–3.09)	*Posidonia*	5655 ± 35	6114–5991	−14.13
Lyon-15118(SacA-53037)	LB3 (3.12)	*Posidonia*	5770 ± 30	6249–6171	Unavailable
Lyon-15119(SacA-53038)	LB3 (3.30–3.34)	Wood	5775 ± 30	6637–6541	Unavailable
Lyon-15114(SacA-53033)	LB3 (3.44–3.48)	Charcoal	6500 ± 30	7460–7336	Unavailable
Lyon-15115(SacA-53034)	LB3 (3.86–3.89)	Shell	6345 ± 30	6862–6763	Unavailable
Lyon-14233(sacA-49741)	LB3 (4.33–4.39)	Charcoal	6335 ± 35	7316–7182	Unavailable
Lyon-14234(sacA-49742)	LB4 (1.00–1.03)	Plant material	1085 ± 35	1045–963	Unavailable
Lyon-15116(SacA-53035)	LB4 (1.12–1.15)	Plant material	5190 ± 30	5987–5918	Unavailable
Lyon-14235(sacA-49743)	LB4 (1.18–1.21)	Wood	4970 ± 30	5727–5657	Unavailable

Table 3. OSL (Optically stimulated Luminescence) datings [52].

Sample	Depth (cm)	K (Bq/kg)	Th (Bq/kg)	U (Bq/kg)	W_{meas} (%)	W_{eff} (%)	D-Q (Gy ka^{-1})	Grain size (μm)	N	Od	D_e OSL (Gy)	Age OSL (ka)
LB2-1	840	1320 ± 80	75.0 ± 4.9	46.4 ± 2.8	15.4	20 ± 5	5.22 ± 0.40	150–200	24	0.28	118.86 ± 7.52	22.8 ± 1.7
LB2-2	1115	1580 ± 160	69.9 ± 3.9	55.0 ± 3.0	13.0	20 ± 5	5.92 ± 0.47	150–200	25	0.22	276.15 ± 14.01	46.7 ± 3.7
LB3-1	478	1400 ± 80	68.3 ± 3.8	47.1 ± 2.9	12.5	20 ± 5	5.28 ± 0.31	200–250	27	0.17	142.98 ± 5.89	27.1 ± 1.6
LB3-2	585	1340 ± 130	67.2 ± 3.9	46.1 ± 2.7	13.7	20 ± 5	5.09 ± 0.44	200–250	21	0.25	193.97 ± 12.42	38.1 ± 3.3
LB4	210	1470 ± 120	96.1 ± 5.8	64.1 ± 4.0	11.6	20 ± 5	6.26 ± 0.33	150–200	24	0.60	70–400	12–64

4.3. Paleovegetational Context

Pollen data from BU1 core provide the floristic and vegetation history of Burano Lake during the last 8000 years (Figure 4). Thermophilous deciduous forest composed especially of *Quercus cerris* and *Q. pubescens* followed by other sub-Mediterranean taxa such as *Carpinus orientalis/Ostrya*, *Corylus*, *Quercus suber*, dominate the landscape in the lowlands and hilly areas. Here, Mediterranean vegetation shows less extension than the sub-Mediterranean one; its main taxon is *Quercus ilex*, followed by Ericaceae, *Olea*, *Phillyrea*, *Pistacia*, etc. Pollen data indicate the presence although with low percentages of *Abies*, *Betula* and *Fagus*, typical taxa of high altitude; among them, *Fagus* shows the more continuous record. Plants growing under greater control of local edaphic conditions include prevalently herbaceous taxa such as Amaranthaceae and Cyperaceae which possibly expanded especially on the lake shoreline.

Mediterranean taxa show some phases of increase, the most evident of them occurred between 8000 and 5860 BP. However, over the previous fluctuations, no dramatic vegetational changes were pointed out throughout the pollen record also before 3775 BP, when a strong increase in herbaceous taxa occurred. In this context, Amaranthaceae exhibited at least two main phases of increase which possibly correspond to a decrease of the extension of the Burano paleo-lagoon.

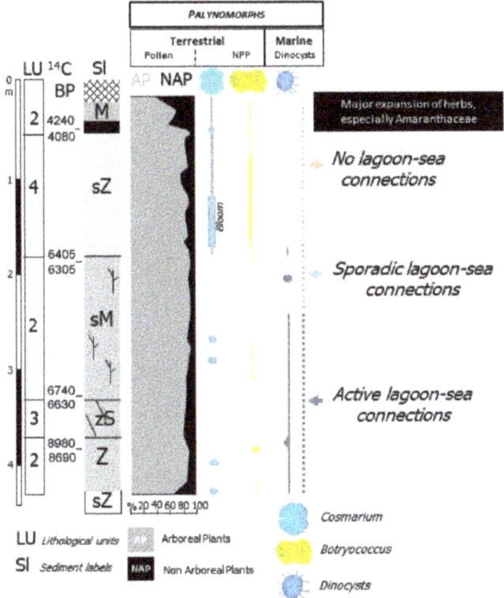

Figure 4. Summary of the palynological evidence from BU1 core. For the legend of sediment labels see Figure 3a. Age calibrated year BP.

4.4. Bio-Indicators of the Aquatic Paleoenvironmental Conditions in Each Units

Unit 1—*Ostracoda*: samples from LB cores are generally devoid of ostracods except for a few LB4 samples showing low ostracod density. The assemblages are highly dominated by the euryhaline taxon *Cyprideis torosa* monospecific or associated with scarce valves of brackish (*Loxoconcha elliptica*) or lagoon/coastal (*Xestoleberis dispar*) species. These assemblages, dominated by *C. torosa*, have low densities in semi-permanent waters of lagoon marginal areas [60,61].

Foraminifera: only oxidized shallow water foraminifera and rare molluscs fragments characterize this unit.

Unit 2—*Ostracoda*: in the LB cores, the samples are mainly characterized by a high ostracod density and a relatively low diversity (2 to 6 taxa). Ostracods are mainly composed of both brackish (*C. torosa* and *L. elliptica*) and lagoon/coastal (*X. dispar*) taxa with variable proportions, all three of which can be dominant or co-dominant. They are occasionally associated with a few phytal coastal (*Aurila* spp. and *Leptocythere* spp.) and/or freshwater to low brackish specimens (*Candona angulata* and *Darwinula stevensoni*); this assemblage is characteristic of an open lagoon environment. The abundance of *X. dispar* indicates polyhaline conditions with substantial seawater inputs [53,62]. In the LB2 core, between 915 and 970 cm, the samples have a low ostracod density and a high diversity (7 to 17 taxa). These assemblages are dominated by *C. torosa*, *X. dispar* or the marine species *Paracytherois mediterranea*, mainly associated with *L. elliptica*, lagoon/coastal (*Loxoconcha* sp. and *Xestoleberis communis*) and phytal coastal taxa (main species are *Cushmanidea turbida* and *Urocythereis favosa*). Sporadic specimens (1 to 3 valves) of about 20 other coastal and marine taxa are reported. The abundance and diversity of marine and coastal taxa is common in euhaline to polyhaline lagoon environment with a strong seawater influence typical of sandy bar inlets surroundings [61,63]. In core LB1, between 20–40 cm, 160–190 cm and 520–530 cm occurs a bispecific assemblage dominated by *C. torosa* associated with *L. elliptica* characteristic of confined lagoons [61,63–66]. The association of these two euryhaline species suggests intra- and inter-annual variations in salinity. In the LB4, two samples (165–168 cm and 233–238 cm) are monospecific with scarce valves of *C. torosa*, which could indicate semi-permanent waters [60,61]. In core LB1, between 40 and 60 cm each sample contains only one reworked valve.

Foraminifera: they are mainly represented by brackish lagoonal assemblage containing an oligospecific fauna characterized by the high dominance of *Ammonia tepida* (75–91%) accompanied by few species (mainly *Haynesina germanica* 4–8% and *Porosononion granosum* 1–19%). The low diversity (S: 3 to 7 taxa, Fisher α-index: 0.4 to 5, H: 0.3 to 1.7) and the high dominance of *A. tepida*, a euryhaline species, are indicative of an enclosed brackish lagoon [67–71]. This assemblage is recorded in BU5 core, from 330 to 300 cm and from 122 to 81 cm, and in BU7 core, from 216 to the top. Towards the top of the unit (BU7: from 166 cm to the top) it is characterized by a clear transition towards freshwater conditions highlighted by a decrease of foraminiferal content. Similar environmental condition is recognized also in the upper part of the BU1 core (from 61 to 30 cm) where the assemblage is similar to that written above for the upper part and it is associated with abundant gastropods like *Hydrobia* spp. and *Planorbis* spp.; while at the lower part (bottom to 386 cm), more marine taxa (*Nonion* spp., *Triloculina* spp. and *Quinqueloculina* spp.) indicate a significant sea influence. This unit in BU1 core is also characterized by the presence of Characeae oogones and monassone spicules of porifers. In BU3 core, this unit is barren.

Palynology (NPPs): palynology was studied in core BU1. Between 426 and 371 cm depth, marine dinocysts are usually present although in low abundance compared to the terrestrial palynomorphs. In particular, at 387 cm dinocysts increase and are well represented by *Spiniferites mirabilis*, an indicator of warm-temperate conditions and *Lingulodinium machaerophorum*, often found in estuarine, coastal/neritic environments in modern sediments. *L. machaerophorum* thrives in nutrient-rich conditions and can reach high abundances in river-dominated marine areas [72]. Blooms of its motile form can cause toxic red tides. The other species found are typical of neritic to coastal warm-temperate environments, such as *Operculodinium centrocarpum*, *Operculodinium israelianum*, *Selenopemphix quanta*, *Spiniferites belerius*, *Spiniferites bulloides*, *Spiniferites membranaceous*, *Spiniferites ramosus*. Notably, typical oceanic taxa such as *Impagidinium* spp. are absent. Some foraminifer linings are also present; such organic test linings probably belong to calcareous benthic foraminifers, the occurrence of which indicates a shallow depositional environment, nutrient-rich waters and relatively more fully marine salinities. *Botryococcus* (Chlorophyceae, Chlorococcales/Tetrasporales), a fresh–brackish water colonial green algae is quite abundant. Throughout this interval, *Pseudoschizaea* (Incertae sedis) and some hemi-cells of the prevalent freshwater Desmidiaceae *Cosmarium* have also scattered occurrences. Terrestrial Fungi are always abundant. Between 332 and 188 cm, Fungi spores, *Pseudoschizaea*, *Botryococcus* and especially *Cosmarium* decrease markedly. Dinocysts also decrease toward the top of the interval. Some foraminifer linings are present. From 56 to 42 cm, marine dinocysts are absent whereas *Cosmarium*, *Botryococcus* and fungal spores are present.

Unit 3—*Ostracoda*: samples from LB2 core are devoid of ostracod fauna. Samples from LB3 core have a high ostracod density and a low specific diversity (9 to 14 taxa). The assemblages are slightly dominated by the euryhaline species *C. torosa* or the phytal coastal species *C. turbida* associated with brackish (*L. elliptica*), lagoon/coastal (*Loxoconcha* sp. and *Xestoleberis* spp.) and phytal coastal taxa (mainly *Aurila* spp., *Bairdia mediterranea*, *Cytheretta* spp., *Hiltermannicythere rubra*, *Leptocythere* spp., *Neocytherideis* spp., *Semicytherura* spp. and *U. favosa*). It can be associated with an open lagoon environment strongly influenced by seawater and the closeness of a sandy bar inlets. In the LB1 core, the samples are devoid of ostracods or contain one or two reworked valves except for one sample (720–730 cm) which has similar characteristics to the LB3 samples.

Foraminifera: in this unit the foraminiferal assemblage shows, in respect to the Unit 2, a clear decrease of lagoonal taxa (e.g., *Ammonia tepida* displays a decreasing trend from BU5-285 to BU5-283) and an increase of typical shallow marine species like miliolids (>22% *Quinqueloculina* spp. and *Triloculina* spp.), *Elphidium* spp. (4–8%) and *Rosalina* spp. (7–40%) [49,58,73]. This assemblage presents very few specimens but high diversity (S: 18 to 38 taxa, Fisher α-index: 6.0 to 20.4, H: 1.8 to 3.1). The environment related to Unit 3 can be associated with a lagoon affected by occasional marine connection. This assemblage is recorded in BU5 core (from 300 to 132 cm) and in BU1 (from 370 to 320 cm) and BU3 (from 405 to 369 cm) cores where siliceous spicules of poriferas were found too.

Moreover, it is characterized by abundant remains of molluscs (*Cardium* spp., *Glycimeris* spp., *Venus* spp. and tellinids) vermetids, bryozoans, coral algae and remains of echinoids. In BU3 core, the upper part of the unit, contains only reworked molluscs.

Palynology (NPPs): data have been collected from the BU1 core between 371 and 332 cm. Samples are mostly barren in NPP but also pollen probably due to the unfavourable lithology marked by a high content in sand. Marine dinocysts such as *S. belerius, S. membranaceous, S. ramosus* are present along foraminifera linings and *Pseudoschizaea*.

Unit 4—*Ostracoda*: samples from LB cores have a low ostracod density. In the LB1 samples, ostracods are low diversified (2 to 4 taxa) and highly dominated by the freshwater to low brackish species *C. angulata* mainly associated with a freshwater to low brackish (*D. stevensoni*) and/or a euryhaline species (*C. torosa*). *C. angulata* tolerate limnetic to mesohaline waters but clearly prefers slightly salty waters and are common in slightly brackish coastal ponds [70]. In the LB3 samples, ostracods are moderately diversified (5 to 6 taxa) and dominated by *C. torosa* mainly associated with freshwater to low brackish (*C. angulata, D. stevensoni* and *Limnocythere inopinata*) and brackish taxa (*L. elliptica*). These two assemblages are common in low brackish lagoon environment. In the LB3 core, *L. elliptica* which tolerates oligohaline to polyhaline waters and the dominance of *C. torosa* for which maximum productivity occurs at salinities between 2 and 16.5‰ [70] indicates slightly higher salinities than in the LB1 core.

Foraminifera: unlike ostracods, in this unit benthic foraminifera are rare or absent with a very low diversity (S: 3 to 4 taxa, Fisher α-index: <1, H: 0.3 to 0.7), advocating freshwater or low brackish conditions. It is recorded in BU1 core from 164 to 72 cm and in BU5 core from 91 cm to the top. Where it is associated with abundant gastropods (*Hydrobia* spp. and *Planorbis* spp.) and frequent oogones.

Palynology (NPPs): the interval between 188 and 56 cm of core BU1 shows some sporadic/rare presence of dinocysts (*Spiniferites* spp.) along with few hemi-cells of the freshwater *Cosmarium* and fresh–brackish water *Botryococcus* at the base. However, in the interval from 170 to 64 cm, a sudden change in the NPP assemblages occurs, marked by the increase of *Cosmarium* and *Botryococcus* as well as by the disappearance of dinocysts. In more detail, *Cosmarium,* shows an acme phase between 170 and 123 cm and *Botryococcus* is also well represented although in lower occurrence.

Unit 5—The samples include scarce reworked valves of coastal or lagoon ostracods. Only one sample, at bottom of LB1 core, present a low ostracod density with a high diversity (12 taxa). This assemblage is slightly dominated by the phytal coastal species *C. turbida* associated with lagoon/coastal (*Loxoconcha* sp. and *Xestoleberis* spp.), brackish (*C. torosa* and *L. elliptica*) and phytal coastal taxa (*B. mediterranea, Cytheretta* sp., *H. rubra, Semicytherura sulcata* and *U. favosa*). Rare tellinids and oxidized foraminifera are also present.

5. Discussion

5.1. Facies and Reciprocal Stratigraphic Relationships (Figure 5):

- Coastal Pleistocene facies (CP)—It is characterized by the lithological Unit 1 in which rare fragments of bivalves, oxidized shallow waters foraminifera and rare brackish and euryhaline ostracods have been found. The measured age varies between about 45 to about 20 ka BP. The lithology, the faunal content and datings, identify a generic coastal plain developed during the Upper Pleistocene.
- Lagoon Holocene facies (LgH)—It includes sediments that refer prevalently to the Unit 2 in which the faunal content consists of predominant brackish and lagoonal/coastal taxa. Sometimes these taxa are accompanied by species more suitable for freshwater environment. It is confirmed by the absence of foraminiferal assemblages in the upper part of the unit. The decrease of marine dinocysts confirm a minor lagoon-sea connection in the time. According to the composition and stratigraphical distribution of dinocysts a main seawater input is documented close to about 8 ka BP (BU1 426 to 371 cm) whereas the marine influence progressively decreases until it disappears towards the more recent times. This facies develops mostly until 3.5 ka BP, even if locally, it is

also present in more recent times (2.8–1.0 ka BP, Table 1). LgH facies is attributable to a coastal Holocene lagoon significantly influenced by the sea in the period before 6 ka BP even if phases of greater or less marine connection alternate over time. Using the calibrated ages in the BU10 core, where the organic sediments of this facies are more continuous and lacking clastic intercalations, the peat sedimentation rates were estimated (Figure 3a). These progressively decrease from 1.1 to 0.27 mm/year.

- Beach-dune Holocene facies (BdH)—It includes Unit 3 deposits characterized by the occurrence of prevalent marine bio-indicators; all previous evidences support the development of foreshore/dune, sand barrier environment and subordinately washover fans.

- Lacustrine Holocene facies (LH)—It corresponds to the whitish $CaCO_3$ enriched Unit 4. The faunal component is essentially made up of freshwater gastropods and freshwater/low brackish ostracods. NPPs as a whole, confirm a prevalent freshwater environment; indeed, the main components of the NPPs assemblages are *Cosmarium*, indicative of prevalent freshwater conditions and *Botryoccoccus*, which thrives in fresh–brackish waters. Notably, their abundances (especially that of *Cosmarium*) indicates a bloom around 5770 year BP (in BU1). Such bloom may indicate a sudden change in one or more environmental parameters such as nutrients (e.g., phosphorus and nitrogen), temperature or pH. It is also necessary to understand if the phenomenon was natural or anthropic-induced. *Cosmarium* and *Botryococcus* proliferate in oligo-to mesotrophic conditions, and therefore a sudden eutrophication episode can be excluded. The absence of other taxa such as *Pediastrum*, aquatic algae commonly present in freshwater habitats rich in mineral and organic nutrients confirms the oligotrophic conditions. Desmids are sensitive organisms and act as an indicator of water quality, concentration of chemical oxygen demand, nitrate and turbidity; *Botryococcus* can dominates (or be present) in relatively extreme environments which prevents for example the occurrence of *Pediastrum*; it is often present when water is clear, oligotrophic, eventually dystrophic [74–76]. All this evidence permits to define a freshwater depositional basin with clear and oligotrophic waters. Moreover, the abundance of calcite suggests medium-high pH waters and probably microbially induced calcium carbonate precipitation a phenomenon also associated with the so-called "Whitings" (e.g., [77,78]). The development of this facies starts after 6 ka BP and ends at about to 4 ka BP. The origin of this facies is not clear at present. Due to the absence of significant river inputs, it might assume a rise of fluids linked to the late hydrothermal phases of the Vulsino volcanism. Similar evidences were observed in the Specchio di Venere lake, a coastal basin in the volcanic island of Pantelleria (southern Sicily) [79].

- Shoreface Holocene facies (SH)—It is expressed by Unit 5. The presence of coastal malaco-ostracofauna partly reworked, shallow water oxidized foraminifera, as well as arenaceous pebbles and gravel levels, suggests that it is related to a shoreface environment. Locally (LB1), this environment could also be supplied by the remobilization of the Pleistocene sediments (CP facies) present at the inner edge of the depressed area.

The CP facies outcrops along the inner edge of the depression. At ESE of the Burano Lake, the CP facies deepens seawards (Figure 5). At WNW, the CP facies was not intercepted in the cores and therefore its seawards trend is not known. Stratigraphically overlapping on the CP facies, the LgH facies is present always with sharp contact. The chronological difference between the two facies highlights a sedimentary hiatus. The LgH facies shows a plane-concave geometry that tapers both seawards and landwards. The BdH facies, outcropping along the outer edge of the depression, constitutes the current Holocene beach-dune system. In the subsoil of south-eastern area, it displays a wedge geometry closed landward into LgH facies constituting a typical sand barrier. In the north-western area, the wedge geometry is only partially evident. The BdH and LgH facies are frequently heteropic and the relative contact is generally sharp or locally transitional. The LH facies presents a lenticular geometry and is always embedded in the LgH one resulting heteropic with this. Regarding to SH facies, there is not enough information to define its geometry.

The characters of the facies and their mutual stratigraphic relations highlight the presence of two depositional sequences pro-parte (Figure 6) [80,81]. The oldest sequence (A) is represented only by the Upper Pleistocene CP facies. A second sequence (B), represented by Holocene facies, overlies sequence A. The overlap is identified along the surface separating the CP and LgH facies; this surface constitutes a *Basal Unconformity (BU)* (red line in Figure 6). The oldest sediments of LgH facies (about 8.0 to 6.0 ka BP), together with the heteropic sediments of BdH, show a typically transgressive setting associated with the Transgressive System Tract (TST) of the sequence B (Figure 6). Therefore, the *BU* surface therefore constitutes partly a *Transgressive Surface (TS)* in blue line in Figure 6. Most recent deposits of facies LgH and BdH display a regressive setting constituting the Highstand System Tract (HST) of the B sequence that develops above the *maximum flooding surface (mfs)* (yellow line in Figure 6). The HST of B sequence includes the LH facies embedded into LgH facies. Beyond the junction point of the *mfs* on the *BU*, the HST lagoon sediments of the B sequence lay directly on the A sequence.

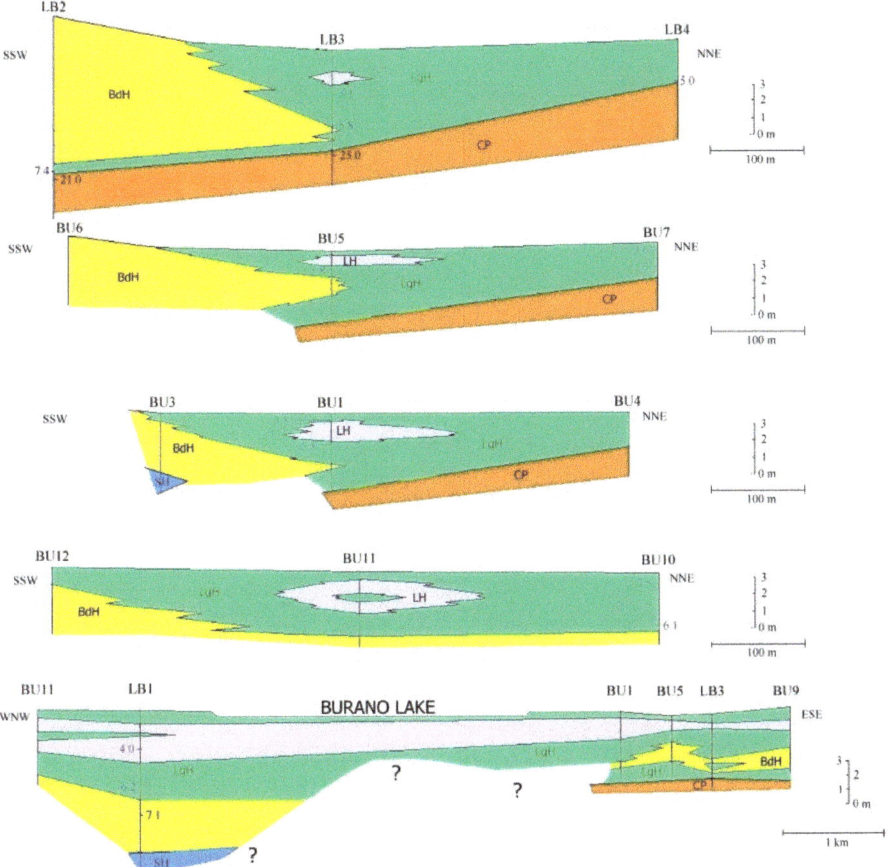

Figure 5. Schematic sedimentological cross-sections. Colours highlight the facies marked by relative label. For chronological reference, some ages approximate in ka BP was inserted.

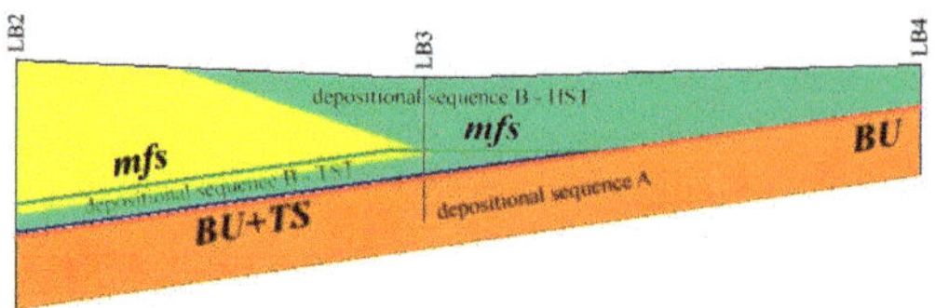

Figure 6. Sequence stratigraphy schematic setting. *BU* (red line): Basal Unconformity; *TS* (blue line): Transgressive Surface; *mfs* (green line): maximum flooding surface; TST: Transgressive System Tract; HST: High Stand System Tract. Depositional sequence A (brown) is constituted by coastal Pleistocene sediments. Depositional sequence B (TST + HST) is constituted by beach, bar and dune sediments (yellow) and lagoon/lacustrine sediments (light green).

5.2. Holocene Evolution

The paleoenvironmental history of the Burano paleo-lagoon can be summarized in 5 main phases starting from about 8000 years BP. This time interval includes the upper portion of the Neolithic up to the High Middle Ages, which largely centre the period characterized by increasing aridification phenomena (e.g., [9,82,83]). The vegetational and climate responses at the Mediterranean scale are documented by several palynological studies (e.g., [84–87] and references therein). Pollen data from the Burano paleo-lagoon show a prevalent warm temperate climate characterized by a quite humid weather conditions at least until about 4000 years BP despite the occurrence of greater seasonality phases. After 4000 years BP the significant increases of herbaceous cover, if not climate-induced, could be interpreted as an anthropogenic indicator of the impact of the Etruscan activities.

The RSL curve was drawn based on the calibrated ages from the sediments of the LgH and LH facies. Since the area is considered to be stable on average in the late-Quaternary, the curve was not corrected for the land vertical movements [88]. For each used data, the RSL was determined based on the equation RSL = A − RWL where: A is the altitude of the data (respect to the MSL) and RWL is a value equal to −0.5 (data from LgH facies) and 0 (data from LH facies) [89]. In Figure 7, the Burano RSL curve is compared with the RLS model curves of Sardinia [89] and Versilia plain [88]. These two locations have been considered as relatively close to the Burano area and considered on average stable.

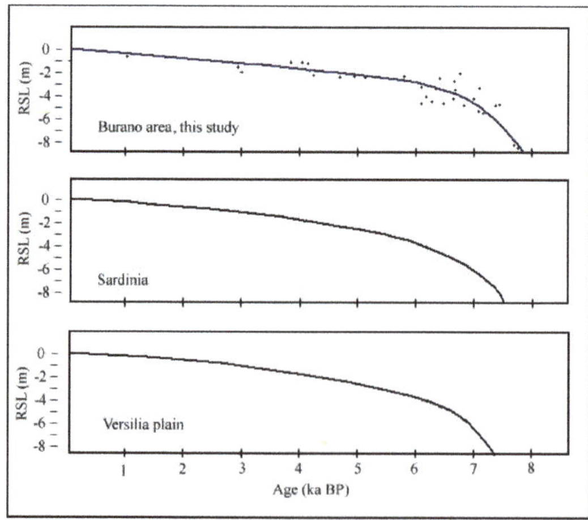

Figure 7. Comparison between the Burano RSL curve and the curves for Sardinia and Versilia plain modified from [88,89].

Phase 1 (Figure 8a)—In the easternmost sector of the study area, during the late transgression phase and before 7500 years BP, a sandy barrier (BdH) was located about 250 m seaward from the innermost shoreline consisting of Pleistocene sediments (CP). A narrow lagoon developed between the Pleistocene sediments and the sandy barrier (LgH) prone to seawater inflow. In the most western sector, the lagoon was not present and a large beach (facies BdH + SH) extended from the Pleistocene deposits seawards. The sandy barrier and beach were fed by the longshore current, carrying towards north-west the sediments from the Fiora, Arrone and Marta river mouths [90,91]. At this time, the sea level was close to −7 m when compared to the current one.

Phase 2 (Figure 8b)—About 6000 years ago, approximately at sea level still-stand, a barrier-lagoon system developed throughout the area. The boundary between the Pleistocene and Holocene deposits migrated to the northeast. The width of the lagoon varied from about 150 m in the eastern part to about 500 m in the western sector. There are no data that allow us to accurately evaluate the width of the barrier that even in the presence of washover events it began to limit the intrusion of seawater into the lagoon. During this phase, the sea level was about at −3 m compared to the current one.

Phase 3 (Figure 8c)—Even after the still stand, the sea level rose but with minor rates, and at around 5000 years BP, it was close to −2.5 m. In this phase, the width of the lagoon ranged from 500 to 700 m. The inner edge of the barrier shows a partial seaward migration in the eastern part. The LH facies occurs along the entire axis of the coastal basin.

Phase 4 (Figure 8d)—About 4000 years ago, the sea level was very close to −2 m. The planimetric features of the lagoon do not show a significant change in respect to the previous phase except for a limited seaward migration of the inner edge of the barrier in the westernmost part. However, the LH facies shows a more discontinuous distribution than the previous phase.

Phase 5—During the last 4000 years, LH facies disappeared and was overlain by limited thicknesses of LgH facies. Successively, the basin rapidly dried up, or, at least, turned into a wetland only locally and periodically submerged so that part of the organic sediments was subject to pedogenesis forming a thin brown soil. The reclamation made during the first part of the twentieth century, definitively dried up the entire area except the Burano Lake zone whose depth is currently not over a meter.

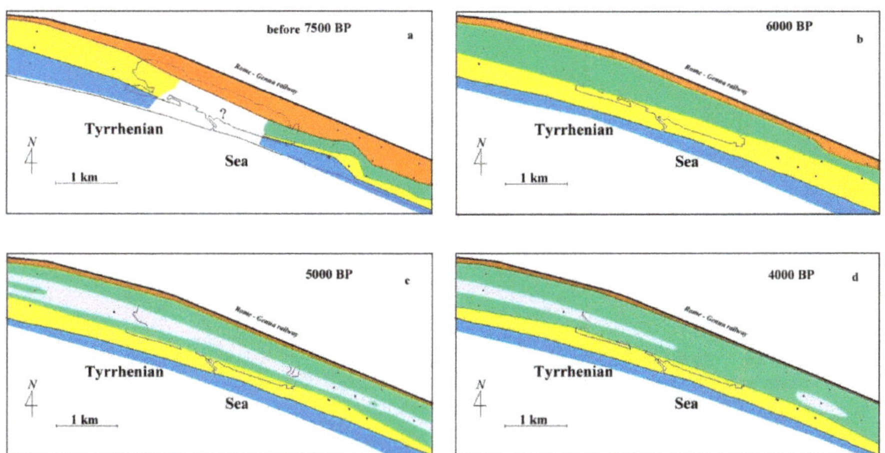

Figure 8. Schematic bidimensional landscape reconstructions of the Burano paleo-lagoon from 7500 to 4000 years BP (**a–d**). Coloured areas indicate: Pleistocene deposits (brown), lagoon/marsh Holocene deposits (green), Holocene sandy barrier and foreshore deposits (yellow), Holocene lacustrine deposits (grey), Holocene shoreface deposits (blue). The location of Burano Lake and the present shoreline is indicated.

6. Conclusions

A beach-dune-lagoon system evolved in the study area between about 8.0 and 4.0 ka BP, and it rests unconformably on the previous coastal sediments of Pleistocene age. In a first phase, the lagoon shows a significant marine influence and then, from 6000 years ago, it progressively changes to more closed environment with brackish water. Lagoon sediments are particularly rich in peat. Between 6.0 and 4.0 ka BP, along the central axis of the lagoon, a sedimentation particularly rich in $CaCO_3$ and bio indicators of freshwater-low brackish environment developed. From 4.0 ka BP ago, the lagoon quickly dried up becoming a locally and periodically submerged wetland. The area was reclaimed at the beginning of the 20th century.

Author Contributions: Conceptualization, M.D., P.B., A.B., L.D.B., C.V. and L.D.; Data curation, A.B., G.C., L.D.B., L.M.F., M.A.L., L.R., C.V., J.-P.G., L.S., P.C., F.P., C.O. and F.S.; Investigation, M.D., A.B., P.C.N., D.F., C.V. and J.-P.G.; Methodology, J.-P.G.; Supervision, P.B., L.D.B. and L.D. All authors have read and agreed to the published version of the manuscript.

Funding: This research received no external funding.

Acknowledgments: Fanny Guerin and Nora Dalal for their help in sieving samples from LB cores. Stoil Chapkanski (CNRS, Archéorient-UMR5133) and Camille Goncalves for the analysis of the Cosa samples. A special thanks to Maurizio Cacopardo for the collaboration during the drillings. We would like to thank SACRA Spa, in particular the Director Piergiorgio Santi, for the availability and collaboration shown, as well as for the authorization to carry out the mechanical drillings on the lands of the Company.

Conflicts of Interest: The authors declare no conflict of interest.

References

1. Kjerfve, B.; Magill, K.E. Geographic and hydrodynamic characteristics of shallow coastal lagoons. *Mar. Geol.* **1989**, *88*, 187–199. [CrossRef]
2. Dalrymple, R.W.; Leckie, D.A.; Tillman, R.W. Incised Valleys in time and space. *SEPM Publ.* **2006**, *85*, 343.
3. Marco-Barba, J.; Holmes, J.A.; Mesquita-Joanes, F.; Miracle, M.R. The influence of climate and sea level change on the Holocene evolution of a Mediterranean coastal lagoon: Evidence from ostracod palaeoecology and geochemistry. *Geobios* **2013**, *46*, 409–421. [CrossRef]
4. Marra, F.; Bozzano, F.; Cinti, F.R. Chronostratigraphic and lithologic feature soft the Tiber River sediments (Rome, Italy): Implications on the post-glacial sea-level rise and Holocene climate. *Glob. Planet. Chang.* **2013**, *107*, 157–176. [CrossRef]
5. Milli, S.; D'Ambrogi, C.; Bellotti, P.; Calderoni, G.; Carboni, M.G.; Celant, A.; Di Bella, L.; Di Rita, F.; Frezza, V.; Magri, D.; et al. The transition from wave-dominated estuary to wave -dominated delta: The Late Quaternary stratigraphic architeture of Tiber River deltaic succession (Italy). *Sediment. Geol.* **2013**, *284*, 159–180. [CrossRef]
6. Benallack, K.; Green, A.N.; Humphries, M.S.; Cooper, J.A.G.; Dladla, N.N.; Finch, J.M. The stratigraphic evolution of a large back-barrier lagoon system with a non-migrating barrier. *Mar. Geol.* **2016**, *379*, 64–77. [CrossRef]
7. Bortolin, E.; Weschenfelder, J.; Cooper, A. Holocene Evolution of Patos Lagoon, Brazil. *J. Coast. Res.* **2019**, *35*, 357–368. [CrossRef]
8. Dladla, N.N.; Green, A.N.; Cooper, J.A.G.; Humphries, M.S. Geological inheritance and its role in the geomorphological and sedimentological evolution of bedrock-hosted incised valleys, lake St Lucia, South Africa. *Estuar. Coast. Shelf Sci.* **2019**, *222*, 154–167. [CrossRef]
9. Mayewski, P.A.; Rohling, E.E.; Stager, J.C.; Karlen, W.; Maasch, K.A.; Meeker, L.D.; Meyerson, E.A.; Gasse, F.; Van Kreveld, S.; Holmgren, K.; et al. Holocene climate variability. *Quat. Res.* **2004**, *62*, 243–255. [CrossRef]
10. Vella, C.; Fleury, T.J.; Raccasi, G.; Provansal, M.; Sabatier, F.; Bourcier, M. Evolution of The Rhône delta plain in the Holocene. *Mar. Geol.* **2005**, *222*, 235–265. [CrossRef]
11. Mulligan, R.P.; Mallinson, D.J.; Clunies, G.J.; Rey, A.; Culver, S.J.; Zaremba, N.; Leorri, E.; Mitra, S. Estuarine Responses to Long-Term Changes in Inlets, Morphology, and Sea Level Rise. *J. Geophys. Res. Ocean.* **2019**, *124*, 12. [CrossRef]

12. Bellotti, P. Sedimentologia ed evoluzione olocenica della laguna costiera un tempo presente alla foce del Tevere. In Proceedings of the Atti del X Congresso Della Associazione Italiana di Oceanologia e Limnologia, Alassio, Italy, 4–6 November 1994.

13. Di Rita, F.; Celant, A.; Magri, D. Holocene environmental instability in the wetland north of the Tiber delta (Rome, Italy): Sea–lake–man interactions. *J. Paleolimnol.* **2010**, *44*, 51–67. [CrossRef]

14. Giraudi, C. The sediments of the 'Stagno di Maccarese' marsh (Tiber river delta, central Italy): A late-Holocene record of natural and human-induced environmental changes. *Holocene* **2011**, *21*, 1233–1243. [CrossRef]

15. Bellotti, P.; Calderoni, G.; Di Rita, F.; D'Orefice, M.; D'Amico, C.; Esu, D.; Magri, D.; Preite Martinez, M.; Tortora, P.; Valeri, P. The Tiber River delta plain (central Italy): Coastal evolution and implications for the ancient Ostia Roman settlement. *Holocene* **2011**, *21*, 1105–1116. [CrossRef]

16. Pannuzi, S. La laguna di Ostia: Produzione del sale e trasformazione del paesaggio dall'età antica all'età moderna. Available online: http://www.efrome.it/publications/resure-en-ligne.html (accessed on 25 November 2013).

17. Vittori, C.; Mazzini, I.; Salomon, F.; Goiran, J.P.; Pannuzi, S.; Rossa, C.; Pellegrino, A. Palaeoenvironmental evolution of the ancient lagoon of Ostia Antica (Tiber delta, Italy). *J. Archaeol. Sci.* **2015**, *54*, 374–384. [CrossRef]

18. Davoli, L.; Raffi, R.; Baldassarre, A.M.; Bellotti, P.; Di Bella, L. New maps relative to the «Palude di Torre Flavia» (Central Tyrrhenian Sea-Italy) prone to severe coastal erosion. *IJEGE* **2019**, *2*, 13–21.

19. D'Orefice, M.; Graciotti, R.; Bertini, A.; Fedi, M.; Foresi, L.M.; Ricci, M.; Toti, F. Latest Pleistocene to Holocene environmental changes in the Northern Tyrrhenian area (central Mediterranean). A case study from southern Elba Island. *AMQ* **2020**, *33*, 1–25.

20. Goiran, J.P.; Pavlopoulos, K.P.; Fouache, E.; Triantaphyllou, M.; Etienne, R. Piraeus, the ancient island of Athens: Evidence from Holocene sediments and historical archives. *Geology* **2011**, *39*, 531–534. [CrossRef]

21. Stanley, D.J.; Bernasconi, M.P. Holocene depositional pattern and evolution in Alexandria's Eastern Harbor, Egypt. *J. Coast. Res.* **2006**, *22*, 283–297. [CrossRef]

22. Stanley, D.J.; Bernasconi, M.P. Sybaris-Thuri-Copia trilogy: Three delta coastal sites become landlocked. *Méditerranée* **2009**, *112*, 75–88. [CrossRef]

23. Bellotti, P.; Caputo, C.; Dall'Aglio, P.L.; Davoli, L.; Ferrari, K. Human settlement in an evolvine landscape. Man-environment interaction in the Sibari Plain (Ionian Calabria). *AMQ* **2009**, *22*, 61–72.

24. Bini, M.; Brückner, H.; Chelli, A.; Pappalardo, M.; Da Prato, S.; Gervasini, L. Palaeogeographies of the Magra Valley coastal plain to constrain the location of the Roman harbour of Luna (NW Italy). *Palaeogeogr. Palaeoclimatol. Palaeoecol.* **2012**, *337*, 37–51. [CrossRef]

25. Amorosi, A.; Bini, M.; Giacomelli, S.; Pappalardo, M.; Ribecai, C.; Rossi, V.; Sammartino, I.; Sarti, G. Middle to late Holocene environmental evolution of the Pisa plain (Tuscany, Italy) and early human settlements. *Quat. Int.* **2013**, *303*, 93–106. [CrossRef]

26. Anthony, E.J.; Marriner, N.; Morhange, C. Human influence and the changing geomorphology of Mediterranean deltas and coasts over last 6000 years: From progradation to destruction phase? *Earth Sci. Rev.* **2014**, *139*, 336–361. [CrossRef]

27. Ghilardi, M.; Istria, D.; Curras, A.; Vacchi, M.; Contreras, D.; Vella, C.; Dussouillez, P.; Crest, Y.; Guiter, P.; Delanghe, D. Reconstructing the landscape evolution and the human occupation of the Lower Sagone River (Western Corsica, France) from the Bronze Age to the Medieval period. *J. Archaeol. Sci.* **2017**, *12*, 741–754. [CrossRef]

28. Giaime, M.; Magne, G.; Bivolaru, A.; Gandouin, E.; Marriner, N.; Morhange, C. Halmyris. Geoarchaeology of a fluvial harbour on the Danube Delta (Dobrogea, Romania). *Holocene* **2018**, *28*, 1–15. [CrossRef]

29. Giaime, M.; Marriner, N.; Morhange, C. Evolution of ancient harbours in deltaic contexts: A geoarchaeological typology. *Earth Sci. Rev.* **2019**, *191*, 1–290. [CrossRef]

30. Regione Toscana. Carta Geologica Della Regione Toscana in Scala 1:10.000. 2015. Available online: http://www502.regione.toscana.it/geoscopio/cartoteca.html (accessed on 28 November 2014).

31. Bartole, R. Caratteri sismostratigrafici, strutturali e paleogeografici della piattaforma continentale tosco-laziale; suoi rapporti con l'Appennino settentrionale. *Boll. Soc. Geol. Ital.* **1990**, *109*, 599–622.

32. Servizio Geologico d'Italia—Carta Geologica d'Italia alla scala 1:100.000—Foglio 135 "Orbetello". 1968. Stab. L. Salomone, Roma. Available online: http://193.206.192.231/carta_geologica_italia/tavoletta.php?foglio=13501.02.2012 (accessed on 27 March 2020).

33. Carmignani, L.; Decandia, F.A.; Fantozzi, P.L.; Lazzaretto, A.; Lotta, D.; Meccheri, M. Tertiary extensional tectonics in Tuscany (Northern Apennines, Italy). *Tectonophysics* **1994**, *238*, 295–315. [CrossRef]

34. Bartole, R. The North Tyrrhenian-Northern Apennines post-collisional system: Constrain for a geodynamic model. *Terra Nova* **1995**, *7*, 7–30. [CrossRef]

35. Principe, C.; Malfatti, A.; Rosi, M.; Ambrosio, M.; Fagioli, M.T. Metodologia innovativa di carotaggio microstratigrafico: Esempio di applicazione alla tefrostratigrafia di prodotti vulcanici distali. *Geol. Tec. Ambient.* **1997**, *4*, 39–50.

36. Ambrosio, M.; Dellomonaco, G.; Fagioli, M.T.; Giannini, F.; Pareschi, M.T.; Pignatelli, L.; Rosi, M.; Santacroce, R.; Sulpizio, R.; Zanchetta, G. Utilizzo di fioretto meccanico e carotiere microstratigrafico inguainante per la valutazione degli spessori e della stratigrafia delle coltri vulcanoclastiche soggette a fenomeni di colata rapida di fango. *Geol. Tec. Ambient.* **1999**, *4*, 23–32.

37. Stuiver, M.; Polach, H.A. Discussion: Reporting of ^{14}C data. *Radiocarbon* **1977**, *19*, 355–363. [CrossRef]

38. Ramsey, B. *Oxcal V.3.10*; Research Laboratory for Archaeology: Oxford, UK, 2005.

39. Preusser, F.; Degering, D.; Fuchs, M.; Hilgers, A.; Kadereit, A.; Klasen, N.; Krbetschek, M.R.; Richter, D.; Spencer, J. Luminescence dating: Basics, methods and applications. *Quat. Sci. J.* **2008**, *57*, 95–149.

40. Murray, A.S.; Wintle, A.G. Luminescence dating of quartz using an improved single-aliquot regenerative-dose protocol. *Radiat. Meas.* **2000**, *32*, 57–73. [CrossRef]

41. Richter, D.; Richter, A.; Dornich, K. Lexsyg—A new system for luminescence research. *Geochronometria* **2013**, *40*, 220–228. [CrossRef]

42. Galbraith, R.F.; Roberts, R.G.; Laslett, G.M.; Yoshida, H.; Olley, J.M. Optical dating of single grains of quartz from Jinmium rock shelter, northern Australia. Part I: Experimental design and statistical models. *Archaeometry* **1999**, *41*, 339–364. [CrossRef]

43. Preusser, F.; Kasper, H.U. Comparison of dose rate determination using high-resolution gamma spectrometry and inductively coupled plasma-mass spectrometry. *Anc. tL* **2001**, *19*, 19–23.

44. Folk, R.L. Distinction between Grain Size and Mineral Composition in Sedimentary-Rocks Nomenclature. *J. Geol.* **1954**, *62*, 344–359. [CrossRef]

45. Murray, J.W. *Ecology and Applications of Benthic Foraminifera*; Cambridge University Press: Cambridge, UK, 2006; pp. 1–426. ISBN 978-051-153-552-9.

46. Hammer, O.; Harper, D.A.T.; Ryan, P.D. PAST: Paleontological Statistics Software Package for Education and Data Analisys. *Palaeontol. Electron.* **2001**, *4*, 1–9.

47. Loeblich, A.R., Jr.; Tappan, H. *Foraminiferal Genera and Their Classification*; Van Nostrand Reinhold Company: New York, NY, USA, 1988; pp. 1–71. ISBN 978-1-4899-5760-3.

48. Cimerman, F.; Langer, M. Mediterranean foraminifera. In *Slovenska Akademija Znanosti Umetnosti, Academia Scientiarum Artium Slovenica*; Classis IV: Ljubljana, Slovenia, 1991; pp. 1–118.

49. Sgarrella, F.; Moncharmont-Zei, M. Benthic foraminifera of the Gulf of Naples (Italy): Systematics and autoecology. *Boll. Soc. Paleontol. Ital.* **1993**, *32*, 145–264.

50. Milker, Y.; Schmiedl, G. A taxonomic guide to modern benthic shelf foraminifera of the western Mediterranean Sea. *Palaeontol. Electron.* **2012**, *15*, 1–134. [CrossRef]

51. WoRMS Editorial Board. World Register of Marine Species. Available online: http://www.marinespecies.orgatVLIZ (accessed on 2 December 2019).

52. Vittori, C. Trajectoires temporelles des environnements fluvio-lagunaires littoraux de la péninsule italique et sociétés anciennes. Ph.D. Dissertation, University of Strasbourg, Strasbourg, France, 2020.

53. Bonaduce, G.; Ciampo, G.; Masoli, M. Distribution of ostracoda in the Adriatic Sea. *Pubbl. Stn. Zool. Napoli* **1975**, *40*, 1–304.

54. Bonaduce, G.; Masoli, M.; Pugliese, N. Ostracodi bentonici dell'alto Tirreno. *Sci. Nat. Acta Biol.* **1977**, *54*, 243–261.

55. Athersuch, J.; Horne, D.J.; Whittaker, J.E. *Marine and Brackish Water Ostracods (Superfamilies Cypridacea and Cytheracea): Key and Notes for Identification of Species*; Synopsis of the British Fauna (New Series) 43; The Linnean Society of London and the Estuarine and Brackish-Water Sciences Association: Leiden, UK, 1989.

56. Meisch, C. Freshwater ostracoda of Western and Central Europe. In *Süsswasserfauna von Mitteleuropa 8/3*; Schwoerbel, J., Zwick, P., Eds.; Spektrum Akademischer Verlag: Heidelberg, Germany, 2000; pp. 1–522. ISBN 978-1-4020-6417-3.

57. Faranda, C.; Gliozzi, E. The ostracod fauna of the Plio-Pleistocene Monte Mario succession (Roma, Italy). *Boll. Soc. Paleontol. Ital.* **2008**, *47*, 215–267.

58. Fuhrmann, R. *Atlas Quartärer und Rezenter Ostrakoden Mitteldeutschlands*; Altenburger Naturwissenschaftliche Forsc-hungen: Altenburg, Germany, 2012.

59. Aiello, G.; Barra, D.; Parisi, R.; Isaia, R.; Marturano, A. Holocene benthic foraminiferal and ostracod assemblages in a paleo-hydrothermal vent system of Campi Flegrei (Campania, South Italy). *Palaeontol. Electron.* **2018**, *21*, 41. [CrossRef]

60. Ruiz, F.; Abad, M.; Galán, E.; González, I.; Aguilá, I.; Olías, M.; Gómez Ariza, J.L.; Cantano, M. The present environmental scenario of El Melah Lagoon (NE Tunisia) and its evolution to a future sabkha. *J. Afr. Earth Sci.* **2006**, *44*, 289–302. [CrossRef]

61. Salel, T.; Bruneton, H.; Lefèvre, D. Ostracods and environmental variability in lagoons and deltas along the north-western Mediterranean coast (Gulf of Lions, France and Ebro delta, Spain). *Rev. Micropaléontologie* **2016**, *59*, 425–444. [CrossRef]

62. Arbulla, D.; Pugliese, N.; Russo, A. Ostracods from the National Park of La Maddalena Archipelago (Sardinia, Italy). *Boll. Soc. Paleontol. Ital.* **2004**, *43*, 91–99.

63. Ruiz, F.; González-Regalado, M.L.; Baceta, J.I.; Menegazzo-Vitturi, L.; Pistolato, M.; Rampazzo, G.; Molinaroli, E. Los ostrácodos actuales de la laguna de Venecia (NE de Italia). *Geobios* **2000**, *33*, 447–454. [CrossRef]

64. Mansouri, R.; Bobier, C.; Carbonel, P.; Tastet, J.P. Contribution à la connaissance des systèmes lagunaires en domaine méditerranéen: Les milieux actuels et les paléoenvironnements du lac de Ghar El Melh et de la sebkha de l'Ariana (Tunisie). *Boll. Oceanol. Teor. Appl.* **1985**, *3*, 167–195.

65. Mazzini, I.; Rossi, V.; Da Prato, S.; Ruscito, V. Ostracods in archaeological sites along the Mediterranean coastlines: Three case studies from the Italian peninsula. In *The Archaeological and Forensic Applications of Microfossils: A Deeper Understanding of Human Histo*; Williams, M., Hill, T., Eds.; The Micropalaeontological Society, Special Publications. Geological Society: London, UK, 2017; pp. 121–142. ISBN 178-620-305-7.

66. Triantaphyllou, M.V.; Kouli, K.; Tsourou, T.; Koukousioura, O.; Pavlopoulos, K.; Dermitzakis, M.D. Paleoenvironmental changes since 3000 BC in the coastal marsh of Vravron (Attica, SE Greece). *Quat. Int.* **2010**, *216*, 14–22. [CrossRef]

67. Koukousioura, O.; Dimiza, M.D.; Kyriazidou, E.; Triantaphyllou, M.V.; Syrides, G.; Aidona, E.; Vouvalidis, K.; Panagiotopoulos, I.; Papadopoulou, L. Environmental evolution of the Paliouras coastal lagoon in the eastern Thermaikos gulf (Greece) during Holocene. *Environ. Earth Sci.* **2019**, *78*. [CrossRef]

68. Debenay, J.P.; Guillou, J.J. Ecological Transitions Indicated by Foraminiferal Assemblages in Paralic Environments. *Estuaries* **2002**, *25*, 1107–1120. [CrossRef]

69. Laut, L.; Silva, F.S.; Figueiredo, A.G., Jr.; Laut, V. Assembleias de foraminíferos e tecamebas associadas a análises sedimentológicas e microbiológicas no delta do rio Paraíba do Sul, Rio de Janeiro, Brasil. *Pesqui. Geociências* **2011**, *38*, 251–267. [CrossRef]

70. Martins, A.M.V.; Zaaboub, N.; Aleya, L.; Frontalini, F.; Pereira, E.; Miranda, P.; Mane, M.; Rocha, F.; Laut, L.; El Bour, M. Environmental Quality Assessment of Bizerte Lagoon (Tunisia) Using Living Foraminifera Assemblages and a Multiproxy Approach. *PLoS ONE* **2015**, *10*, e0137250. [CrossRef]

71. Belart, P.; Frontalini, F.; Laut, V.; Fortes, R.; Clemente, I.; Raposo, D.; Martins, V.; Lorini, M.L.; Rocha Fortes, R.; Laut, L. Living benthic Foraminifera from the Saquarema lagoonal system (Rio de Janeiro, southeastern Brazil). *Check List* **2017**, *13*, 1–2062. [CrossRef]

72. Zonneveld, K.A.F.; Marret, F.; Versteegh, G.J.M.; Bogus, K.; Bonnet, S.; Bouimetarhan, I.; Crouch, E.; de Vernal, A.; Elshanawany, R.; Edwards, L.; et al. Geographic distribution of dinoflagellate cysts in surface sediments. *Pangaea* **2013**. [CrossRef]

73. Dimiza, M.D.; Koukousioura, O.; Triantaphyllou, M.V.; Dermitzakis, M.D. Live and dead benthic foraminiferal assemblages from coastal environments of the Aegean Sea (Greece): Distribution and diversity. *Rev. Micropaléontologie* **2016**, *59*, 19–32. [CrossRef]

74. Batten, D.J.; Grenfell, H.R. Green and blue-green algae. *Palynol. Princ. Appl.* **1996**, *36*, 205–214.

75. Guy-Ohlson, D. *Botryococcus* as an aid in the interpretation of palaeoenvironment and depositional processes. *Rev. Palaeobot. Palynol.* **1992**, *71*, 1–15. [CrossRef]

76. Tyson, R.V. Distribution of the palynomorph group: Phytoplankton subgroup, chlorococcale algae. In *Sedimentary Organic Matter*; Springer: Dordrecht, The Netherlands, 1995.

77. Castro Alonso, M.J.; Montañez Hernández, L.E.; Sanchez Muñoz, M.A.; Franco, M.; Rubi, M.; Narayanasamy, R.; Balagurusamy, N. Microbially Induced Calcium carbonate Precipitation (MICP) and its potential in Bioconcrete: Microbiological and molecular concepts. *Front. Mater.* **2019**, *6*, 1–126. [CrossRef]
78. Karami, F.; Balci, N.; Guven, B. A modeling approach for calcium carbonate precipitation in a hypersaline environment: A case study from a shallow, alkaline lake. *Ecol. Complex.* **2019**, *39*, 100–774. [CrossRef]
79. Cangemi, M.; Bellanca, A.; Borin, S.; Hopkinson, L.; Mapelli, F.; Neri, R. The genesis of actively growing siliceous stromatolites: Evidence from Lake Specchio di Venere, Pantelleria Island, Italy. *Chem. Geol.* **2010**, *276*, 318–330. [CrossRef]
80. Posamentier, H.W.; Vail, P.R. Eustatic Controls on Clastic Deposition II: Sequence and Systems Tract Models. In *Sea Level Change: An Integrated Approach*; Wilgus, C.K., Hastings, B.S., Eds.; SEPM Publ.: Tulsa, OK, USA, 1988; pp. 125–154. ISBN 978-091-898-574-3.
81. Posamentier, H.W.; Allen, G.P. Siliciclastic sequence stratigraphy—Concepts and applications. *Sedimentol. Paleontol.* **1999**, *7*, 210.
82. Roberts, N.; Brayshaw, D.; Kuzucuoğlu, C.; Perez, R.; Sadori, L. The mid-Holocene climatic transition in the Mediterranean: Causes and consequences. *Holocene* **2011**, *21*, 3–13. [CrossRef]
83. Bini, M.; Zanchetta, G.; Perşoiu, A.; Cartier, R.; Català, A.; Cacho, I.; Dean, J.R.; Di Rita, F.; Drysdale, R.N.; Finnè, M.; et al. The 4.2 ka BP Event in the Mediterranean region: An overview. *Clim. Past* **2019**, *15*, 555–577. [CrossRef]
84. Currás, A.; Ghilardi, M.; Peche-Quilichini, K.; Fagel, N.; Vacchi, M.; Delanghe, D.; Dussouillez, P.; Vella, C.; Bontempi, J.M.; Ottaviani, J.C. Reconstructing past landscapes of the eastern plain of Corsica (NW Mediterranean) during the last 6000 years based on molluscan, sedimentological and palynological analyses. *J. Archaeol. Sci. Rep.* **2017**, *12*, 755–769. [CrossRef]
85. Di Rita, F.; Fletcher, W.J.; Aranbarri, J.; Margaritelli, G.; Lirer, F.; Magri, D. Holocene forest dynamics in central and western Mediterranean: Periodicity, spatio-temporal patterns and climate influence. *Sci. Rep.* **2018**, *8*, 8929. [CrossRef]
86. Melis, R.T.; Depalmas, A.; Di Rita, F.; Montisa, F.; Vacchi, M. Mid to late Holocene environmental changes along the coast of western Sardinia (Mediterranean Sea). *Glob. Planet. Chang.* **2017**, *155*, 29–41. [CrossRef]
87. Sadori, L.; Koutsodendris, A.; Panagiotopoulos, K.; Masi, A.; Bertini, A.; Combourieu-Nebout, N.; Francke, A.; Kouli, K.; Joannin, S.; Mercuri, A.M.; et al. Pollen data of the last 500 ka BP at Lake Ohrid (south-eastern Europe). PANGAEA. Supplement to 2016, Pollen-based paleoenvoronmental and paleoclimatic change at Lake Ohrid (south-eastern Europe) during the past 500 ka. *Biogeosciences* **2018**, *13*, 1423–2016. [CrossRef]
88. Lambeck, K.; Antonioli, F.; Anzidei, M.; Ferranti, L.; Leoni, G.; Scicchitano, G.; Silenzi, S. Sea level change along the Italian coast during the Holocene and projections for the future. *Quat. Int.* **2011**, *232*, 250–257. [CrossRef]
89. Vacchi, M.; Ghilardi, M.; Melis, R.T.; Spada, G.; Giaime, M.; Marriner, N.; Lorscheid, T.; Morhange, C.; Burjachs, F.; Rovere, A. New relative sea-level insights into the isostatic history of the Western Mediterranean. *Quat. Sci. Rev.* **2018**, *201*, 396–408. [CrossRef]
90. Bartolini, C.; Corda, L.; D'Alessandro, L.; La Monica, G.B.; Regini, E. Studi di Geomorfologia costiera: III—Il tombolo di Feniglia. *Boll. Soc. Geol. Ital.* **1979**, *96*, 117–157.
91. Evangelista, S.; Full, W.E.; Tortora, P. Provenance and dispersion of fluvial, beach and shelf sands in the bassa Maremma coastal system (central Italy): An integrated approach using Fourier shape analysis, grain size and seismic data. *Boll. Soc. Geol. Ital.* **1996**, *115*, 195–217.

Article

"Continuous" Backstepping of Holocene Coastal Barrier Systems into Incised Valleys: Insights from the Ofanto and Carapelle-Cervaro Valleys

Vincenzo De Santis *, Massimo Caldara and Luigi Pennetta

Department of Earth and Environmental Science, University of Bari, 70125 Bari, Italy; massimoangelo.caldara@uniba.it (M.C.); luigi.pennetta@uniba.it (L.P.)
* Correspondence: vincenzo.desantis@uniba.it

Received: 25 May 2020; Accepted: 21 June 2020; Published: 24 June 2020

Abstract: Two recently recognised incised valleys in the Manfredonia Gulf are described. The first (CCV) is correlated with the current Carapelle and Cervaro streams. The second (OSFV) is correlated mostly with the current Ofanto River. Six seismic facies and seven unconformity-bounded seismic units have been identified, which infilled CCV and OSFV. In CCV, during the sea-level ranges from −29 to −18 and from −18 to −4.7 m b.s.l., two barrier/spit-backbarrier systems formed in the most landward sector of the valley. The lower system was attributed to a time interval between 9.2 ka BP and ca. 8.3 ka BP, chronologically constrained by the ZS2 borehole. In OSFV, during the sea level ranges from −39 to −29, and from −29 and to −18 m b.s.l., two beach/spit-backbarrier systems, arranged in a "continuous" landward backstepping pattern, formed. The phase that contributed most to the beach/spit-backbarrier systems formation is that which is coeval with the formation of the sapropel S1 in the Mediterranean. The conservation of barrier/spit-backbarrier systems arranged in a "continuous" landward backstepping pattern, is due to a strong and continued sediment supply that occurred during the sapropel S1 formation, coupled with low-gradient settings and a regime of slow sea-level rise.

Keywords: coastal barrier; continuous backstepping pattern; incised valley; sapropel S1; sea-level rise

1. Introduction

The term coastal barrier refers to different geomorphological features at different times. Otvos [1] reviewed the nomenclature and classification of coastal barriers and considered two main types: (1) barrier islands, parallel to the shore and located between the inshore and open marine environments, and (2) barrier spits, which are linked to the mainland shore. Modern barrier islands are found worldwide, located mainly along wave-dominated coasts, and these islands were subjected to rising sea levels of the late Holocene [2]. Barrier-lagoon systems include the shoreface sector, the barrier, and the backbarrier sector, where wash-over fans and lagoons develop [1,3].

The formation of these systems is controlled by the interplay of wave energy, sediment supply, accommodation and sea-level oscillations [4–6]. In particular, barriers and backbarrier-lagoons can develop when the substrate gradient is less than 0.8° [7].

The responses of coastal barrier and backbarrier systems to sea-level rise are crucial issues, in the light of the future sea level rise due to global warming. Studies of evolving barrier morphology indicate several responses to sea-level rise, all of which are strongly influenced by local factors. Most low-gradient coasts form transgressive barriers during sea-level rise. These barriers were composed almost entirely of tidal delta and wash-over deposits that retrograded into estuarine/lagoonal environments as sea-level rose [7]. The seabed exposed by the retreating barrier is a ravinement surface [7,8]. Typically, the two main responses of coastal barriers to sea-level rise are identified as [9] rollover and overstepping. Rollover is the dominant retreat process and consists of a continuous migration of the barrier systems following shoreline retreat, with almost complete reworking of shoreface and barrier deposits [10–12]. The overstepping process involves the partial preservation of transgressive deposits or barrier morphology and is generally associated with a fast rate of sea-level rise [13,14]. Overstepping involves near-instantaneous drowning of the barrier by the sea and translation of the marine, wave-influenced shoreline to the landward margin of the former lagoon, which may be several kilometres away [15]. Barrier overstepping has been attributed to a combination of rapid sea-level rises, such as those associated with meltwater pulses [16], with other particularly favourable circumstances, such as early cementation (beachrock and aeolianite formation) or gravelly sediments [12,17–20].

The preservation potential of drowned barriers (only possible in the case of overstepping) and related transgressive deposits is considered to be relatively low in comparison to the deposits associated with regressive coasts [13,21].

In this study, we describe the formation and landward migration of coastal barrier/spit and backbarrier systems during the last post-Würm transgression. This migration affected an incised valley created by the Carapelle and Cervaro streams and a shallow and very wide valley created by the Ofanto River and its tributary San Ferdinando stream, which have been progressively infilled. The aim of this study is to provide an example of how coastal environments migrate landwards in response to slow transgression, on low-gradient substrates and in the presence of shallow incised valleys.

2. Regional Setting

The Manfredonia Gulf is the natural offshore continuation of the Tavoliere di Puglia Plain, the second largest plain in Italy characterised by several Quaternary marine and alluvial terraces [22–26]. The coastal area is subsiding, at a maximum rate of 0.21 mm/year between MIS 5.5 and the present day [22].

The Gulf extends ca. 83 km seaward from the coastline. The seabed has an average slope of approximately 0.07° towards the east [27]. The gulf has a microtidal regime.

In the Gulf, Trincardi et al. [28,29], Maselli and Trincardi [30], Maselli et al. [31], and De Santis and Caldara [32] recognised an extensive unconformity (today buried) representing the low-stand surface of subaerial exposure (LSSE) formed during the last glacial maximum (LGM). The Manfredonia Incised Valley (MIV) cuts into the LSSE (Figure 1); it is a sinuous valley elongated for more than 60 km in a W-E direction from the inner shelf to the mid-outer shelf. On the LSSE, the transgressive system tract (TST) and the high-stand system tract (HST) were deposited from the beginning of the last sea-level rise. The TST also constitutes most of the infilling of the MIV.

The Manfredonia Gulf was also affected by deposition of the so-called Gargano subaqueous delta [33]. This delta was formed on the eastern and southeastern sides of the Gargano promontory (Figure 1). This subaqueous deposit represents the southernmost portion of the Holocene HST developing along the western side of the Adriatic. The HST rests above a regional down-lap surface (maximum flooding surface; mfs) that marks the time that the maximum landward shoreline shift occurred, approximately 5.5 cal. ka BP during the end of the late-Pleistocene to Holocene sea-level rise [30,33–35].

Following the work of Maselli et al. [36] in the central Adriatic Sea, the TST is found to consist of a tripartite deposit, recording the step-wise nature of sea-level rise. The three distinct units, each deposited during a specific interval of the last sea-level rise, are the lower TST unit (lTST), middle TST (mTST) and upper TST (uTST). The lTST formed as sea-level rose between −120 and −95 m, the mTST between −95 and −60 m, and the uTST between −60 m and the position of the shoreline that was reached at the time of maximum marine ingression (5.5 cal. ka BP). The boundary between the lTST and mTST consists of the S1 surface, whereas the boundary between the mTST and the uTST consists of the S2 surface. S1 and S2 are regional erosional surfaces generated by reduced sediment inputs and enhanced marine reworking as a consequence of Meltwater Pulses 1A and 1B (MWP-1A and MWP-1B), respectively [37]. Within the mTST, another regional erosional surface (Si), can be detected, thus dividing the mTST into two different subunits (mTST-1 below and mTST-2 above the Si, respectively). The stratigraphic complexity of the mTST unit, recording the Bölling-Allerød and Younger Dryas events, can best be explained by sea-level oscillations, including a minor sea-level fall during the Younger Dryas event, which likely formed the Si surface. This conclusion is consistent with the formation of a Younger Dryas shoreline deposit ca. −75 m depth, which is within the mTST unit.

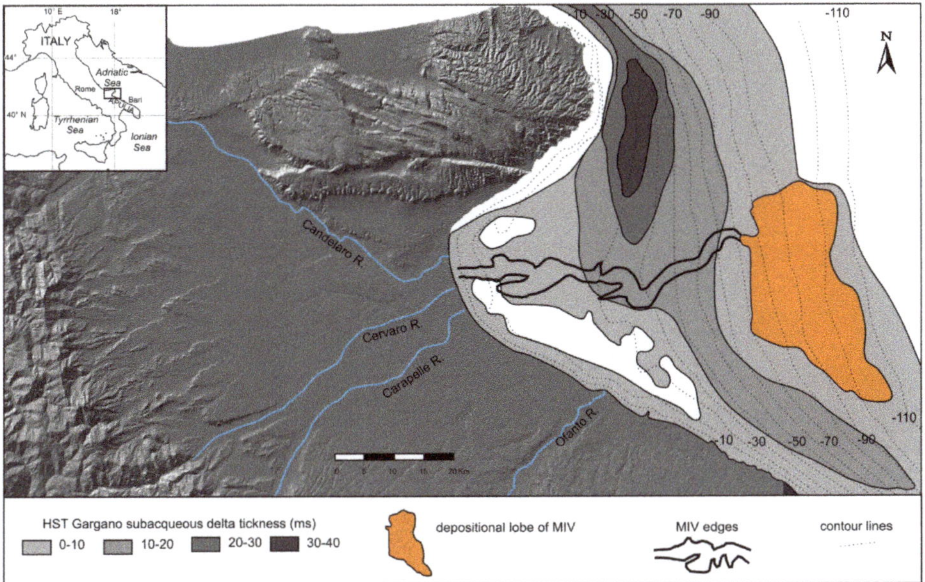

Figure 1. Simplified map showing position of the Manfredonia Incised Valley (MIV) and of its depositional lobe, high-stand system tract (HST) Gargano sub-acqueous delta (gray tones) and bathymetry of the Manfredonia Gulf shelf with 10-m-spaced contours. Redrawn from literature data [29,31,32].

3. Methods

3.1. Seismic Survey

Seismic surveys were carried out in the Manfredonia Gulf (southern Adriatic) in 2016 and 2017 using the motorboat "ISSEL", which is owned by the Interuniversity Consortium for Sea Sciences (Con.I.S.Ma.). The following instruments and sensors were mobilised: (i) AppliedAcoustic Squid CSP-P 350J "sparker" source; and (ii) Geo-Resources Geo-Sense single channel seismic acquisition system with 8 hydrophone streamer and amplification/filtering stage.

The data was processed using the IXSEA DELPH software. Through the use of the system, all the necessary steps for filtering, gaining and interpreting the data have been performed.

Track-line positioning is based on D-GPS navigation, assuring a sub-metric position accuracy using the WGS84 datum.

The seismic survey comprised 15 sparker profiles oriented perpendicular to the coastline and 12 sparker profiles subparallel to the coastline for a total of ca. 580 km (Figure 2). The minimum and maximum water depths were <8 m and ca. 97 m, respectively.

The data were subsequently converted into a digital format/image for interpretation. Each seismic profile was imported into TEI Delph Map with Seismic GIS, allowing for immediate georeferencing and interpretation of acquired data. Finally, a buried erosional surface (ES1) was reconstructed using a triangular interpolation method in GIS.

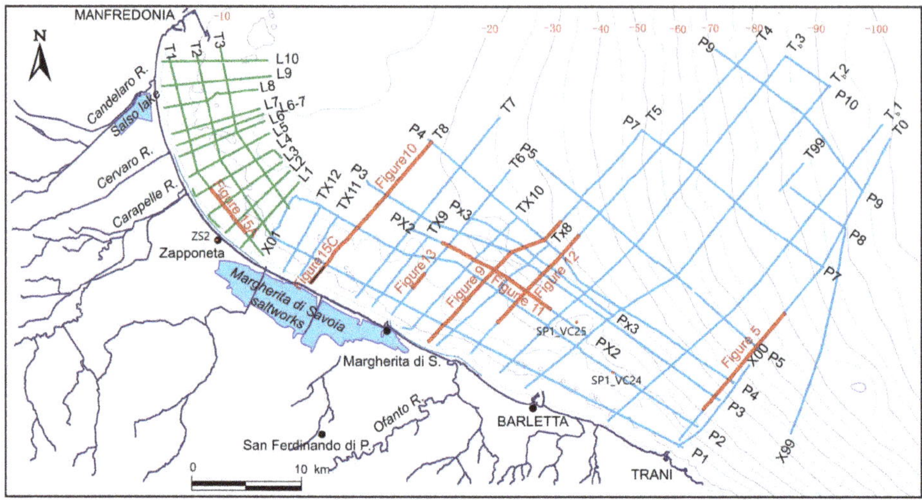

Figure 2. Study area with seismic profiles acquired in the two surveys: in green colour those of the 2013–2014, in blue colour those of the 2016–2017; the red parts of the seismic profiles correspond to the sections reported in the figures indicated. The black part in seismic profile T8 is reported in Figure 13C. Contour line of the present sea bottom and positions of ZS2 borehole and SP1_VC24 and SP1_VC25 vibro-cores are also shown. The abbreviations of seismic profiles are shown at their ends.

In addition to the surveys described above, this work reports the seismic profile T1, which was acquired and processed within other surveys carried out between 2013 and 2014 using sub-bottom profiler Benthos CHIRP III in the Manfredonia Gulf, immediately NW of the area investigated in the present study (Figure 2). For the technical details of that survey see [32,38].

3.2. Seismic Facies and Unconformity-Bounded Seismic Units

For this work, seismic facies and unconformity-bounded seismic units (UBSUs) belonging to the TST have been identified (Figure 3).

By seismic facies we mean groups of reflections whose parameters (bedform geometry, lateral continuity, amplitude, frequency, and interval velocity) differ from those of adjacent groups of reflections [39]. We also used the spatial extension of seismic facies, along with the reciprocal position between facies, to infer depositional environments and formative processes. The seismic facies interpretation is also supported by previous study, e.g., some general seismic facies characteristics of the latest Quaternary transgressive deposits and incised valley infillings were established (i.e., [13,15,40–43]). More specifically for the southern Adriatic, Maselli and Trincardi [30], Maselli et al. [36], and Maselli et al. [31] provide a complete description of the seismic facies that characterise the TST and HST.

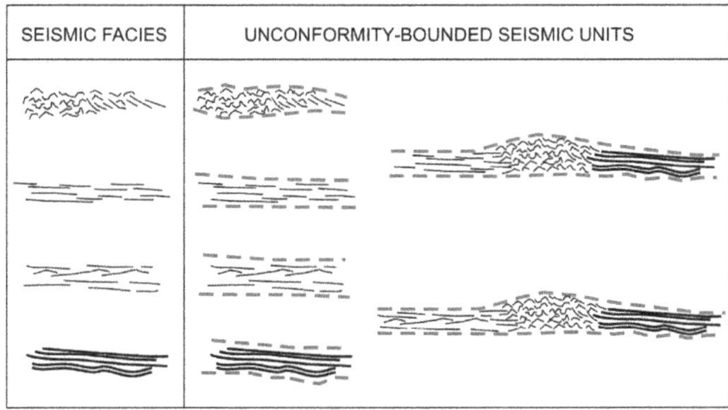

Figure 3. Conceptual scheme of relationships between seismic facies and unconformity-bounded seismic units (UBSUs) applied in this study. The red dashed lines are the unconformity. The symbols of the seismic facies used in this figure are indicative only.

The seismic facies interpretations are also based on two vibro-cores drilled in 2015 in the Manfredonia Gulf: SP1_VC24 (intercepted by profile P2); and SP1_VC25 (close to profile PX2) on behalf of the Puglia Basin Authority (AdB) for a sand resource study of the Apulian continental shelf. The chronological constraint of the transgressive UBSUs is based on a borehole drilled in Zapponeta town (ZS2) in 2012 by the AdB during a seismic microzonation project; this borehole was directly analysed by the authors. Seismic facies, coupled with the identification of discontinuity surfaces, were used to separate the TST sediments into UBSUs. More precisely, UBSUs have been distinguished based on unconformities, interpretable as surfaces of erosion or discontinuity in sedimentation. The UBSUs in some cases are made up of a single seismic facies, in other cases of multiple seismic facies (Figure 3) that transit into one another (i.e., the seismic units described in [17] and [44]).

In this work, the seismic facies have been recognised based only on seismic sparker profiles, because they are almost all of the profiles present in the study area. In the single sub-bottom profile T1, only the unconformity bounded seismic units have been recognised.

3.3. Age Range Evaluation and Chronological Constraints

For an approximate assessment of the age of the UBSUs, we have modified the method based on the current depth of the units, already applied in the southern Adriatic (see for example the age estimation made by Maselli and Trincardi [30] for the units 2C and 2D of area 3 of the MIV or the age estimation made by Pellegrini et al. [45] for the Palaeo Gargano compound delta).

One of the seismic facies (seismic facies 2) was found to be helpful in an approximated evaluation of the age range of the UBSUs. Seismic facies 2 has been interpreted in this study as a high-energy marine deposit and therefore deposited at (or close to) sea level; this interpretation is validated also because the seismic facies 2 is intercepted by vibro-cores SP1_VC24 and SP1_VC25, where it is present between the sea bottom and the bottom of the two vibro-cores. In addition, the interpretation of seismic facies 2 is in accordance with Trincardi et al. [28] (p. 86) which defined the sediments cropping out at the sea bottom in the area were the two vibro-cores have been realised as "proximal transgressive sandy deposits of beach or paralic environment".

Having said that, because seismic facies 2 forms sedimentary bodies within different UBSUs arranged in a landward backstepping pattern [46] on the ES1, it follows that each of these bodies indicates an approximated sea-level range if we consider the base and the top of the body. Then, we considered the local sea-level curve (site 25 in [47]) to infer two ages corresponding to two sea levels that we associated with the base and top of each sedimentary body formed by seismic facies 2. These two sea levels have been corrected considering a possible subsidence calculated for the Holocene deposits of our study area. In conclusion, we associate each sedimentary body formed by seismic facies 2 with two approximated palaeo sea-levels and therefore with an approximated age range (Figure 4).

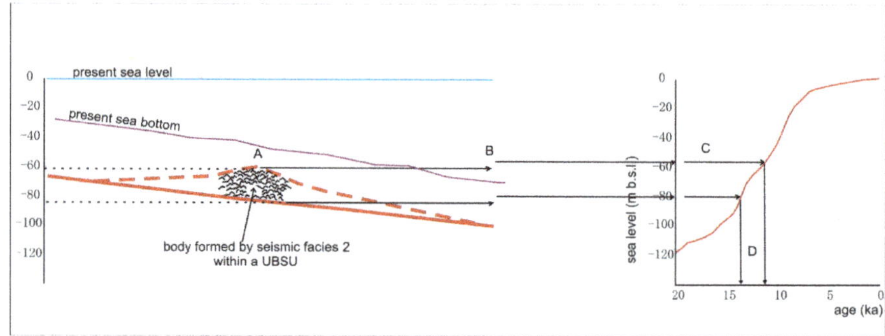

Figure 4. Conceptual sketch illustrating the method used for the approximated evaluation of the age range of the discontinuity-bounded seismic units. A: recognition of the depth below present sea level of the base and the top of the bodies formed by seismic facies 2; B: correction of these depths in order to remove the subsidence; C: projection of these corrected depths on the sea level curve considered; D: reading of the corresponding age range.

If the body formed by seismic facies 2 belongs to an UBSUcontaining multiple seismic facies, we extend to the whole unit the approximated age range obtained from the facies 2 body. This procedure allowed us to link the UBSUs at some sea-level ranges and therefore at corresponding age ranges.

The timing of the deposition of transgressive UBSUs was chronologically constrained by the ZS2 borehole, which intercepts the transgressive fill of the inland sector of an incised valley recognised in this study. Three samples of this perforation were dated using the 14C method at DirectAMS Laboratories, Bothell, Washington (laboratory ID: D-AMS). The radiocarbon dates were calibrated using the CALIB 7.1 program. The IntCal13, Marine/IntCal13, and Marine13 datasets [48] were used for terrestrial, transitional and marine samples, respectively (Table 1). The percentage of marine carbon was calculated based on the $\delta13C$ value [49]. A ΔR correction ($\Delta R =121 \pm 60$) calculated for the southern Adriatic Sea off Barletta (http://intcal.qub.ac.uk/marine/) was used. Furthermore, the ZS2 borehole intercepted a pumice layer, which has been identified.

Table 1. Samples dated with the 14C method coming from ZS2 borehole. For the location of the samples in the core, see Figure 4.

Lab. ID.	Submitter ID	Material	Calibration	$\delta(13C)$ ‰	Radiocarbon Age (Year BP)	Calibrated Age (Year BP)
D-AMS 007669	ZS2/4.80	*Donax semistriatus*	marine13	5.8	2090 ± 23	1527 ± 161
D-AMS 007670	ZS2/7.60–7.70	wood/organic	intcal13	−25.1	3668 ± 25	4002 ± 82
D-AMS 007672	ZS2/28.65	wood/organic	marine/intcal13	−14.3	8401 ± 37	9160 ± 137

4. Data and Results

We present the following seismic and stratigraphic data regarding the most inner part of the Manfredonia Gulf continental shelf: (1) the erosional surface that formed during the last sea-level fall, sub-aerial exposure of the shelf and erosional reworking during subsequent sea level rise; (2) the main landforms described by the erosional surface; (3) borehole and vibro-cores; (4) seismic facies recognizable within uTST; and (5) UBSUs belonging to the uTST.

4.1. Erosional Surface ES1

The seismic survey allowed us to identify a regional erosional surface (ES1). In some cases, the ES1 appears as a clearly visible reflector, with moderate to high amplitude, coinciding with a more or less evident angular unconformity between the underlying reflectors and the overlying reflectors; in other cases, the ES1 appears instead as an abrupt contact between very different seismic facies. However, in both cases, ES1 is identified as a surface that separates the substrate from younger sediments having a very different seismic architecture. The substrate is almost always characterised by seaward-inclined reflectors, referable to oblique-tangential clino-forms [39] which dip at low angles (usually fractions of a degree) and are truncated up-dip by the ES1. Sediments above ES1, on the other hand, are highly variable in that they are characterised by numerous seismic facies (Figure 5).

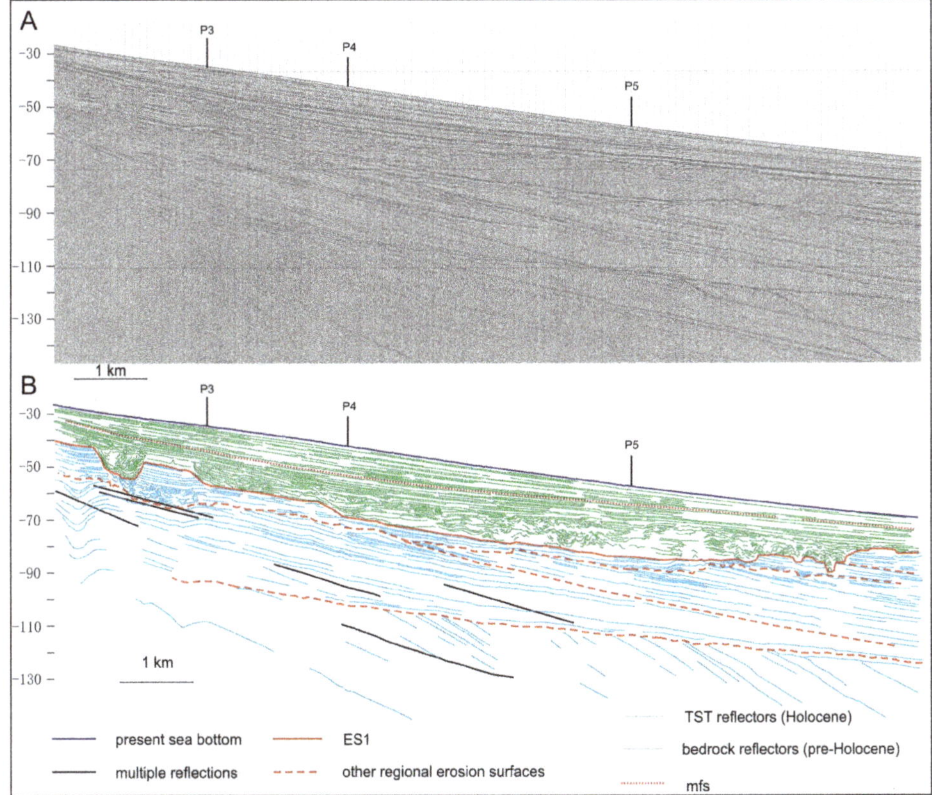

Figure 5. Part of seismic profile T0 (**A**) and its interpretation (**B**), showing the erosional surface (ES1) on a regional scale. Note that ES1 is identified as a surface that separates the substrate from younger sediments having a different seismic architecture.

4.2. Erosive Landform: Incised Valley A and B

ES1 presents three main erosive landforms (Figure 6). The north-western, not the subject of this work, is already known in the literature as Manfredonia incised valley (MIV [30,32]).

The central is an incised valley (hereinafter incised valley 1: IV1; Figure 6), that reaches a depth (in profiles T8 and P1) of ca. −37 m b.s.l. and ca. 18/22 m below the surrounding ES1. In the P2 profile, the maximum depth is ca. −38 m. b.s.l., which reaches −50 m. b.s.l. in profile P4.

This valley has a meandering course at the turn of the current coastline.

The south-eastern landform is another incised valley (hereinafter incised valley 2: IV2; Figure 6), which appears, in the sector closest to the coast, as two different sub-incisions. These two sub-incisions merge in a single wide and shallow valley with a very rugged and irregular bottom. The IV2 reaches a maximum width of 11.9 km (profile P2) and narrows to approximately 2.5 km in correspondence with the Barletta strait (Figure 6). The IV2 reaches depths between ca. −13/−15 m b.s.l. (profile T4) and ca. −50 m b.s.l. at the Barletta strait; a topographic low is present at the intersection between profiles P3 and T5, where the bottom reaches −48 m b.s.l. With respect to the surrounding ES1, the IV2 shows decreasing depths towards the open sea, between ca. 17 m and less than 10 m b.s.l.

Above the ES1 surface, a series of units were deposited during the last transgressive phase; they constitute the infilling of the IV1 and IV2.

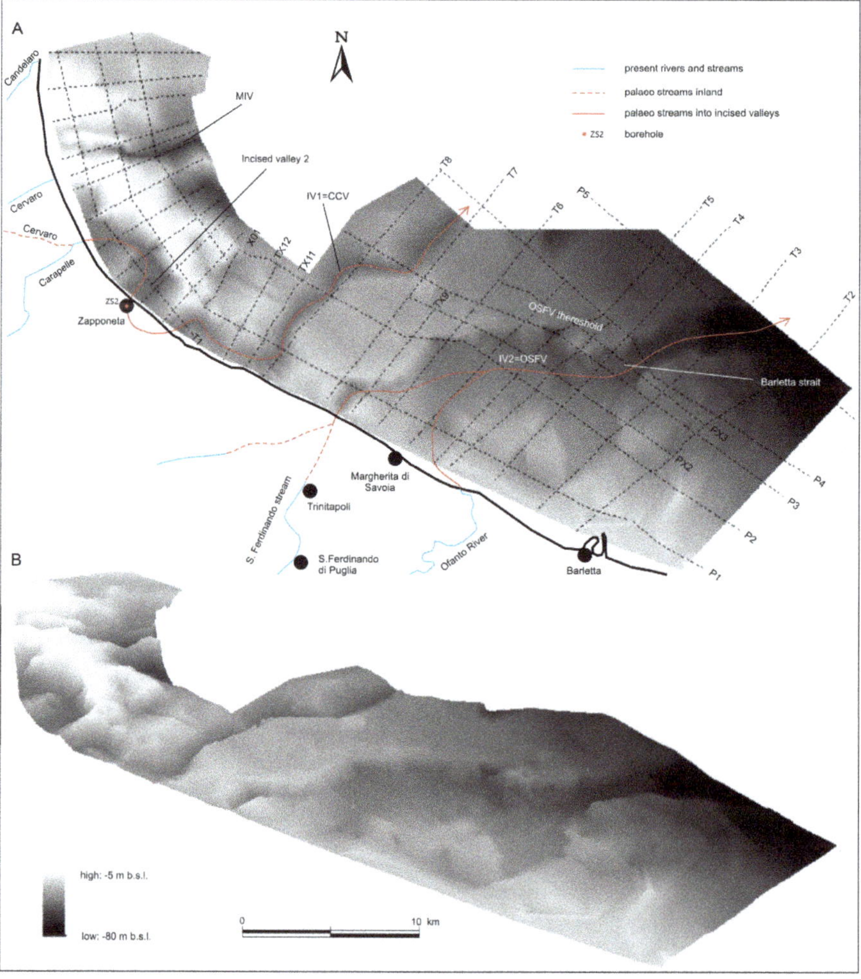

Figure 6. Digital Elevation Model (DEM) of the erosional surface ES1 both in plan view (**A**) and in oblique view (**B**). For the former, seismic profiles and the reconstructed drainage network are also shown. The seismic profiles showed in the following figures are indicated with their names.

4.3. Borehole and Vibrocores

The borehole and vibro-cores intersect only transgressive sediments lying above ES1.

4.3.1. Borehole ZS2

The borehole ZS2 (Figure 7), drilled in the town of Zapponeta, is 30 metres deep. The ground level is 1.2 m above sea level. From the bottom up, the recognised lithological units are as follows:

1. from 30 to 29 m below ground level (−28.8 to −27.8 m b.s.l.), light grey clay with *Cyprideis torosa*, Pulmonata, Characeae and vegetal remnants;
2. from 29 to 28.5 m (−27.8 to −27.3 m b.s.l.), dark grey clay rich in *Cerastoderma glaucum* (Bruguiere), *Abra segmentum* (Recluz), *Turboella lineolata* (Michaud), and Hydrobiidae spp.;
3. from 28.5 to 27.8 m (−27.3 to −26.6 m b.s.l.), graded light grey pumice, from very coarse to medium-fine;
4. from 27.8 to 24.5 m (−26.6 to −23.3 m b.s.l.), very dark and dark grey clay rich in *C. glaucum* and *Bittium reticulatum* (Da Costa) in the lowest 30 cm. In the remaining part of the unit, the fauna is different: first, the fauna is composed of species such as *Ostea edulis* (Linneo), *Modiolus barbatus* (Linneo) and *Mysella bidentata* (Montagu); then, the fauna comprises *M. bidentata*, *Nucula nitidosa* (Winckworth), *Abra nitida* (Müller), and *Dentalium inaequicostatum* (Dautz);

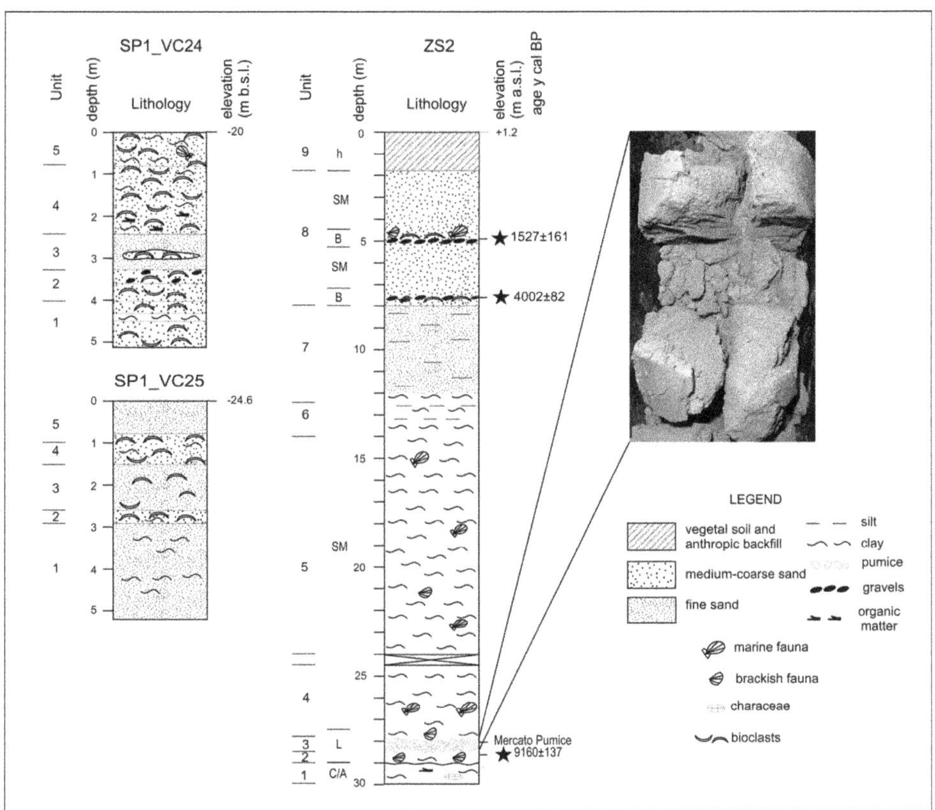

Figure 7. Stratigraphic logs of ZS2 borehole and SP1_VC24 and SP1_VC25 vibro-cores; a photo showing a particular of the Mercato pumice layer in ZS2 borehole has been reported. C/A = continental/alluvial; L = lagoonal; SM = shallow marine; B = beach; h = anthropic.

There is an interval from 24.5 to 24.0 m (−23.3 to −22.8 m b.s.l.) where no sediment recovery was obtained. Subsequently, the stratigraphy continues as follows;

5. from 30 to 29 m below ground level (−28.8 to −27.8 m b.s.l.), light grey clay with *Cyprideis torosa*, Pulmonata, Characeae and vegetal remnants;

6. from 29 to 28.5 m (−27.8 to −27.3 m b.s.l.), dark grey clay rich in *Cerastoderma glaucum* (Bruguiere), *Abra segmentum* (Recluz), *Turboella lineolata* (Michaud), and Hydrobiidae spp.;

7. from 28.5 to 27.8 m (−27.3 to −26.6 m b.s.l.), graded light grey pumice, from very coarse to medium-fine;

8. from 27.8 to 24.5 m (−26.6 to −23.3 m b.s.l.), very dark and dark grey clay rich in *C. glaucum* and *Bittium reticulatum* (Da Costa) in the lowest 30 cm. In the remaining part of the unit, the fauna is different: first, the fauna is composed of species such as *Ostea edulis* Linneo, *Modiolus barbatus* Linneo and *Mysella bidentata* (Montagu); then, the fauna comprises *M. bidentata*, *Nucula nitidosa* (Winckworth), *Abra nitida* (Müller), and *Dentalium inaequicostatum* (Dautz).

There is an interval from 24.5 to 24.0 m (−23.3 to −22.8 m b.s.l.) where no sediment recovery was obtained. Subsequently, the stratigraphy continues as follows:

9. from 24.0 to 14.0 m (−22.8 to −12.8 m b.s.l.), grey, dark grey, or leaden clay, generally with abundant and diversified fauna: *Tellina distorta* (Poli), *Corbula gibba* (Olivi), *M. bidentata*, *Pitar rudis* (Poli), *Gouldia minima* (Montagu), *A. nitida*, *Nuculana pella* (Linneo), *N. nitidosa*, and *D. inaequicostatum*. Many ossicles of Asteroidea and vertebrae of Ophiuroidea are present;

10. from 14.0 to 12.45 m (−12.8 to −11.25 m. b.s.l.), dark grey clay layers alternating with a thin layer of light grey silt containing rare specimens of *A. nitida*, *C. gibba*, *N. pella*, *N.a nitida*, and *Odostomia* sp.;

11. from 12.45 to 8.0 m (−11.25 to −6.8 m b.s.l.), dark grey to light grey fine silty sand, containing rare specimens of *A. nitida* and *N. pella*;

12. from 8.0 to 1.75 (−6.8 to −0.55 m b.s.l.), grey brown to dark grey medium-coarse sand; two thin layers of gravels in a sandy matrix with abundant shell fragments (*Glycymeris* sp., *Chamelea gallina* (Linneo), *Donax* sp.) are present at approximately −6.5 and −3.5 m, respectively; and

13. from 1.75 to 0.0 m (−0.55 to +1.2 m b.s.l.), vegetal soil and anthropic backfill.

Three samples from this borehole have been 14C dated (Table 1).

In addition, the level of pumice present from 28.5 to 27.8 m (−27.3 to −26.6 m b.s.l.) was attributed to the eruption of the Mercato Pumice [50] of the Somma-Vesuvius volcanic complex, dated to 8890 ± 90 cal year BP [51,52].

4.3.2. Vibro-cores SP1_VC24 and SP1_VC25

Vibro-core SP1_VC24 was obtained 5 km offshore at a water depth of 20 m, and this core was intercepted by the seismic profile P2 (Figure 2). Coring ceased at 5.13 m below the seabed, and from the bottom up the core consists of the following units (Figure 7):

1. from 5.13 to 3.59 m, bioclastic coarse material with some pelitic intercalation;

2. from 3.59 to 3.27 m, coarse black sand with abundant bioclasts and pebbles;

3. from 3.27 to 2.43 m, fine and very fine light olive brown sand. Pockets of coarse bioclastic black sand are present;

4. from 2.43 to 0.79 m, bioclastic gravel and coarse sand in a grey pelitic matrix, with abundant vegetal remnants towards the lower part of the level; and

5. from 0.79 m to the sea floor, bioclastic gravel of bivalves and gastropods in a clayey-sandy, dark grey matrix.

Vibro-core SP1_VC25 was obtained 8 km offshore at a water depth of 24.6 m and is located very close to seismic profile PX2 (Figure 2). The coring ceased at 5.20 m below the seabed, and from the bottom upwards, the core consists of the following units (Figure 7):

1. from 5.20 to 2.92 m, very fine ochre pelitic sand, with many clay intercalations of various greyish shades;
2. from 2.92 to 2.61 m, bioclastic gravel in an ochre sandy/pelitic matrix;
3. from 2.61 to 1.52 m, very fine ochre sand with rare small bioclasts;
4. from 1.52 to 0.78 m, bioclastic gravel in an ochre sandy/pelitic matrix; and
5. from 0.78 m to the sea floor, fine and very fine ochre sand.

4.4. Seismic Facies

In this section, we describe the seismic facies recognised in the sediments belonging to the TST, which cover ES1 and infill the IV1 and IV2.

- Facies 1 (Figure 8A,B). This facies is only present at the bottom of the incised valleys and is characterised by high amplitude down-lapping reflector or sets of concave and/or convex and/or wavy reflectors; reflectors also show frequent changes in dip direction; the lateral continuity is low-medium. In the seismic profiles normal to the narrow valley axes, the reflectors present a conformable channel infill pattern.
- Facies 2. These seismic facies can be divided into two subtypes: 2a and 2b. Facies 2a (Figure 8C): this facies, in general, has a chaotic, non-symmetrical internal configuration with reflectors of low or very low amplitude and very low lateral continuity. Commonly, these facies contains V and Λ-shaped reflectors. Alternatively, the reflectors are undulating, concave, convex or oblique-parallel, but always with low amplitude and very low lateral continuity. Facies 2b (Figure 8D): the reflectors are sigmoidal and/or clinoform with moderate-high amplitude and medium-high lateral continuity, which alternate with low amplitude reflectors or with chaotic and non-symmetrical packages, and all are inclined landward. Facies 2 overlies ES1 on the shelf and previous infilling units in IV1 and IV2. From offshore to onshore, a transition between facies 2 and facies 3 or 4 is often observed.
- Facies 3 (Figure 8E,F). This facies is characterised by moderate amplitude and plane-parallel reflectors with medium lateral continuity; sometimes, reflectors are weakly undulating with low lateral continuity. This facies overlies ES1 or seismic facies 1. The relationship with seismic facies 2 is twofold: in some cases, facies 2 transitions landwards into facies 3; in other cases, facies 3 is surmounted by facies 2.
- Facies 4 (Figure 8G,H). This facies is characterised by moderate to high amplitude, medium-high lateral continuity plane-parallel reflectors alternating with wavy or gently inclined reflectors. The relationship of facies 4 with seismic facies 2 is twofold: in some cases, facies 2 transitions landwards into facies 4; in other cases, facies 5 is surmounted by facies 2.
- Facies 5 (Figure 8I). This facies is characterised by moderate amplitude and medium lateral continuity reflectors, with the presence of a topset, fore-set and bottomset; thus, it is a pro-gradational seismic facies. Downdip, this facies transitions into facies 3.
- Facies 6 (Figure 8). This facies is characterised by moderate to high amplitude and maximum observed lateral continuity reflectors that onlap onto inclined substrate and/or draping over uneven substrate. Reflectors are gently seaward-inclined (Figure 8L) or sub-horizontal (Figure 8M).

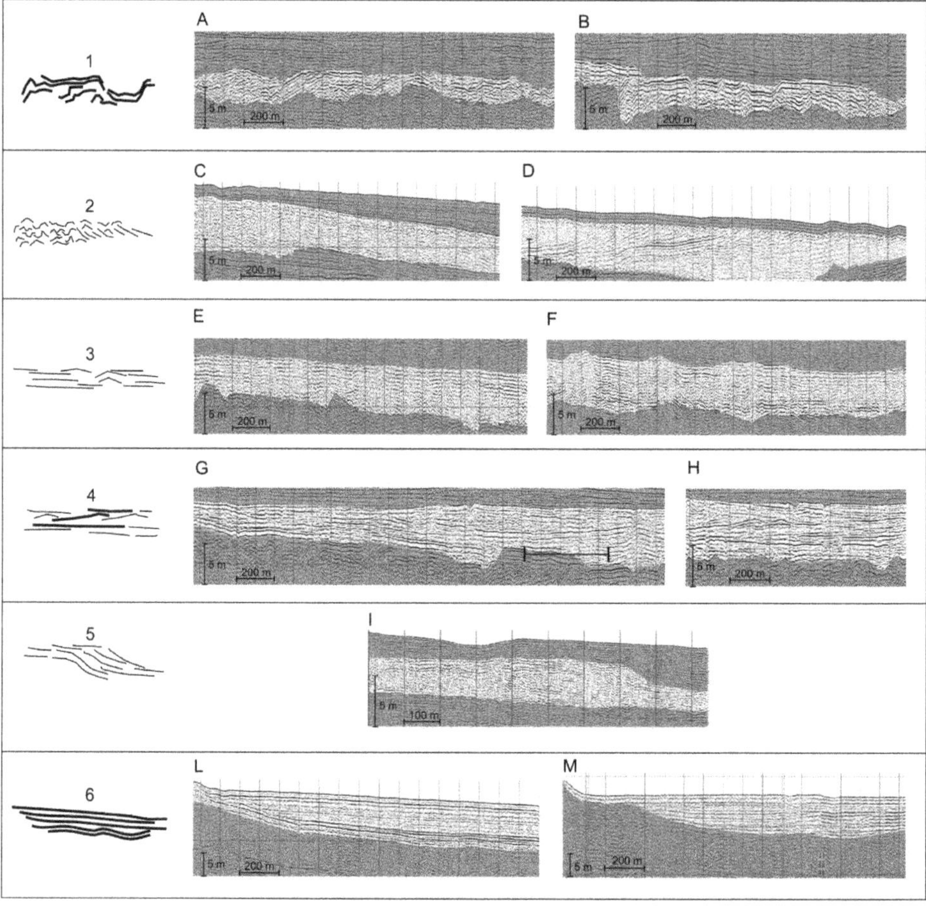

Figure 8. Examples of the seismic facies recognised in the study area in the sparker profiles. On the left the six seismic facies are drawn, with the corresponding number; the drawings roughly summarize the main characteristics of the facies. On the right are examples of each seismic facies, taken from the studied profiles. (**A,B**): facies 1; (**C,D**): facies 2a and 2b, respectively; (**E,F**): facies 3; (**G,H**): facies 4; (**I**): facies 5; (**L,M**): facies 6. The shaded areas cover seismic facies different from that described in each row. The red dashed line represents the ES1. All the profile segments shown are along valleys except that of box A, which is across the IV2.

4.5. Unconformity-Bounded Seismic Units

We identified seven UBSUs that lie above ES1 and/or infilled the IV1 and IV2.

4.5.1. Unit A_u

Unit A_u (the subscript letter u in the abbreviation of the units indicates that all units belong to upper part of TST) consists of seismic facies 1, occupies discontinuous portions of the IV1 and IV2 bottom and lies always above ES1 (Figures 9–12). Unit A_u reaches a maximum thickness of ca. 9 m.

4.5.2. Unit B_u

Unit B_u extends between the offshore boundary of the study area and an ES1 depth of ca. -40 m b.s.l. and it consists exclusively of seismic facies 2a. This unit does not affect IV1 or IV2 and always lies above ES1 (Figures 9 and 12). Unit B_u reaches maximum thickness of ca. 5 m.

4.5.3. Unit C_u

Unit C_u occupies only a small lower area of the IV2 at the intersection of profiles T5 and P3 (Figures 6 and 9), and it never exceeds a depth of -38 b.s.l. Overall, this unit has a lenticular shape, lies above unit A_u and ES1, reaches a maximum thickness of ca. 5 m and consists exclusively of seismic facies 3.

4.5.4. Unit D_u

Unit D_u is present between depths of ca. -40 m b.s.l. and ca. -30 b.s.l., and lies on unit A_u, unit B_u, and ES1.

In the IV1 sector (profile T8, Figure 10), unit D_u is present exclusively seawards and outside the IV1 and lies on the ES1. Here, unit D_u is composed of seismic facies 2a, with some sets of landward-dipping reflectors, and the unit reaches a maximum thickness of ca. 7 m.

Passing to IV2 sector, unit D_u changes proceeding from the NW to the SE. At the NW (profile T6), unit D_u consists only of the seismic facies 2a, lies on ES1 and is present exclusively seawards and outside the IV2. Moving to the SE (profile T5; Figure 9), unit D_u centres on the threshold that separates the IV2 from the continental shelf and consists of a central body formed by seismic facies 2a which transitions into seismic facies 3 inside the IV2 (landwards) and seismic facies 6 on the shelf (seawards). The central body is characterised by two sets of opposing inclined reflectors towards the NE and the SW (Figure 9). This internal organisation of the seismic facies within unit D_u remains even further to the SE (profile T4; Figure 12), with the difference that the central body of seismic facies 2a appears narrower than that in profile T5 and transitions into seismic facies 4 inside the IV2.

Into the IV2, unit D_u extends landwards much further in the NW area (where it appears with seismic facies 3) than in the SW area, where unit D_u it is not present (profile PX2; Figure 11A–C).

The central body formed by seismic facies 2 stops at the Barletta strait. This is visible in profile PX3 which is elongated along the external threshold of the IV2 (Figure 6) and then it runs across the central body of the unit D_u.

4.5.5. Unit E_u

Unit E_u is present between depths of ca. -30 m and ca. -16 b.s.l., and lies on ES1, unit A_u, unit B_u, and unit D_u.

In the IV1 sector, unit E_u is formed by seismic 2a both on the shelf and in the seaward part of the valley (profile T8, Figure 10). In the landward part of the valley, unit E_u consists of seismic facies 3 at the base, surmounted by seismic facies 2b (Figure 10).

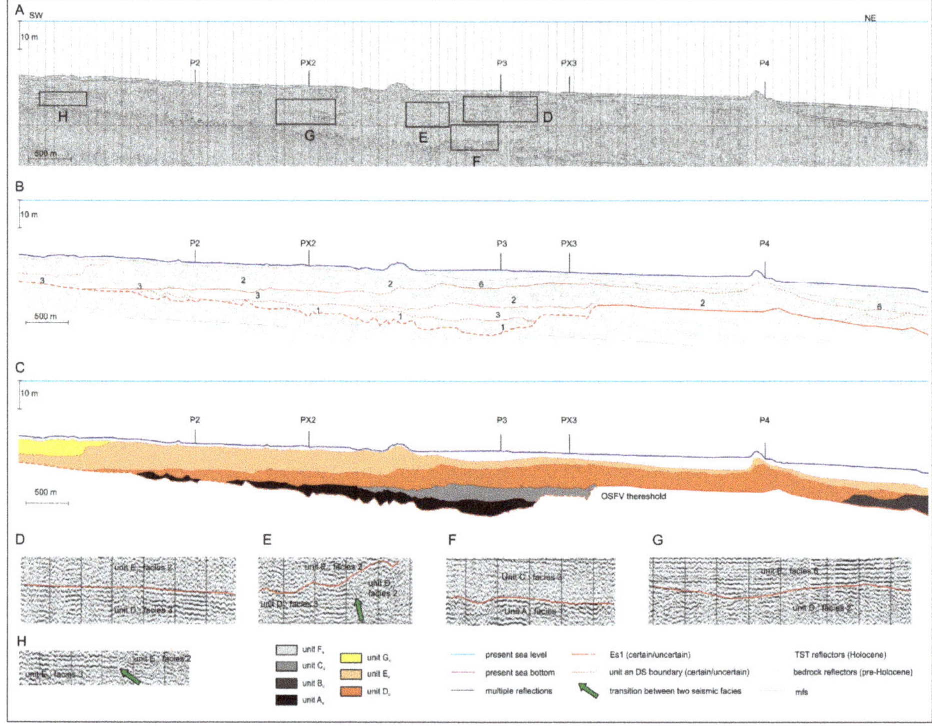

Figure 9. Part of seismic profile T5. Seismic section (**A**), interpretation (**B**), seismic unit recognised (**C**), enlargement of some significant sectors of the unconformities between units and some transitions between different seismic facies (**D–H**). In this profile, in particular, the two sets of opposite inclined reflectors that characterise the central body (seismic facies 2) of the unit D_u, the position of the central body of unit E_u (seismic facies 2) above the backbarrier deposits (seismic facies 3) of the former unit D_u, and the landward backstepping pattern of unit D_u and unit E_u are evident. The legend reported below applies until Figure 13 and for Figure 15; the numbers reported in the interpretations of Figures 9–13 indicate the seismic facies.

Passing to IV2 sector, unit E_u appears to be the most complex system in terms of internal organisation, and it changes proceeding from the NW to the SE. To the NW (profile T6), unit E_u is composed of a very thin central body of seismic facies 2a, which seaward transitions into seismic facies 6 and landwards transitions first into facies 3 and then into seismic facies 5; this latter forms a body up to 10 m tick with an evident rollover point between the fore-set and the topset at ca. −19 m (Figure 13). Moving to the SE (profile T5; Figure 9), the central body reaches a thickness of ca. 8 m and transitions landward into seismic facies 3 and seaward into seismic facies 6. Even further to the SE (profile T4; Figure 12), unit E_u is constituted by a body formed by seismic facies 2a, up to ca. 5 m thick, which transitions landwards into seismic facies 4.

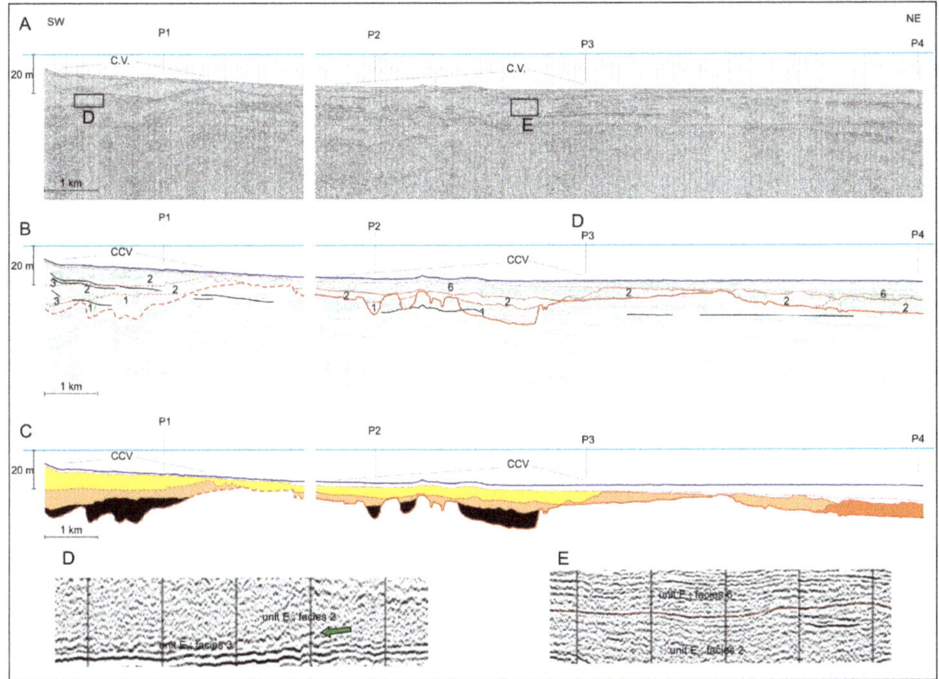

Figure 10. Part of seismic profile T8. Seismic section (**A**), its interpretation (**B**), seismic units recognised (**C**), and enlargement of some significant sectors of the unconformities between units, with some transitions between different seismic facies (**D,E**). This profile intercepts two parts of the CCV. In the landward side of the profile, note the superimposition of unit E_u and unit G_u, which are two successive barrier/spit-backbarrier systems. In both systems, as the transgression progressed, the barrier/spit retreated landwards and covered the backbarrier sediments through over-wash and/or tidal inlet deposits, as evidenced by the reflectors in seismic facies 2b down-lapping landwards onto the plane-parallel reflector of seismic facies 3.

The central body of unit E_u, constituted by seismic facies 2a, is much wider in the NW sector of the IV2 compared to the SE sector. In fact, facies 2a reaches landwards the profile PX2 (Figure 11A–C) only in the NW sector, while, in the SE sector, it gives way to facies 4 (Figure 11A–C).

The deposition of unit E_u involves most of the IV2 infilling and its division into two residual sub-basins, which consist of two areas not completely infilled. The first is a smaller sub-basin located to the NW in front of the sedimentary body formed by seismic facies 5 visible in profile T6 (Figure 13). The second is a larger sub-basin located to the SE in the sector behind the Barletta strait and is visible in profiles PX2 and T4 (Figures 11 and 12).

4.5.6. Unit F_u

Unit F_u infills the NW sub-basin and lies always on unit E_u. Unit F_u formed after deposition of unit E_u, in front of the sedimentary body formed by seismic facies 5 (Figure 13). Unit F_u consists of seismic facies 3 whose reflectors onlap both onto the down-lapping reflectors of the sedimentary body formed by seismic facies 5 (Figure 13) on the landward side and on the chaotic reflectors (facies 2a) of unit E_u on the seaward side of the sub-basin.

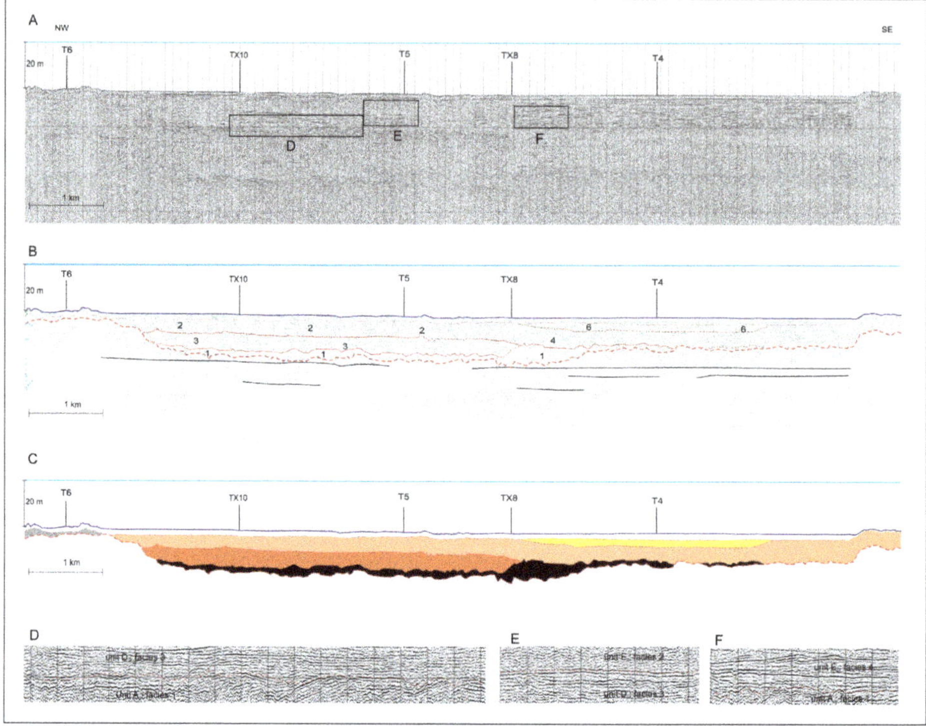

Figure 11. Part of seismic profile PX2. Seismic section (**A**), its interpretation (**B**), seismic units recognised (**C**), enlargement of some significant sectors of the unconformities between units (**D–F**). The evidence that the unit D_u is intercepted by this profile only in the NW sector of the OSFV indicates that this system is much more extended into the OSFV in this sector than that of the SE. Note also the position of the central body of unit E_u (seismic facies 2) above the backbarrier deposits of the previous unit D_u (seismic facies 3) and the transition, within unit E_u, between the central body at the NW (seismic facies 2) and the seismic facies 4 in the SE sector.

4.5.7. Unit G_u

Unit G_u is present between depths of ca. −22 m b.s.l. and ca. −4.7 b.s.l., and lies on ES1, unit F_u, unit D_u, and unit E_u. The unit G_u infills the last residual depressions of the IV1 and IV2 and cancels them completely.

In the IV1 sector, unit G_u is composed of seismic facies 2b, 3 and 6. Proceeding from the SW to the NE (profile T8; Figure 10), unit G_u shows a sedimentary body consisting of seismic facies 3 which is surmounted by seismic facies 2b and has a maximum thickness of ca. 10 m; seawards, seismic facies 2 transitions into seismic facies 6, which have a maximum thickness of ca. 7 m (profile T8; Figure 10).

In the IV2 sector, unit G_u is composed of seismic facies 2a and 6 and fills the residual accommodation within the sub-basin left in the SE sector in front of Barletta strait after deposition of unit E_u. In some cases, thin beds of seismic facies 2a, overlaid by the draping reflectors of seismic facies 6, occur (profiles T6 and T5; Figure 9). In other cases, only seismic facies 6, which drapes or onlaps onto underlying units and systems, occur (profile T4; Figure 12). A depo-centre of up to ca. 6 m thickness is observed at the turn of the intersection between profiles T4 and PX2 (Figures 11 and 12).

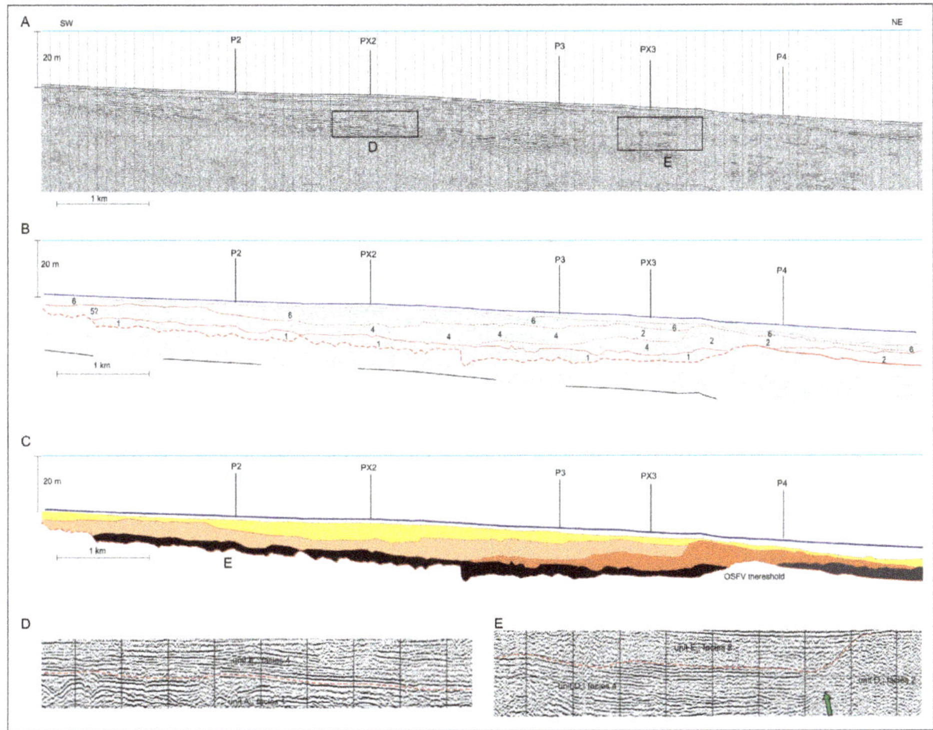

Figure 12. Part of seismic profile T4. Seismic section (**A**), interpretation (**B**), seismic unit recognised (**C**), and enlargement of some significant sectors of the unconformities between units and, with some transitions between different seismic facies within some units (**D,E**). In this profile we note the seismic facies 4 which characterises the backbarrier deposits of both unit D_u and unit E_u, as well as the smaller width of the central bodies (seismic facies 2a) of unit D_u and unit E_u, with respect to seismic profile T5 (Figure 9).

5. Discussion

5.1. Erosional Surface ES1

Based on data previously collected in the Manfredonia Gulf [30,32] and in the adjacent areas [36], the ES1 is interpreted as the surface of subaerial exposure formed during the last sea-level low stand attained during the LGM.

The ES1 describes the two incised valleys IV1 and IV2. Regarding the IV1, we believe that this valley is the seaward continuation of "incised valley 2" (Figure 6) from De Santis and Caldara [32]. Our data further support that it is unified valley of the Carapelle and Cervaro streams. Historical maps up to and including that of Marzolla [53] shows the two streams merged more or less at the current position of the Carapelle mouth. After 1851 the lower stretch of the Cervaro stream was diverted to the north, bringing its mouth into Salso Lake, and then the stream was diverted to its current position [54]. For this reason, we refer to this landform as the Cervaro-Carapelle valley (CCV).

The IV2 is an incised valley (Figure 6) which appears, in the sector closest to the current coastline, as two different incisions. The main incision corresponds to the present Ofanto River mouth. The second incision is placed in correspondence with the probable mouth of an inactive stream; this mouth is masked today by the Margherita di Savoia saltworks. This stream is not indicated with its proper name in recent and historical maps; however, because of its proximity to the town of San Ferdinando di Puglia, we named it San Ferdinando stream. As a consequence, we indicated the IV2 as Ofanto and San Ferdinando valley (OSFV), where the San Ferdinando stream is a tributary of the Ofanto river.

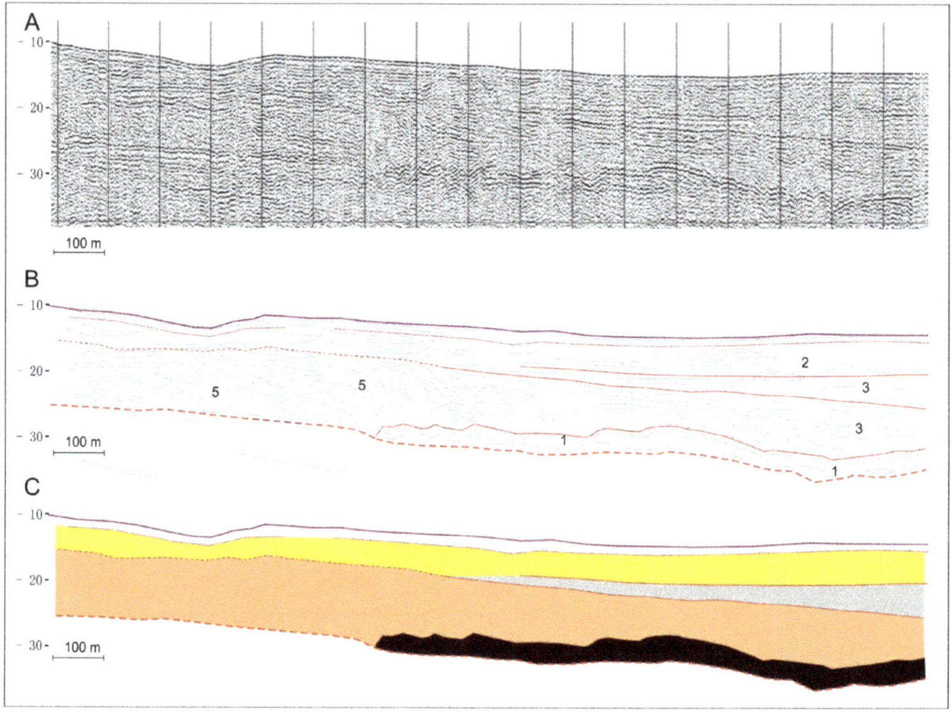

Figure 13. Part of seismic profile T6 showing the small deltaic body D1=San Fernando delta (SFD). Seismic section (**A**), interpretation (**B**), seismic unit recognised (**C**).

5.2. Borehole ZS2 and Vibrocores SP1_VC24 and SP1_VC52

Since borehole ZS2 intercepts the transgressive filling of the CCV (Figure 6), it allows the reconstruction of its sedimentary palaeoenvironments. From 30 to 29 m, the environment is continental and characterised by the presence of stagnant freshwater. From 29 to 27.5 m, a euryhaline and eurytherm lagoon (LEE, *sensu* [55]) is established, which was initially confined and then became more open. Lagoon sedimentation was interrupted by a pyroclastic episode. Starting from 27.5 m, there is a rapid marine transgression characterised by a strong sedimentary instability typical of the heterogeneous palaeo-community [56]. This palaeo-community developed at the passage between the infralittoral zone and the terrigenous mud biocoenosis (VTC, *sensu* [55]) of the circalittoral zone. After a transition phase (from 12.45 to 8.0 m), starting at 8.0 m, a regressive phase occurred characterised by fine well-sorted sand biocoenosis (SFBC, *sensu* [55]) with some beach-layers. Finally, the succession is closed by an anthropic deposit.

The borehole ZS2 also provides two important chronological constraints for the transgressive filling of CCV and allows for an assessment of the subsidence rate of the area during the Holocene.

The chronological constraints are: (i) sample ZS2 28.65, located at −27.45 m b.s.l., consisting of lagoonal environment clays rich in *Cerastoderma glaucum*, which provided an age of 9160 ± 137 cal year BP; and (ii) the level of coarse pumice, whose base is located at −27.30 m b.s.l., which is attributed to the Mercato pumice eruption that occurred at 8890 ± 90 cal year BP [51,52].

Above the pumice level, the lagoonal clays rich in *Cerastoderma glaucum* reoccur. When the Mercato eruption occurred, local sea level was ca. −24.3 m b.s.l. [47]. If we consider the base of pumice level (currently at −27.30 m b.s.l.) as the bottom of the palaeo lagoon on the eve of the eruption, it follows that the palaeo lagoon was max. ca. 3 m deep if no subsidence occurred. However, if we consider that *Cerastoderma glaucum* generally lives up to a maximum depth of −2 m [57–60], we hypothesize that, at the date of the Mercato pumice eruption, the lagoon bottom (base of the pumice layer) was at ca. −26.3 m b.s.l. The current position of the base of the pumice at −27.3 m b.s.l. implies a total subsidence of one meter, with a rate of 0.11 mm/a.

The subsidence value of one meter was used to correct the palaeo sea-levels obtained from the current position of the base and the top of the bodies formed by seismic facies 2.

The vibro-cores SP1_VC24 and SP1_VC25 show a predominantly sandy deposit rich in coarse bioclasts, with a minor content of silt and/or clay, which we interpret as a high energy beach deposited at (or close to) sea level (beach/shoreface).

5.3. Interpretation of Seismic Facies

Facies 1. The characteristics of this seismic facies, its position, and the similarity with other seismic facies of fluvial origin [40,42] suggest a fluvial lag which was most likely deposited under high-energy conditions. The deposition of this facies occurred when streams incised the substrate and developed during both the low-stand and transgression until the arrival of the sea.

Facies 2. The interpretation of the seismic facies 2 was based also on the vibro-cores SP1_VC24 and SP1_VC25 (Figure 2), which intercept only this facies from the sea bottom to borehole bottom (Figure 14). Overall, the data allows us to interpret seismic facies 2a as a predominantly sandy deposit rich in coarse bioclasts, with a minor content of silt and/or clay. Based on this evidence, we interpret this facies as the expression of high-energy coastal sediments in a beach/shoreface environment. This interpretation is in accordance with Trincardi et al. [28], who defined the transgressive sediments cropping out at the sea bottom in the area where the two vibro-cores are located as "proximal transgressive sandy deposits of beach or paralic environment". Facies 2b is interpreted as a beach/shoreface deposit within a transgressive barrier or spit that migrated landward by wave erosion and over-washing. In fact, landward-dipping reflectors are dominant in transgressive barriers, and they can be interpreted as former flood-tidal deltas and wash-over fans, which generally overlie sandy-muddy deposits that formed in a backbarrier environment [61,62]. In conclusion, seismic facies 3 represents deposits that formed following arrival of the sea and wave action: this wave action reworked ES1 and created a coastal high-energy palimpsest deposit due to mixing among reworked sediments from the substrate with coeval sediments.

Facies 3. The position of facies 3 with respect to facies 2 is a fundamental element for its interpretation: the transition between facies 2 and 3 is interpreted as a passage from a beach/barrier/spit environment to a low-energy environment; therefore, facies 3 is interpreted as a backbarrier deposit (perhaps a semi-enclosed bay/microtidal estuarine lagoon). The cases where facies 2 lies above facies 3 are interpreted as resulting from the migration of coastal environments landwards, when the beach passed over the backbarrier during progression of the marine transgression.

Facies 4. As in the case of facies 3, the transition between facies 2 and 4 is interpreted as a passage from a beach/barrier/spit environment to a lower energy environment. In addition, the position of facies 4 in the area of the IV2, near the opening towards the sea, allows us to interpret this seismic facies as a deposit of open backbarrier (perhaps bay/ microtidal estuarine lagoon), probably of silty nature. The proximity to the opening implies the presence of currents and channels, which create a moderate energy environment. Thus, the wavy undulations which characterise this facies can be interpreted in some cases as channel features, in other cases as deformation of the estuarine/lagoon bottom due to waves or currents.

Facies 5. This seismic facies is interpreted as a deltaic body. Thus, the body formed by seismic facies 5 within unit E_u represents a small delta (D1; Figure 13) up to ca. 10 m thick.

Facies 6. This seismic facies is interpreted as a lower shoreface/offshore, probably silty-clay deposit that lies around or below the wave action base. It is formed due to the rapid landward shift in the shoreline, which was in turn due to the low-gradient substrate.

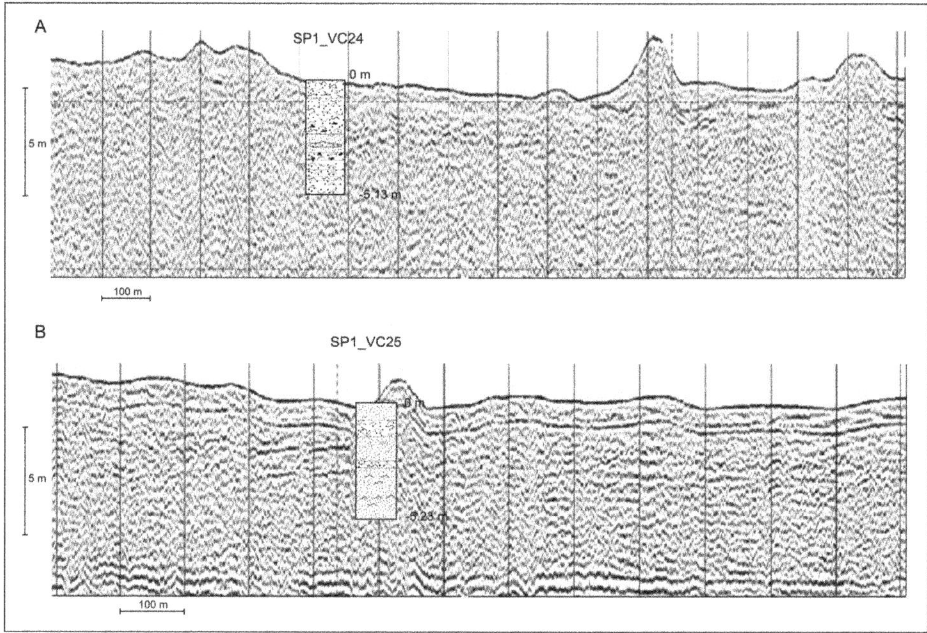

Figure 14. Seismic facies 2 within profiles P2 (**A**) and PX2 (**B**), with stratigraphic logs of the two vibrocores SP1_VC24 and SP1_VC25.

5.4. Interpretation of UBSUs and Transgressive Architecture (Sea Level Ranges and Deposition of UBSUs)

We have linked the UBSUs at four sea-level ranges and therefore at four corresponding age ranges. The sea-levels which define the ranges are: −39 m, −29 m, −18 m, −4.7 m. The first two values are derived from upper and/or lower limit of bodies formed by seismic facies 2, corrected considering a subsidence of 1 m. The level −18 m is derived from the small coastal deltaic body (D1; seismic facies 5) visible within the unit E_u, by assuming its rollover point between the fore-set and the topset as a proxy of sea level (according to [30,45,63]), after a correction to remove the subsidence of 1m. The level −4.7 m is that coinciding with the maximum landward shift of the shoreline (ca. 5.5 ka BP; [47] and marks the end of the late-Pleistocene to Holocene transgression [30,33–35].

5.4.1. Sea-Level Range 1: up to ca. −39 m b.s.l.

This phase coincides with a sea-level rise up to ca. −39 m b.s.l., i.e., until ca. 9.8 ka BP [47]. In this phase, the sea entered the OSFV occupying only the lowest reaches, corresponding to the Barletta strait and the northernmost corner (intersection between profiles P3 and T5; Figure 6). In this phase, unit A_u, unit B_u, and unit C_u were deposited.

Unit A_u contains only seismic facies 1, corresponding to a fluvial lag. In the OSFV, this unit extends both below and above −39 m b.s.l., while in the CCV, it extends only above the −39 m b.s.l. Thus, it was deposited in a subaerial environment on the bottom of the two valleys until the arrival of the sea.

Unit B_u consists exclusively of seismic facies 2a and is interpreted as a beach deposit formed on the shelf above the ES1 in conjunction with a sea-level rise of up to −39 m.

Unit C_u consists exclusively of seismic facies 3, and its deposition is correlated to the final phases of sea-level rise up to −39 m. We interpret this unit as an enclosed bay/lagoon deposit that formed when the sea began to enter the OSFV through the Barletta strait, without having yet submerged the IV2 threshold.

5.4.2. Sea-Level Range 2: from ca. −39 to ca. −29 m b.s.l.

This range coincides with a sea-level rise from ca. −39 to ca. −29 m b.s.l., i.e., from ca. 9.8 ka BP to ca. 9.2 ka BP [47].

In CCV sector, this phase involves the deposition of part of unit A_u and unit D_u. Unit A_u maintains the same characteristics as those described in the previous phase. Unit D_u is constituted only of seismic facies 2a; thus, unit D_u is interpreted as a simple beach deposit of high-energy environment deposited on the ES1.

In the OSFV sector, this phase involves the deposition of unit D_u; the OSFV threshold is not jet submerged in the NW sector of the valley (profile T6), while it is fully submerged in the central and south-eastern sector (profiles T5 and T4; Figures 9 and 12). Therefore, unit D_u is confined to the outside of the valley in the first sector, while it centres on the threshold in the central and south-eastern sector. Considering the overall geometry of unit D_u, as well as the seismic facies that constitute it, we believe that it consists of a central body composed of a beach/spit that, into the OSFV, transitions into backbarrier deposits (perhaps semi-enclosed bay/microtidal estuarine lagoon) and towards the shelf into the lower shoreface/offshore deposits. In particular, the presence of a spit is inferred from the two sets of opposing inclined reflectors that characterise the central body consisting of seismic facies 2a (Figure 9), from the progressive narrowing of the central body proceeding from the NW to the SE (compares profiles T5 and T4; Figures 9 and 12), and from its terminations at the correspondence of the Barletta strait (profile PX3).

5.4.3. Sea-Level Range 3: from ca. −29 to ca. −18 m b.s.l.

This range coincides with sea-level rising from ca. −29 m to ca. −18 m b.s.l., which occurred from ca. 9.2 ka BP to ca. 8.3 ka BP [47].

In the CCV sector, deposition of unit E_u only occurred. We interpret the infilling of the landward part by unit E_u (profile T8; Figure 10) as a transgressive barrier/spit that migrated landward and covered the backbarrier deposits [64,65]. This system formed at the unified mouth of the Carapelle and Cervaro streams.

Passing to OSFV, deposition of units E_u and F_u occurred. In the NW sector (profile T6) rising sea levels submerged the threshold only in this phase, albeit slightly. The small delta D1 recognisable within unit E_u is located seaward of the probable position of the San Ferdinando stream mouth before it was covered by the Margherita di Savoia saltworks. Thus, we named this feature the San Ferdinando delta (SFD).

The deposition of unit E_u involves most of the OSFV infilling and its division into two residual sub-basins, the most northwest of which, placed in front of SFD, is rapidly filled by unit F_u in the last part of this same phase, after the SFD formation (Figure 13). Considering the overall geometry of unit E_u, as well as the seismic facies that constitute it, we interpret unit E_u as a complex coastal system consisting of lower shoreface/offshore deposits that pass landwards into to beach/spit deposits, then to backbarrier deposits (perhaps a semi-enclosed bay/ microtidal estuarine lagoon) in which a small fluvial delta flow.

5.4.4. Sea-Level Range 4: from ca. −18 to ca. −4.7 m b.s.l

This range coincides with a sea-level rise from ca. −18m to ca. −4.7 m b.s.l. This phase involves the deposition of unit G_u both in the CCV and OSFV sectors.

In the CCV sector, based on the seismic facies and their mutual positions, unit G_u is interpreted as a transgressive spit-barrier sandy deposit that overlies backbarrier deposits (perhaps a semi-enclosed bay/microtidal estuarine lagoon), which passes seaward into the lower shoreface/offshore deposits.

In the OSFV sector, based on the seismic facies and their mutual positions, unit G_u is interpreted as a sandy beach deposit of high energy at the bottom (seismic facies 2a) dating back only to the earliest moments of the phase. Then, the beach deposit was quickly replaced by lower shoreface/offshore deposits (seismic facies 6), which are probably silty-clayey, dating back to when the shoreline transgressed rapidly across the gentle slope of the shelf.

5.5. Focus on Beach/Spit-Backbarrier Systems in CCV and OSFV Infilling

In this section, we propose models for the formation and evolution dynamics of beach/spit-backbarrier systems in the infilling of CCV and OSFV.

5.5.1. CCV Infilling

We interpret the landward sector of CCV as the superimposition of two successive barrier/spit-backbarrier systems at the unified mouth of the Carapelle-Cervaro streams, with the lower system belonging to unit E_u, and the upper systems belonging to unit G_u (Figure 10). In both systems, as the transgression progressed, the barrier/spit retreated landwards and covered the backbarrier sediments through over-wash and/or tidal inlet deposits, as evidenced by the reflectors in seismic facies 2b down-lapping landwards onto the plane-parallel reflector of seismic facies 3 (Figure 10).

The ZS2 borehole (Figure 7) intercepts the infilling of CCV (Figures 6 and 15) and allows the correlation between sparker profile T8 and CHIRP profile T1, and a chronological validation of the age range attributed to unit E_u. In particular, in profile T1 we recognise three seismic units correlated with units A_u, E_u, G_u (Figure 15B).

The lagoonal deposit with the enclosed Mercato pumice (dated at 9160 ± 137 and 8890 ± 90 cal year BP) in ZS2 borehole (Figure 15B) is correlated with a series of very high-amplitude reflectors visible roughly at the same depth at the base of unit E_u in the CCV sector shown by profile T1 (Figure 15A,B) and, in terms of both depth and sedimentary environment, with seismic facies 3 visible at the base of unit E_u in the near profile T8 (Figure 15C,D). This confirms the age range attributed to unit E_u according to our method of age range evaluation (from ca. 9.2 ka BP to ca. 8.3 ka BP) which, in addition, is within the time of sapropel S1a formation in Mediterranean [66,67].

About the presence of two barrier/spit-backbarrier systems that characterise the inner part of the CCV (units E_u and G_u), we hypothesise that it is due to the abrupt sea-level rise that occurred during the so-called 8.2 event [68–70]. This sea-level jump may have drowned the lower barrier/spit-backbarrier system (unit E_u); subsequently, the conditions were appropriate for construction of a new system over the previous one (unit G_u). This hypothesis is consistent with a scheme in which unit E_u is attributed to the sapropel S1a event, and unit G_u can be attributed to sapropel S1b, as these units are separated by a discontinuity dating back to the 8.2 event [67].

5.5.2. OSFV Infilling

With rising sea level, the inherited morphology of the OSFV favors the creation of beach/spit-backbarrier systems associated with the Ofanto river mouth. The first evidence of the creation of a real beach/spit-backbarrier system (unit D_u) occurs during sea-level range 2 (Figure 16). Sea-level rise flooded the threshold separating the OSFV from the shelf. The threshold appears to play an important role in triggering the formation of the system. In fact, the first spit that forms (within unit D_u) is perched on the bedrock high of the threshold (Figures 9 and 12). The presence of the high provided material for the construction of the spit (as marine erosion affected the high) and promoted the deposition of material transported by the littoral drift. This feature appears to be like other barriers, such as barrier system of the Miquelon-Langlade island [71]. The landward expansion of the unit D_u was enabled by the initial infilling of the OSFV (unit C_u).

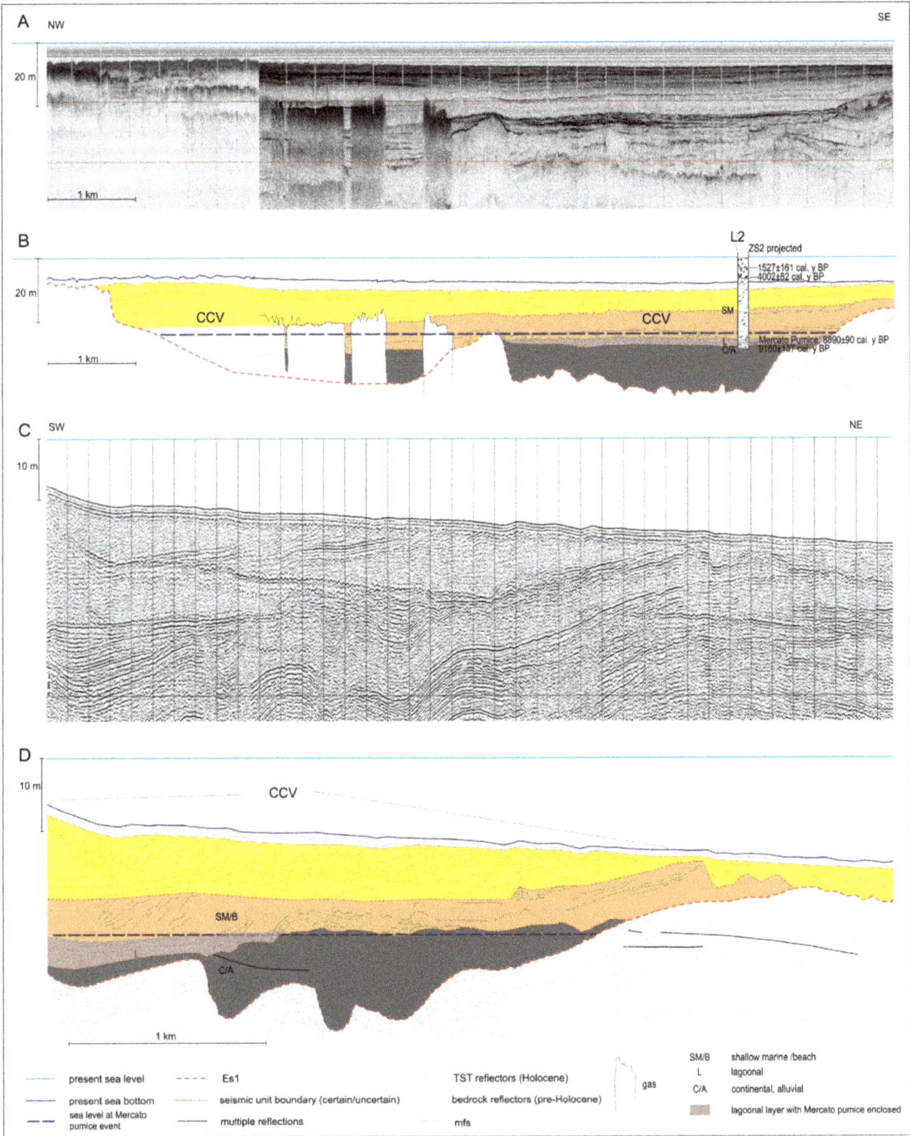

Figure 15. Southeastern sector of the profile T1 (**A**) and its interpretation (**B**), in which the stratigraphic log of ZS2 borehole has been projected. Inland sector of the T8 profile (**C**) and its interpretation (**D**). The lagoonal level containing the Mercato pumice in borehole ZS2 has been correlated both to a series of very high-amplitude reflectors placed roughly at the same depth in the nearby profile T1 at the base of unit E_u and to seismic facies 3 reflectors placed roughly at the same depth at the base of unit E_u in profile T8.

The accretion of the unit D_u spit left an opening in the Barletta strait (profile PX3). This landform, inherited from the time of the OSFV incision, is only modified by spit accretion but remains as an opening between the OSFV and the open sea. This opening influenced the different hydrodynamics that affect the backbarrier deposits. In the NW sector (profile T5, Figure 9), these deposits are composed of seismic facies 3, which indicates a low-energy environment. In the SE sector (profile T4, Figure 12), these deposits are composed of the wavy facies 4, which indicate relatively high energy due to the presence of channels, currents and/or waves in the sector closest to the opening to the sea.

During sea-level range 3, a general landward shift in the beach/spit-backbarrier system (unit E_u) occurs, together with an increase in sediment supply, resulting in a greater thickness and extension of the unit E_u (Figure 16). Compared to the sea level range 2, the different hydrodynamics within the backbarrier is conserved between the NW and the SE sectors, with the same characteristics and same causes. In its final stages, this phase resulted in the near-complete infilling of the OSFV.

The small delta (SFD) that formed during sea level range 3 at the inner edge of the OSFV within unit E_u is important to test our age model; in fact, the sea level of −18 m, reconstructed from the rollover point of SFD, corresponds to an age of ca. 8.3 ka BP [47], that is, within the formation period of sapropel S1a [66,72]. This data supports the attribution of the unit E_u to the sapropel S1a, as in the CCV sector.

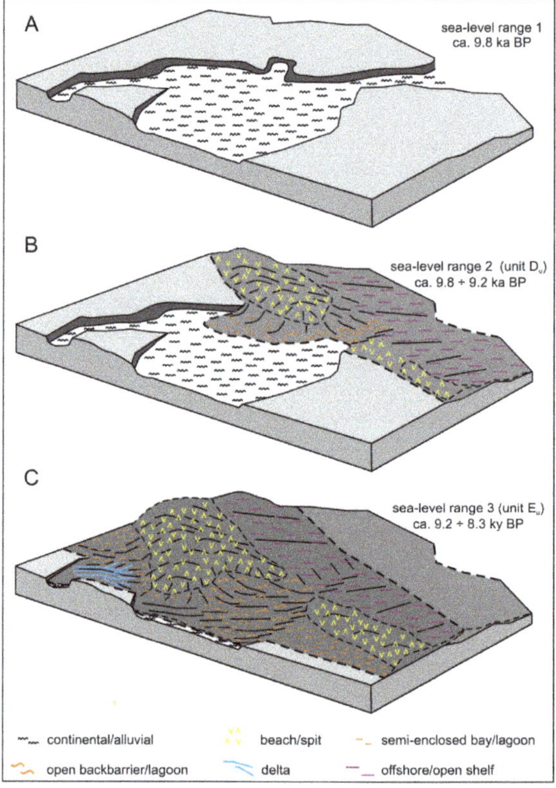

Figure 16. Three-dimensional model of the main phases of OSFV infilling, involving the deposition of unit D_u and unit E_u. (**A**) sea-level range 1: up to ca. −39 m b.s.l.; (**B**) sea-level range 2: from ca. −39 to ca. −29 m b.s.l.: deposition of unit D_u; (**C**) sea-level range 3: from ca. −29 m to ca. −18 m b.s.l.: deposition of unit E_u.

5.6. Palaeoenvironmental and Palaeoclimatic Considerations

The overall characteristics of the OSFV infilling suggest sediment movement proceeded from the NW to the SE. All the units decrease in size in this direction, and the OSFV infilled is faster in the NW sector than in the SE sector. In particular, the construction of a sandy spit points to a littoral drift from the N-NW to the E-SE; that is, in the opposite direction of the current littoral drift in the Manfredonia Gulf.

The direction of sediment transport reconstructed in this work suggests a scenario such as that described by De Santis and Caldara [73] for the period preceding the Holocene climatic transition (ca. 5.5–4.5 cal ka BP). For that period, the authors recognised a strong southward and south-eastward littoral drift in the Manfredonia Gulf, which was promoted by the prevalence and dominance of the N, E and NE winds, in turn promoted by more frequent cyclogenesis to the E, S and SE of the Italian Peninsula. Therefore, this littoral drift pattern in the Manfredonia Gulf seems to have been active from at least ca. 9.8 ka BP, when the first beach/spit-backbarrier system started to form (unit D_u).

We consider units D_u, E_u, and G_u as beach/spit-backbarrier systems that were arranged in a landward backstepping pattern, which involved partial preservation of transgressive deposits or spit morphology. According to the literature, the preservation of relict barrier within a backstepping sequence is rare [12,21] and has been attributed to combinations of rapid sea-level rise, such as those associated with meltwater pulses [16], and accompanied by particularly favourable circumstances, such as early cementation (beach-rock and aeolianite formation) or gravelly sediments [12,14,17].

In our case, the barrier system overstepping occurred after MWP-1B and is therefore in a regime of sea-level rise slower than the previous period [13,74]. Furthermore, the overstepping affected loose and mostly sandy sediments, as demonstrated by the vibro-core collected in this study and by the previous studies [28,30,31]. Therefore, to justify the partial preservation of two barrier systems arranged in a landward backstepping pattern, we believe that a strong and continued supply of sediments attenuated or compensated for the erosive actions. In this context, subsequent units or systems may partially cover and protect previous systems (Figure 17A–C).

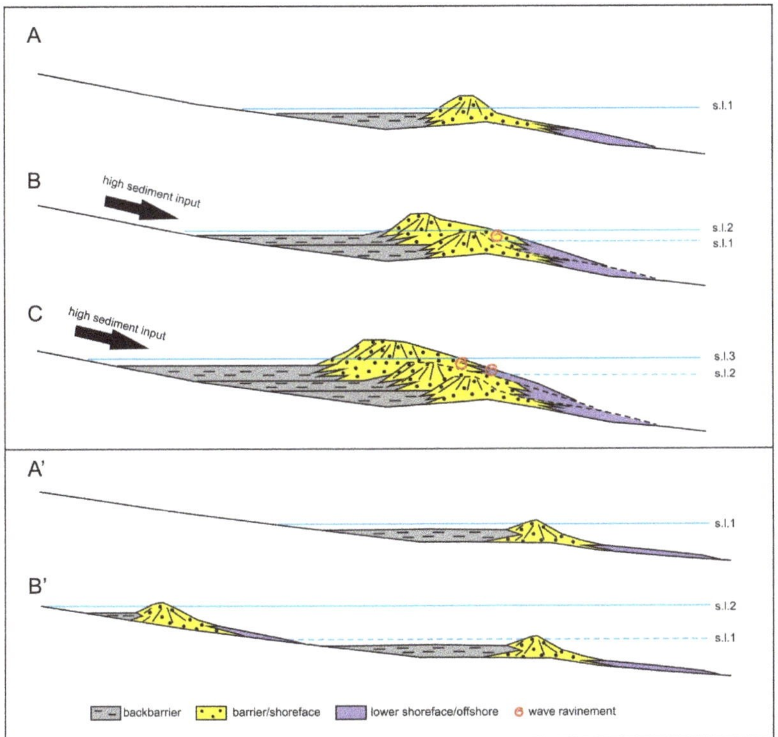

Figure 17. Comparison between our observed "continuous" landward backstepping pattern (**A–C**) and the model of backstepping of Sanders and Kumar [75] and Rampino and Sanders [76] (**A'–B'**).

We correlate this strong and continued sediment supply with the sapropel S1 event. In fact, the chronological constraint provided by the borehole ZS2 and the sea level reconstructed thanks to SFD allows the consideration of unit E_u as coeval with formation of sapropel S1a. As this conclusion confirms and validates our age range evaluation method based on seismic facies 2, it appears that unit D_u and, probably, at least part of unit G_u, were also formed simultaneously with the deposition of sapropel S1 in the Mediterranean. Thus, units D_u, E_u, and G_u represent the shallow-water equivalent of the cm-thick sapropel layers that accumulated offshore in the deeper southern Adriatic basin.

During the minima of the precession of the equinoxes, the northward shift in the African monsoon led to increased precipitation and river runoff in northern African and eastern European regions, which favoured water column stratification and reduced deep-water ventilation of the eastern Mediterranean Sea and Adriatic. This allowed for the accumulation and preservation of organic matter in deep marine sediments and the formation of sapropels [77–79]. The youngest of such events, sapropel S1, occurred from ca. 10/9.0 to ca. 6.8 cal ka BP [66,67] and formed in two distinct pulses (S1a and S1b), which are separated by an interruption in sapropel formation, dating to ca. 8.3–7.8 cal ka BP in the Adriatic Sea [67] due to a short-lived episode of cooler and drier conditions [66].

Overall, units D_u, and E_u demonstrate a period of increased sediment influx to the sea, which has promoted the formation of complex coastal systems. In particular, the formation of the small delta within unit E_u shows a period of increased precipitation during the S1a sapropel event in this area of southern Italy [80] and not only in the northern African and eastern European regions.

Therefore, the roles of the S1 event in the creation and partial preservation of our complex beach/spit-backbarrier systems (units D_u, E_u, and G_u) seem to be as follows: i) a direct role in the formation of the beach/spit-backbarrier systems of units D_u, E_u, and G_u, due to the increased sediment supply; and ii) an indirect role in the partial conservation of previous beach/spit-backbarrier systems because the same increased sediment supply that created a subsequent system also allowed partial protection and conservation of a previous system from wave ravinement action.

The "in-place drowning" transgressive model for low-gradient shelves [75,76], although questioned later [11], predicts that a barrier island-lagoon system develops during conditions of slow relative sea-level rise (Figure 17A′); the barriers can be drowned and preserved due to a very rapid pulse of relative sea-level rise, such as the MWP-1A and MWP-1B, which are recognized in most of the sea-level curves. Later, a new barrier forms landwards during newly created conditions of slower relative sea-level rise (Figure 17B′). This is the case, for example, in the barriers systems of the northern Adriatic [13], a low-gradient setting [81]; due to the episodic occurrence of very high rates of sea-level rise up to 60 mm/a [74], some barriers were overstepped and drowned. As a result, two barriers can be several kilometres apart [81]. A similar case is that described by Cooper et al. [15] in Tijucas Bay (southern Brazil), where a former sandy shoreface-barrier was overstepped during the 8.2 ka event and decoupled from the contemporary shoreline. Upon overstepping, the wave-influenced shoreline was shifted from the barrier to the landward margin of the former backbarrier (over 7 km cross-shore distance).

Our study instead suggests that, in situations of high and continued sedimentary inputs, low-gradient settings and sea-level rise lower than those recorded during MWP-1A and MWP-1B, a stacking pattern of the barrier systems that we define as "continuous" backstepping may be generated, in which the next barrier system partially covers the previous one (Figure 17A–C). Continued sediment supply is a key factor in our model. In an extreme situation, a very strong and continuous sediment supply, along with a slow sea-level rise, can generate a prograding barrier system [82]. Instead, in our case, a fine balance between SLR and sediment supply causes a continuous landward backstepping.

6. Conclusions

Holocene sea-level rise has affected most of the LGM surface (ES1) in our study area between the last phases of the MWP−1B [13,74,83,84] and the subsequent slower rise until the maximum landward shift in the shoreline occurred. This implies that in our study area, the upper part of the TST was deposited (uTST [36]), as well as the subsequent HST, from 5.5 ka to present day.

In the innermost area of the Manfredonia Gulf, the post-Würm transgression resulted in the backfilling of two erosive landforms incised within the ES1 which we documented for the first time.

The first landform (CCV), located further to the NW, is correlated with the current Carapelle and Cervaro streams and consists of a meandering incised valley with stretches parallel to the current coastline in the most landward sector and a straighter route on the shelf, where it flows as a tributary into the so-called MIV.

The second erosive landform (OSFV), which is present farther to the SE, is correlated with the current San Ferdinando stream and Ofanto River. The OSFV consists of two separate incisions that join to form a very wide and shallow valley.

Six different seismic facies and seven UBSUs have been identified, which have covered the ES1 and/or infilled the CCV and OSFV. The UBSUs in some cases are made up of a single seismic facies, in other cases of multiple seismic facies that transit into one another.

We have identified some approximated age ranges for the different units, starting from the current position of one of the seismic facies identified, seismic facies 2. It has been interpreted as deposited at (or close to) contemporary sea level. Since seismic facies 2 forms several sedimentary bodies (belonging to different units) in a landward backstepping pattern on ES1, we associated two sea levels with the base and top of each sedimentary body. Then, we considered the local sea-level curve to infer two ages corresponding to these two sea levels. Thus, several sedimentary bodies formed by seismic facies 2 can be correlated to many sea-level ranges and therefore to age ranges.

Both in the CCV and OSFV, some barrier/spit-backbarrier systems, arranged in a landward backstepping pattern, were formed. One of these (unit E_u) was attributed to the sea level range between -29 to -18 m b.s.l. and, thus, to a time interval between 9.2 ka BP and ca. 8.3 ka BP. This age range attribution, obtained from the current position of sedimentary bodies formed by seismic facies 2 included in unit E_u, was chronologically constrained both in CCV and OSFV.

In CCV the chronological constraint was obtained thanks to a dated lagoonal deposit with enclosed Mercato pumice present in the ZS2 borehole, which intercepts unit E_u. In OSFV, the chronological constraint was obtained thanks to a small delta included in unit E_u, in that the rollover point of the delta was assumed as a sea-level indicator, and thus as an age indicator.

The chronological constraint identified for unit E_u validated the age range evaluation method based on the current position of sedimentary bodies formed by seismic facies 2. Thus, it results that most of the OSFV and CCV infilling is coeval with the formation of the sapropel S1a in the Mediterranean.

The role of the S1 event in the creation and partial preservation of our complex beach/spit-backbarrier systems seem to be both direct and indirect: direct in the formation of the systems, due to the increased sediment supply; indirect in the partial conservation of previous systems because the same increased sediment supply, coupled with the low gradient setting and slow relative sea-level rise, promoted what we call a "continuous" backstepping. This latter, in turn allowed the partial protection and conservation of the previous barrier system from wave ravinement action.

The progression of the infilling units into CCV and OSFV indicates a prevalent littoral drift from the N-NW to the E-SE; that is, in a direction contrary to the current littoral drift of the Manfredonia Gulf. This evidence allows us to extend the climatic phase of frequent cyclogenesis to the E, S and SE of the Italian Peninsula, identified by De Santis and Caldara [73] before the middle-late Holocene transition (5.5–4.5 ka cal BP), backwards in time to at least ca. 9.8 ka BP.

Author Contributions: Conceptualization, V.D.S., M.C., L.P.; methodology, V.D.S.; software, V.D.S., M.C.; formal analysis, V.D.S., M.C.; investigation, V.D.S., M.C.; data curation, V.D.S., M.C.; writing—original draft preparation, V.D.S.; writing—review and editing, V.D.S., M.C.; supervision, V.D.S., M.C., L.P.; funding acquisition, V.D.S., L.P. All authors have read and agreed to the published version of the manuscript.

Funding: This research received no external funding.

Acknowledgments: We are very grateful to (1) Puglia Basin Authority (AdB) for cores of borehole ZS2 obtained within the "Studio di fattibilità per il monitoraggio e la messa in sicurezza delle aree urbane a rischio di stabilità statica e vulnerabilità strutturale Comune e Provincia di Foggia, Codice DT 011, affidato dalla Regione Puglia all'Autorità di Bacino della Puglia e finanziato con fondi assegnati con delibera CIPE n. 20/2004"; (2) Puglia Basin Authority (AdB) for stratigraphic descriptions of vibrocores SP1_VC24 and SP1_VC25, obtained within the *"Piano operativo per l'individuazione di giacimenti di sabbia sottomarini utilizzabili per il ripascimento artificiale dei litorali sabbiosi in erosione della regione. P.O.R. Puglia 2000–2006-FESR-Asse IV sistemi locali di sviluppo, misura 4.16"*; (3) Con.I.S.Ma. (Interuniversity Consortium for Sea Sciences) for use of the motorboat "ISSEL", within the research agreement n. 270-2017 "Rilievi sismici sulla piattaforma continentale pugliese"; (4) Ionic University section of the University of Bari for instruments and sensors used for navigation and seismic campaign; (5) En.Su (Environmental Survey) and Francesco De Giosa for data processing; and (6) Roberto Sulpizio for the anticipation of some results of the pumice analyses in borehole ZS2.

Conflicts of Interest: The authors declare no conflicts of interest.

References

1. Otvos, E.G. Coastal barriers-nomenclature, processes, and classification issues. *Geomorphology* **2012**, *139–140*, 39–52. [CrossRef]

2. Stutz, M.I.; Pilkey, O.H. Open-ocean barrier islands. Global influence of climatic, oceanographic, and depositional settings. *J. Coast. Res.* **2011**, *27*, 207–222. [CrossRef]

3. Otvos, E.G. Coastal barriers, northern Gulf - Last Eustatic Cycle; genetic categories and development contrasts. A review. *Quat. Sci. Rev.* **2018**, *193*, 212–243. [CrossRef]

4. Hesp, P.A.; Short, A.D. Barrier morphodynamics. In *Handbook of Beach and Shoreface Morphodynamics*; Short, A.D., Ed.; John Wiley & Sons: Hoboken, NJ, USA, 1999; pp. 307–333. ISBN 978-0-471-96570-1.

5. De Santis, V.; Caldara, M.; Torres, T.; Ortiz, J.E.; Sánchez-Palencia, Y. A review of MIS 7 and MIS 5 terrace deposits along the Gulf of Taranto based on new stratigraphic and chronological data. *Ital. J. Geosci.* **2018**, *137*, 349–368. [CrossRef]

6. De Santis, V.; Caldara, M.; Torres, T.; Ortiz, J.E.; Sánchez-Palencia, Y. The role of beach ridges, spits, or barriers in understanding marine terraces processes on loose or semiconsolidated substrates: Insights from the givoni of the Gulf of Taranto (southern Italy). *Geol. J.* **2020**, *55*, 2951–2975. [CrossRef]

7. Roy, P.S.; Cowell, P.J.; Ferland, M.A.; Thom, B.G. Wave dominated coast. In *Coastal Evolution, Late Quaternary Shoreline Morphodynamics*; Carter, R.W.G., Woodroffe, C.D., Eds.; Cambridge University Press: Cambridge, UK, 1994; pp. 121–186. ISBN 0-521-59890-7.

8. Zecchin, M.; Catuneanu, O.; Caffau, M. Wave-ravinement surfaces: Classification and key characteristics. *Earth-Sci. Rev.* **2019**, *188*, 210–239. [CrossRef]

9. Mellett, C.L.; Plater, A.J. Drowned Barriers as Archives of Coastal-Response to Sea-Level Rise. In *Barrier Dynamics and Response to Changing Climate*; Moore, L.J., Murray, A.B., Eds.; Spinger: Berlin/Heidelberg, Germany, 2018; pp. 57–89, ISBN 978-3-319-68086-6.

10. Swift, D.J.P.; Moslow, T.F. Holocene transgression in south-central Long Island, New York-discussion. *J. Sediment. Petrol.* **1982**, *53*, 1014–1019.

11. Leatherman, S.P.; Rampino, M.R.; Sanders, J.E. Barrier island evolution in response to sea level rise; discussion and reply. *J. Sediment. Res.* **1983**, *53*, 1026–1033. [CrossRef]

12. De Falco, G.; Antonioli, F.; Fontolan, G.; Lo Presti, V.; Simeone, S.; Tonielli, R. Early cementation and accommodation space dictate the evolution of an overstepping barrier system during the Holocene. *Mar. Geol.* **2015**, *369*, 52–66. [CrossRef]

13. Storms, J.E.A.; Weltje, G.J.; Terra, G.J.; Cattaneo, A.; Trincardi, F. Coastal dynamics under conditions of rapid sea-level rise: Late Pleistocene to Early Holocene evolution of barrier-lagoon systems on the northern Adriatic shelf (Italy). *Quat. Sci. Rev.* **2008**, *27*, 1107–1123. [CrossRef]

14. Mellett, C.L.; Hodgson, D.M.; Mauz, B.; Lang, A.; Selby, I.; Plater, A.J. Preservation of a drowned gravel barrier complex: A landscape evolution study from the northeastern English Channel. *Mar. Geol.* **2012**, *315–318*, 115–131. [CrossRef]

15. Cooper, J.A.G.; Green, A.N.; Meireles, R.P.; Klein, A.H.F.; Souza, J.; Toldo, E.E. Sandy barrier overstepping and preservation linked to rapid sea level rise and geological setting. *Mar. Geol.* **2016**, *382*, 80–91. [CrossRef]

16. Green, A.N.; Cooper, J.A.G.; Salzmann, L. Geomorphic and stratigraphic signals of postglacial meltwater pulses on continental shelves. *Geology* **2014**, *24*, 151–154. [CrossRef]

17. Green, A.N.; Cooper, J.A.G.; Leuci, R.; Thackeray, R. Formation and preservation of an overstepped segmented lagoon complex on a high-energy continental shelf. *Sedimentology* **2013**, *60*, 1755–1768. [CrossRef]

18. Belknap, D.F.; Kraft, J.C. Preservation potential of transgressive coastal lithosomes on the U.S. Atlantic Shelf. *Mar. Geol.* **1981**, *41*, 419–442.

19. Belknap, D.F.; Kraft, J.C. Influence of antecedent geology on evolution of barrier systems. *Mar. Geol.* **1985**, *63*, 235–262. [CrossRef]

20. Belknap, D.F.; Kraft, J.C.; Dunn, R.K. Transgressive valley-fill lithosomes: Delaware and Maine. In *Incised-Valley Systems: Origin and Sedimentary Sequences*; Dalrymple, R., Boyd, R., Zaitlin, B., Eds.; SEPM: Broken Arrow, OK, USA, 1994; Volume 51, pp. 303–320.

21. Storms, J.E.A.; Swift, D.J.P. Shallow marine sequences as the building blocks of stratigraphy: Insights from numerical modelling. *Basin Res.* **2003**, *15*, 287–303. [CrossRef]

22. De Santis, V.; Caldara, M.; Torres, T.; Ortiz, J.E. Stratigraphic units of the Apulian Tavoliere Plain: Chronology, correlation with marine isotope stages and implication regarding vertical movements. *Sediment. Geol.* **2010**, *228*, 255–270. [CrossRef]

23. De Santis, V.; Caldara, M.; Pennetta, L.; Torres, T.; Ortiz, J.E. Unconformity bounded stratigraphic units in an Italian alluvial plain area: Recognising and dating. *J. Sediment. Res.* **2013**, *83*, 96–114. [CrossRef]

24. De Santis, V.; Caldara, M.; Torres, T.; Ortiz, J.E. Two middle Pleistocene warm stages in the terraced deposits of the Apulia region (southern Italy). *Quat. Int.* **2014**, *332*, 2–18. [CrossRef]

25. De Santis, V.; Caldara, M.; Pennetta, L. The marine and alluvial terraces of Tavoliere di Puglia plain (southern Italy). *J. Maps* **2014**, *10*, 114–125. [CrossRef]

26. Caldara, M.; Pennetta, L. Pleistocenic buried abrasion platforms in Southeastern "Tavoliere" (Apulia, South Italy). *Alp. Mediterr. Quat.* **1991**, *4*, 303–309.

27. Fabbri, A.; Gallignani, P. Ricerche geomorfologiche e sedimentologiche nell'Adriatico meridionale. *Giorn Geol.* **1972**, *38*, 453–498.

28. Trincardi, F.; Argnani, A.; Correggiari, A. *Note Illustrative della Carta Geologica dei Mari Italiani, alla scala 1:250,00, Foglio NK33-6 Vieste, e Foglio NK33-8/9 Bari*; ISPRA: Roma, Italy, 2011; pp. 1–194.

29. Trincardi, F.; Correggiari, A.; Ridente, D.; Verdicchio, G.; Cattaneo, A.; Minisini, D.; Remia, A.; Taviani, M.; Asioli, A.; Piva, A.; et al. *Carta Geologica dei Mari Italiani, Foglio NK33-8/9 Bari*; ISPRA: Roma, Italy, 2011.

30. Maselli, V.; Trincardi, F. Large-scale single incised valley from a small catchment basin on the western Adriatic margin (central Mediterranean Sea). *Glob. Planet Chang.* **2013**, *100*, 245–262. [CrossRef]

31. Maselli, V.; Trincardi, T.; Asioli, A.; Ceregato, A.; Rizzetto, F.; Taviani, M. Delta growth and river valleys: The influence of climate and sea level changes on the South Adriatic shelf (Mediterranean Sea). *Quat. Sci. Rev.* **2014**, *99*, 146–163. [CrossRef]

32. De Santis, V.; Caldara, M. Evolution of an incised valley system in the southern Adriatic Sea (Apulian margin): An onshore-offshore correlation. *Geol. J.* **2016**, *51*, 263–284. [CrossRef]

33. Cattaneo, A.; Correggiari, A.; Langone, L.; Trincardi, F. The late-Holocene Gargano subaqueous delta, Adriatic shelf: Sediment pathways and supply fluctuations. *Mar. Geol.* **2003**, *193*, 61–91. [CrossRef]

34. Trincardi, F.; Asioli, A.; Cattaneo, A.; Correggiari, A.; Langone, L. Stratigraphy of the late-Quaternary deposits in the Central Adriatic basin and the record of short-term climatic events. *Memorie-Istituto Italiano di Idrobiologia* **1996**, *55*, 39–70.

35. Correggiari, A.; Trincardi, F.; Langone, L.; Roveri, M. Styles of failure in heavily-sedimented highstand prodelta wedges on the Adriatic shelf. *J. Sediment. Res.* **2001**, *71*, 218–236. [CrossRef]

36. Maselli, V.; Hutton, E.W.; Kettner, A.J.; Syvitski, J.P.M.; Trincardi, F. High-frequency sea level and sediment supply fluctuations during Termination I: An integrated sequence-stratigraphy and modeling approach from the Adriatic Sea (Central Mediterranean). *Mar. Geol.* **2011**, *287*, 54–70. [CrossRef]

37. Cattaneo, A.; Trincardi, F. The late-Quaternary transgressive record in the Adriatic epicontinental sea: Basin widening and facies partitioning. In *Isolated Shallow Marine Sand Bodies: Sequence Stratigraphic Analysis and Sedimentologic Interpretation*; Bergman, K., Snedden, J., Eds.; SEPM: Broken Arrow OK, USA, 1999; Volume 64, pp. 127–146.

38. Caldara, M.; Capolongo, D.; Del Gaudio, V.; De Santis, V.; Pennetta, L.; Maiorano, P.; Simone, O. *Note Illustrative Della Carta Geologica d'Italia alla Scala 1:50.000 Foglio, 409 "Zapponeta"*; Istituto Superiore per la Protezione e la ricerca Ambientale: Roma, Italy, in press.

39. Mitchum, R.M.; Vail, P.R.; Sangree, J.B. Seismic stratigraphy and global changes in sea-level, part 6. Stratigraphic interpretation of seismic reflection patterns in depositional sequences. In *Seismic Stratigraphy-Applications to Hydrocarbon Exploration*; Payton, C.E., Ed.; AAPG Memoires: Boulder, Tulsa, OK, USA, 1977; Volume 26, pp. 117–133, ISBN 117-133.

40. Esker, D.; Eberli, G.P.; McNeill, D.F. The Structural and Sedimentological Controls on the Reoccupation of Quaternary Incised Valleys, Belize Southern Lagoon. *AAPG Bull.* **1998**, *82*, 2075–2109.

41. Maio, C.V.; Gontz, A.M.; Sullivan, R.M.; Madsen, S.M.; Weidman, C.R.; Donnelly, J.P. Subsurface evidence of storm-driven breaching along a transgressing barrier system, Cape Cod, U.S.A. *J. Coast. Res.* **2015**, *32*, 264–279. [CrossRef]

42. Nordfjord, S.; Goff, J.A.; Austin, J.A.; Gulick, S.P.S. Seismic facies of incised-valley fills, New Jersey continental shelf: Implications for erosion and preservation processes acting during latest Pleistocene-Holocene transgression. *J. Sediment. Res.* **2006**, *76*, 1284–1303. [CrossRef]

43. Labaune, C.; Tesson, M.; Gensous, B.; Parize, O.; Imbert, P.; Delhaye-Prat, V. Detailed architecture of a compound incised valley system and correlation with forced regressive wedges: Example of Late Quaternary Têt and Agly rivers, western Gulf of Lions, Mediterranean Sea, France. *Sediment. Geol.* **2010**, *223*, 360–379. [CrossRef]

44. Raynal, O.; Bouchette, F.; Certain, R.; Sabatier, P.; Lofi, J.; Séranne, M.; Dezileau, L.; Briqueu, L.; Ferrer, P.; Courp, T. Holocene evolution of a Languedocian lagoonal environment controlled by inherited coastal morphology (northern Gulf of Lions, France). *Bull. Soc. Geol. France* **2010**, *181*, 211–224. [CrossRef]

45. Pellegrini, C.; Maselli, V.; Cattaneo, A.; Piva, A.; Ceregato, A.; Trincardi, F. Anatomy of a compound delta from the post-glacial transgressive record in the Adriatic Sea. *Mar. Geol.* **2015**, *362*, 43–59. [CrossRef]

46. Catuneanu, O.; Galloway, W.E.; Kendall, C.G.S.t.C.; Miall, A.D.; Posamentier, H.W.; Strasser, A.; Tucker, M.E. Sequence stratigraphy: Methodology and Nomenclature. *Newsl. Stratigr.* **2011**, *44/3*, 173–245. [CrossRef]

47. Lambeck, K.; Antonioli, F.; Anzidei, M.; Ferranti, L.; Leoni, G.; Silenzi, S. Sea level change along the Italian coasts during Holocene and prediction for the future. *Quat. Int.* **2011**, *232*, 250–257. [CrossRef]

48. Reimer, P.J.; Bard, E.; Bayliss, A.; Warren Beck, J.; Blackwell, P.G.; Bronk Ramsey, C.; Buck, C.E.; Cheng, H.; Lawrence Edwards, R.; Friedrich, M.; et al. IntCal13 and Marine13 radiocarbon age calibration curves 0-50000 years cal BP. *Radiocarbon* **2013**, *55*, 1869–1887. [CrossRef]

49. Di Rita, F.; Simone, O.; Caldara, M.; Gehrels, W.R.; Magri, D. Holocene environmental changes in the coastal Tavoliere Plain (Apulia, southern Italy): A multiproxy approach. *Palaeogeogr. Palaeoclimatol. Palaeoecol.* **2011**, *310*, 139–151. [CrossRef]

50. Sulpizio, R.; (University of Bari, Bari, Italy). Personal communication, 2019.

51. Andronico, D.; Calderoni, G.; Cioni, R.; Sbrana, A.; Sulpizio, R.; Santacroce, R. Geological map of Somma-Vesuvius volcano. *Period Miner.* **1995**, *64*, 77–78.

52. Santacroce, R.; Cioni, R.; Marianelli, P.; Sbrana, A.; Sulpizio, R.; Zanchetta, G.; Donahue, D.J.; Joron, J.L. Age and whole rock—glass compositions of proximal pyroclastics from the major explosive eruptions of Somma-Vesuvius: A review as a tool for distal tephrostratigraphy. *J. Volcanol. Geotherm. Res.* **2008**, *177*, 1–18. [CrossRef]

53. Marzolla, B. *Provincia di Capitanata*; Real Officio Topografico della Guerra: Napoli, Italy, 1851.

54. Caldara, M.; Pennetta, L.; Simone, O. Holocene Evolution of the Salpi Lagoon (Puglia, Italy). *J. Coast. Res.* **2002**, *36*, 124–133. [CrossRef]

55. Pérès, J.M. The Mediterranean benthos. *Oceanogr. Mar. Biol.* **1967**, *5*, 449–533.

56. Picard, J. Recherches qualitatives sur les Biocoenoses marines des substrats meubles dragables de la région marseillaise. *Recueil Trav. Stat. Mar. Endoume* **1965**, *36*, 1–102.

57. Gravina, M.F.; Ardizzone, G.D.; Scaletta, F.; Chimenz, C. Descriptive analysis and classification of benthic communities in some Mediterranean coastal lagoons (central Italy). *Mar. Ecol.* **1989**, *10*, 141–216. [CrossRef]

58. Lambeck, K.; Antonioli, F.; Purcell, A.; Silenzi, S. Sea-level change along the Italian coast for the past 10,000 yr. *Quat. Sci. Rev.* **2004**, *23*, 1567–1598. [CrossRef]

59. Primavera, M.; Simone, O.; Fiorentino, G.; Caldara, M. The Palaeo-environmental study of the Alimini Piccolo Lake enables a reconstruction of Holocene sea-level changes in south-east Italy. *Holocene* **2011**, *21*, 553–563. [CrossRef]

60. Vacchi, M.; Marriner, N.; Morhange, C.; Spada, G.; Fontana, A.; Rovere, A. Multiproxy assessment of Holocene relative sea-level changes in the western Mediterranean: Sea-level variability and improvements in the definition of the isostatic signal. *Earth-Sci. Rev.* **2016**, *155*, 172–197. [CrossRef]

61. Billy, J.; Robin, N.; Hein, C.J.; FitzGerald, D.M.; Certain, R. Impact of relative sea-level changes since the last deglaciation on the formation of a composite paraglacial barrier. *Mar. Geol.* **2018**, *400*, 76–93. [CrossRef]

62. Tillmann, T.; Wunderlich, J. Barrier rollover and spit accretion due to the combined action of storm surge induced washover events and progradation: Insights from ground-penetrating radar surveys and sedimentological data. *J. Coast. Res.* **2013**, *65*, 600–605. [CrossRef]

63. Patruno, S.; Hampson, G.J.; Jackson, C.A.-L. Quantitative characterisation of deltaic and subaqueous clinoforms. *Earth-Sci. Rev.* **2015**, *142*, 79–119. [CrossRef]

64. Montes, A.; Bujalesky, G.G.; Paredes, J.M. Geomorphology and internal architecture of Holocene sandy-gravel beach ridge plain and barrier spits at Río Chico area, Tierra del Fuego, Argentina. *J. S. Am. Earth Sci.* **2018**, *84*, 172–183. [CrossRef]

65. Timmons, E.A.; Rodriguez, A.B.; Mattheus, C.R.; DeWitt, R. Transition of a regressive to a transgressive barrier island due to backbarrier erosion, increased storminess, and low sediment supply: Bogue Banks, North Carolina, USA. *Mar. Geol.* **2010**, *278*, 100–114. [CrossRef]

66. Ariztegui, D.; Asioli, A.; Lowe, J.J.; Trincardi, F.; Vigliotti, L.; Tamburini, F.; Chondrogianni, C.; Accorsi, C.A.; Bandini Mazzanti, M.; Mercuri, A.M.; et al. Palaeoclimate and the formation of sapropel S1: Inferences from Late Quaternary lacustrine and marine sequences in the central Mediterranean region. *Palaeogeogr. Palaeoclimatol. Palaeoecol.* **2000**, *158*, 215–240. [CrossRef]

67. Tesi, T.; Asioli, A.; Minisini, D.; Maselli, V.; Dalla Valle, G.; Gamberi, F.; Langone, L.; Cattaneo, A.; Montagna, P.; Trincardi, F. Large-scale response of the Eastern Mediterranean thermohaline circulation to African monsoon intensification during sapropel S1 formation. *Quat. Sci. Rev.* **2017**, *159*, 139–154. [CrossRef]

68. Cronin, T.M.; Vogt, P.R.; Willard, D.A.; Thunell, R.C.; Halka, J.; Berke, M.; Pohlman, J. Rapid sea level rise and ice sheet response to 8200-year climate event. *Geophys. Res. Lett.* **2007**, *34*. [CrossRef]

69. Kendall, R.A.; Mitrovica, J.X.; Milne, G.A.; Törnqvist, T.E.; Li, Y. The sea-level fingerprint of the 8.2 ka climate event. *Geology* **2008**, *36*, 423–426. [CrossRef]

70. Hijma, M.P.; Coen, K.M. Timing and magnitude of the sea-level jump preluding the 8200 yr event. *Geology* **2010**, *38*, 275–278. [CrossRef]

71. Billy, J.; Robin, N.; Hein, C.J.; FitzGerald, D.M.; Certain, R. Dominance of inherited geologic framework on the development of coastal barrier system. *J. Coast. Res.* **2018**, *85*, 406–410. [CrossRef]

72. De Rijk, S.; Hayes, A.; Rohling, E.J. Eastern Mediterranean sapropel S1 interruption: An expression of the onset of history of climatic deterioration around 7 ka B.P. *Mar. Geol.* **1999**, *153*, 337–343. [CrossRef]

73. De Santis, V.; Caldara, M. The 5.5-4.5 kyr climatic transition as recorded by the sedimentation pattern of coastal deposits of Apulia region, southern Italy. *Holocene* **2015**, *25*, 1313–1329. [CrossRef]

74. Liu, J.P.; Milliman, J.D. Reconsidering melt-water pulses 1A and 1B: Global impacts of rapid sea-level rise. *J. Ocean U China* **2004**, *3*, 183–190. [CrossRef]

75. Sanders, J.E.; Kumar, N. Evidence of shoreface retreat and in-place "drowning" during Holocene submergence of barriers, shelf off Fire Island, New York. *Geol. Soc. Am. Bull.* **1975**, *86*, 65–76. [CrossRef]

76. Rampino, M.R.; Sanders, J.E. Holocene transgression in south-central Long Island, New York. *J. Sediment. Petrol.* **1980**, *50*, 1063–1080.

77. Grant, K.M.; Grimm, R.; Mikolajewicz, U.; Marino, G.; Ziegle, M.; Rohling, E.J. The timing of Mediterranean sapropel deposition relative to insolation, sea-level and African monsoon changes. *Quat. Sci. Rev.* **2016**, *140*, 125–141. [CrossRef]

78. Rossignol-Strick, M.; Nesteroff, W.; Olive, P.; Vergnaud-Grazzini, C. After the deluge: Mediterranean stagnation and sapropel formation. *Nature* **1982**, *295*, 105–110. [CrossRef]

79. Rohling, E.J. Review and new aspects concerning the formation of eastern Mediterranean sapropels. *Mar. Geol.* **1994**, *122*, 1–28. [CrossRef]

80. Piccarreta, M.; Caldara, M.; Capolongo, D.; Boenzi, F. Holocene geomorphic activity related to climatic change and human impact in Basilicata, southern Italy. *Geomorphology* **2011**, *128*, 137–147. [CrossRef]

81. Cattaneo, A.; Steel, R.J. Transgressive deposits: A review of their variability. *Earth-Sci. Rev.* **2003**, *62*, 187–228. [CrossRef]

82. Billy, J.; Robin, N.; Hein, C.J.; Certain, R.; FitzGerald, D.M. Insight into the late Holocene sea-level changes in the NW Atlantic from a paraglacial beach-ridge plain south of Newfoundland. *Geomorphology* **2015**, *248*, 134–146. [CrossRef]
83. Bard, E.; Hamelin, B.; Fairbanks, R.G. U-Th ages obtained by mass spectrometry in corals from Barbados: Sea level during the past 130,000 years. *Nature* **1990**, *346*, 456–458. [CrossRef]
84. Bard, E.; Hamelin, B.; Arnold, M.; Montaggioni, L.; Cabioch, G.; Faure, G.; Rougerie, F. Deglacial sea-level record from Tahiti corals and the timing of global Meltwater discharge. *Nature* **1996**, *382*, 241–244. [CrossRef]

Article

Ancient Coastal Changes Due to Ground Movements and Human Interventions in the Roman Portus Julius (Pozzuoli Gulf, Italy): Results from Photogrammetric and Direct Surveys

Pietro P. C. Aucelli [1], Gaia Mattei [1], Claudia Caporizzo [1,*], Aldo Cinque [2], Salvatore Troisi [1], Francesco Peluso [1], Michele Stefanile [3] and Gerardo Pappone [1]

[1] Dipartimento di Scienze e Tecnologie, Università degli Studi di Napoli Parthenope, Centro Direzionale Is. C4, 80121 Napoli, Italy; pietro.aucelli@uniparthenope.it (P.P.C.A.); gaia.mattei@uniparthenope.it (G.M.); salvatore.troisi@uniparthenope.it (S.T.); francesco.peluso@uniparthenope.it (F.P.); gerardo.pappone@uniparthenope.it (G.P.)

[2] Dipartimento di Scienze della Terra, dell'Ambiente e delle Risorse, Università degli Studi di Napoli Federico II, Largo San Marcellino, 10, 80138 Napoli, Italy; aldocinque@hotmail.it

[3] Dipartimento Asia Africa e Mediterraneo, Università degli Studi di Napoli L'Orientale, Piazza S. Domenico Maggiore, 12, 80134 Napoli, Italy; michelestefanile@gmail.com

* Correspondence: claudia.caporizzo@uniparthenope.it

Received: 30 January 2020; Accepted: 25 February 2020; Published: 29 February 2020

Abstract: This research aims to evaluate the amount of vertical ground movements during Roman times inside the archaeological area of Portus Julius (Gulf of Pozzuoli) using high-precision surveys on the most reliable archaeological sea-level markers. Measuring the submersion of ancient floors, structural elements belonging to a former fish tank, and several roman *pilae*, two different relative sea levels (RSLs), related to the beginning and the end of the first century BCE, respectively, −4.7/−5.20 m and −3.10 m MSL (mean sea level), were detected. A photogrammetric survey was carried out in order to produce a 3D model of the fish tank. The results in terms of the RSL variations have enabled us to reconstruct a morpho-evolution of the ancient coastal sector during the last 2.1 kyBP. At the beginning of the first century BCE, the area was characterized by a sheltered gulf with numerous maritime *villae* located along the coast. In 37 BCE, the construction of the military harbour of Portus Julius strongly modified the paleogeography of the sector, which was also affected by a prevailing subsidence at least until the end of the first century BCE (year 12 BCE), when the port was converted into a commercial hub.

Keywords: coastal changes; vertical ground movements; geoarchaeology; 3D model; relative sea level changes; Campi Flegrei

---

1. Introduction

The study of past geomorphological events may reveal the vulnerability of a specific area to dangerous phenomena, even if they have very long return times and/or very slow rates of action [1]. Moreover, such studies are fundamental to build site-specific process response models, which are useful to predict the consequences of future environmental changes (climatic, tectonic, etc.) [2–4].

Regarding particular volcanic coastal areas, vertical ground movements and related changes of the relative sea level [5–20] represent crucial information for hazard forecasting based on the style and rate of landscape changes over the time.

The present study concerning the Gulf of Pozzuoli, located in the wide caldera of the Campi Flegrei volcanic complex (S. Italy), can be framed in this context.

The study area is known worldwide for the striking evidence of vertical ground movements, many of which are related to the ruins (both emerged and submerged) of ancient buildings and coastal structures scattered throughout this territory [21–28].

Around 770 BCE, the establishment of a commercial base (*emporion*) on the island of *Pithekoussai* (Ischia) and the later foundation of the first Greek colony of *Kyme* (now Cuma) in Southern Italy marked the beginning of the human presence in the territory in the historical period.

However, the great urbanization of Campi Flegrei occurred during the Roman age, after the settlement in 194 BCE of the Roman colony of *Puteoli* (now Pozzuoli) in the same place of the old Samian colony of *Dicearchia*. *Puteoli* soon became the main commercial port of Rome and, therefore, a centre of maritime traffic so intense and important that it earned the appellation of *Delus minor* (i.e., small Delo, from the important Aegean free port) [29–31]. Moreover, this territory had a strategic military value hosting two important military harbours. Portus Julius was built in 37 BCE by M. Vipsanius Agrippa and served as a naval military base during the war against Sextus Pompeius at the end of the Roman republican era. The *Misenum* port, instead, was built in 12 BCE after the abandonment of Portus Julius, which became a commercial hub.

Since the end of the II century BCE, the CF coasts hosted the largest and most luxurious seaside villas built especially in the second half of the first century BCE. The numerous thermal springs, connected with the volcanism of the region, the proximity to Rome and the *amoenitas* of the natural landscape attracted many wealthy people from the Roman aristocracy of the Late Republican period. This resulted in the consequence of quick occupation, with private properties, of the whole coastal sector. The almost uninterrupted sequence of luxurious villas, visible by the sea in the Gulf of Naples, was precisely described in a famous quotation by Strabone [32].

Currently, the largest part of the ancient coastal strip, including all the buildings and maritime structures, is submerged due to the overall subsidence starting at the end of the Roman period [25,33].

When better preserved and carefully investigated, the remains of those structures also reveal the evidence of relative sea level (RSL) changes that occurred during the Roman period. The typologies and dimensions of these remains may document the strong human capacity to modify the coastal physiography of the time, such as the opening of navigable channels to access the coastal lakes (see Lucrino and Averno lakes during the construction of the Portus Julius) or the artificial infill by which a 2 km long reach of Puteoli coast (*ripa puteolana*) was advanced tens of meters [30].

The purpose of this study is to present the results of high-precision surveys on the most reliable archaeological markers of the ancient sea level, aimed at the evaluation of the ground movements affecting the area of Portus Julius during the Roman period.

This data can be considered very relevant to issues regarding the relative sea-level positions at the time of the construction of the harbour [23,25–28,34]. In addition, the ancient Portus Julius is one of the largest underwater archaeological sites in Italy and represents a great point of interest for marine archaeology studies [35–39]. This site-located in the Underwater Archeological Park of Baia is a perfect example of submerged archaeological heritage that records fundamental traces of the ancient coastal morphologies, highly vulnerable to present coastal processes because of recent submersion [40,41]. Indeed, a further aim of this research is to document, through a high-precision photogrammetric survey, the fish tank detected at Portus Julius, which represents the most significant evidence of the relative sea-level variations that occurred in the area in the few decades between the first century BCE and the first century CE. In fact, with respect to the historical sources, the area is well-dated and documented and can be used as a key site in the RSL studies.

However, the study of Mediterranean submerged archaeological sites as indicators of sea-level oscillations is a challenge of great scientific interest [14,15,42–48].

In the framework of caldera modelling studies, these results can be considered crucial to evaluate the exacerbated effects of vertical ground movements on the accelerated sea-level rise due to the ongoing climate change, in terms of coastal modification and consequent human adaptations.

2. Geological Setting

The Gulf of Pozzuoli (Figure 1) belongs to the northern sector of the Gulf of Naples and covers an area of about 130 km^2 [49]. The coasts of the Gulf are mainly exposed to the waves coming from the SW sector (fetch amplitude 45° and average distance of 485 km, [50]), and were modified in the last millennia due to both the physiography and the volcano-tectonic ground movements affecting the whole area. In fact, Pozzuoli Gulf is located along the northern margin of the Campi Flegrei caldera ([51] and references therein).

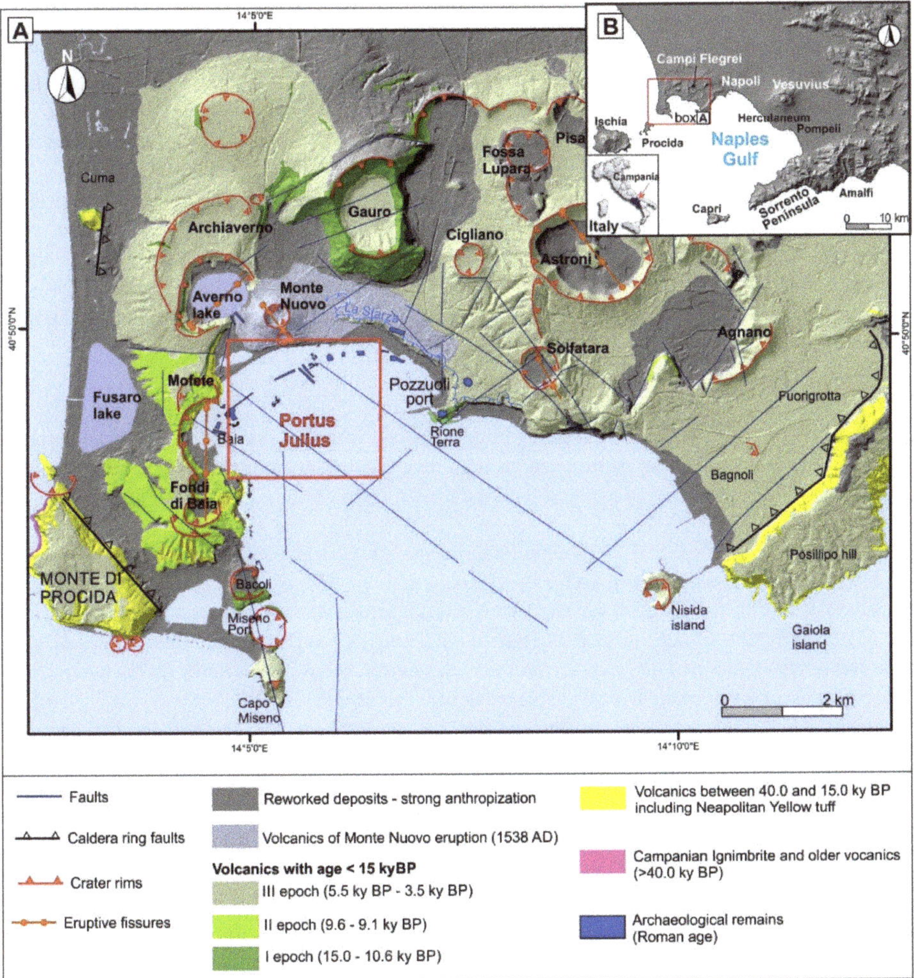

Figure 1. Geological map of the Campi Flegrei volcanic area with the chronologic classification of the main volcanic events (after Isaia et al. [52]) and the positioning of the main archaeological remains of Roman age (after Camodeca [30]).

The Campi Flegrei volcanic area is a poly-calderic system made of structural depressions covering an area of about 230 km^2. The volcanism of the area can be divided into pre- and post-caldera activities [53–55].

The pre-caldera activity is characterized by three main different eruptive events [50], interspersed with periods of volcanic quiescence, starting with the Campanian Ignimbrite (CI) super eruption, which occurred 40 ky BP [56,57].

After the CI eruption, the northern part of the just-formed caldera was invaded by the sea and progressively filled with tuffites and subordinate submarine flows. The second volcanic event, which led to the formation of the Masseria del Monte Tuff, occurred 29.3 ky BP [52,58] and it preceded the last eruption that has contributed to the formation of the caldera 15 ky BP, called the Neapolitan Yellow Tuff (NYT) Eruption [59,60].

The volcanic activity following the NYT eruption was characterized by about 72 explosive and effusive eruptions, recognized in stratigraphic records over an area of about 1000 km^2 [61]. These eruptions, both explosive and effusive, have produced lava flows and pyroclastic deposits and they occurred in three epochs of volcanic activity, between 15 and 10.6 ka BP, between 9.6 and 9.1 ka BP, and between 5.5 and 3.5 ka BP [62]. After a quiescence of about 3000 years, the last volcanic event has deeply modified the morphology of the area with the formation of the Monte Nuovo tuff cone in 1538 CE [21,28,63,64].

The study area, located in the central sector of Pozzuoli Gulf, is characterized by a low sandy coast at the feet of the Monte Nuovo Lithosome. This unit, with a height of 100 m, is made of a grey cineritic deposit rich in blackish scoriae in its upper part ([52] Figure 1). In particular, the coastal sector of Portus Julius is bordered southward by La Starza marine terrace, uplifted 30–40 m MSL about 5 ky BP [61,65], whose depositional sequence recorded the older coastal changes in the area, with an alternation of marine sediments and pyroclastic levels.

3. Archeological Setting

During the Roman period, Portus Julius (Figure 2) experienced at least two different construction phases: a brief military phase during which the entry channel was built and a long commercial phase during which the area was restored with the construction of numerous warehouses (*horrea*) and fish tanks.

The first phase started in 37 BCE, when the coastal sector enclosed between the Lucrino and the Averno Lakes was chosen by Agrippa for the construction of a new military harbour system, named precisely Portus Julius. At that time, the Lucrino Gulf appeared as a lake due to the construction of *via Herculanea*, built on a spit formed between *Baiae* and *Puteoli* [32]. When the military port was positioned into the Lucrino lake, the narrow opening linking Averno and Lucrino lakes was enlarged and fortified by walls to create a sheltered landing for warships [37,39,66].

During this period, the whole area was renovated, as testified by the construction of a new main entrance to *Portus Julius*, formed by a channel with two 300 m long banks and protected by pier structures with huge *pilae* in *opus caementicum*. These were realized following the principles well described by Vitruvius [67].

However, the use of the area for military purposes, related to the quick and victorious war against the pirate fleet of Sextus Pompeius, was very short and the harbour was converted in a commercial basin around the year 12 BCE, together with the relocation of the military port in the nearby Misenum, favoured by the presence of a natural triple-basin port and by a less problematic tendency to sand infilling. With the change of purposes, Portus Julius was rethought and restructured, with the transformation and the adaptation of military environments into warehouses and fish tanks, also along the entry channel to Lake Lucrino [37–39].

In this study, the fish tank cut into the embankment of the entry channel was deeply surveyed and analysed, as it can be considered as good evidence of the renovation that occurred after 12 BCE. This fish tank is particularly interesting as it represents one of the few examples of fish tanks related to commercial activities and not to private maritime villas. The existence of a tank within the entry channel can be related to the presence of a fish market or areas intended for the production of the *garum.* According to the historical sources, Cassiodoro [68] in particular, the harbour was abandoned during the IV century CE due to a bradyseismic crisis.

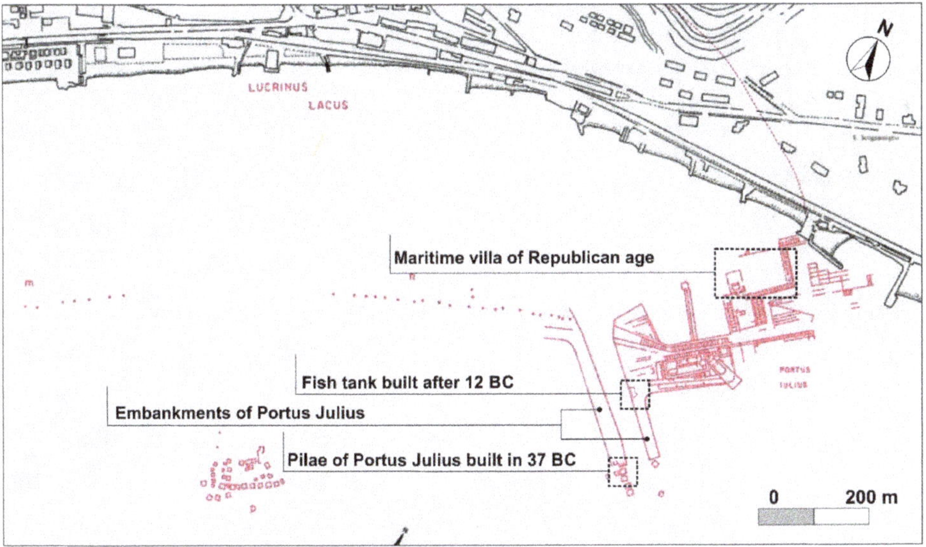

Figure 2. Archaeological map of Portus Julius (after Scognamiglio [69]) with the archaeological sea level markers used in this study highlighted with black rectangles.

4. Methods

4.1. Direct Survey on Archaeological Features

In this study, direct surveys were carried out on three different types of archaeological sea level markers. Two markers are directly related to the former sea level, and therefore considered high-precision sea level index points (SLIPs) [70], i.e., a fish tank and a port structure. The third marker is a floor belonged to a maritime villa built on the ancient sandy coast, therefore considered a terrestrial limiting point (TLPs), intended as a marker positioned above the ancient sea level [70].

The survey was planned by evaluating the bibliographic, cartographic, and videographic documentation describing the archaeological area. It was carried out to measure the size and submersion of the archaeological structures investigated with direct methods by a team of scuba divers composed of two geomorphologists and an archaeologist. The divers were assisted by two surveyors on a support boat, equipped with a GIS-GPS cartographic station used to georeference the position of the targets.

The submersion measurements were corrected with respect to the tidal level and the barometric pressure taken from the tide gauge in the Port of Naples.

The following describes the main features of the archaeological sea-level markers selected for the reconstruction of the ancient sea-levels.

4.1.1. Fish Tanks

The Roman fish tanks (*piscinae*) consisted of single or multiple tanks used for fish farming [69], interconnected with the open sea through channels. This direct connection of constructional elements with the sea level, and specifically with the tidal range, makes these structures excellent indicators of ancient sea levels.

In their treaties, Varro and Columella [71,72] described the different construction techniques for fish tanks, which varied according to the coastal type [73–76].

There are essentially three constructional elements directly linked to the sea level at the time of fish tank's construction [73,75–78]: *crepidines* (a foot-walk border surrounding the tank and the internal pools); channel system (canals which allowed tidally controlled water exchange); *cataractae* (closing gates located at the access of the canal into the basin or at the communication passage between each tank).

In this study, the submersion measurements concerned the *Cataractae*, considering that these structural elements were perfectly preserved in their original position. The measurements were carried out on top of the vertical posts cut by grooves, originally excavated in order to guide the vertical movement of the old gate, made of a sheet of lead or stone with several small holes (diameter between ~3 and ~9 centimetres, Figure 3).

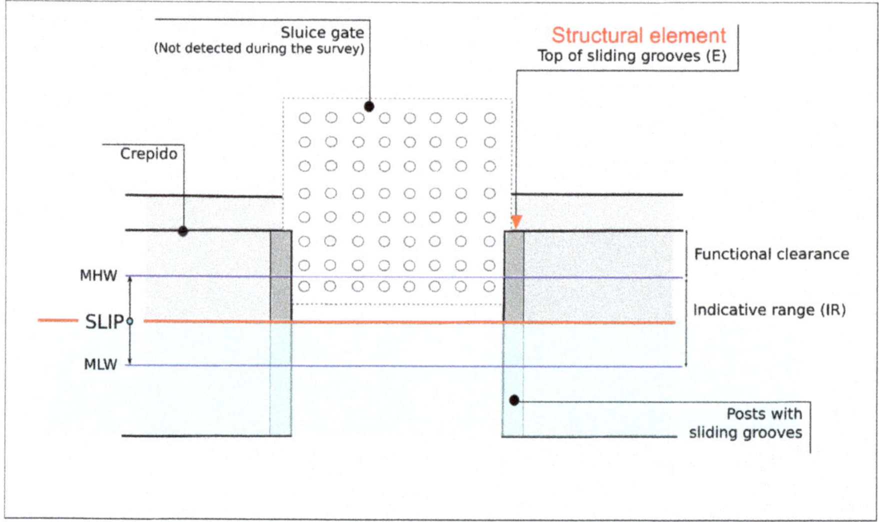

Figure 3. Measuring point in the Portus Julius fish tank (MHW: Mean High Water; MLW: Mean Low Water; SLIP: Sea Level Index Point).

These constructional elements are considered high-resolution SLIPs of the second construction phase of Portus Julius. The submersion measurements (elevation bsl, E) were carried out on the top of the sliding grooves because these structural elements were always built above the highest tide level (mean high water, MHW, sensu Shennan et al. [70], obtaining at least 0.2 m of functional clearance (fc) (after Lambeck et al. [76]). Moreover, the original elevation of the structural elements above the ancient MSL, indicative meaning (IM) sensu Shennan et al. [70], can be evaluated considering that the IM describes the relationship between the marker and the former sea-level and that it is composed of two elements:

- Indicative range (IR), the elevation range over which the marker forms;
- Reference water level (RWL), the mid-point of the above-mentioned range [14,70].

In the case of fish tanks, the indicative meaning and the related IR can be defined with high precision as reported in the ancient texts of Varro and Columella [71,72]. Thereby, while the functional clearance is considered a constant amount according to the archaeological interpretations, in the case of a fish tank, the IR varies between the mean high water (MHW) and the mean low water (MLW) (Figure 3).

Finally, by correcting the E measurements with respect to the functional clearance (fc $_{cataracta}$ = 0.2 m), the RSL (SLIP) can be evaluated [78,79] using the equation:

$$\text{SLIP}_{\text{cataracta}} \ (m) = E - fc_{\text{cataracta}} - IR/2 \tag{1}$$

where IR = (MHW–MLW) is the indicative range.

Concerning the vertical uncertainty, the functional clearance derived from archaeological interpretations cannot be included in its calculation (after Vacchi et al. [14]). Consequently, this uncertainty is equal to ±(MHW to MLW)/2. Taking into account that the tidal range (MHW to MLW) in the Gulf of Naples is about 0.4 m [80], the vertical uncertainty of our measurement is ±0.2 m.

4.1.2. Ancient Pier Structures

Port structures are excellent sea-level indicators, as they were built in direct functional relation with the ancient sea level [77], even if it is not always easy to identify a precise measuring point.

During the planning phases of the survey, the *pilae* protecting the embankments of Portus Julius were chosen as high-precision archaeological sea-level markers related to the first construction phase of Portus Julius. The submersion measurements were carried out on the upper part of the *pila* made in hydraulic concrete, here considered a SLIP according to Mattei et al. [16].

The *pilae* are Roman square construction made of hydraulic concrete and built on the sea bottom with their upper part generally emerged [81]. These cubes were built forming single or multiple lines, often connected by arches, and called *opus pilarum*. This kind of structure was largely distributed in the Mediterranean area after the discovery of Roman hydraulic concrete during the II century BCE [16,82,83].

Due to the importance in Roman engineering, this construction material, mainly made of *pulvis puteolanus* (Pozzolana), was mentioned by several ancient authors, including Vitruvius [67], Strabo [32], and Plinius [84]. This volcanic ash—largely detectable in Campi Flegrei caldera—is composed of reactive aluminosilicates, mixed with irregular stones or tuff aggregate and it has the main characteristic of being hardened by the sea to a strength "which neither the waves nor the force of the water can dissolve" (Vitruvius, [67]). When seawater infiltrates in the concrete, it dissolves some of the ash and, rather than undermining the structure, the alkali fluid that is left allows minerals to strengthen it [85].

Since its discovery, the concrete became the most used material for the construction of ports and coastal structures [16,82,83], as confirmed by the analyses of several *pilae* located in Italy (Portus Julius, Anzio, Cosa, Santa Liberata, Brindisi), Israel (Caesarea), Egypt (Alexandria), and Greece (Crete) [76] carried out during the ROMACONS project since 2002.

The Roman *pilae* were built on the seabed using an innovative building technique, consisting of wooden cofferdams with an open upper surface and sides made of vertical oaken poles (*destinae*) held together by horizontal beams (*catenae*). The cofferdam technique, described by Vitruvius [67] in 15 BCE, required that the hydraulic concrete be cast and set directly underwater [16,82,83].

The emerged parts of these port facilities were often built with the same concrete but set in sub-aerial environments, and were therefore less resistant due to the absence of seawater infiltration.

The port structures were often refined with the typical roman *opus reticulatum* (covering made up of small square blocks forming a grid) or *opus testaceum* (covering made up of simple brickwork, [16,82,83,86,87].

In this study, the *pilae* protecting the embankments of Portus Julius were used as SLIP (sensu Shennan et al. [70]) by detecting the concrete change (i.e., the limit between the areas in hydraulic concrete and the areas in concrete totally laid in a subaerial environment) (Figure 4) [16,88].

In fact, as described by Vitruvius [67], the cofferdams filled with hydraulic concrete had to emerge from the ancient mean sea level by an amount equal to a wooden board (about 0.50 m). Consequently, according to this archaeological interpretation, the elevation range over which the marker forms (IR) is directly related to the tidal range and varies between the mean high water (MHW) and the mean low water (MLW).

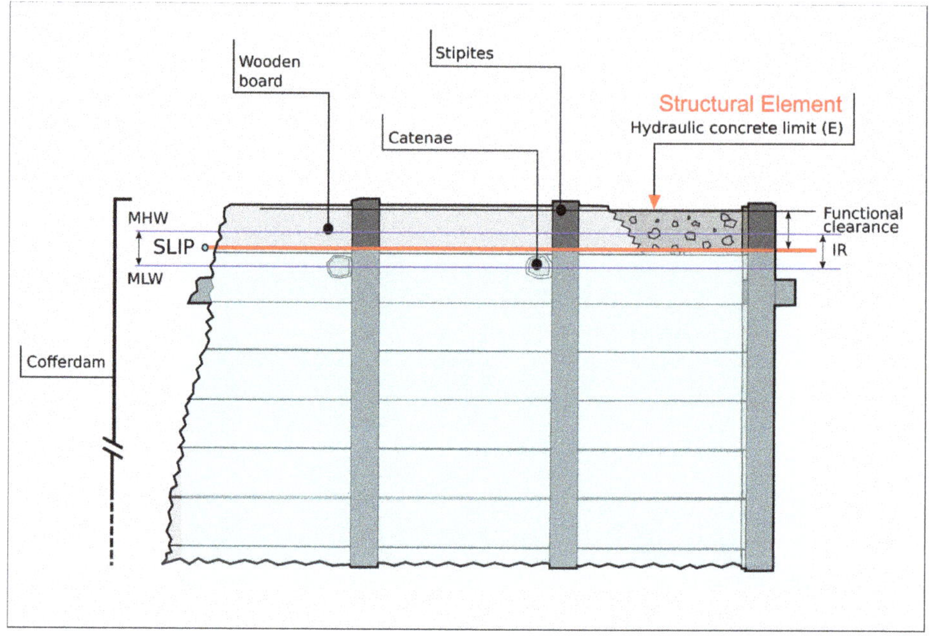

Figure 4. Measuring point in the Portus Julius *pila* (MHW: Mean High Water; MLW: Mean Low Water; SLIP: Sea Level Index Point).

Knowing the submersion measurement (emersion bsl, E) and the amount of the functional clearance (fc_{pila}), by determining the original elevation above the ancient MSL (indicative meaning, IM, (sensu Shennan et al. [70])) it is possible to evaluate the SLIP (RSL) by using the following equation:

$$SLIP_{pila} \ (m) = E - fc_{\,pila} - IR/2 \tag{2}$$

where IR = (MHW to MLW) is the indicative range, and $fc_{\,pila}$ (m) = 0.50 − MHW is the functional clearance with respect to the high tide.

As in the previous case of the fish tank, concerning the vertical uncertainty, the functional clearance derived from archaeological interpretations cannot be included in its calculation (after Vacchi et al. [14]). Consequently, this uncertainty is equal to ±(MHW–MLW)/2. Considering that the tidal range (MHW–MLW) in the Gulf of Naples is about 0.4 m [80], the vertical uncertainty of our measurement is ±0.2 m.

During the surveys, we observed that the hydraulic concrete cast underwater and immersed in the sea water during the last 2000 years effectively strengthened its hardness [83].

Indeed, the part of the *pila* built in the emerged environment, with the same hydraulic concrete or with another material, is currently deteriorated more than the part cast underwater (Figure 5).

The divers recognized some areas where the hydraulic concrete was visible, thanks to its typical characteristics perfectly coincident with those described for the *pila* of Nisida by Mattei et al. [16]. These sectors detected on the upper face of the pila—where the subaerial concrete was totally eroded—appear as deeper areas with a flat and sub-horizontal morphology (Figure 5). These areas were always positioned underneath the areas in sub-aerial concrete that appeared very eroded with a rugged morphology (Figure 5).

Figure 5. Comparison between the Nisida and Portus Julius *pilae*: the black line identifies the concrete change limit.

4.1.3. Ancient Buildings

To detect the elevation of an ancient RSL, the ruins of coastal buildings can be profitably studied, particularly the remnants of ground floor pavements and entrance thresholds. Their relations to the coeval sea level is not so direct and simple so as to be able to use a universal rule to derive a past marine level from it. However, they permit to fix a maximum height limit of the RSL (terrestrial limiting point, TLP) by considering at what minimum height it was safe to build a residence or a warehouse to minimize the risks of invasion by storm waves and/or the penetration of groundwater (Figure 6) [77,89]. Concerning the local wave climate, during winter and autumn, the storm events are characterized by mean wave height values up to 4.8 m, associated with atmospheric low-pressure systems; the spring and summer are characterized by low wave height values ranging from 0.4 to 0.6 m [90–92].

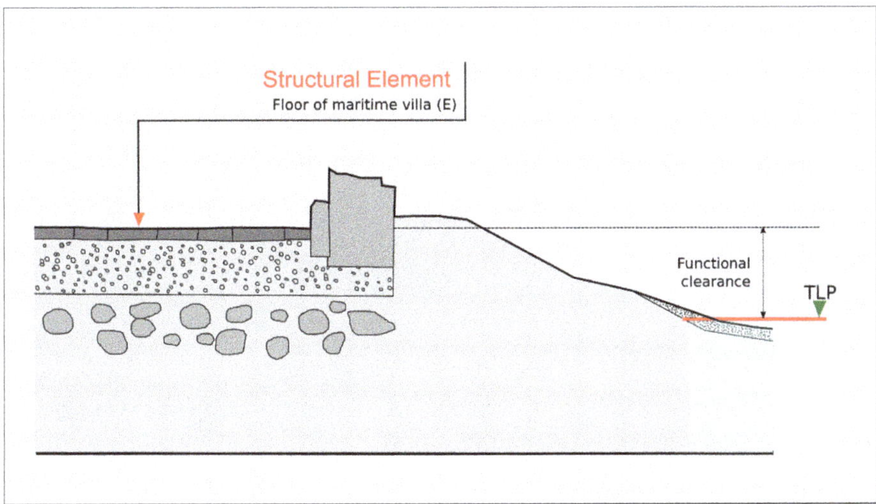

Figure 6. Terrestrial limiting point in the villa close to Portus Julius (TPL: Terrestrial Limiting Point).

In naturally sheltered coastal plains, such as the one where Portus Julius is located (see section geological setting), we assume that, for a site located right behind the backshore, the minimum height necessary to minimize the effects of storm waves (i.e., functional clearance) ranged between 1.5 and 2 m MSL) [77,89].

4.2. Photogrammetric Survey

The fish tank here studied represented the key element of our geoarchaeological interpretations in terms of ground movements occurred during the Roman times, due to its strategic position. The sluice gate of this fish tank was built directly by cutting the channel embankment. This led to the assumption that it was therefore built in the second construction phase of the Portus Julius, according to Benini and Giacobelli [37]. Therefore, in this area, a high precision photogrammetric survey has been carried out both to obtain a detailed morphometric analysis of this functional element and to preserve a 3D documentation of this important and unique historical testimony, also considering the extreme fragility of this underwater landscape.

The survey was planned using a photogrammetric system consisting of three high-resolution cameras, two parallel to the vertical optical axis in the nadiral position with respect to the sea-bottom and a third inclined with respect to the bottom by about 30° (Figure 7).

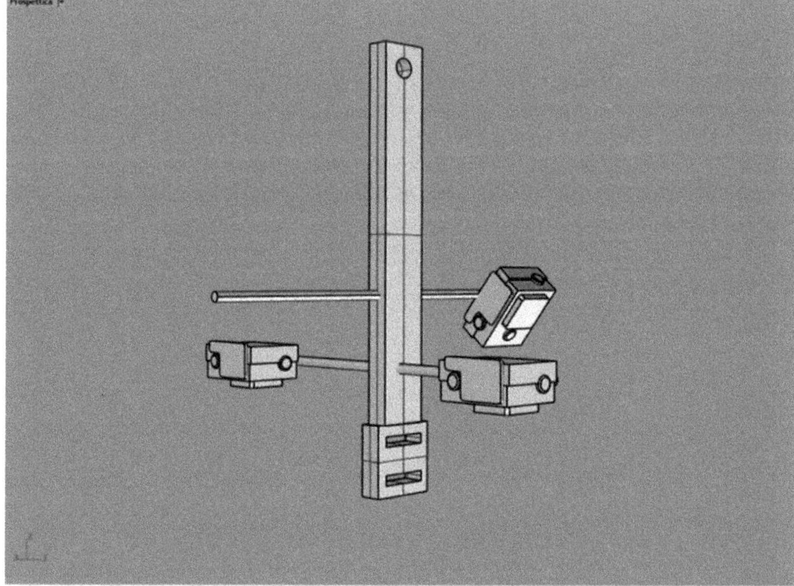

Figure 7. 3D reconstruction of the photogrammetric system used during the survey.

The stereoscopic base (b) and the submersion of the system were chosen in relation to the bathymetry of the study area and the level of the resolution of the three-dimensional model of the submerged findings, with our settings for the cameras ensuring a minimum overlap of 80% during the survey (Figure 8). Taking into account the characteristics of the video cameras, to ensure a transversal overlap of at least 80%, the stereoscopic base was chosen equal to 30 centimeters. The depth of the photogrammetric system was modified during the survey to guarantee both an adequate completeness of the three-dimensional model and its resolution (see Figure 8).

As shown in Figure 8a, the photographic system was towed by a rope connected to an emerged float and moved by divers. This system provides two basic positions, namely position A at −1 m and position B at −2 m below the sea surface.

During the survey, two videoshots were taken in sequence, the first with the photogrammetric system at an altitude of −1 m MSL (Positon A in Figure 8a) and the second at an altitude of −2 m MSL (Positon B in Figure 8a) and therefore close to the target.

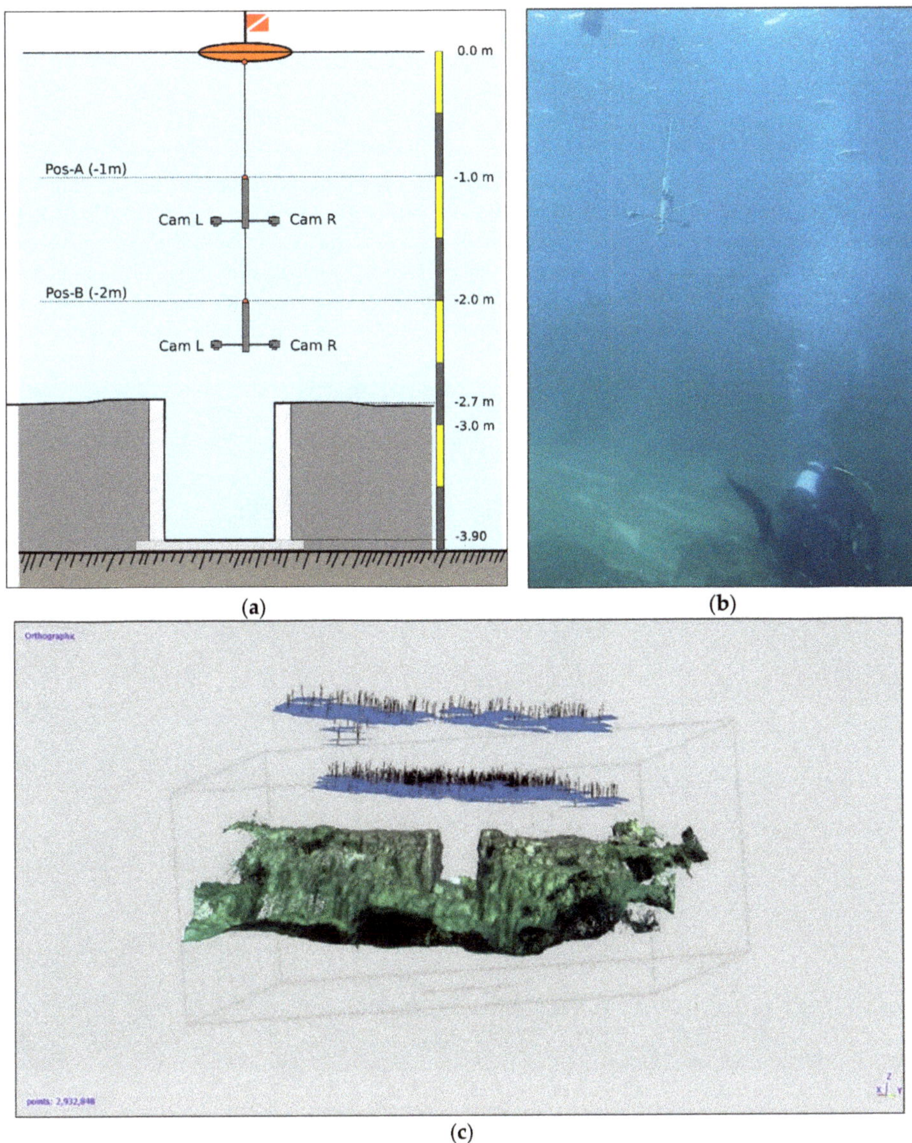

Figure 8. (**a**) Reconstruction of the photogrammetric system used during the survey; (**b**) The photogrammetric system used during the survey; (**c**) Position of the cameras during the survey.

The photogrammetric 3D model of the surveyed fish tank was obtained in three two steps [88]:

1. The videos at 30 fps recorded by the two Xiaomi cameras (previously calibrated in an underwater environment close to the study area to achieve the inner orientation parameters) were synchronized

by the use of the trigger system and the images were extracted using a frame every six seconds. More than one thousand 1920 × 1080 images were thus obtained.

2. The alignment procedure of the images and the dense point cloud extraction was performed using the Agisoft Metashape software. The 3D model was not georeferenced as it was not possible to determine the coordinates of the control points; however, some linear measurements made by the diver directly on the underwater structure made it possible to scale the dense point cloud.

5. Results

5.1. Maritime Villa (I Century BCE)

In the first instance, the survey involved an area near to Portus Julius but already existing before its construction, where a maritime villa was built at the end of the Republican Age (I century BCE) [39]. The submersion of the better-preserved mosaic floor was measured at −3.2 m MSL (Figure 9) by using a metric roll and depth gauge. Consequently, by correcting this measurement with respect to the functional clearance, a SLIP (RSL) at −4.70/−5.20 m MSL was deduced. This amount of submersion was also measured in the best-preserved parts from the nearby *via Herculanea* (dated I century BCE).

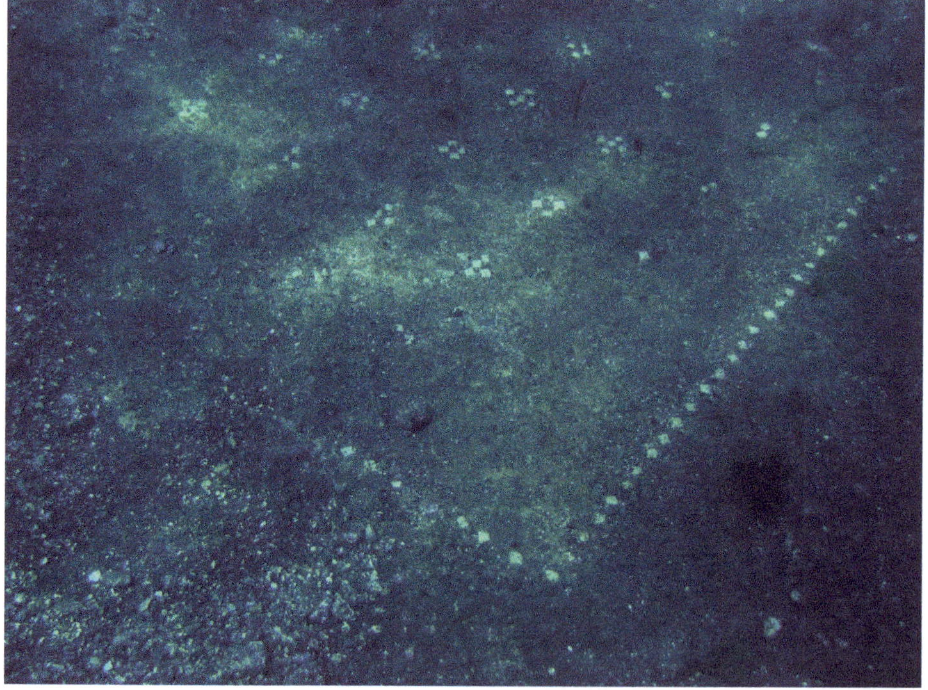

Figure 9. One of the opus signinum floors of the maritime villa of Republican age.

5.2. Fish Tank (12 BCE)

The second archaeological structure, surveyed with direct and indirect methods, was the fish tank of Portus Julius, the main archaeological sea level markers related to the second constructive stage of the harbour after 12 BCE, when its usage switched from a military to a commercial one. This archaeological interpretation was endorsed by the position of the fish tank cutting the embankment of the entry channel of Portus Julius, as clearly shown in the photogrammetric reconstruction (Figure 8a). In fact, according to Benini and Giacobelli [37], the relationship between the fish tank and the military port seems incongruous. It is more likely that the embankment was cut afterward—when the port became commercial—to allow the conversion of the rooms adjacent to the embankment in the fish tank for commercial use.

To better describe the fish tank, a morphometric analysis of the photogrammetric point cloud was performed (Figure 10b–d). The photogrammetric reconstruction was crucial both for the evaluation of the elements related to the study of the past relative sea levels and for the documentation of this singular fish tank built for commercial use, with a great cultural value.

The triangular fish tank, with a size of $20 \times 22 \times 16$ m, is entirely made of tuff and it appears to be carved into the embankment of the entry channel [37], constructed in the year 37 BCE by the order of Agrippa. Along the external wall, two different ancient *cataractae* are clearly identifiable (Figures 10 and 11) and, while the southern one is still in situ, the northern one is partially destroyed and shows one of the two vertical posts collapsed.

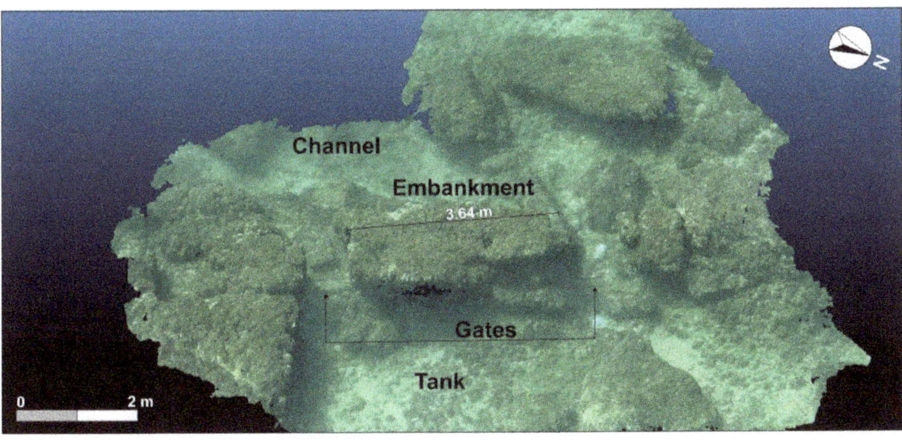

(**a**)

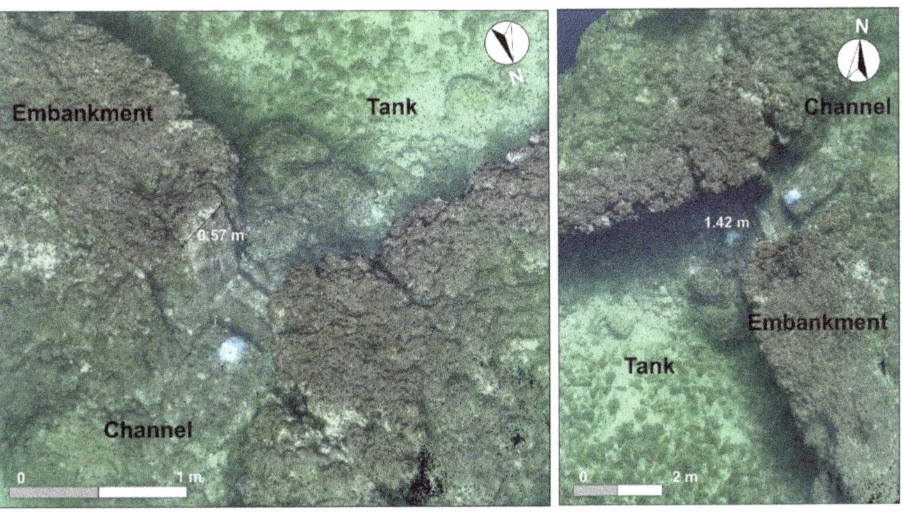

(**b**)

Figure 10. *Cont.*

Figure 10. (**a**) Dense cloud point of the fish tank where the two gates cutting the embankment are clearly visible; (**b**) Zoom of the 3D reconstruction of the well preserved gate; (**c**) Zoom of the 3D reconstruction of two perimeter walls of the fish tank; (**d**) 3D sketch of the surveyed fish tank.

The submersion of the top of the sliding grooves located along the posts still in loco was measured at −2.70 m MSL and, by correcting this measurement by using the Equation (1), a SLIP (RSL) at −3.10 m MSL was deduced.

(a)

(b)

(c)

(d)

Figure 11. (**a**,**b**) details of the sluice gate of the fish tank still *in situ*; (**c**,**d**) details of the collapsed sluice gate of the fish tank.

5.3. Pilae of Portus Julius (37 BCE)

During the same survey, the Roman fish tank of Portus Julius and the five *pilae* located at the entry of the main channel were analyzed. The *pilae* are coeval with the entry channel, and belong to the first construction phase of the military harbour precisely dated at 37 BCE.

The submersion value of the concrete change limit in the better-preserved *pila* has been determined (Figure 12), according to Mattei et al. [16]. The submersion measurement of the concrete change limit is currently located at −2.6 m MSL, and by correcting this measurement by using Equation (2) a SLIP (RSL) is determined at −3.1 m MSL.

(a)

(b)

Figure 12. *Cont.*

Figure 12. Underwater photos of the *pilae* of the Portus Julius entry channel (**a**) vertical slide of one of the *pilae*; (**b–d**) limit of the concrete change measured during the survey on different *pilae*; (**e**) detail of the remain of a wooden stake (*destina*) used for the construction of the *pilae*; (**f**) measuring the concrete change submersion.

All the submersion measurements carried out in this study were analyzed to evaluate the RSL changes that occurred between the first century BCE and the first half of the first century CE (Table 1).

Table 1. The archaeological sea level markers and the related RSLs. Where: E = submersion measurements; fc = functional clearance; RWL = Reference Water Level (IR/2); RSL = Relative sea level.

	Marker Type	Age	E (m)	fc (m)	RWL (m)	RSL (m)	Uncertainty
Piscina	Top Sluice Gate	I cen. CE	−2.70	−0.20	−0.20	−3.10	±0.20
Pilae	Concrete Change	I cen. CE	−2.60	−0.30	−0.20	−3.10	±0.20
Maritime Villa	Floor	I cen. BCE	−3.20	−1.50/−2.00	-	−4.70/−5.20	-

6. Discussion

The results achieved in terms of the RSL variations allowed us to reconstruct the evolution of the coastal sector over the last 2100 years.

According to the narrations of Strabo [32], it is possible to affirm that, at the end of the Republican Age, the study area was shaped as by a low coastal sector located next to a sandy spit, above which, the former Via Herculanea was constructed. At that time, the littoral was characterized by shallow water and the Lucrino Gulf was not deep enough to permit navigation. For this reason, it was intended for oyster farming.

The maritime villa, analyzed in this paper, was located near to the above-mentioned spit in a flat area at a depth of about −3.5 m MSL. Therefore, considering its functional clearance, an RSL at −4.70/−5.2 m MSL can be presumed for that period.

The second evolutionary phase of the area can be posed right before 37 BCE, when the coastal sector was chosen by Agrippa for the construction of a new military harbour system and the access channel to the Lucrino Lake was enlarged and fortified. During this period, the older warehouses were still used for the supply of raw materials for the troops. The relative sea level at −3.2 m MSL measured from the *pilae* demonstrates that some subsidence had already occurred when the military port was built in 37 BCE. As well, we can suppose that the subsidence, which produced an increase in the water depth, encouraged the construction of the harbour facility (Figure 11).

The standing of the RSL from the fish tank at about −3.2 m MSL, even after 12 BCE, demonstrates that the area experienced a period of volcano-tectonic stability (Figure 13).

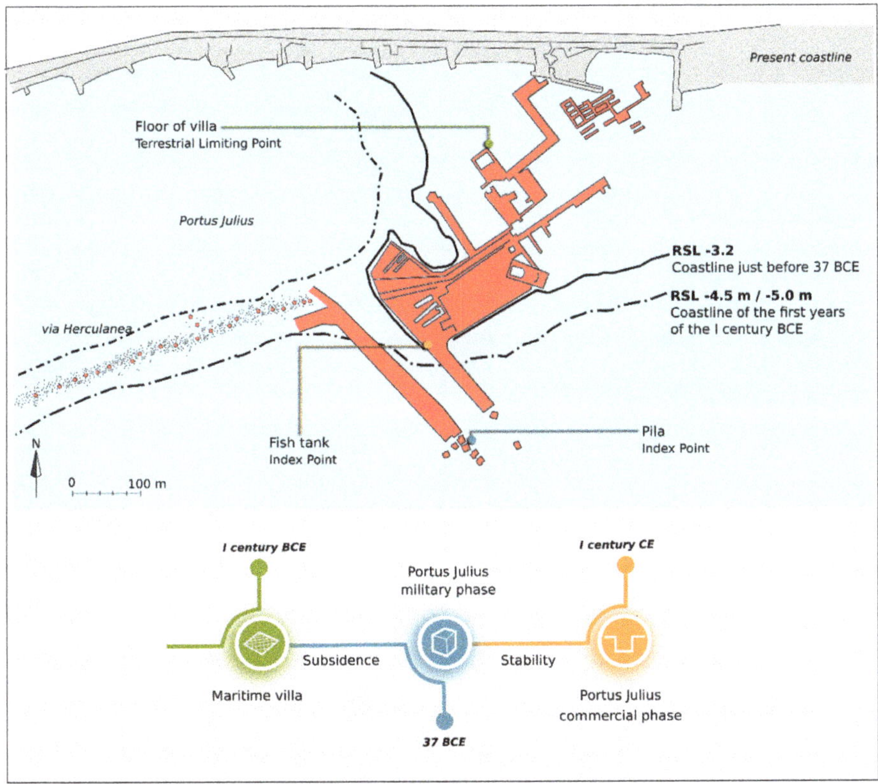

Figure 13. Reconstruction of the probable coastline during the first years of the first century BCE (continuous black line) when the maritime villa was built and the coastline in 37 BCE (stretch and point black line), after the 2 m of subsidence, when the Portus Julius was built.

Until now, several authors have assumed that the military port was abandoned due to subsidence [28,93] but the results of our study support the idea that the most of subsidence occurred before the port was built.

Coming to more recent times, around the year 500 CE, it is well known that the area was affected by a new and stronger phase of subsidence which led to an RSL up to +7.00 m MSL (Morhange et al. [25] and reference therein). This is also supported by the boreholes carried out by Welter-Schultes and Richling [94], claiming that, in the Middle Ages, the Averno Lake became a completely marine environment, with the lithodome holes located at a height of +7.00 m MSL on the marble columns of the Serapeo in Pozzuoli [25,33,65].

Moreover, by comparing our measurements with the Glacio-Hydro-Isostatic Adjustment (GIA) models proposed by several authors [95,96], it is possible to determine that the only contribution of the vertical ground movements (VGMs) occurred between the early first century BCE and the early first century CE. In this period, the GIA produced an SL variation in our study area of a centimetric amount (from −1.05 to −1.09 m MSL). Consequently, at the same time and in terms of VGMs, the study area suffered a subsidence of about 4.00 meters, and experienced a rapid increase (bradyseismic crisis) of another 2.00 m before 37 BCE.

By extending the geoarchaeological observations to the entire Gulf of Pozzuoli, it is interesting to compare the results obtained in the area of PJ with those of the Baia-Miseno sector studied by Aucelli et al. [97]. Along that costal sector, the authors recognized a period of volcano-tectonic stability that occurred between 60 BCE and the early years of the first century CE. This trend was testified by a stable RSL at −4 ± 0.2 m MSL, as deduced by studying three Roman fish tanks dotted along the coastline. Additionally, by comparing the RSL data with the eustatic models [95,96], the authors evaluated in an overall subsidence of 3 m for the VGMs that affected the coastal area in the last 2000 years.

Consequently, during the time span ranging between the first century BCE and the beginning of the first century CE, the NE sector (Portus Julius) and the SW sector (Baia-Miseno) of Pozzuoli Gulf showed different behaviors. Both sectors went through a period of tectonic volcano stability between the mid-late first century BCE and the beginning of the first century CE; however, the overall subsidence that followed up to the current state, was 2 m in the NE sector and about 3 m in the SW sector (Figure 12).

Concerning the beginning of the detected VGMs stability, the data on Portus Julius led us to suppose that it started before 37 BCE, when the Port was designed and built. A better constraint is available in the SW sector, were the fish tank of *Hortensius Hortalus* demonstrates that this stability was already set around 60 BCE (Figure 14).

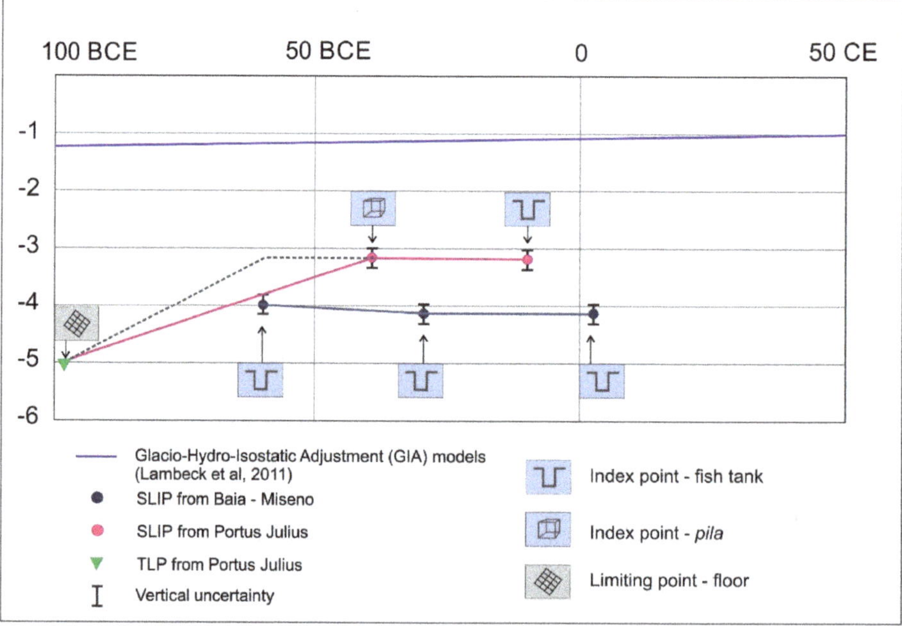

Figure 14. Comparison between the RSL data derived from the geoarchaeological study of the Baia-Miseno sector Aucelli et al. [97] and those obtained for the Portus Julius sector.

7. Concluding Remarks

In this study, two different RSLs related to the same study area were recognized using archaeological sea-level markers related to different construction phases of Portus Julius.

The ancient RSL of −3.20 m, determined from the fish tank and the *pilae*, is related to the time between the construction of the military port (*pilae*) in 37 BCE and its conversion into a commercial hub (fish tank) at the end of the first century BCE (after 12 BCE). The result is innovative as all the studies carried out in the area before always proposed RSLs for the same period ranging between −6.00 m and −10 m MSL [25–27,34] and none had measured high-precision sea-level markers.

On the other hand, in this study, a relative sea level during the first half of the first century BCE of about −4.70/−5.20 m MSL was evaluated using the submersion measurements of the ground floor pavements of an ancient villa in the same area, well-dated, thanks to the mosaics covering the floors. Consequently, a subsidence of metric entity and volcano-tectonic nature has been deduced from the comparison between the submersion measurements of the floors of the Late Republican villa and the most recent (37 BCE) *pilae* of the military port. Furthermore, the high precision measurements of the fish tank (built after 12 BCE) lead us to suppose that this subsiding trend stopped before the construction of the military port, and that the area experienced a period of volcano-tectonic stability at least until the first century CE.

A further result of this work is related to the use of port structures in hydraulic concrete as SLIPs. Until now, the fish tanks were the only highly accurate archaeological indicators for sea-level reconstruction [74–78,98,99]. In this paper, we demonstrated that it is also possible to use port facilities as high precision markers, as long as the concrete change is recognizable. To be acceptable, this indicator must meet the following requirements:

- A visible horizontal and flat surface (concrete change limit), which is detached from the upper planes;
- The surface (concrete change limit) must have an extension equal to at least 20% of the total upper face of the *pila*;
- The surface (concrete change limit) must be harder and more compact than that of the upper layers and must be free or almost free of cover vegetation, while the upper layers must be rough with vegetation.

In conclusion, it can be affirmed that the roman *pilae*, when showing the above-mentioned characteristics, represent a new SLIP of great importance considering their dissemination across all of the Mediterranean Sea.

Author Contributions: Conceptualization, P.P.C.A., A.C., G.M., G.P., M.S., S.T.; methodology, P.P.C.A., G.M., C.C., A.C., F.P., S.T.; software, C.C., G.M., F.P., S.T.; validation, P.P.C.A., A.C., G.M., G.P., M.S., S.T.; investigation, C.C. and F.P.; data curation, P.P.C.A., C.C., G.M., F.P., S.T.; writing—original draft preparation, C.C., A.C., G.M., M.S.; writing—review and editing, P.P.C.A., C.C., A.C., G.M., G.P., M.S., S.T. All authors have read and agreed to the published version of the manuscript.

Funding: This research was financially supported by research found of Parthenope University granted to Pietro P.C. Aucelli and Gerardo Pappone.

Acknowledgments: The authors extend their sincere thanks to Francesco Salvatore Ruggiero and Gennaro Sorrentino of Meno 100 Underwater TEK ASD for their willingness and their important support during the diving surveys. The coastal LIDAR data used in this paper were kindly provided by the Ministry of the Environment in March 2013.

Conflicts of Interest: The authors declare no conflict of interest.

References

1. Derbyshire, E.; Hails, J.R.; Gregory, K.J. *Geomorphological Processes: Studies in Physical Geography;* Butterworth-Heinemann: London, UK, 1979.
2. Eybergen, F.A.; Imeson, A.C. Geomorphological processes and climatic change. *Catena* **1989**, *16*, 307–319. [CrossRef]

3. Alberico, I.; Amato, V.; Aucelli, P.P.C.; Di Paola, G.; Pappone, G.; Rosskopf, C.M. Historical and recent changes of the Sele River coastal plain (Southern Italy): Natural variations and human pressures. *Rend. Lincei* **2012**, *23*, 3–12. [CrossRef]

4. Pappone, G.; Aucelli, P.P.C.; Alberico, I.; Amato, V.; Antonioli, F.; Cesarano, M.; Di Paola, G.; Pelosi, N. Relative sea-level rise and marine erosion and inundation in the Sele river coastal plain (Southern Italy): Scenarios for the next century. *Rend. Lincei* **2012**, *23*, 121–129. [CrossRef]

5. Antonioli, F.; Ferranti, L.; Fontana, A.; Amorosi, A.; Bondesan, A.; Braitenberg, C.; Dutton, A.; Fontolan, G.; Furlani, S.; Lambeck, K.; et al. Holocene relative sea-level changes and vertical movements along the Italian and Istrian coastlines. *Quat. Int.* **2009**, *206*, 102–133. [CrossRef]

6. Cinque, A.; Irollo, G.; Romano, P.; Ruello, M.R.; Amato, L.; Giampaola, D. Ground movements and sea levelchanges in urban areas: 5000 years of geological and archaeological record from Naples (Southern Italy). *Quat. Int.* **2011**, *232*, 45–55. [CrossRef]

7. Aucelli, P.; Cinque, A.; Mattei, G.; Pappone, G. Historical sea level changes and effects on the coasts of Sorrento Peninsula (Gulf of Naples): New constrains from recent geoarchaeological investigations. *Palaeogeogr. Palaeoclimatol. Palaeoecol.* **2016**, *463*, 112–125. [CrossRef]

8. Aucelli, P.; Cinque, A.; Giordano, F.; Mattei, G. A geoarchaeological survey of the marine extension of the Roman archaeological site Villa del Pezzolo, Vico Equense, on the Sorrento Peninsula, Italy. *Geoarchaeology* **2016**, *31*, 244–252. [CrossRef]

9. Aucelli, P.; Cinque, A.; Mattei, G.; Pappone, G. Late Holocene landscape evolution of the gulf of Naples (Italy) inferred from geoarchaeological data. *J. Maps* **2017**, *13*, 300–310. [CrossRef]

10. Aucelli, P.P.C.; Cinque, A.; Mattei, G.; Pappone, G.; Stefanile, M. Coastal landscape evolution of Naples (Southern Italy) since the Roman period from archaeological and geomorphological data at Palazzo degli Spiriti site. *Quat. Int.* **2018**, *483*, 23–38. [CrossRef]

11. Aucelli, P.P.C.; Cinque, A.; Mattei, G.; Pappone, G.; Stefanile, M. First results on the coastal changes related to local sea level variations along the Puteoli sector (Campi Flegrei, Italy) during the historical times. *Alp. Mediterr. Quat.* **2018**, *31*, 13–16.

12. Aucelli, P.P.C.; Cinque, A.; Mattei, G.; Pappone, G.; Rizzo, A. Studying relative sea level change and correlative adaptation of coastal structures on submerged Roman time ruins nearby Naples (southern Italy). *Quat. Int.* **2019**, *501*, 328–348. [CrossRef]

13. Rovere, A.; Raymo, M.E.; Vacchi, M.; Lorscheid, T.; Stocchi, P.; Gómez-Pujol, L.; Harris, D.L.; Casella, E.; O'Leary, M.J.; Hearty, P.J. The analysis of the Last Interglacial (MIS 5e) relative sea-level indicators: Reconstructing sea-level in a warmer world. *Earth Sci. Rev.* **2016**, *159*, 404–427. [CrossRef]

14. Vacchi, M.; Marriner, N.; Morhange, C.; Spada, G.; Fontana, A.; Rovere, A. Multiproxy assessment of Holocene relative sea-level changes in the western Mediterranean: Sea-level variability and improvements in the definition of the isostatic signal. *Earth Sci. Rev.* **2016**, *155*, 172–197. [CrossRef]

15. Benjamin, J.; Rovere, A.; Fontana, A.; Furlani, S.; Vacchi, M.; Inglis, R.H.; Galili, E.; Antonioli, F.; Sivan, D.; Miko, S.; et al. Late Quaternary sea-level changes and early human societies in the central and eastern Mediterranean Basin: An interdisciplinary review. *Quat. Int.* **2017**, *449*, 29–57. [CrossRef]

16. Mattei, G.; Troisi, S.; Aucelli, P.P.C.; Pappone, G.; Peluso, F.; Stefanile, M. Sensing the Submerged Landscape of Nisida Roman Harbour in the Gulf of Naples from Integrated Measurements on a USV. *Water* **2018**, *10*, 1686. [CrossRef]

17. De Giosa, F.; Scardino, G.; Vacchi, M.; Piscitelli, A.; Milella, M.; Ciccolella, A.; Mastronuzzi, G. Geomorphological Signature of Late Pleistocene Sea Level Oscillations in Torre Guaceto Marine Protected Area (Adriatic Sea, SE Italy). *Water* **2019**, *11*, 2409. [CrossRef]

18. Khan, N.S.; Horton, B.P.; Engelhart, S.; Rovere, A.; Vacchi, M.; Ashe, E.L.; Törnqvist, T.E.; Dutton, A.; Hijma, M.P.; Shennan, I. Inception of a global atlas of sea levels since the Last Glacial Maximum. *Quat. Sci. Rev.* **2019**, *220*, 359–371. [CrossRef]

19. Kolaiti, E.; Mourtzas, N.; Kissas, K.; Antonioli, F.; Lambeck, K. New insights into the uplifted Roman harbour at Mavra Litharia (N Peloponnese, Greece) in the geodynamic context of the southern margin of the Corinth Gulf. *Quat. Int.* **2019**, *508*, 23–35. [CrossRef]

20. Vacchi, M.; Russo Ermolli, E.; Morhange, C.; Ruello, M.; Di Donato, V.; Di Vito, M.; Giampaola, D.; Carsana, V.; Liuzza, V.; Cinque, A.; et al. Millennial variability of rates of sea-level rise in the ancient harbour of Naples (Italy, western Mediterranean Sea). *Quat. Res.* **2019**, 1–15. [CrossRef]

21. Dvorak, J.J.; Mastrolorenzo, G. *The Mechanisms of Recent Vertical Crustal Movements in Campi Flegrei Caldera, Southern Italy*; The Geological Society of America Spec. Pap.: Boulder, CO, USA, 1991; Volume 263, p. 49.

22. Cinque, A.; Aucelli, P.P.C.; Brancaccio, L.; Mele, R.; Milia, A.; Robustelli, G.; Romano, P.; Russo, F.; Santangelo, N.; Sgambati, D. Volcanism, tectonics and recent geomorphological change in the bay of Napoli. *Suppl. Geogr. Fis. Din. Quat.* **1997**, *3*, 123–141.

23. Morhange, C.; Bourcier, M.; Laborel, J.; Giallanella, C.; Goiran, J.P.; Crimaco, L.; Vecchi, L. New data on historical relative sea level movements in Pozzuoli, Phlegrean Fields, southern Italy. *Phys. Chem. Earth* **1999**, *24*, 349–354. [CrossRef]

24. Morhange, C.; Blanc, F.; Schmitt-Mercury, S.; Bourcier, M.; Carbonel, P.; Oberlin, C.; Prone, A.; Vivent, D.; Hesnard, A. Stratigraphy of late-Holocene deposits of the ancient harbour of Marseilles, southern France. *Holocene* **2003**, *13*, 539–604. [CrossRef]

25. Morhange, C.; Marriner, N.; Laborel, J.; Todesco, M.; Oberlin, C. Rapid sea-level movements and non-eruptive crustal deformation in the phlegrean Fields caldera, Italy. *Geology* **2006**, *34*, 93–96. [CrossRef]

26. Bellucci, F.; Woo, J.; Kilburn, C.R.; Rolandi, G. Ground deformation at Campi Flegrei, Italy: Implications for hazard assessment. *Geol. Soc. Lond. Spec. Publ.* **2006**, *269*, 141–157. [CrossRef]

27. Passaro, S.; Barra, M.; Saggiorno, R.; Di Giacomo, S.; Leotta, A.; Uhlen, H.; Mazzola, S. Multi-resolution morpho-bathymetric survey results at the Pozzuoli-Baia underwater archaeological site (Naples, Italy). *J. Archaeol. Sci.* **2013**, *40*, 1268–1278. [CrossRef]

28. Todesco, M.; Costa, A.; Comastri, A.; Colleoni, F.; Spada, G.; Quareni, F. Vertical ground displacement at Campi Flegrei (Italy) in the fifth century: Rapid subsidence driven by pore pressure drop. *Geophys. Res. Lett.* **2014**, *41*, 1471–1478. [CrossRef]

29. Sommella, P. Forma e urbanistica di Pozzuoli Romana. In *Puteoli, Studi di Storia Antica Vol. II*; Azienda Autonoma di Soggiorno, Cura e Turismo: Pozzuoli, Italy, 1980.

30. Camodeca, G. Puteoli porto annonario e il commercio del grano in età imperiale. *Publ. École Fr. Rome* **1994**, *196*, 103–128.

31. Camodeca, G. *Puteoli Romana: Istituzioni e Società*; UniorPress: Napoli, Italy, 2018.

32. Strabone. *Geografia V*; 23 CE.

33. Parascandola, A. *I Fenomeni Bradisismici del Serapeo di Pozzuoli, Napoli*; Guida: Napoli, Italy, 1947.

34. Pappalardo, U.; Russo, F. Il Bradisismo dei Campi Flegrei (Campania): Dati Geomorfologici ed Evidenze Archeologiche. In *Forma Maris, Atti Della Rassegna Internazionale di Archeologia Subacquea (Pozzuoli 1998)*; Gianfrotta, P.A., Maniscalco, F., Eds.; Massa Editore: Pozzuoli, Italy, 2001.

35. Gianfrotta, P.A. I porti dell'area flegrea. *Pt. Approdi Rotte Mediterr. Antico Studi Filol. Lett.* **1998**, *4*, 155–168.

36. Benini, A. *Storia, Archeologia e Tutela dei Beni Archeologici Sommersi: L'esempio dei Campi Flegrei*; Edipuglia: Bari, Italy, 2004; pp. 35–43.

37. Benini, A.; Giacobelli, M. Peschiera o Pescheria? Curiosità flegree. In Proceedings of the Atti del III Convegno di Archeologia Subacquea, Manfredonia, Italy, 4–6 October 2007.

38. Miniero, P. Baia sommersa e portus Iulius. Il rilievo con strumentazione integrata Multibeam. In *Ricoveri Per Navi Militari Nei Porti del Mediterraneo Antico e Medievale*; Blackman, D.J., Lentini, M.C., Eds.; Edipuglia: Bari, Italy, 2010; pp. 101–108.

39. Gianfrotta, P.A. Ricerche nell'area sommersa del "Portus Iulius" (1988-'90 e successive): Un riepilogo. *Atlante Temat. Topogr. Antica* **2012**, *22*, 1–20.

40. Mattei, G.; Rizzo, A.; Anfuso, G.; Aucelli, P.P.C.; Gracia, F.J. A tool for evaluating the archaeological heritage vulnerability to coastal processes: The case study of Naples Gulf (southern Italy). *Ocean Coast. Manag.* **2019**, *179*, 104876. [CrossRef]

41. Mattei, G.; Rizzo, A.; Anfuso, G.; Aucelli, P.P.C.; Gracia, F.J. Enhancing the protection of archaeological sites as an integrated coastal management strategy: The case of the Posillipo Hill (Naples, Italy). *Rend. Lincei Sci. Fis. Nat.* **2020**. [CrossRef]

42. Caputo, M.; Pieri, L. Eustatic variation in the last 2000 years in the Mediterranean. *J. Geophys. Res.* **1976**, *81*, 5787–5790. [CrossRef]

43. Flemming, N.C. *Archaeological Evidence for Eustatic Change of Sea Level and Earth Movements in the Western Mediterranean during the Last 2000 Years*; The Geological Society of America: Boulder, CO, USA, 1969.

44. Pirazzoli, P.A. Sea level variations in the northwest Mediterranean during Roman times. *Science* **1976**, *194*, 519–521. [CrossRef] [PubMed]

45. Antonioli, F.; Anzidei, M.; Auriemma, R.; Gaddi, D.; Furlani, S.; Lambeck, K.; Orrù, P.; Solinas, E.; Gaspari, A.; Karinja, S.; et al. Sea level change during Holocene from Sardinia and northeastern Adriatic (Central Mediterranean sea) from archaeological and geomorphological data. *Quat. Sci. Rev.* **2007**, *26*, 2463–2486. [CrossRef]

46. Lambeck, K.; Woodroffe, C.D.; Antonioli, F.; Anzidei, M.; Gehrels, W.R.; Laborel, J.; Wright, A.J. Paleoenvironmental Records, Geophysical Modeling, and Reconstruction of Sea-Level Trends and Variability on Centennial and Longer Timescales. In *Understanding Sea-Level Rise and Variability*; Wiley-Blackwel: Chichester, UK, 2010.

47. Morhange, C.; Marriner, N.; Excoffon, P.; Bonnet, S.; El-amouri, M.; Zibrowius, H. Relative sea level changes during Roman times in the NW Mediterranean. The first century Ad fish tank of Forum Julii (Fréjus, France). *Geoarchaeology* **2013**, *28*, 363–372. [CrossRef]

48. Anzidei, M.; Lambeck, K.; Antonioli, F.; Furlani, S.; Mastronuzzi, G.; Serpelloni, E.; Vannucci, G. Coastal structure, sea-level changes and vertical motion of the land in the Mediterranean. In *Sedimentary Coastal Zones from High to Low Latitudes: Similarities and Differences. Special Publications 388*; Martini, I.P., Ed.; The Geological Society: London, UK, 2014. [CrossRef]

49. D'Argenio, A.; Pescatore, T.; Senatore, M.R. Sea-level change and volcano-tectonic interplay. The Gulf of Pozzuoli (Campi Flegrei, Eastern Tyrrhenian Sea) during the last 39 ka. *J. Volcanol. Geotherm. Res.* **2004**, *133*, 105–121.

50. De Pippo, T.; Donadio, C.; Pennetta, M.; Terlizzi, F.; Vecchione, C.; Vegliante, M. Seabed morphology and pollution along the Bagnoli coast (Naples, Italy): A hypothesis for environmental restoration. *Mar. Ecol.* **2002**, *23*, 154–168. [CrossRef]

51. Somma, R.; Iuliano, S.; Matano, F.; Molisso, F.; Passaro, S.; Sacchi, M.; Troise, C.; De Natale, G. High-resolution morpho-bathymetry of Pozzuoli Bay, southern Italy. *J. Maps* **2016**, *12*, 222–230. [CrossRef]

52. Isaia, R.; Vitale, S.; Marturano, A.; Aiello, G.; Barra, D.; Ciarcia, S.; Iannuzzi, E.; Tramparulo, F. High-resolution geological investigations to reconstruct the long-term ground movements in the last 15 kyr at Campi Flegrei caldera (southern Italy). *J. Volcanol. Geotherm. Res.* **2019**, *385*, 143–158. [CrossRef]

53. Rosi, M.; Sbrana, A.; Principe, C. The Phlegraean Fields: Structural evolution, volcanic history and eruptive mechanisms. *J. Volcanol. Geotherm. Res.* **1983**, *17*, 273–288. [CrossRef]

54. Rosi, M.; Sbrana, A. *Carta Geologica e Gravimetrica dei Campi Flegrei*; Consiglio Nazionale Delle Ricerche: Rome, Italy, 1986.

55. Isaia, R.; Iannuzzi, E.; Sbrana, A.; Marianelli, P. *Note Illustrative Della Carta Geologica d'Italia Alla Scala 1: 50.000*; Foglio 446-477 Napoli (aree emerse); Regione Campania: Napoli, Italy, 2016.

56. Giaccio, B.; Isaia, R.; Fedele, F.; Di Canzio, E.; Hoffecker, J.; Ronchitelli, A.; Sinitsyn, A.; Anikovich, M.; Lisitsyn, S.; Popov, V. The Campanian Ignimbrite and Codola tephra layers: Two temporal/stratigraphic markers for the Early Upper Palaeolithic in southern Italy and eastern Europe. *J. Volcanol. Geotherm. Res.* **2008**, *177*, 208–226. [CrossRef]

57. Costa, A.; Folch, A.; Macedonio, G.; Giaccio, B.; Isaia, R.; Smith, V. Quantifying volcanic ash dispersal and impact of the Campanian Ignimbrite super-eruption. *Geophys. Res. Lett.* **2012**, *39*, L10310. [CrossRef]

58. Albert, P.G.; Giaccio, B.; Isaia, R.; Costa, A.; Niespolo, E.M.; Nomade, S.; Pereira, A.; Renne, P.R.; Hinchliffe, A.; Mark, D.F.; et al. Evidence for a large-magnitude eruption from Campi Flegrei caldera (Italy) at 29 ka. *Geology* **2019**, *47*, 595–599. [CrossRef]

59. Orsi, G.; D'Antonio, M.; de Vita, S.; Gallo, G. The Neapolitan Yellow Tuff, a large-magnitude trachytic phreatoplinian eruption: Eruptive dynamics, magma withdrawal and caldera collapse. *J. Volcanol. Geotherm. Res.* **1992**, *53*, 275–287. [CrossRef]

60. Deino, A.L.; Orsi, G.; Piochi, M.; de Vita, S. The age of the Neapolitan Yellow Tuff caldera-forming eruption (Campi Flegrei caldera – Italy) assessed by 40Ar/39Ar dating method. *J. Volcanol. Geotherm. Res.* **2004**, *185*, 48–56. [CrossRef]

61. Di Vito, M.A.; Isaia, R.; Orsi, G.; Southon, J.; de Vita, S.; D'Antonio, M.; Pappalardo, L.; Piochi, M. Volcanism and deformation in the past 12 ka at the Campi Flegrei caldera (Italy). *J. Volcanol. Geotherm. Res.* **1999**, *91*, 221–246. [CrossRef]

62. Smith, V.C.; Isaia, R.; Pearce, N.J.C. Tephrostratigraphy and glass compositions of post-15 kyr Campi Flegrei eruptions: Implications for eruption history and chronostratigraphic markers. *Quat. Sci. Rev.* **2011**, *30*, 3638–3660. [CrossRef]

63. Di Vito, M.; Lirer, L.; Mastrolorenzo, G.; Rolandi, G. The 1538 Monte Nuovo eruption (Campi Flegrei, Italy). *Bull. Volcanol.* **1987**, *49*, 608–615. [CrossRef]

64. D'Antonio, M.; Civetta, L.; Orsi, G.; Pappalardo, L.; Piochi, M.; Caradente, A.; de Vita, S.; Di Vito, M.A.; Isaia, R. The present state of the magmatic system of the Campi Flegrei caldera based on a reconstruction of its behavior in the past 12 Ka. *J. Volcanol. Geotherm. Res.* **1999**, *91*, 247–268. [CrossRef]

65. Cinque, A.; Rolandi, G.; Zamparelli, V. L'estensione dei depositi marini olocenici nei Campi Flegrei in relazione alla vulcano-tettonica. *Boll. Soc. Geol. Ital.* **1985**, *104*, 327–348.

66. Reddé, M. *Mare Nostrum. Les Infrastructures, le Dispositif et L'histoire de la Marine Militaire Sous L'empire Romain*; Ecoles Françaises d'Athènes et de Rome: Rome, Italy, 1986; Volume 260, pp. 164–171.

67. Vitruvius. *De Architectura II*; 15 BCE; Volume 6, p. 1.

68. Cassiodorus. *Chronica*; 519 CE.

69. Scognamiglio, E. Aggiornamenti per la topografia di Baia sommersa. *Archeol. Sub.* **1997**, *2*, 35–45.

70. Shennan, I.; Long, A.J.; Horton, B.P. *Handbook of Sea-Level Research*; John Wiley & Sons: Oxford, UK; New York, NY, USA, 2015.

71. Varro. *De Re Rustica III*; 37 BCE.

72. Columella. *De Re Rustica XVII*; 30 CE.

73. Leoni, G.; Dai Pra, G. *Variazioni del Livello del Mare nel Tardo Olocene, Ultimi 2500 Anni, Lungo la Costa del Lazio in Base ad Indicatori Geo-Archeologici: Interazioni fra Neotettonica, Eustatismo e Clima*; ENEA, Unità Comunicazione e Informazione: Rome, Italy, 1997.

74. Antonioli, F.; Leoni, G. Siti archeologici e loro utilizzazione quali indicatori per lo studio delle variazioni recenti del livello del mare. *Quat. Ital. J. Quat. Sci.* **1998**, *11*, 122–139.

75. Evelpidou, N.; Pirazzoli, P.A.; Saliegec, J.F. Submerged notches and doline sediments evidence for Holocene subsidence. *Cont. Shelf Res.* **2011**, *31*, 1273–1281. [CrossRef]

76. Lambeck, K.; Anzidei, M.; Antonioli, F.; Benini, A.; Verrubbi, V. Tyrrhenian sea level at 2000 BP: Evidence from Roman age fish tanks and their geological calibration. *Rend. Lincei* **2018**, *29*, 69–80. [CrossRef]

77. Auriemma, R.; Solinas, E. Archaeological remains as sea level change markers: A review. *Quat. Int.* **2009**, *206*, 134–146. [CrossRef]

78. Lambeck, K.; Anzidei, M.; Antonioli, F.; Benini, A.; Esposito, E. Sea level in Roman time in the central Mediterranean and implications for recent change. *Earth Planet. Sci. Lett.* **2004**, *224*, 563–575. [CrossRef]

79. Morhange, C.; Marriner, N. Archeological and biological relative sea-level indicators. *Handb. Sea Level Res.* **2015**, 146–156. [CrossRef]

80. Capuano, P.; Buonocore, B.; Tammaro, U.; Obrizzo, F.; La Rocca, A.; Pinto, S.; Pingue, F. Caratteristiche spettrali delle variazioni del livello marino delle baie di Napoli e Pozzuoli. *ASITA Roma* **2004**, *1*, 615–620.

81. Stefanile, M. The Project PILAE, For an Inventory of the Submerged Roman Piers. A Preliminary Overview. *Int. J. Environ. Geoinform.* **2015**, *2*, 34–39. [CrossRef]

82. Gianfrotta, P.A. Structures of the Augustan Age in Italy. In *Caesarea Maritima, a Retrospective after Two Millennia*; Raban, A., Holum, K.G., Eds.; Brill: Leiden, The Netherlands, 1996; pp. 65–76.

83. Brandon, C.J.; Hohlfelder, R.L.; Jackson, M.D.; Oleson, J.P. *Building for Eternity: The History and Technology of Roman Concrete Engineering in the Sea*; Oxbow Books: Oxford/Barnsley, UK, 2014.

84. Plinius. *Naturalis Historia*; 78 DC.

85. Oleson, J.P.; Brandon, C.; Cramer, S.M.; Cucitore, R.; Gotti, E.; Hohlfelder, R.L. The ROMACONS Project: A Contribution to the Historican and Engineering Analysis of the Hydrauilc Concrete in Roman Maritime Structures. *Int. J. Naut. Archaeol.* **2004**, *33*, 199–229. [CrossRef]

86. Scognamiglio, E. Baia sommersa: Gli sviluppi di ricerca. In *Forma Maris, Atti Della Rassegna Internazionale di Archeologia Subacquea (Pozzuoli 1998)*; Gianfrotta, P.A., Maniscalco, F., Eds.; Massa Editore: Pozzuoli, Italy, 2001; pp. 43–50.

87. Stefanile, M.; Mattei, G.; Troisi, S.; Aucelli, P.P.C.; Pappone, G.; Peluso, F. The Pilae of Nisida. Some geological and archaeological observations. *Archaeol. Marit. Mediterr.* **2018**, *15*, 81–100.

88. Mattei, G.; Troisi, S.; Aucelli, P.P.C.; Pappone, G.; Peluso, F.; Stefanile, M. Multiscale reconstruction of natural and archaeological underwater landscape by optical and acoustic sensors. In Proceedings of the 2018 IEEE International Workshop on Metrology for the Sea; Learning to Measure Sea Health Parameters (MetroSea), Bari, Italy, 8–10 October 2018; pp. 46–49.

89. Sivan, D.; Wdowinski, S.; Lambeck, K.; Galili, E.; Raban, A. Holocene sea-level changes along the Mediterranean coast of Israel, based on archaeological observations and numerical model. *Palaeogeogr. Palaeoclimatol. Palaeoecol.* **2001**, *167*, 101–117. [CrossRef]

90. Benassai, G.; de Maio, A.; Sansone, E. Altezze e periodi delle onde significative nel Golfo di Napoli dall'aprile 1986 al giugno 1987. *Ann. IUN* **1994**, *61*, 3–9.

91. Buonocore, B.; Sansone, E.; Zambardino, G. Rilievi ondametrici nel Golfo di Napoli. *Ann. IUN* **2003**, *67*, 203–211.

92. Saviano, S.; Kalampokis, A.; Zambianchi, E.; Uttieri, M. A year-long assessment of wave measurements retrieved from an HF radar network in the Gulf of Naples (Tyrrhenian Sea, Western Mediterranean Sea). *J. Oper. Oceanogr.* **2019**, *12*, 1–15. [CrossRef]

93. Maniscalco, F. Gli edifici sommersi del Palatium dei Severi a Baia: Nuovi dati per la definizione del bradisismo flegreo. In *Forma Maris, Atti Della Rassegna Internazionale di Archeologia Subacquea (Pozzuoli 1998)*; Gianfrotta, P.A., Maniscalco, F., Eds.; Massa Editore: Pozzuoli, Italy, 2001.

94. Welter-Dchultes, F.W.; Richling, I. Palaeoenvironmental history of the Holocene volcanic crater lake Lago d'Averno (central southern Italy) inferred from aquatic mollusc deposits. *J. Quat. Sci.* **2000**, *15*, 805–812. [CrossRef]

95. Lambeck, K.; Antonioli, F.; Anzidei, M.; Ferranti, L.; Leoni, G.; Scicchitano, G.; Silenzi, S. Sea level change along the Italian coasts during Holocene and prediction for the future. *Quat. Int.* **2011**, *232*, 250–257. [CrossRef]

96. Vacchi, M.; Ghilardi, M.; Melis, R.T.; Spada, G.; Giaime, M.; Marriner, N.; Lorscheid, T.; Morhange, C.; Burjachs, F.; Rovere, A. New relative sea-level insights into the isostatic history of the Western Mediterranean. *Quat. Sci. Rev.* **2018**, *201*, 396–408. [CrossRef]

97. Aucelli, P.P.C.; Caporizzo, C.; Cinque, A.; Mattei, G.; Pappone, G.; Stefanile, M. Reconstructing the relative sea level and the palaeo-shoreline during the first century BC in the unstable Campi Flegrei caldera (S. Italy). 2020; under review.

98. Amato, V.; Aucelli, P.P.C.; Mattei, G.; Pennetta, M.; Rizzo, A.; Rosskopf, C.M.; Schiattarella, M. A geodatabase of Late Pleistocene-Holocene palaeo sea-level markers in the Gulf of Naples. *Alp. Mediterr. Quat.* **2018**, *31*, 5–9.

99. Pappone, G.; Aucelli, P.P.C.; Mattei, G.; Peluso, F.; Stefanile, M.; Carola, A. A Detailed Reconstruction of the Roman Landscape and the Submerged Archaeological Structure at "Castel dell'Ovo islet" (Naples, Southern Italy). *Geosciences* **2019**, *9*, 170. [CrossRef]

*water*

Article

Coastal Modification in Relation to Sea Storm Effects: Application of 3D Remote Sensing Survey in Sanremo Marina (Liguria, NW Italy)

Ilaria Ferrando [1], Pierluigi Brandolini [2,*], Bianca Federici [1], Antonio Lucarelli [3], Domenico Sguerso [1], Danilo Morelli [2] and Nicola Corradi [2,4]

[1] Geomatics Laboratory, Department of Civil, Chemical and Environmental Engineering (DICCA), University of Genoa, 16145 Genoa, Italy; ilaria.ferrando@edu.unige.it (I.F.); bianca.federici@unige.it (B.F.); domenico.sguerso@unige.it (D.S.)

[2] Department of Earth, Environment and Life Sciences (DiSTAV), University of Genoa, 16132 Genoa, Italy; danilo.morelli@unige.it (D.M.); nicola.corradi@unige.it (N.C.)

[3] Drafinsub Survey, 16126 Genoa, Italy; alucarelli@dssurvey.com

[4] National Research Group for Coastal Environment (GNRAC), 16132 Genoa, Italy

* Correspondence: brando@unige.it

Citation: Ferrando, I.; Brandolini, P.; Federici, B.; Lucarelli, A.; Sguerso, D.; Morelli, D.; Corradi, N. Coastal Modification in Relation to Sea Storm Effects: Application of 3D Remote Sensing Survey in Sanremo Marina (Liguria, NW Italy). *Water* **2021**, *13*, 1040. https://doi.org/10.3390/w13081040

Academic Editors: Giorgio Anfuso and Elias Dimitriou

Received: 1 March 2021
Accepted: 6 April 2021
Published: 9 April 2021

Publisher's Note: MDPI stays neutral with regard to jurisdictional claims in published maps and institutional affiliations.

Abstract: Integrated remote sensing techniques, such as photogrammetry from unmanned aerial vehicles (UAV), mobile laser scanners (MLS) and multibeam echosounders (MBES), are particularly effective in detecting and measuring coastal and seabed features and their modifications over time (4D analysis) induced by sea storms. In fact, these techniques allow the production of very high-resolution 3D models, with a continuum between above and below sea level. The present research is focused on the area of Portosole Marina (Sanremo, Western Liguria), affected by a severe sea storm in October 2018 and the following restoration. Two integrated 3D surveys were performed in February 2019 and in November 2019, obtaining accurate and reliable high-definition digital surface models (DSMs) in both emerged and submerged areas. The comparison between the two surveys highlighted volumetric changes in the seabed induced by the sea storm and the effects of a temporary worksite on the emerged and submerged breakwater. In particular, a total deficit of sediments of about 5000 m^3 caused an average lowering of about 4 cm over the entire area, concurring with the breakwater instability. This study aims to contribute to the understanding of coastal system resilience within ongoing global climate changes, that is, increasing the intensity of extreme events in the Mediterranean area.

Keywords: geomorphological coastal changes; sea storm effects; integrated 3D remote sensing surveys; sedimentary dynamics; western Ligurian sea

1. Introduction

The most advanced integrated remote sensing techniques, such as photogrammetry from unmanned aerial vehicle (UAV), mobile laser scanner (MLS) and multibeam echosounder (MBES), applied to a geomorphological survey of coastal areas and to topographic measurements of coastal infrastructures, allow creating very high-resolution 3D models [1–13]. These methods are particularly effective in detecting and measure seabed features and their modifications induced by extreme events, such as severe storm surges and short-term local variations in sea level [14–16]. Detailed reconstructions of sea storms effects, such as seabed and beach instability other than damage to port infrastructure, provide coastal scenarios assessment of the impact of extreme marine events, such as the 2018 Vaia Storm [17], which are intensifying in the context of ongoing global warming/climate changes [18–20]. The application of remote sensing integrated systems also represents support in assessing resilience and vulnerability, in the monitoring of geo-risk and in the evaluation of proper mitigation measures within land management in coastal/fluvial areas [21–31].

The present research is focused on Portosole Marina (Sanremo, Western Liguria), affected by a severe storm in October 2018 [32–34]. After this event, the breakwater of Portosole Marina was restored. Two integrated 3D surveys were performed: the first one (survey 1) carried out in February 2019 [35] and the second one (survey 2) in November 2019, before and after the restoration, respectively. A system integrating laser scanner in mobile mode (MLS) and multibeam echosounder (MBES) mounted on a survey boat was employed to get a continuous metrically reliable 3D model. The good quality of data coming from such integrated survey has already been verified, both in terms of continuity and coherence of the MLS and MBES point clouds on the overlapping area and of metrical accuracy, which was centimetric for both emerged and submerged areas referred to the 0 mean-sea levels (MSL) [35].

The aim of the present research is the 4D very high-resolution morpho-dynamic analysis through the comparison of the two surveys of the study area to highlight the seabed changes both from metric and geomorphological points of view.

The comparison between the two surveys was performed employing multiple procedures. First, a simple difference in elevation between the two entire DSMs was performed. Then, the test area was divided into three regions, characterized by homogeneous morphology, applying such difference in elevation to each region. Hence, the total deposited and removed volumes were computed for each homogeneous region. Finally, a comparison was performed along vertical sections.

Then, the high-resolution bathymetric data were analyzed both for verifying the state of conservation of the maritime structures from a structural point of view and for highlighting erosion and accumulation of sediments. The detailed and quantitative analysis of these elements represents a useful tool for assessing the coastal area's vulnerability to intense storm surge.

The paper follows with a brief description of the study area and the employed survey techniques (Section 2). Subsequently, details on the storm that affected Sanremo in October 2018 are reported (Section 3). A description of the analysis procedure and a discussion on the obtained results are outlined in Sections 4 and 5, respectively.

2. Study Area

The study area is located in the central sector of the Sanremo coastline (Liguria, northwest of Italy). It is included in a wide bay of about 8 km extent, located between the Cape Nero promontory to the west and the Cape Verde promontory to the east (Figure 1), respectively featured by outcrops of marly arenaceous flysch (Campanian–Eocene) and polygenic conglomerates (Pliocene) [36].

The area is characterized by a narrow fluvial-coastal plain formed by the alluvial deposits of the Foce, Mafalda, San Romolo, San Francesco, San Lazzaro, San Martino and Val D'olivi streams and by the sediments deriving from sea cliff erosion and landslide phenomena affecting the nearby rocky coast [37–44]. The coast is mainly exposed to storms of Libeccio (225°) and Scirocco (135°).

The former coastal morphology has been almost totally modified by the expansion of urban settlements and by the construction of port infrastructures. Consequently, today this coastal zone appears as a "techno-coast", as occurs in many other Ligurian and Italian areas [45–51].

The natural geomorphological dynamics, together with the anthropic transformations that have occurred over time, have marked different conditions of the coastline and its beaches here summarized:

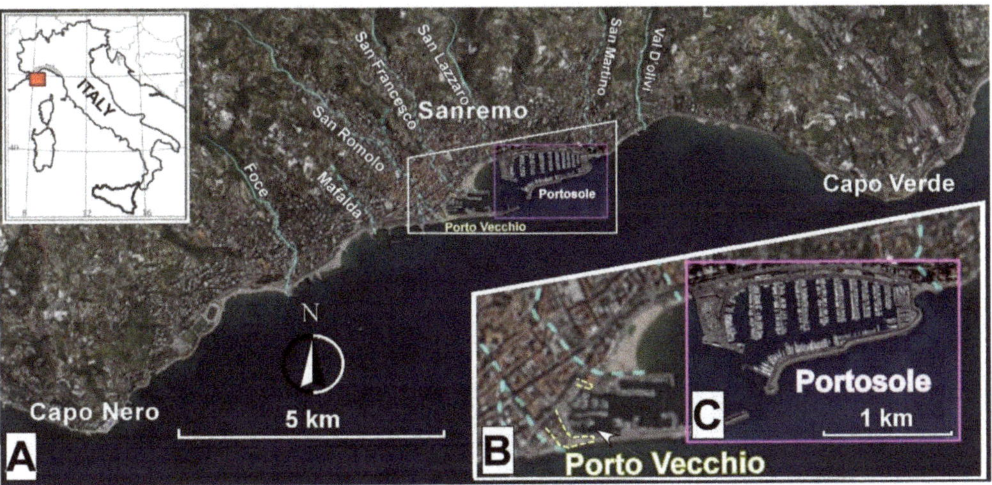

Figure 1. Location of Sanremo: Portosole study area. (**A**) The orthophoto of the Liguria Region (2016) shows how the narrow coastal plain of Sanremo and the course of the main streams crossing it (light blue dotted and continuous lines) are totally occupied by urban settlements. The zoom (**B**) shows the area of the Porto Antico and its ancient piers (dotted yellow line) and the Portosole area, where the morpho-bathymetric surveys were realized (**C**).

- Up to the middle of the 19th century, the coast was affected by a progradation of the beaches, thanks to the sedimentary input by the different watercourses and the erosional retreat of the sea cliffs. It is demonstrated by the historical map of Matteo Vinzoni [52], showing the former conditions of the coast in 1753 (Figure 2);
- After the second half of the 19th century, the beaches were generally affected by erosion, albeit with alternating phases of advancement and retreat. Since the beginning of the 20th century, to counteract the effects of sediments deficit connected to the port works on sediment transport, the construction of numerous defense works was necessary;
- Since 1965 up to the mid-seventies, several works of sea-embankments (reclamation area) and beach nourishment have been carried out;
- Between 1975 and 1980, the Portosole Marina was built between the Porto Vecchio (west) and the mouth of the San Martino stream (east). This has determined the current coastal layout, together with recent and further expansion and reinforcement work. The construction of Portosole caused the disappearance of pre-existing beaches, in erosion (about 13 m in the period 1944–1973) and protected by six groins. The artificial advancement of the coastline due to the construction of the infrastructures (techno-coast) was more than 350 m;
- The small beach about 90 m in extent, located between the mouth of the San Martino Stream and the root of the breakwater, assumed the characteristics of an anthropic pocket-beach [53–55]. In fact, it is between the maritime works and the jettied mouth of the stream, which reached its current layout between 2013 and 2016. Figure 3 shows the variations of the shoreline position over the years. Overall, the multi-temporal coastal line comparison shows a fairly stable equilibrium of the shoreline, with advancement of about 3 m (1983–2016). The alternating stages of advancement and retreat were presumably due to the restoration works at the river mouth and to small maintenance interventions of beach nourishment [56–58]. Since 1983 (Figure 3), the eastern side of the jetted mouth is constituted by the seawall.

The study here presented concerns the breakwater of the Portosole Marina and the facing seabed (Figure 1, box C), with a total extent of about 850 m, together with the small artificial pocket beach immediately to the east.

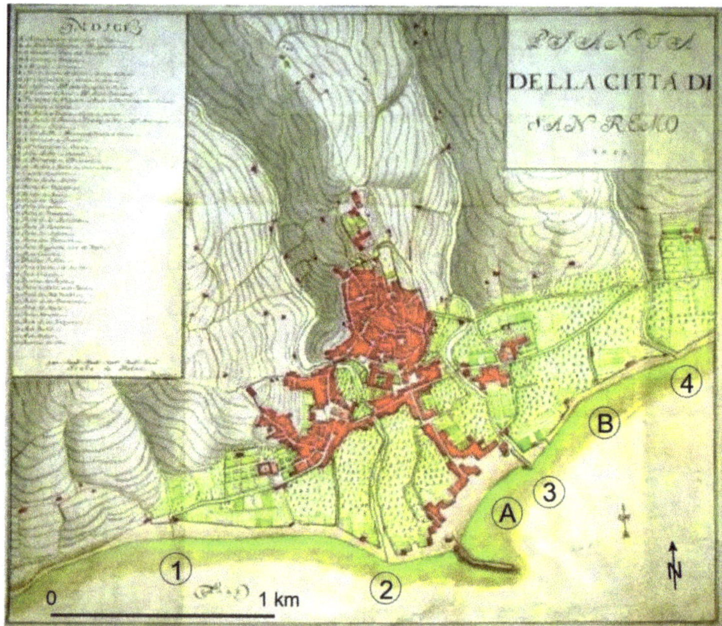

Figure 2. Historical map of Sanremo coast (1753): (**1**) Mafalda, (**2**) San Romolo, (**3**) San Francesco, (**4**) San Lazzaro streams; (**A**) and (**B**) represent the current locations of Porto Vecchio and Portosole, respectively (modified after [52]).

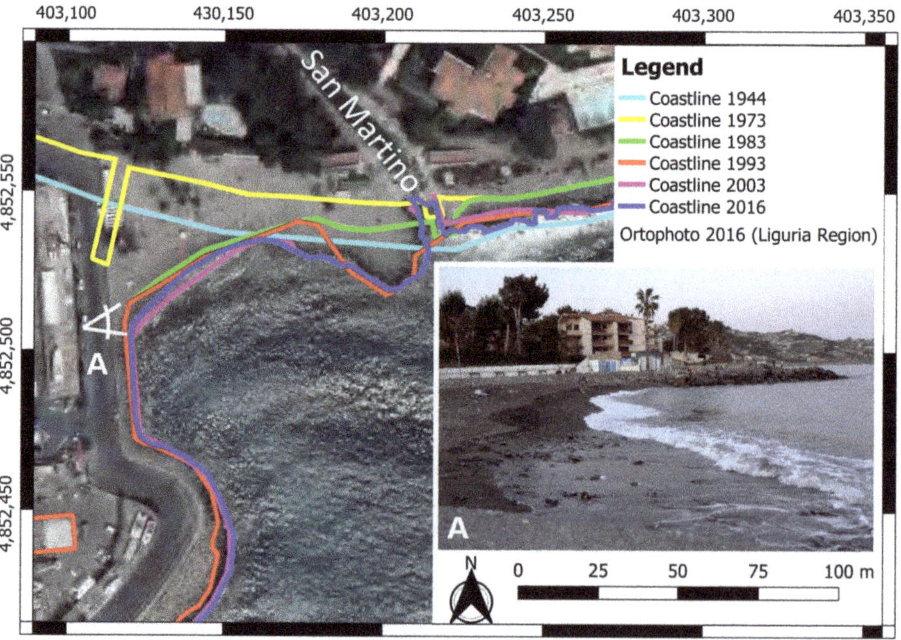

Figure 3. Variations of the position of the shoreline over the years 1944–2016 [58], Liguria Region orthophoto of 2016 used as background. (**A**) Photo taken from the west (the point of view is indicated in white on the main map) shows the features of the shoreline deposits. The coordinates are expressed in the ETRF2000–2008.0 reference system with UTM 32 projection.

3. The Event of 29 October 2018

On 29 October 2018, the Portosole Marina and the entire coastline were affected by an intense storm that caused severe damage to the breakwater and surrounding areas.

This event was part of a wider meteorological event, known as the Vaia storm, which affected the entire Mediterranean basin with particular intensity on the Ligurian and North Adriatic Sea. The storm developed from 27 to 29 October 2018, generating a cyclonic circulation centered on the west of Corsica, with a drop in pressure up to 17 hPa in 18 h. This depression, moving towards Italy's northeast, caused extreme southerly winds with gusts up to 119 km/h at Genoa airport and 171 km/h at La Spezia weather stations [17]. This cyclogenesis, together with the collision of warm air masses from the south and the cold front over the Alps from the north, led to developing heavy rainfall, winds and storms (Figure 4).

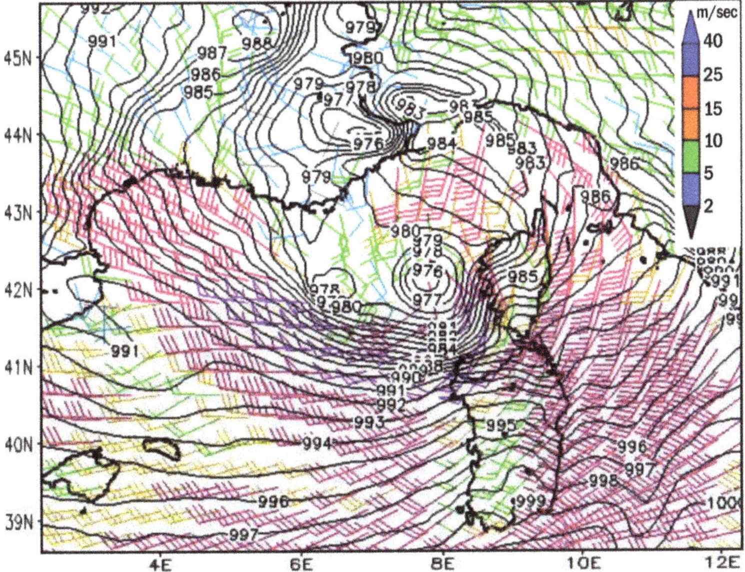

Figure 4. Wind at 10 m (m/s) and mean sea-level pressure (hPa) during Vaia Storm on 29 October 2018, 00 UTC (modified after ARPAL (Regional Agency for Ligurian Environment Protection) [34]).

Referring to the study area, at the Marina of Loano weather station, located about 52 km northeast of Sanremo, southerly winds from 230°, with gusts of 180 km/h and average velocity of 82 km/h, were recorded on 29 October 2018 [33,34]. It resulted from a baric minimum of 976 hPa and a gradient of 8 hPa between Provence and Corsica (Figure 4).

The wave buoy at Capo Mele, located about 35 km northeast of Sanremo, recorded a significant wave height (Hs) of 6.5 m with a period of 11–12 s and a maximum wave peak of 10.3 m, with the sea initially coming from the southeast and then in rotation from the southwest. A sea-level rise of about 50–60 cm was recorded [34], ascribed to a storm surge due to a depression minimum, with a wave tide of about +/− 15–20 cm.

4. Materials and Methods

4.1. Portosole

The site was surveyed in February 2019 (survey 1) using a survey boat equipped with a multibeam echosounder (MBES) and a mobile laser scanner (MLS). On that occasion, the emerged part of the breakwater was surveyed also using a camera mounted on an unmanned aerial vehicle (UAV) and the 3D model was derived by Agisoft Metashape© [59] Structure for Motion processing. To correctly set the parameters of such elaborations, refer to [60] while adopting the tool coming from [61] to plan the UAV photogrammetric survey in a realistic way, obtaining a rigorous evaluation of the precision. The UAV 3D model resolution and accuracy are comparable with MLS ones [35]. Note that the Laser Scanner accuracy could be higher, on the order of a few millimeters in static conditions [62], but the results of an MLS on a boat, hence supported by IMU measurements, could be on the order of a few centimeters.

A second survey was performed in November 2019 (survey 2) with the same survey boat and the same survey criteria.

The survey boat was equipped with the Teledyne Reson PDS2000 platform (Teledyne RESON B.V., Rotterdam, Netherlands [63]) for the simultaneous acquisition of an MBES R2Sonic 2024 (R2Sonic, Austin, TX, USA), a sound velocity profiler (SVP) RESON mod. SVP-15 (RESON B.V., Rotterdam, The Netherlands), an inertial measurement unit (IMU) IXBLUE mod. HYDRINS III (IXBLUE, Paris, France), a MLS RIEGL mod. LMS-Z420i (RIEGL, Horn, Austria) used in profiler mode and a GPS 5700 TRIMBLE (Trimble Navigation Limited, Dayton, OH, USA) receiver in "rover" real-time kinematic configuration (GPS-RTK).

The MBES R2Sonic 2024 is characterized by 256 beams of $0.5° \times 1.0°$, along and across track beamwidth, at 400 kHz.

To investigate the portions of the sea bottom close to the free surface, the transducer was mounted on a joint angled at $25°$, thus physically tilting the transducer to exploit the entire swath of the MBES.

The MLS RIEGL mod. LMS-Z420i is a time of flight (TOF) instrument, characterized by a 1 km maximum range and a repeatability of 8 mm on a single measurement and 4 mm on average in static 3D configuration. In mobile profiler mode, accuracy varies between 2 and 5 cm, depending on the distance of targets and objects measurement and on the quality of IMU (angular precisions). In the present case, precisions in the order of 2–3 cm were obtained, measured and compared both with Total Station and GPS-RTK data.

The MBES/MLS systems required a preventive calibration phase both to synchronize the time scale of each instrument (including IMU) and to compensate the roll, pitch and yaw angles of the system concerning the theoretical (0,0,0) point, called common reference point (CRP). Data of both instruments were processed by the software platform Teledyne Reson PDS2000.

The two performed surveys are framed in the same reference system and cartographic projection, i.e., ETRF2000–2008.0 projected in UTM 32. Three ground control points (GCPs) located along the breakwater structure permit to to register the two point clouds so that they are in the same reference frame.

For survey 1, a (10×10) cm digital surface model (DSM) was produced for the forthcoming elaborations. The DSM reproduces both the emerged and submerged parts of the study area; the first one was obtained from the MBES survey, while the latter was obtained from the MLS survey.

On the contrary, for survey 2, the single point clouds deriving from the MBES and MLS surveys were made available. A (10×10) cm DSM of the entire study area, including both MBES and MLS, surveyed areas, was realized using the Rasterize tool of the free and open-source CloudCompare ver. 2.10 [64].

4.2. Resulting Products

As a preliminary step, particular attention was paid to the removal of the background noise (despiking) and to data filtering. This is essential for automatic/semi-automatic noise removal in the water column, typically caused by navigation motion and reflections of some types of structures geometries.

The quality of the original point clouds related to survey 1 was assessed as follows: First, the MBES, MLS and UAV point clouds have been compared at the head of the Portosole breakwater through the free and open-source software CloudCompare ver. 2.10, to estimate their overlapping area. From these point clouds, three DSMs with (10×10) cm cell resolution were computed using the CloudCompare Rasterize tool. Besides the DSM cell height, the Rasterize tool allows computing the per-cell population (i.e., the number of points falling in a cell) and the heights standard deviation (i.e., the dispersion of height values inside a cell) for each cell of the DSM. Thus, these two parameters can give a rough indication of the point cloud density and of the distribution of height values, respectively. It resulted that the UAV DSM had a higher average value of per-cell population (48), whereas the MBES and MLS DSMs have lower and comparable values; moreover, the MBES and MLS DSMs have similar values of average standard deviation (0.074 m and 0.067 m), showing a substantial comparable distribution of heights in the point clouds, while the UAV DSM has a lower value (0.036 m), probably due to the flatter surveyed area (the breakwater emerged part, mainly constituted by a flat service area, other than the rocky blocks forming the breakwater). These results confirmed the good quality and the reliability of the point clouds derived by the integration of the employed survey techniques. Further details on point clouds quality assessment are available in [35].

The DSMs, derived by MBES and MLS integrated surveys, describes both the breakwater and the sea bottom facing it. The difference in time between the two surveys was enhanced employing different procedures here described, thought to be applied as much automatically as possible and to different case studies.

A simple difference in elevation between the DSMs was performed using the Cloud-Compare M3C2 plugin [2], which computes the signed distances between input point clouds, i.e., the obtained distances have positive or negative signs. The M3C2 output is a new point cloud where each point represents the distance along a defined direction (the vertical direction in the present case, i.e., along the z-axis of a conventional right-handed orthogonal Cartesian triplet) between the reference and the compared point clouds. The M3C2 output cloud follows this convention: positive sign if the compared DSM has a higher height concerning the reference DSM, negative sign vice versa.

To better quantify the metrical difference, a histogram was created, dividing the distance values into 256 classes of 0.05 m of amplitude and computing the number of occurrences for each class. Therefore, the histogram displays the number of occurrences of each class along the y-axis and the central value (v_c) of the 256 classes along the x-axis, obtained as

$$v_c = \frac{v_f - v_i}{2} \tag{1}$$

where v_f and v_i are the final and the initial value of each class.

The average value μ and standard deviation σ of this distribution were computed as follows:

$$\mu = \frac{\sum v_c \times n}{\sum n} \tag{2}$$

$$\sigma = \sqrt{\frac{\sum((v_c - \mu) \cdot n)^2 / n}{\sum n}} \tag{3}$$

where v_c and n are the central value and the number of occurrences of each class, respectively.

This procedure was first applied to compare the entire surveyed area, then on smaller areas, characterized by homogeneous morphology. To divide the entire surveyed area into regions characterized by homogeneous morphology, a manual re-allocation was necessary using the Segment tool of CloudCompare. It allows the assignment of the boundary between areas with near-zero slopes (the sea bottom) and areas characterized by slopes different from zero (the rocky blocks constituting the breakwater). Moreover, the breakwater was divided into two regions, the submerged and emerged areas, respectively.

A comparison was also performed along vertical sections oriented orthogonally to the pier. The sections were extracted starting from a directrix polyline following the pier development using the CloudCompare Extract sections along polylines tool, with a 10 m step between two consecutive sections, for a total of 75 sections. Each section trace is 250 m long orthogonally to the directrix, and it has a thickness of 0.2 m in the xy plane. The section thickness is a required parameter for the Extract sections along polylines tool in case, besides the linear sections (profiles), the corresponding point cloud portions (section clouds) are needed, as in the present case. For this purpose, a thickness of 0.2 m guarantees that the section includes at least one point of the DSM, as its spacing is 0.1 m. Thus, following the tool terminology, the section cloud is a set of points falling inside a parallelepiped of undefined height and base given by the section trace length and its thickness, whereas the profile is a line that approximates the points joining them. Since the profile, for its definition and construction, describes the points configuration using approximation, the section clouds were analyzed. Figure 5 graphically represents the difference between the profile (depicted as a black line) and the section cloud (depicted as red points), referring to a portion of a representative section on the head of the breakwater of Portosole.

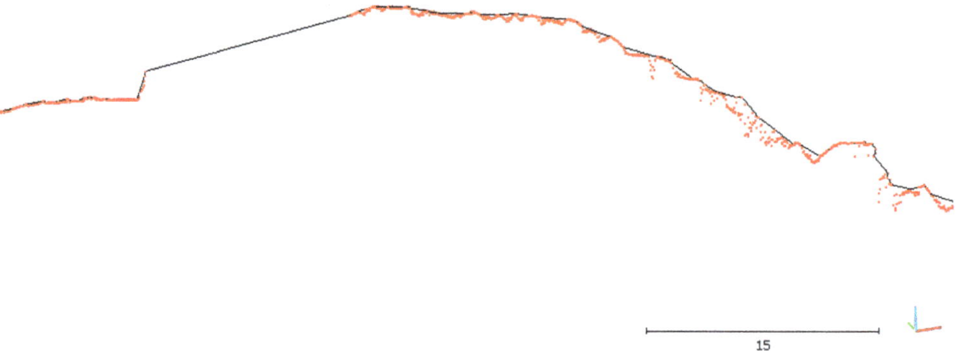

Figure 5. Profile (**black line**) and section cloud (**red points**) of a portion of a representative section on the head of the breakwater of Portosole.

Finally, the total deposited and removed volumes were computed for each homogeneous region using the CloudCompare Compute 2.5D volume tool. It requires a reference point cloud (DSM 1) as input, which is referred to as the initial epoch of study, and a second surface (DSM 2), referred to as the final epoch. The computed volumes and the ratio q between deposited and eroded volumes (in absolute value) were computed for the three portions.

Then, to give an overall idea of the average height variation (Δh), the ratio between the volume variation ΔV and the total extent S was computed for the sea bottom, the submerged and the emerged breakwater portions, using the relation in Equation (4), where V_d and V_e are the deposited and the eroded volumes, respectively.

$$\Delta h = \frac{\Delta V}{S} = \frac{|V_d| - |V_e|}{S} \tag{4}$$

5. Results

A comparison between the DSMs derived by MBES and MLS integrated surveys, performed on February 2019 and November 2019, was carried out for the entire Portosole breakwater extent, revealing breakwater modifications (Section 3). Moreover, the analysis revealed some peculiar aspects of the morphological features and sedimentary dynamics of the seabed facing the breakwater and of the beach located between the root of the breakwater and the mouth of the Rio San Martino.

5.1. Results on DSMs Comparison

The difference in time between the DSMs relative to surveys 1 and 2, hereafter DSM 1 and DSM 2, were enhanced employing different procedures, whose results are presented here.

First, a simple difference in elevation between the two entire DSMs was performed. The DSM 1 was taken as a reference, whereas the DSM 2 was assumed than the surface.

The result is depicted in Figure 6. The distances between DSM 1 and DSM 2 range between −7.5 and 5 m, but most of the values are centered around 0 m, as shown in the histogram in the top left part of Figure 6. The resulting average value and standard deviations of different values are 0.06 m and 0.43 m, respectively.

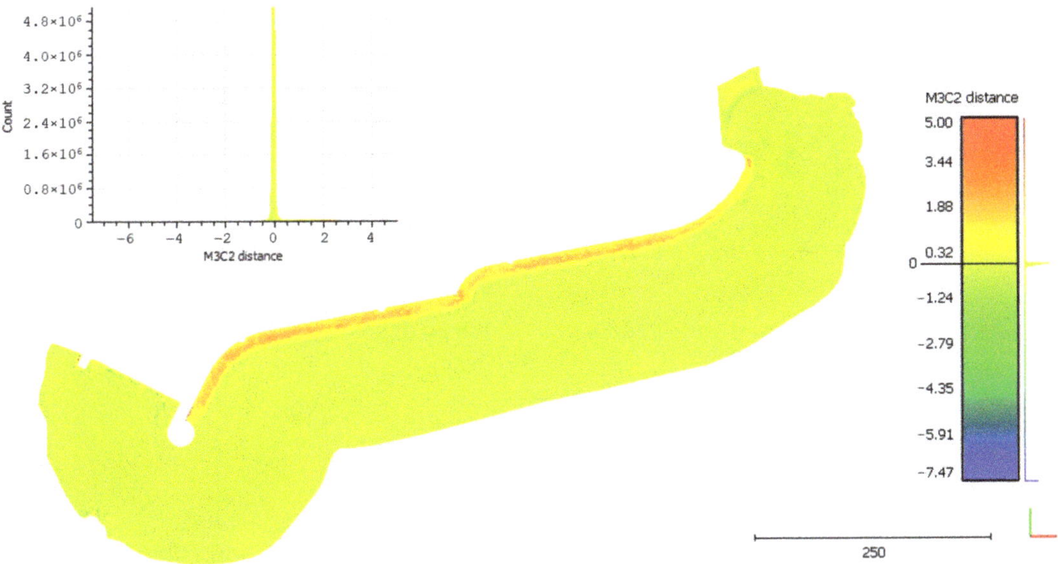

Figure 6. Comparison between digital surface model (DSM) 1 (reference surface) and DSM 2 (compared surface). The histogram shows the distribution of distance values.

Assuming a tolerance interval of ± 3σ in amplitude, the difference values must range between ±1.3 m, centered on the average value (μ), and, at least theoretically, the distance values outside the tolerance interval should be considered as outliers. It is evident that this criterion is not applicable to the area surveyed as a whole because height differences due to changes over time, both in the sea bottom and in the breakwater, can be higher than 1.3 m, not being outliers.

For this reason, three homogeneous regions were identified and analyzed separately: sea bottom, submerged breakwater and emerged breakwater. The obtained homogeneous regions of DSM 1 and DSM 2 are shown in Figure 7a,b, respectively.

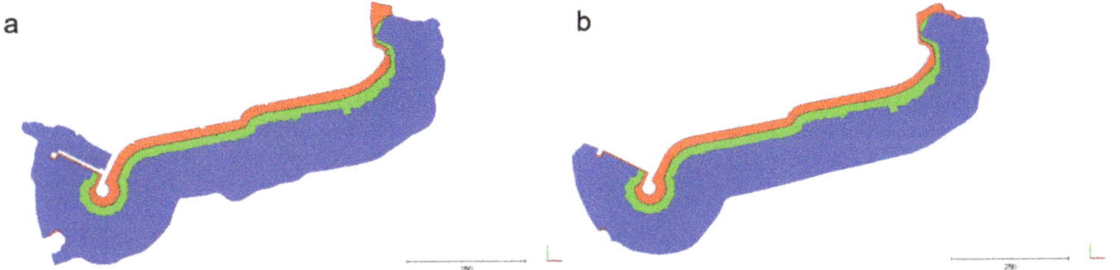

Figure 7. Sea bottom (blue), submerged breakwater (green) and emerged breakwater (red) for DSM 1 (**a**) and DSM 2 (**b**).

The differences in elevation between DSM1 and DSM2 relative to the three homogeneous areas (Figures 8–10), together with their average values (μ) and standard deviations (σ), were computed.

The Gaussian functions $N(\mu, \sigma^2)$ fitting the distribution of DSMs differences of the three homogeneous portions are represented in Figure 11. The differences between DSMs and Gaussian functions are represented by blue histograms and red lines, respectively, with their values, reported on the left and right y-axes, respectively. For all the three cases, the class values are limited within their confidence intervals (amplitude of 3σ) to improve the readability of the graphs. The average values (μ) and standard deviations (σ) of DSMs distances in the three homogeneous portions are also reported in the table inside Figure 11.

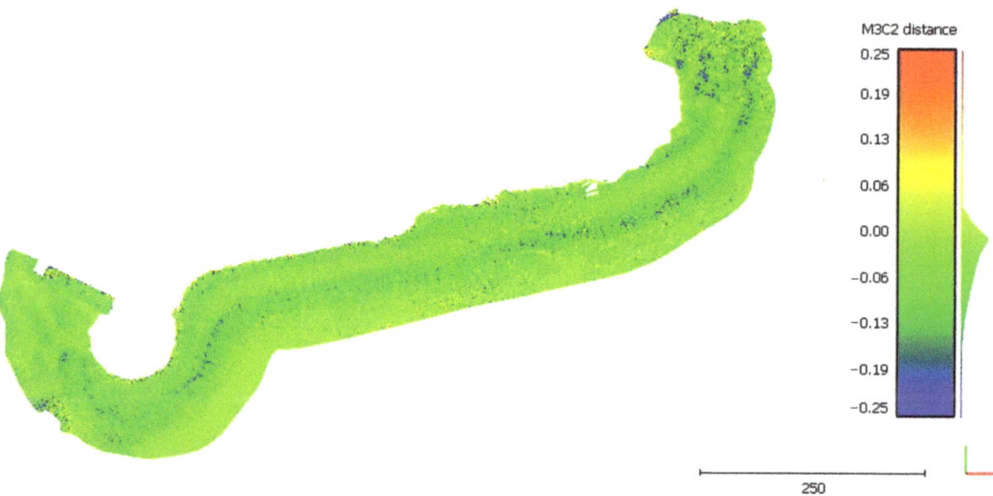

Figure 8. Comparison between DSM 1 (reference) and DSM 2 (compared) on the sea bottom, with distance values limited within the interval $(-0.25; 0.25)$ m.

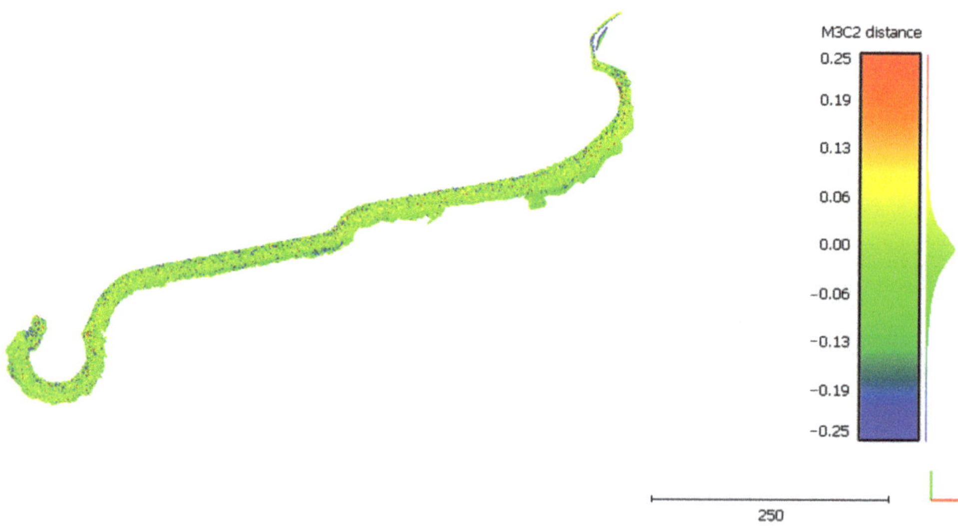

Figure 9. Comparison between DSM 1 (reference) and DSM 2 (compared) on the submerged breakwater, with distance values limited within the interval $(-0.25; 0.25)$ m.

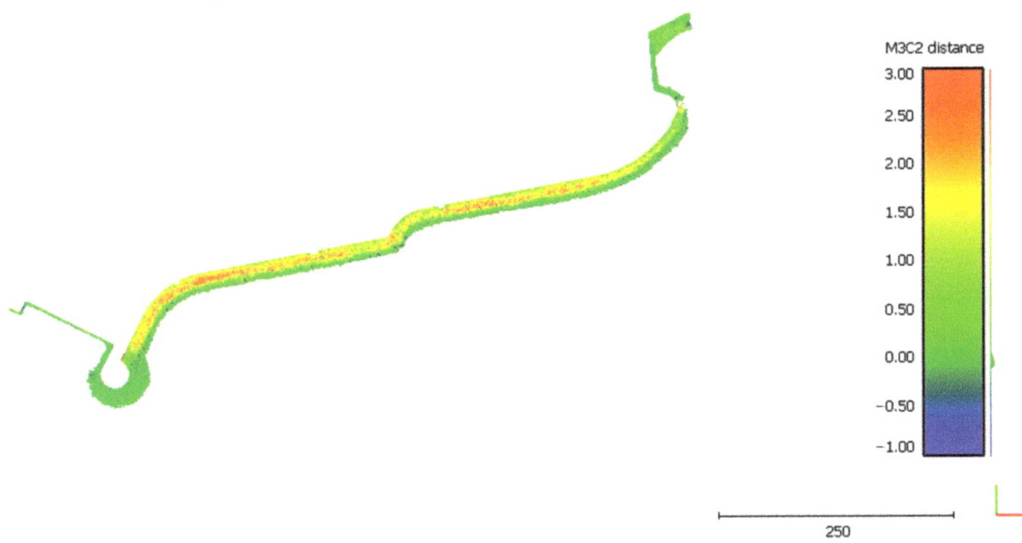

Figure 10. Comparison between DSM 1 (reference) and DSM 2 (compared) on the emerged breakwater, with distance values limited within the interval (−1; 3) m.

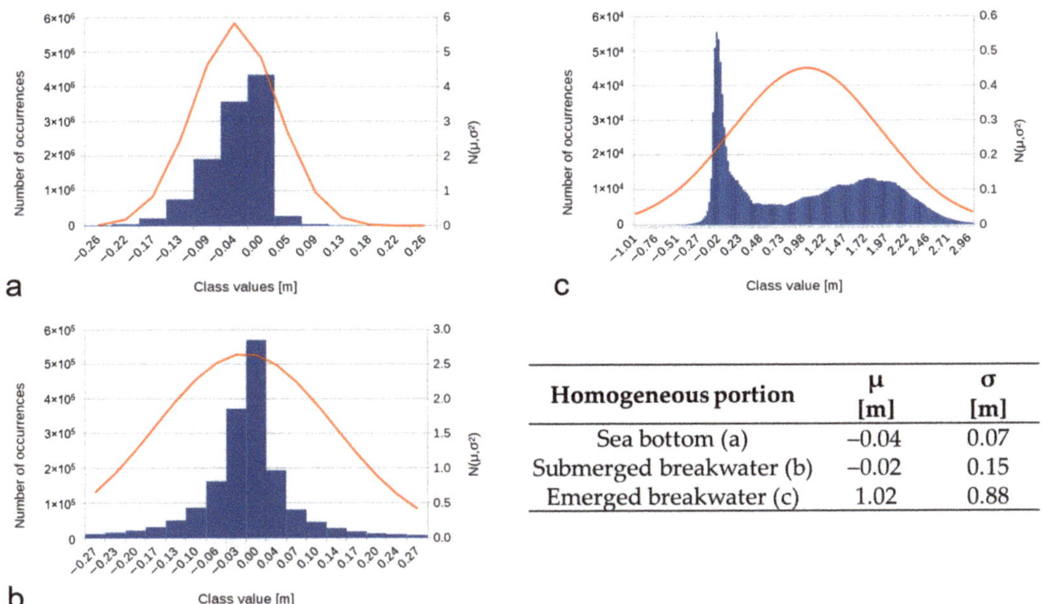

Homogeneous portion	μ [m]	σ [m]
Sea bottom (a)	−0.04	0.07
Submerged breakwater (b)	−0.02	0.15
Emerged breakwater (c)	1.02	0.88

Figure 11. Fitting between a Gaussian function (red line) and the distribution of DSMs differences of sea bottom (**a**), submerged breakwater (**b**), and emerged breakwater (**c**). The distance values are limited within the interval (−1; 3) m, corresponding to the μ ± 3σ confidence interval.

For a more accurate comparison between the DSMs, 250 m long vertical sections and with a 10 m, for a total of 75 sections, were extracted. These section traces, represented in pink in Figure 12, are oriented orthogonally to the breakwater, whose directrix is represented as a black line in Figure 12.

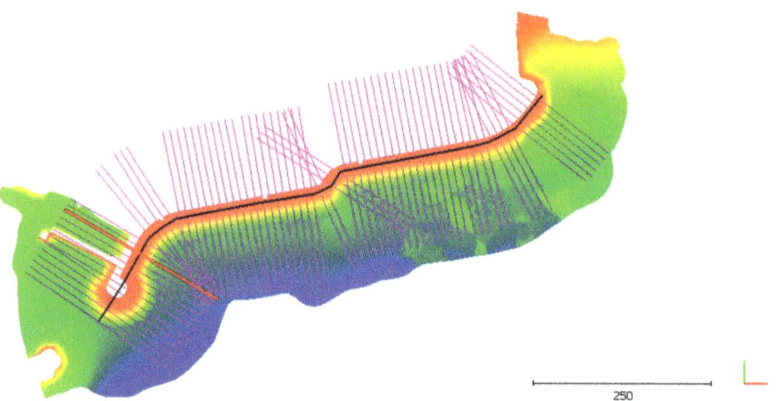

Figure 12. Sections traces (**pink**) and directrix (**black**) along the pier. Section trace number 12 is highlighted in red for considerations in the text.

Figure 13 represents a view of the section clouds (from the head of the breakwater, thus almost from southwest to northeast). The red-magenta and blue-cyan section clouds depict the submerged and emerged breakwater points for DSM 1 and DSM 2, respectively. To perform this, the CloudCompare Filter points by values tool was used to distinguish the points below 0 m from the points above it. Figure 14 represents a zoom of the extracted section clouds along the section trace number 12, highlighted in red in Figure 12, to better underline and quantify the distances between DSM 1 and 2.

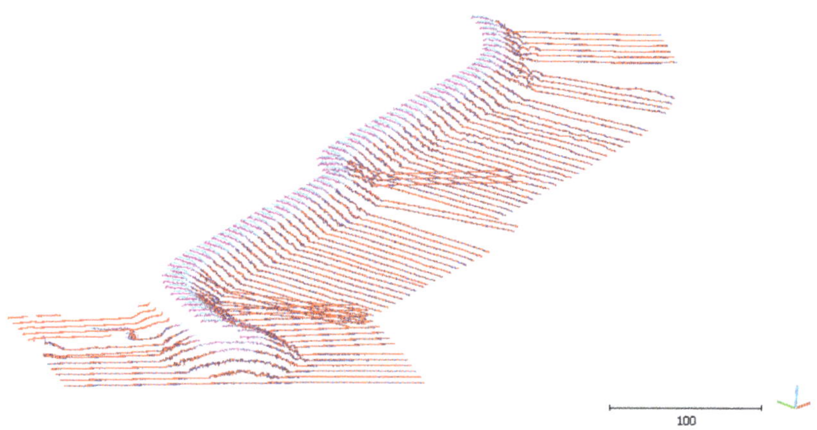

Figure 13. View of the section clouds of DSM 1 and DSM 2. The red and blue sections refer to the submerged portion of DSM 1 and 2, respectively, whereas the magenta and cyan sections are relative to the emerged portion.

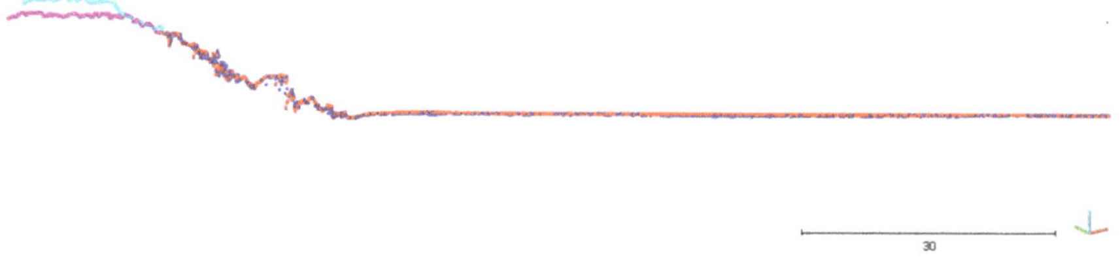

Figure 14. Zoom of the section cloud relative to section trace number 12.

As highlighted in the overall view of the section clouds (Figure 13) and in the zoom reported in Figure 14, and as expected from the previous analysis, the section clouds relative to DSMs 1 and 2 are almost undistinguishable over sea bottom and submerged breakwater areas, whereas a distance between them is noticeable concerning the emerged part of the breakwater.

In conclusion, the DSM 2 is globally slightly lower than the DSM 1 over the sea bottom and the submerged breakwater areas, due to the erosive phenomena that took place; on the contrary, its average height is higher for the emerged breakwater area due to the construction of a temporary service track, which was built to allow the transit of the vehicles transporting rocky blocks and other materials for the restoration of the breakwater.

Moreover, it is noteworthy to underline how the values of average values μ and standard deviations σ reported in the table inside Figure 11 are compatible with the ones computed using an independent method and contained in the following Table 1. Furthermore, in this case, the most pronounced variation is found in the emerged breakwater portion.

Table 1. Deposited and removed volumes between DSMs 1 and 2, surface extensions, ratios (q) between deposited and removed volumes, and average height variations (Δh) for the three homogeneous portions.

Homogeneous Portion	Deposited Volume (m^3)	Eroded Volume (m^3)	Surface (m^2)	q (–)	Δh (m)
Sea bottom	950	−5964	113,500	0.16	−0.04
Submerged breakwater	1094	−1483	18,068	0.74	−0.02
Emerged breakwater	15,022	−221	13,688	67.9	1.08

The total deposited and eroded volumes, the ratio q between them and the ratio Δh between the volume variation ΔV and the total extent S were computed for each homogeneous portion (Table 1).

5.2. Results on Morphological and Sedimentary Dynamics of the Seabed

The survey conducted in February 2019 (Figure 15A), before the restoration of the breakwater, and the one in November 2019 (Figure 15B), when the restoration was completed, have allowed highlighting variations both in the emerged structure of the breakwater and in the morpho-sedimentary features of the seabed (Figure 15C).

The seabed morphology facing the breakwater shows a regular deepening (about 4%) of the bathymetry, between 10 m up to 14 m (section A–A′ in Figure 15C).

Section A–A′ in Figure 15C shows the erosion operated at the base of the breakwater by the reflection of the waves, highlighted by a depression between 0.5 and 0.8 m in-depth. The section B–B′ in Figure 15C shows the presence of *Posidonia oceanica* matte partially affected by erosion [65–68]. The *Posidonia oceanica* upper limit, reported after [58,69], is mapped in the zoom of Figure 15.

Between the root of the breakwater and the mouth of the Rio San Martino, there is a small anthropic pocket beach, confined between the maritime infrastructure and the jetted river mouth with quays and embankments, whose foot extends into the submerged beach.

As shown in the sections C–C′ in Figure 15C, D–D′ and F–F′ in Figure 16, ripple marks of different sizes can be recognized, whose wavelengths range between 0.4 and 1 m, according to the hydraulic regime. Their structures indicate a vergence towards the coast.

The section E–E′ in Figure 16 indicates two erosive channels due to the backflow currents following the sea wave runup on the beach [51].

As shown in Table 1, the ratio between deposited and eroded volumes highlights that the sea bottom was affected by a total deficit of about 5000 m^3 of sediment, with an average lowering of about 4 cm over the entire area. In particular, as shown in Figure 15C, erosional features were found close to the base of the submerged breakwater (section A–A′) and along the sea bottom (e.g., section C–C′); no significant change was detected in the matte areas (section B–B′).

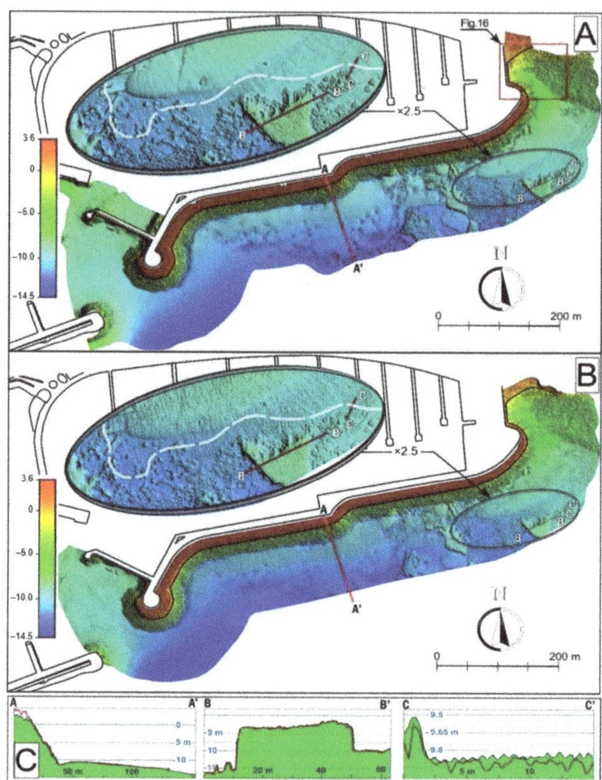

Figure 15. Morpho-bathymetry of the study area related to the two surveys: (**A**) February 2019; (**B**) November 2019. The zooms in (**A**) and (**B**) show the high detail of the acquisition; the white dotted line indicates the upper limit of the *Posidonia oceanica* seagrass after [58]. The red lines indicate the sections reported in (**C**), where the black and red lines refer to February and November 2019, respectively. The red box in (**A**) indicates the location of the zoom reported in Figure 16.

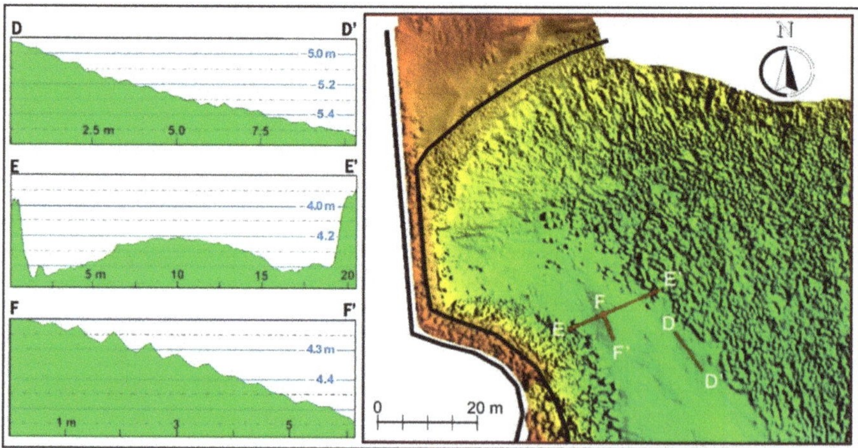

Figure 16. Zoom of the eastern sector of the study area (for location, see Figure 15): the red lines indicate the location of sections shown in the left box.

Analyzing the height variations along the entire breakwater, an average lowering of 2 cm was found in the submerged breakwater due to the adjustment of stone and concrete blocks, while the emerged breakwater shows an average increase of 1 m due to the restoration works.

6. Discussion

The comparison between the DSMs obtained from the survey immediately after the storm (survey 1) that affected Sanremo in October 2018 and the second one (survey 2) after the restoration of the breakwater allowed the detection of 3D variations of the study area over time (4D analysis) with centimetric accuracy. In particular, significant breakwater modifications due to the temporary worksite, the block adjustment of the submerged breakwater and a globally slight erosive phenomena along the sea bottom were revealed.

Concerning the comparison on the sea bottom, the average value of the difference between the surve1 and survey 2 was very close to zero, and the standard deviation was very low, as expected, due to the almost 2D structure, with limited variation along the z-axis. The observed differences, highlighted in blue in Figure 8, could be due to different meteorological and marine conditions encountered during the two surveys. In fact, the blue "stripes" highlight the roll artifacts of the MBES, so they are not effective differences concerning the previous survey, i.e., accumulation/erosion of sediments. In the area where such blue "stripes" are not present, the comparison between the two surveys provided quantitative data of the effects of the October 2018 storm on the geomorphological-sedimentary processes on the sea bottom in front of the breakwater, confirming the reliability of high-resolution survey data within coastal hazard evaluation [70–72]. The comparison on the submerged breakwater area shows an average value of differences close to zero but a higher standard deviation concerning the value of sea bottom, probably due to the 3D structure of the blocks forming it. In fact, even in the case of small movements/rotations of the blocks forming the breakwater, the standard deviation of the differences can vary significantly. The comparison in the emerged breakwater area presents the highest average value and standard deviation. In particular, the height difference of approximately 1 m is due to the temporary service track built over the breakwater between the two surveys to allow the transition of material-handling vehicles to transport rocky blocks. The new road construction also affects the standard deviation value, which is higher than the ones of the sea bottom and submerged breakwater.

The employed survey methods and the resulting products allowed to detect and measure bottom forms related to sedimentary structures and subcritical/supercritical hydraulic flows, such as ripple marks and erosion channels. This highlights their usefulness in recognizing the direction of sedimentary transport and the intensity of the hydraulic flow [73–75]. In fact, in both the surveys, the orientation of the ridges of the asymmetrical ripple marks allowed to identify the directions of the provenance of the waves and the refraction waves due to the interaction with the breakwater and the seabed [76,77]. DMS 1 showed the erosive effects produced by the storm, while DMS 2, confirming the persistence of these erosive processes, provided detailed quantitative data on the variation of the seabed that occurred during the 9 months between the two surveys. In fact, a noteworthy result of the comparison between the two surveys was the measurement of volumetric changes in the seabed and the submerged and emerged breakwater. The surveys showed that the erosion produced by the October 2018 storm, especially at the base of the breakwater, was not compensated by sufficient sedimentary contributions after 9 months. This is evidence of an ongoing sedimentation deficit, resulting in the instability of the rock and concrete blocks at the base of the breakwater. As found in other coastal sectors with high anthropogenic pressure, insufficient sediment supply is attributable to changes in the river network, the obstacle of port infrastructure on the littoral drift and the capture of the sediment by canyon heads [28,42,78–81].

7. Conclusions

Integrated remote sensing techniques have been successfully applied to survey both the breakwater and the neighboring beach of Portosole Marina. Mobile laser scanner (MLS) and multibeam echosounder (MBES), eventually integrated with photogrammetry from (unmanned aerial vehicle) UAV, result to be very effective and reliable to obtain high-definition digital surface models (DSMs) both of the emerged and submerged areas.

The surveys presented in this study were carried out in a coastal stretch totally modified by human intervention consisting of a port infrastructure, an artificial pocket beach and a jetted stream mouth, defined as "techno-coast".

The marine event of 29 October 2018 that affected the area of Portosole Marina represented an outstanding storm associated with the transition of deep low-pressure (976 hPa) over the Ligurian Gulf. This caused a significant sea-level change with a local increase of about 50–60 cm. The phenomenon provoked a great wave penetration on the coasts, as observed in other sectors of Eastern Liguria, with catastrophic effects on port infrastructures and urban facilities.

To verify the coastal impact of the storm, two integrated 3D surveys were performed in February 2019 and in November 2019. The collected data allowed to obtain accurate and reliable high-definition DSMs, both of the emerged and the submerged areas. The comparison between the two surveys highlighted volumetric changes in the seabed and the effects on the emerged and submerged breakwater induced by storm. The estimated deficit of sediment is about 5000 m^3, with a lowering of about 4 cm over the entire submerged area, which contributed to the instability of the breakwater base. The second survey confirmed the sea bottom's erosional state, which is not compensated by the coastal sediment supply. This shows that the storm's effects superimposed ongoing erosional phenomena along the coastal stretch strongly affected by anthropogenic impact.

The recurrence of high-intensity sea storms, as also evidenced by the event that affected Eastern Liguria in November 2000 with similar characteristics (Hs 5.8, period 9–11 s), shows the current increase of storm surges hazard and coastal risk in the Mediterranean area. This highlights the importance of providing adequate coastal zone management measures for risk mitigation through the definition of proper marine flooding hazard scenarios), considering the ongoing climate change and related sea-level rise.

Author Contributions: Remote sensing and field survey, A.L.; Geomatics and numerical elaboration, I.F., B.F. and D.S.; Geomorphology and Sedimentology, P.B., D.M. and N.C. All authors have read and agreed to the published version of the manuscript.

Funding: This research was funded by Research financed with University of Genoa funds (FRA).

Institutional Review Board Statement: Not applicable.

Informed Consent Statement: Not applicable.

Data Availability Statement: Data presented is available under request from the corresponding author.

Acknowledgments: The authors thank Portosole Marina of Sanremo for the support and availability of the data.

Conflicts of Interest: The authors declare no conflict of interest.

References

1. Mills, J.P.; Buckley, S.J.; Mitchell, H.L.; Clarke, P.J.; Edwards, S.J. A geomatics data integration technique for coastal change monitoring. *Earth Surf. Process. Landf. J. Br. Geomorphol. Res. Group* **2005**, *30*, 651–664. [CrossRef]
2. Lague, D.; Brodu, N.; Leroux, J. Accurate 3D comparison of complex topography with terrestrial laser scanner: Application to the Rangitikei canyon (N-Z). *ISPRS J. Photogramm. Remote Sens.* **2013**, *82*, 10–26. [CrossRef]
3. Mancini, F.; Dubbini, M.; Gattelli, M.; Stecchi, F.; Fabbri, S.; Gabbianelli, G. Using unmanned aerial vehicles (UAV) for high-resolution reconstruction of topography: The structure from motion approach on coastal environments. *Remote Sens.* **2013**, *5*, 6880–6898. [CrossRef]
4. Monteys, X.; Harris, P.; Caloca, S.; Cahalane, C. Spatial prediction of coastal bathymetry based on multispectral satellite imagery and multibeam data. *Remote Sens.* **2015**, *7*, 13782–13806. [CrossRef]

5. Troisi, S.; Del Pizzo, S.; Gaglione, S.; Miccio, S.; Testa, R.L. 3D models comparison of complex shell in underwater and dry environments. *Int. Arch. Photogramm. Remote Sens. Spat. Inf. Sci.* **2015**, *40*, 215–222. [CrossRef]

6. Turner, I.L.; Harley, M.D.; Drummond, C.D. UAVs for coastal surveying. *Coast. Eng.* **2016**, *114*, 19–24. [CrossRef]

7. Aguilar, F.J.; Fernandez, I.; Casanova, J.A.; Ramos, F.J.; Aguilar, M.A.; Blanco, J.L.; Moreno, J.C. 3D Coastal Monitoring from Very Sense UAV-Based Photogrammetric Point Clouds. In *Advances on Mechanics, Design Engineering and Manufacturing*; Springer: Cham, Switzerland, 2017; pp. 879–887.

8. Capra, A.; Castagnetti, C.; Dubbini, M.; Gruen, A.; Guo, T.; Mancini, F.T.; Troyer, M. High Accuracy Underwater Photogrammetric Surveying. In Proceedings of the 3rd IMEKO International Conference on Metrology for Archeology and Cultural Heritage, Lecce, Italy, 23–25 October 2017; pp. 696–701.

9. Chen, B.; Yang, Y.; Wen, H.; Ruan, H.; Zhou, Z.; Luo, K.; Zhong, F. High-resolution monitoring of beach topography and its change using unmanned aerial vehicle imagery. *Ocean Coast. Manag.* **2018**, *160*, 103–116. [CrossRef]

10. Mattei, G.; Troisi, S.; Aucelli, P.P.C.; Pappone, G.; Peluso, F.; Stefanile, M. Multiscale Reconstruction of Natural and Archaeological Underwater Landscape by Optical and Acoustic Sensors. In Proceedings of the 2018 IEEE International Workshop on Metrology for the Sea, Learning to Measure Sea Health Parameters (MetroSea), Bari, Italy, 8–10 October 2018; pp. 46–49.

11. Aucelli, P.P.C.; Cinque, A.; Mattei, G.; Pappone, G.; Rizzo, A. Studying relative sea level change and correlative adaptation of coastal structures on submerged Roman time ruins nearby Naples (southern Italy). *Quat. Int.* **2019**, *501*, 328–348. [CrossRef]

12. Federici, B.; Corradi, N.; Ferrando, I.; Sguerso, D.; Lucarelli, A.; Guida, S.; Brandolini, P. Remote sensing techniques applied to geomorphological mapping of rocky coast: The case study of Gallinara Island (Western Liguria, Italy). *Eur. J. Remote Sens.* **2019**, *52*, 123–136. [CrossRef]

13. Guida, S.; Corradi, N.; Federici, B.; Lucarelli, A.; Brandolini, P. Laser Scanner and Multibeam Integrated Survey for The Assessment of Rocky Sea Cliff Geomorphological Hazard. In *Earth Observation Advancements in A Changing World*; Chirici, G., Gianinetto, M., Eds.; AIT: Firenze, Italy, 2019; Volume 1, pp. 162–165.

14. Scicchitano, G.; Pignatelli, C.; Spampinato, C.R.; Piscitelli, A.; Milella, M.; Monaco, C.; Mastronuzzi, G. Terrestrial Laser Scanner techniques in the assessment of tsunami impact on the Maddalena peninsula (south-eastern Sicily, Italy). *Earth Planets Space* **2012**, *64*, 889–903. [CrossRef]

15. Piscitelli, A.; Milella, M.; Hippolyte, J.-C.; Shah-hosseini, M.; Morhange, C.; Mastronuzzi, G. Numerical approach to the study of coastal boulders: The case of Martigues, Marseille, France. *Quat. Int.* **2017**, *439*, 52–64. [CrossRef]

16. Oprandi, A.; Mucerino, L.; De Leo, F.; Bianchi, C.N.; Morri, C.; Azzola, A.; Benelli, F.; Besio, G.; Ferrari, M.; Montefalcone, M. Effects of a severe storm on seagrass meadows. *Sci. Total Environ.* **2020**, *748*, 141373. [CrossRef]

17. Biolchi, S.; Denamiel, C.; Devoto, S.; Korbar, T.; Macovaz, V.; Scicchitano, G.; Vilibic, I.; Furlani, S. Impact of the October 2018 storm Vaia on coastal boulders in the northern Adriatic Sea. *Water* **2019**, *11*, 2229. [CrossRef]

18. Gaertner, M.A.; Jacob, D.; Gil, V.; Domínguez, M.; Padorno, E.; Sánchez, E.; Castro, M. Tropical cyclones over the Mediterranean Sea in climate change simulations. *Geophys. Res. Lett.* **2007**, *34*, 1–5. [CrossRef]

19. Lionello, P.; Galati, M.B.; Elvini, E. Extreme storm surge and wind wave climate scenario simulations at the Venetian littoral. *Phys. Chem. Earth* **2012**, *40*, 86–92. [CrossRef]

20. Gonzáles-Alemán, J.J.; Pascale, S.; Gutierrez-Fernandez, J.; Murakami, H.; Gaertner, M.A.; Vecchi, G.A. Potential Increase in Hazard from Mediterranean Hurricane Activity with Global Warming. *Geophys. Res. Lett.* **2019**, *46*, 1754–1764. [CrossRef]

21. Faccini, F.; Brandolini, P.; Robbiano, A.; Perasso, L.; Sola, A. Instability, precipitation phenomena and land planning: The flood of 2002 in lower Lavagna valley (Eastern Liguria, Italy). *Geogr. Fis. Din. Quat.* **2005**, *7*, 145–153.

22. Brandolini, P.; Faccini, F.; Robbiano, A.; Terranova, R. Geomorphological hazards and monitoring activity along the western rocky coast of the Portofino Promontory (Italy). *Quat. Int.* **2007**, *171*, 131–142. [CrossRef]

23. Brandolini, P.; Canepa, G.; Faccini, F.; Robbiano, A.; Terranova, R. Geomorphological and geo-environmental features of the Graveglia Valley (Ligurian Apennines, Italy). *Geogr. Fis. Din. Quat.* **2007**, *30*, 99–116.

24. Casella, E.; Rovere, A.; Pedroncini, A.; Mucerino, L.; Casella, M.; Cusati, L.A.; Vacchi, M.; Ferrari, M.; Firpo, M. Study of wave runup using numerical models and low-altitude aerial photogrammetry: A tool for coastal management. *Estuar. Coast. Shelf Sci.* **2014**, *149*, 160–167. [CrossRef]

25. Brandolini, P.; Cevasco, A. Geo-Hydrological Risk Mitigation Measures and Land-Management in A Highly Vulnerable Small Coastal Catchment. In *Engineering Geology for Society and Territory*; Springer: Cham, Switzerland, 2015; Volume 5, pp. 759–762.

26. Galve, J.; Cevasco, A.; Brandolini, P.; Piacentini, D.; Azañón, J.M.; Notti, D.; Soldati, M. Cost-based analysis of mitigation measures for shallow-landslide risk reduction strategies. *Eng. Geol.* **2016**, *213*, 142–157. [CrossRef]

27. Carpi, L.; Mucerino, L.; Besio, G.; Corradi, N.; Ferrari, M. MAREGOT Project Experience: Integrated Approach to Understanding Coastal Dynamics Behaviour. In Proceedings of the 2019 IMEKO TC19 International Workshop on Metrology for the Sea: Learning to Measure Sea Health Parameters, MetroSea, Genoa, Italy, 3–5 October 2019; pp. 167–171.

28. Ferrari, M.; Carpi, L.; Pepe, G.; Mucerino, L.; Schiaffino, C.F.; Brignone, M.; Cevasco, A. A geomorphological and hydrodynamic approach for beach safety and sea bathing risk estimation. *Sci. Total Environ.* **2019**, *671*, 1214–1226. [CrossRef]

29. Mucerino, L.; Albarella, M.; Carpi, L.; Besio, G.; Benedetti, A.; Corradi, N.; Firpo, M.; Ferrari, M. Coastal exposure assessment on Bonassola bay. *Ocean Coast. Manag.* **2019**, *167*, 20–31. [CrossRef]

30. Ascari, M.; Baccino, L.; Sanguineti, G. *Ricerche Sulle Variazioni Delle Spiagge Italiane. Le Spiagge Della Riviera Ligure*; C.N.R.: Roma, Italy, 1937; p. 328.

31. Porta, M.; Buosi, C.; Trogu, D.; Ibba, A.; De Muro, S. An integrated sea-land approach for analyzing forms, processes, deposits and the evolution of the urban coastal belt of Cagliari. *J. Maps* **2020**. [CrossRef]

32. Mentaschi, L.; Besio, G.; Cassola, F.; Mazzino, A. Developing and validating a forecast/hindcast system for the Mediterranean Sea. *J. Coast. Res.* **2013**, *65*, 1551–1556. [CrossRef]

33. Iengo, A.; del Giudice, T. Analysis of the 29 October 2018 Sea-Storm in The Ligurian Sea. In Proceedings of the 2019 IMEKO TC 19 International Workshop on Metrology for the Sea, Genoa, Italy, 3–5 October 2019; pp. 145–150.

34. Pedemonte, L.; Corazza, M.; Forestieri, A. Internal Technical Event Report from CMI—ARPAL. Available online: https://www.arpal.gov.it/contenuti_statici//pubblicazioni/rapporti_eventi/2018/Report_speditivo_20181027-30_vers20181122.pdf (accessed on 6 November 2020).

35. Lucarelli, A.; Brandolini, P.; Corradi, N.; De Laurentiis, L.; Federici, B.; Ferrando, I.; Lanzone, A.; Sguerso, D. Potentialities of Integrated 3D Surveys Applied to Maritime Infrastructures and to The Study of Morphological/Sedimentary Dynamics of The Seabed. In Proceedings of the IMEKO TC-19 International Workshop on Metrology for the Sea, Genoa, Italy, 3–5 October 2019; pp. 161–166.

36. Giammarino, S.; Fanucci, F.; Orezzi, S.; Rosti, D.; Morelli, D.; Colbianchi, M.; Di Stefano, A.; Fravega, P.; Vannucci, G.; Piazza, M.; et al. Note Illustrative della Carta Geologica d'Italia alla scala 1:50.000. In *Foglio 258–271 San Remo*; ISPRA Servizio Geologico d'Italia, Regione Liguria: Genova, Italy, 2010; p. 134. Available online: https://www.isprambiente.gov.it/Media/carg/note_illustrative/258_SanRemo.pdf (accessed on 6 November 2020).

37. Zingaro, M.; Refice, A.; Giachetta, E.; D'Addabbo, A.; Lovergine, F.; De Pasquale, V.; Pepe, G.; Brandolini, P.; Cevasco, A.; Capolongo, D. Sediment mobility and connectivity in a catchment: A new mapping approach. *Sci. Total Environ.* **2019**, *672*, 763–775. [CrossRef]

38. Fierro, G.; Berriolo, G.; Ferrari, M. *Le Spiagge Della Liguria Occidentale*; Regione Liguria: Genova, Italy, 2010; p. 174.

39. Corradi, N. La costa della Liguria. In *La Costa d'Italia*; Ginesu, S., Ed.; Carlo Delfino: Sassari, Italy, 2011; pp. 185–200. ISBN 978-88-7138-581-5.

40. Brandolini, P.; Faccini, F.; Robbiano, A.; Terranova, R. Slope instability on rocky coast: A case study of Le Grazie landslides (eastern Liguria, northern Italy). *Geol. Soc. Lond. Spec. Publ.* **2009**, *322*, 143–154. [CrossRef]

41. Scarpati, A.; Pepe, G.; Mucerino, L.; Brandolini, P.; Firpo, M. Rocky cliff landslide hazard: The capo Noli promontory case study (western Liguria, NW Mediterranean Sea). *Rend. Online Soc. Geol. Ital.* **2013**, *28*, 137–141.

42. Fierro, G.; Berriolo, G.; Ferrari, M. *Le Spiagge Della Liguria Centro-Orientale*; Regione Liguria: Genova, Italy, 2015; p. 96.

43. Brandolini, P. The Outstanding Terraced Landscape of The Cinque Terre Coastal Slopes (Eastern Liguria). In *Landscapes and Landforms of Italy*; Springer: Cham, Switzerland, 2017; pp. 235–244.

44. Pepe, G.; Mandarino, A.; Raso, E.; Scarpellini, P.; Brandolini, P.; Cevasco, A. Investigation on farmland abandonment of terraced slopes using multitemporal data sources comparison and its implication on hydro-geomorphological processes. *Water* **2019**, *11*, 1552. [CrossRef]

45. Brandolini, P.; Faccini, F.; Paliaga, G.; Piana, P. Man-made landforms survey and mapping of an urban historical center in a coastal Mediterranean environment. *Geogr. Fis. Din. Quat.* **2018**, *41*, 23–34.

46. Ascione, A.; Aucelli, P.P.; Cinque, A.; Di Paola, G.; Mattei, G.; Ruello, M.; Russo Ermolli, E.; Santangelo, N.; Valente, E. Geomorphology of Naples and the Campi Flegrei: Human and natural landscapes in a restless land. *J. Maps* **2020**. [CrossRef]

47. Brandolini, P.; Mandarino, A.; Paliaga, G.; Faccini, F. Anthropogenic landforms in an urbanized alluvial-coastal plain (Rapallo city, Italy). *J. Maps* **2020**. [CrossRef]

48. Cappadonia, C.; Di Maggio, C.; Agate, M.; Agnesi, V. Geomorphology of the urban area of Palermo (Italy). *J. Maps* **2020**, *16*, 1739154. [CrossRef]

49. Faccini, F.; Giardino, M.; Paliaga, G.; Perotti, L.; Brandolini, P.; Giardino, M.; Paliaga, G.; Perotti, L.; Brandolini, P. Urban geomorphology of Genoa old city (Italy). *J. Maps* **2020**. [CrossRef]

50. Roccati, A.; Mandarino, A.; Perasso, L.; Robbiano, A.; Luino, F.; Faccini, F. Large-scale geomorphology of the Entella River floodplain (Italy) for coastal urban areas management. *J. Maps* **2020**. [CrossRef]

51. Mucerino, L.; Carpi, L.; Schiaffino, C.F.; Pranzini, E.; Sessa, E.; Ferrari, M. Rip current hazard assessment on a sandy beach in Liguria, NW Mediterranean. *Nat. Hazards* **2021**, *105*, 137–156. [CrossRef]

52. Vinzoni, A. Pianta della città di Sanremo (1753). In *Raccolta dei Tipi, disegni e mappe, Mappe e tipi della Repubblica di Genova, involto RR, n. 2.*; Archivio di Stato: Genova, Italy, 1753.

53. Short, A.D.; Masselink, G. Embayed and Structurally Controlled Beaches. In *Handbook of Beach and Shoreface Morphodynamics*; John Wiley & Son: Chichester, UK, 1999; pp. 230–249.

54. Simeoni, U.; Corbau, C.; Pranzini, E.; Ginesu, S. *Le Pocket Beach. Dinamica E Gestione Delle Piccole Spiagge*; FrancoAngeli: Milano, Italy, 2012; p. 177.

55. Mastronuzzi, G.; Aringoli, D.; Aucelli, P.P.; Baldassarre, M.A.; Bellotti, P.; Bini, M.; Biolchi, S.; Bontempi, S.; Brandolini, P.; Chelli, A.; et al. The geomorphological map of the Italian coast: From a descriptive to a morphodynamic approach. *Geogr. Fis. Din. Quat.* **2017**, *40*, 1–36.

56. Berriolo, G.; Sirito, G. *Spiagge e Porti Turistici*; Ulrico Hoepli: Milano, Italy, 1972; p. 430.

57. Lamanna, S. Il Sistema Informativo Geografico Della Costa e Il Censimento Delle Opere Costiere Della Regione Liguria. In *Lo Studio e la Rappresentazione Della Costa Ligure nel Progetto Europeo BEACHMED-E., A Cura Di Corinna Artom*; Regione Liguria: Genova, Italy, 2008; pp. 113–119.
58. SICoast-Regione Liguria. Available online: https://geoportal.regione.liguria.it/catalogo/mappe.html (accessed on 31 August 2020).
59. Agisoft Metashape©. Available online: https://www.agisoft.com/ (accessed on 25 March 2020).
60. Gagliolo, S.; Fagandini, R.; Passoni, D.; Federici, B.; Ferrando, I.; Pagliari, D.; Pinto, L.; Sguerso, D. Parameter optimization for creating reliable photogrammetric models in emergency scenarios. *Appl. Geomat.* **2018**, *10*, 501–514. [CrossRef]
61. Passoni, D.; Federici, B.; Ferrando, I.; Gagliolo, S.; Sguerso, D. The estimation of precision in the planning of UAS photogrammetric surveys. *Int. Arch. Photogramm. Remote Sens. Spat. Inf. Sci.* **2018**, *42*, 837–843. [CrossRef]
62. Fagandini, R.; Federici, B.; Ferrando, I.; Gagliolo, S.; Pagliari, D.; Passoni, D.; Pinto, L.; Rossi, L.; Sguerso, D. Evaluation Of The Laser Response Of Leica Nova Multistation MS60 For 3D Modelling And Structural Monitoring, Computational Science and Its Applications—ICCSA 2017. In *Lecture Notes in Computer Science*; Gervasi, O., Ed.; Springer: Cham, Switzerland, 2017; pp. 93–104.
63. Teledyne Reson PDS2000. Available online: https://www.teledyne-pds.com (accessed on 25 March 2020).
64. CloudCompare (version 2.10) [GPL Software]. Available online: http://www.cloudcompare.org/ (accessed on 25 March 2020).
65. Vacchi, M.; Montefalcone, M.; Bianchi, C.N.; Morri, C.; Ferrari, M. The influence of coastal dynamics on the upper limit of the Posidonia oceanica meadow. *Mar. Ecol.* **2010**, *31*, 546–554. [CrossRef]
66. Ferrari, M.; Montefalcone, M.; Schiaffino, C.F.; Bianchi, C.; Corradi, N.; Morri, C. Geomorphological Constraint and Boundary Effect on Posidonia Oceanica Meadows. In *Rendiconti on Line Della Società Geologica Italiana*; Società Geologica Italiana: Roma, Italy, 2013; Volume 28, pp. 62–65.
67. Montefalcone, M.; Albertelli, G.; Nike Bianchi, C.; Mariani, M.; Morri, C. A new synthetic index and a protocol for monitoring the status of Posidonia oceanica meadows: A case study at Sanremo (Ligurian Sea, NW Mediterranean). *Aquat. Conserv. Mar. Freshw. Ecosyst.* **2006**, *16*, 29–42. [CrossRef]
68. Montefalcone, M. Ecosystem health assessment using the Mediterranean seagrass Posidonia oceanica: A review. *Ecol. Indic.* **2009**, *9*, 595–604. [CrossRef]
69. Diviacco, G.; Coppo, S. *Atlante Degli Habitat Marini Della Liguria: Descrizione Cartografia Delle Praterie di Posidonia Oceanica e dei Principali Popolamenti Marini Costieri*; Grafiche Amadeo: Imperia, Italy, 2006; p. 205.
70. De Pippo, T.; Donadio, C.; Pennetta, M.; Petrosino, C.; Terlizzi, F.; Valente, A. Coastal hazard assessment and mapping in Northern Campania, Italy. *Geomorphology* **2008**, *97*, 451–466. [CrossRef]
71. Aminti, P.L.; Pranzini, E. Indagine sperimentale per la ristrutturazione delle difese di Marina di Pisa. *Studi Costieri* **2000**, *3*, 57–70.
72. Pranzini, E. Shore protection in Italy: From hard to soft engineering and back. *Ocean Coast. Manag.* **2018**, *156*, 43–57. [CrossRef]
73. Tanner, W.F. Ripple mark indices and their uses. *Sedimentology* **1967**, *9*, 89–104. [CrossRef]
74. US Army Corps of Engineers. *Shore Protection Manual*; Department of the Army, US Army Corps of Engineers: Washington DC, USA, 1984; Volume 2, p. 652.
75. Gerhard, E. *Sedimentary Basins Evolution, Facies and Sediment Budget*; Springer: Berlin/Heidelberg, Germany, 2000.
76. Bruun, P. The development of downdrift erosion. *J. Coast. Res.* **1995**, *11*, 1242–1257.
77. Cipriani, L.E.; Ferri, S.; Iannotta, P.; Paolieri, F.; Pranzini, E. Morfologia e dinamica dei sedimenti del litorale della Toscana settentrionale. *Studi Costieri* **2001**, *4*, 119–156.
78. Aminti, P.; Cipriani, L.E.; Pranzini, E. Back to the Beach: Converting Seawalls into Gravel Beaches. In *Soft Shore Protection*; Springer: Dordrecht, The Netherlands, 2003; pp. 261–274.
79. Aucelli, P.C.; Aminti, P.L.; Amore, C.; Artom, C.; Bellotti, P.; Bozzano, A.; Caputo, C.; Castelliti, G.; Cipriani, L.E.; Cocco, E.; et al. Lo stato dei litorali italiani. *Studi Costieri* **2006**, *10*, 174.
80. Migeon, S.; Cattaneo, A.; Hassoun, V.; Larroque, C.; Corradi, N.; Fanucci, F.; Dano, A.; Mercier de Lepinay, B.; Sage, F.; Gorini, C. Morphology, distribution and origin of recent submarine landslides of the Ligurian Margin (North-western Mediterranean): Some insights into geohazard assessment. *Mar. Geophys. Res.* **2011**, *32*, 225–243. [CrossRef]
81. Brandolini, P.; Cappadonia, C.; Luberti, G.M.; Donadio, C.; Stamatopoulos, L.; Di Maggio, C.; Faccini, F.; Corrado, S.; Vergari, F.; Palliaga, G.; et al. Geomorphology of the Anthropocene in Mediterranean urban areas. *Prog. Phys. Geogr. Earth Environ.* **2020**, *44*, 461–494. [CrossRef]

Article

The First Video Witness of Coastal Boulder Displacements Recorded during the Impact of Medicane "Zorbas" on Southeastern Sicily

Giovanni Scicchitano [1], Giovanni Scardino [2,*], Sebastiano Tarascio [1], Carmelo Monaco [3,4,5], Giovanni Barracane [6], Giuseppe Locuratolo [6], Maurilio Milella [6], Arcangelo Piscitelli [6], Gianfranco Mazza [7] and Giuseppe Mastronuzzi [2]

[1]   Studio Geologi Associati T.S.T.–Via Etnea n.169, 95030 Pedara, Italy; scicchitano@studiogeologitst.com (G.S.); tarascio@studiogeologitst.com (S.T.)
[2]   Dipartimento di Scienze della Terra e Geoambientali, Università degli Studi di Bari Aldo Moro, 70121 Bari, Italy; giuseppe.mastronuzzi@uniba.it
[3]   Dipartimento di Scienze Biologiche, Geologiche e Ambientali, Università degli Studi di Catania, 95129 Catania, Italy; cmonaco@unict.it
[4]   CRUST-Interuniversity Center for 3D Seismotectonics with Territorial Applications, 66100 Chieti Scalo, Italy
[5]   Istituto Nazionale di Geofisica e Vulcanologia, Osservatorio Etneo, 95131 Catania, Italy
[6]   Environmental Surveys S.r.l. Academic Spin-Off Università degli Studi di Bari Aldo Moro, 74121 Taranto, Italy; giovannibarracane@ensu.it (G.B.); giuseppelocuratolo@ensu.it (G.L.); mauriliomilella@ensu.it (M.M.); arcangelopiscitelli@ensu.it (A.P.)
[7]   Area Marina Protetta del Plemmirio, 96100 Siracusa, Italy; Gianfranco.Mazza@plemmirio.it
*   Correspondence: giovanni.scardino@uniba.it

Received: 4 May 2020; Accepted: 21 May 2020; Published: 23 May 2020

Abstract: Over the last few years, several authors have presented contrasting models to describe the response of boulders to extreme waves, but the absence of direct observation of movements has hindered the evaluation of these models. The recent development of online video-sharing platforms in coastal settings has provided the opportunity to monitor the evolution of rocky coastlines during storm events. In September 2018, a surveillance camera of the Marine Protected Area of Plemmirio recorded the movement of several boulders along the coast of Maddalena Peninsula (Siracusa, Southeastern Sicily) during the landfall of the Mediterranean tropical-like cyclone (Medicane) Zorbas. Unmanned autonomous vehicle (UAV) photogrammetric and terrestrial laser scanner (TLS) surveys were performed to reconstruct immersive virtual scenarios to geometrically analyze the boulder displacements recorded in the video. Analyses highlighted that the displacements occurred when the boulders were submerged as a result of the impact of multiple small waves rather than due to a single large wave. Comparison between flow velocities obtained by videos and calculated through relationships showed a strong overestimation of the models, suggesting that values of flow density and lift coefficient used in literature are underestimated.

Keywords: boulders; Medicane; flow; UAV; waves

---

1. Introduction

Mediterranean hurricanes, also called Mediterranean Tropical Like Cyclones (TLCs) or medicanes, are warm-core cyclones that develop over the Mediterranean Sea [1–3] with characteristics similar to tropical cyclones. Such storms are constituted by rotating cloud systems characterized by gale force winds, severe precipitation, and a low pressure center, accompanied by a spiral pattern of thunderstorms [4–7]. Two specific areas appear to be the favored locations for medicane genesis: the western Mediterranean [2,8] and the central Mediterranean–Ionian Sea [9,10]. During the last decades, the impacts of medicanes along the coasts of Mediterranean basin have strongly influenced the human settlements, causing a lot of damage and casualties [6,11]. Moreover, several authors predict that, in the next future, climate changes could modify medicanes, decreasing the frequency of their occurrence but increasing the strength of their impacts [4,12,13]. In the last 10 years, two different medicanes made landfall along the coast of Southeastern Sicily; the first one occurred in 2014 and was called Qendresa, and the second occurred in 2018 and was called Zorbas.

In September 2018, Medicane Zorbas impacted the Ionian coasts of Southeastern Sicily and was registered along the coasts of Apulia, Basilicata, and Calabria with minor energy [14]. Evidence of this storm event was observed along the rocky coast of the Maddalena Peninsula (Siracusa, southeastern Sicily; Figure 1), where a surveillance camera of the Marine Protected Area of Plemmirio recorded several boulder displacements that occurred inside an ancient Greek quarry. This coastal sector is characterized by the presence of important boulder fields, interpreted as evidence of the impact of severe storms and tsunami events in the past [15,16]. Since 2009, the ancient quarry has been intensely monitored, using several survey techniques, to identify and analyze boulder displacements with the main purpose to verify if storm waves could be responsible for the movement of boulders which have been attributed to tsunami events [15].

A determination if storm or tsunami waves were responsible for the boulder displacements on coastal area has been the object of considerable debate. Several studies agree that severe storms are generally able to displace most of the small boulders found along coastlines all over the world [17–22]. Some authors proposed that, although boulder displacements occur both during storm and tsunami events, the main cause for movements of the biggest boulders are probably tsunami [15,23]. In contrast, other studies ascribe storms the capability to detach and transport the boulders [24]. A new debate has been recently opened on the number of tsunami events reconstructed for the Mediterranean Sea. Marriner et al. [25] consider this number strongly overestimated, contesting the attribution to tsunami of most of the field evidences described in the literature, and suggesting that cyclical periods of increased storminess, driven by late Holocene climate changes, would be responsible for them. Vött et al. [26] replied to this theory disputing the tsunami DB and the statistical analyses used by Marriner et al. [25], confirming the reliability of the literature in the definition of tsunami events occurred in the Mediterranean Sea.

The Mediterranean basin, due to the lack extreme events such as cyclones, has been considered an excellent area for studies aimed to describe tsunami events [27]. In the absence of observatories, the analytical approach for the study boulder displacements has been the application of different hydrodynamic models (e.g., references [28–32]) to estimate the main features of the wave responsible for the dislocation (Typology, Wave Height, Wave flow velocity). To apply these models it is fundamental to acquire high-resolution data of the boulders (size, volume and mass), but it is also very important to reconstruct (i) the scenario in which the movement occurred (joint bounded or subaerial/submerged); (ii) the typology of movement (sliding, rolling/overturning, saltation/lifting); and (iii) the entity of displacement. These parameters are often very complicated to deduct by field evidence [33,34]. Up to the present, all the studies started from the assumption that the different position of a boulder, after and before an extreme marine event, represents the effect of a single wave impact [20,30,35]. For these reasons, until now, it has been very difficult to test the reliability of the hydrodynamic models in the description of the natural processes.

The aim of this work is to overcome this problem through the analyses of video recorded along the Maddalena Peninsula. Given that video witness has never been used before to analyze boulder displacements, we defined a methodology to reconstruct immersive scenarios, useful for georeferencing the images recorded in the video and to accurately measure, from them, several important parameters such as the wave heights at time of impacts and the wave flows at time of displacements. This kind of approach let us accurately test hydrodynamic equations, showing a significant overestimation of them, probably due to an underestimation of important parameter, such as seawater flow density and lift coefficient [31,32,36,37].

Furthermore, the work provides evidence about the dynamics of boulder movements, suggesting that single wave impacts are unlikely to be responsible for displacements, which normally occur as a result of multiple waves impact, generating a turbulent flow able to nullify frictions. In addition, our work shows that some boulders, previously interpreted as deposited by a tsunami event, have never been displaced in 10 years of monitoring, and so accord with the hypothesis that only tsunamis could be responsible for the deposition of boulders with particular features as dimension and shape.

2. Geological Settings

Southeastern Sicily (Figure 1A) is part of the emerged portion of the Pelagian Block, the foreland domain of the Neogene-Quaternary Sicilian Collision Zone [38]. It is mostly formed by the Hyblean Plateau that along the Ionian coastline is dissected by a system of Quaternary normal faults bounding NNW–SSE oriented horst and graben structures [39]. The Maddalena Peninsula, constituted by a Neogene-Quaternary calcareous succession, is one of the horst structures located along the Ionian coast of southeastern Sicily (Figure 1B). The whole area lies at the footwall of a large normal fault system that, since the Middle Pleistocene, has reactivated the Malta Escarpment [40], a large Mesozoic boundary separating the Pelagian Block from the Ionian oceanic domain to the East [41–43].

Southeastern Sicily is one of the most seismically active areas of the central Mediterranean. It is characterized by a high level of crustal seismicity producing earthquakes with intensities of up to XI–XII Mercalli-Cancani-Sieberg (MCS) and M ~7, such as the 1169, 1693, and 1908 events [44–46] and related tsunami [15,16,47–49]. Being located along a collisional belt, at the footwall of a normal fault system, the analyzed coastal area has experienced vertical deformation that, combined with sea-level changes, has been recorded by several orders of Middle-Upper Quaternary marine terraces and palaeo-shorelines [40,50–52]. However, since the Late Pleistocene the Maddalena Peninsula (Figure 1C) was tectonically stable, as inferred from the elevation of the MIS 5.5 terraces [53] but lightly uplifting in the Holocene [54–57]. According to Anzidei et al. [58], during the last few decades, Global Positioning System (GPS) data and Glacial Isostatic Adjustment (GIA) models indicated current weak subsidence at rates close to 1 mm/yr. This is relevant considering that the area is undergoing heavy coastal retreat and so is exposed to severe storms associated with high-waves, also as a consequence of the global sea-level rise [59–62].

The Ionian coast of Southeastern Sicily, between the towns of Augusta and Siracusa, is characterized by the occurrence of anomalous calcareous and calcarenitic boulders. Scicchitano et al. [15] performed direct observations on these boulders (distance from the coastline, size and weight), together with statistical analysis of the storm regime of the area and hydrodynamic estimations, to verify if tsunami or storm waves were responsible for their detachment and transport. Radiocarbon age determinations of marine organisms encrusting the blocks, compared to historical catalogues, suggested that, in the last 1000 years, the largest earthquakes with local sources (e.g., the 1169, 1693, and 1908 events) could have triggered tsunami waves that displaced the largest boulders occurring in the area. Other evidence of these tsunami events were found along the coasts of Southeastern Sicily, inside several lagoons [63–65] and in coastal deposits [48].

Along Maddalena Peninsula, boulder deposits usually occur on large surfaces gently sloping towards the sea, bordered by high cliffs up to 5 m, formed by Pleistocene depositional terraces or rock platforms dissecting Neogene sub-horizontal, well-stratified, and fractured limestones. The surveillance camera detected boulder displacements inside an ancient Greek quarry located in the northern sector of the Maddalena Peninsula, whose floor has been partially submerged by Holocene sea level rise (up to 40 cm [54]). Boulders located inside the quarry were surveyed with terrestrial laser scanner (TLS) techniques in order to estimate, through the use of a specific hydrodynamic equation [30], the inland penetration limit of the tsunami responsible for the deposition of the block [16]. The two biggest boulders surveyed inside the quarry were represented by a boulder of about 13 ton in weight, previously attributed to a tsunami impact [15], and a boulder of about 41 ton displaced in January 2009 by a storm. Hydrodynamic analyses performed by several authors reveal that coasts of Southeastern Sicily can be affected by severe storm which generate waves able to detach large blocks from the coastal edge and transport them inland [15,16,19,24,66,67].

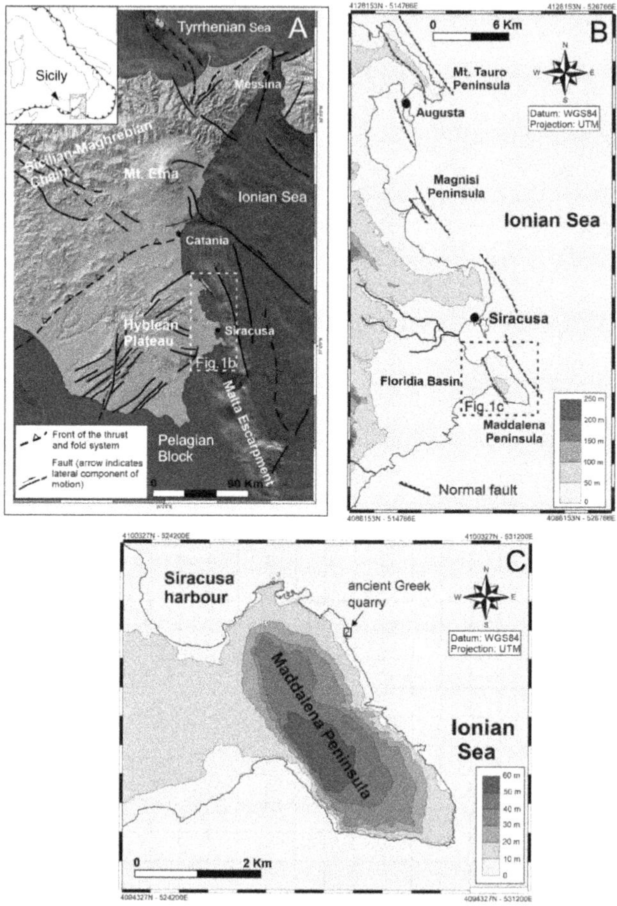

Figure 1. Geological settings of Southeastern Sicily. (**A**) Tectonic sketch map of the Sicilian Collision Zone in eastern Sicily (from Cultrera et al. [39], modified). (**B**) Position of the Maddalena Peninsula area along the Ionian coast of southeastern Sicily. (**C**) Morphological map of the Maddalena Peninsula showing the location of the study area (box).

For the Ionian coast of Sicily, significant spectral wave height (Hm0) and peak period (Tp), recorded during the last 18 years by the Catania buoy (RON—Rete Ondametrica Nazionale; www.idromare.com [68]), are available (Figure 2A). Wave data recorded by the Catania buoy indicated that the most severe storm, which occurred on 2 February 1996, was characterized by significant spectral wave height (Hm0) of about 6.2 m and peak period Tp = 11.3 s. According to Inglesi et al. [69], the return value Hm0 (50) corresponding to a return period of 50 years in the Catania sector do not exceed 6.24 m. A monitoring operated at the quarry located on Maddalena Peninsula indicated that the main boulder displacements have occurred as result of three distinct storms that occurred in 2009, 2014, and 2018. The events of 2014 and 2018 were two medicanes, called Qendresa and Zorbas, respectively.

Qendresa formed on 5 November and rapidly intensified two days later, reaching peak intensity on 7 November. It directly hit Malta in the afternoon and then crossed the Eastern coast of Sicily on 8 November. Later, the cyclone weakened significantly and dissipated over Crete on 11 November. Measurements taken by the ondametric buoy of Crotone and Catania (RON—Rete Ondametrica Nazionale; www.idromare.com [68,70]; Figure 2B), during the passage of Qendresa, show values of significant spectral wave height Hm0 of about 4 m. Medicane Zorbas emerged into the Aegean Sea, moved westward to reach the center of Ionian sea, then inverted its track, moving over northwestern Turkey. Although Zorbas did not affect southeastern Sicily directly, as in the case of Medicane Qendresa, its impact induced similar effects, as recorded by satellite data in off-shore (significant wave height Hs of about 4.1 m) (source AVISO satellite altimetry, credits CLS/CNES [71]). Another relevant effect induced by Zorbas was a storm surge, up to 1m above, detected by the tide gauge sited inside Catania harbor (Figure 2C) and also observed in the video recorded in the Maddalena Peninsula.

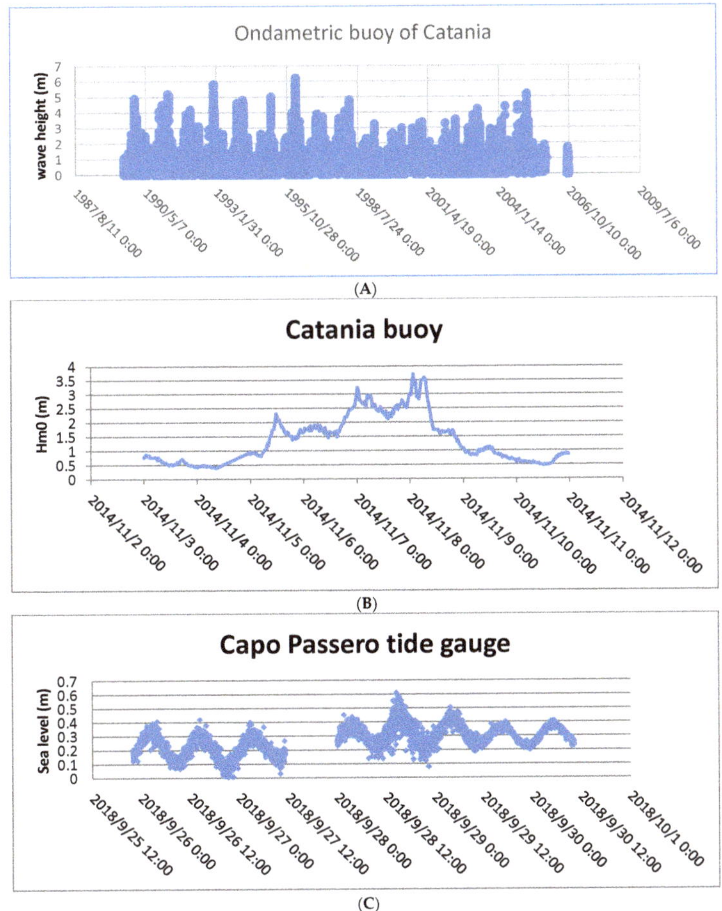

Figure 2. Meteo-marine features recorded in Southeastern Sicily. (**A**) Data of significant spectral wave heights Hm0 recorded by ondametric buoy of Catania (1986–2006). (**B**) Wave height Hm0 recorded by Catania buoy (RON) during Medicane Qendresa. (**C**) Sea level recorded by tide gauge of Capo Passero (ISPRA [72]) during Medicane Zorbas.

3. Material and Methods

In 2003, the technicians of the Marine Protected Area of Plemmirio installed 10 surveillance stations along the coasts of Maddalena Peninsula with the aim to detect illegal fishing operations and to monitoring sea conditions. One of these stations was positioned in proximity of the ancient Greek quarry sited in the northern sector of the Maddalena Peninsula (Figure 3A). The stations are mounted 9 m high on a pole (Figure 3B) and equipped with a HDTV camera (Figure 3C) with 30x optical zoom framing the coastal area (lens 4.3–129 mm; F1.6–4.7; horizontal field of view: 65.6°–2.0°; vertical field of view: 39.0°–1.2°).

Figure 3. (**A**) Orthophoto (Resolution 1cm/pxl) of ancient Greek quarry located in the northern sector of the Maddalena Peninsula, the arrow shows the Monitoring Station. (**B**) Monitoring Station mounted on pole at 9 m height. (**C**) HDTV camera. (**D**) Unmanned Autonomous Vehicle (UAV) survey operations in the studied area, the pole with Monitoring Station is also visible.

During 28 September 2019, the camera recorded about 4 h of video during the Medicane Zorbas. Analyses of this video identified 28 boulder movements inside the quarry. Some of these boulders were already present, and they had been attributed both to tsunami and storm events [15]. The others were detached from the submerged area and transported inland during the Medicane Zorbas. In order to proceed with the analysis of the videos extracted from the Control Center Data Base (resolution 1920 × 1080, frame rate of 25 fps, 50 Hz), it has been necessary to reconstruct a detailed and accurate immersive virtual scenario of the quarry and of the boulders, pre and post the impact of Medicane Zorbas.

In 2009, the northern sector of the Maddalena Peninsula was surveyed using Terrestrial Laser Scanner (TLS) techniques in order to reconstruct the morpho-topographic features of the quarry and of the boulders inside it [16]. The survey was performed by scanning the area from four different points, located on the top of the quarry and from two other points sited inside it. The complete TLS dataset was treated and analyzed by using HDS Cyclone software in order to remove any outliers such as vegetation or anthropogenic features. Then, a dense cloud of points was generated permitting the reconstruction of a 3D model of the quarry and of the boulders present in it. Moreover, since the acquisition of TLS data, we started to monitor the boulder positions inside the quarry by surveying after every storm event, through RTK-GPS techniques, benchmarks located on boulders reconstructed with TLS. In January 2015, a few months after the Medicane Qendresa (8 September 2014), we performed UAV photogrammetric surveys of the quarry to detect the position of the boulders displaced during the Medicane. The results convinced us to replace TLS with UAV photogrammetry techniques (cheaper than TLS and with similar resolution and accuracy) to monitor boulder movements occurring inside the quarry. With this aim, we performed surveys in 2016–2018.

A pre-impact virtual immersive scenario has been modelled combining cloud points reconstructed with TLS in 2009 with the cloud points detected with the UAV photogrammetric surveys performed on 2017. This scenario was used to detect in the video morphological features, such as the edges of the quarry, useful as benchmarks to geometrically analyze, through specific software, the images recorded in the video showing boulder movements. To reconstruct the post-impact virtual immersive scenario we performed, three days after the impact of the Medicane Zorbas, an UAV photogrammetric survey (Figure 3D) of the quarry area together with a proximity photogrammetric survey of the new boulders displaced on the coast.

Geographic Information System (GIS) analyses of the products of previous and recent surveys let us identify most of the boulders appearing in the recorded video (the others have probably been fragmented into smaller blocks). The video has been analyzed by Tracker (https://physlets.org/tracker/ [73]), a video analysis and modelling software built on the Open Source Physics (OSP, Doug Brown, Cabrillo College, California, U.S.) Java framework. Distance between specific features clearly visible in the video, as for example quarry edges, holes, and fractures, were inserted as a spatial reference in Tracker. Once defined, the spatial reference, tagging in each frame of the video objects in motion (boulders, waves, and flows), the software is able to calculate velocity and accelerations. We focused our analyses on the estimation of the flow velocity at time boulder movements occurred.

From the assessment of boulder features, we have calculated, through the hydrodynamic equations of Nandasena et al. [32], the flow needed to start the movements of the blocks analyzed in the videos. Flow values estimated trough video analyses were compared with those calculated with the Nandasena et al. [32] model in order to evaluate the possible discrepancies.

3.1. UAV Data and GIS Analyses

The use of Unmanned Autonomous Vehicles (UAV), better known as drones, in various fields of geoscience has increased during the last five years. In coastal geomorphology studies, UAV photogrammetry techniques have been applied for flood estimation [58,74] and for boulder field monitoring [20,21]. Since 2015, the area of study has been seasonally monitored with UAV photogrammetry surveys in order to detect boulder movements inside the Greek quarry sited in Maddalena Peninsula. Considering the large number of boulders displaced during the impact of the storm generated by the Medicane Zorbas, we performed an UAV survey three days after the storm. The survey was performed with a Multicopter NT4 Airvision (Studio Geologi Associati T.S.T., Catania, Italy) (Figure 3), equipped with an High Definition camera (resolution 24Mpx, lens 16 mm, f3, 5–5, 6), that flew at 30m of altitude, during four distinct flights, with a speed of 1.5 m/s.

In order to obtain an accurate georeferencing of the UAV data, we used the ground control point (GCP) net installed in 2015 and composed of 40 benchmarks regularly spaced along the quarry area (Figure 4A). Benchmarks positions were surveyed with real-time kinematic (RTK) GPS performing 1 h of acquisition for each point of the net. Colored markers, visible from 30 m of altitude, were placed on the benchmarks (Figure 4B) before realizing the flights. A total of 152 pictures were collected and processed using Agisoft Photoscan Professional software version 1.4.0 (St. Petersburg, Russia) to obtain a high-resolution digital elevation model (DEM, 2 cm grid cell; Figure 5A) and orthophotographs (1 cm/pxl; Figure 5B).

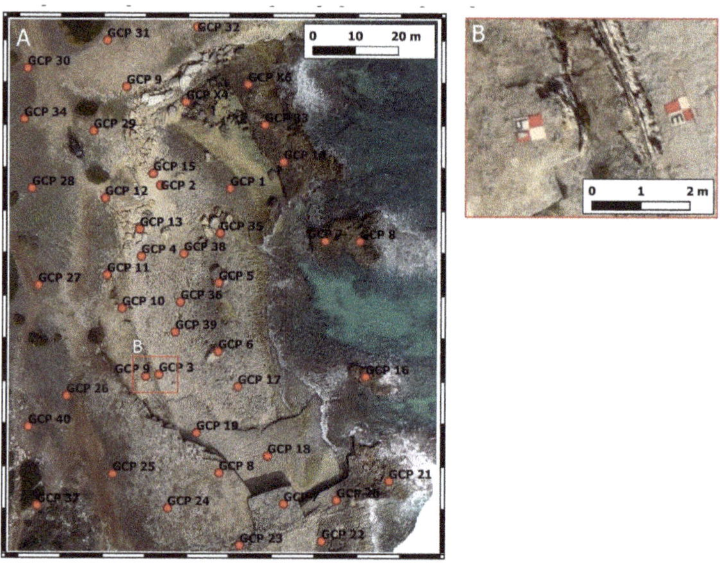

Figure 4. (**A**) Ground control point (GCP) net installed in 2015 and composed of 40 benchmarks regularly spaced along the quarry area. (**B**) Detail of the colored markers positioned on the ground.

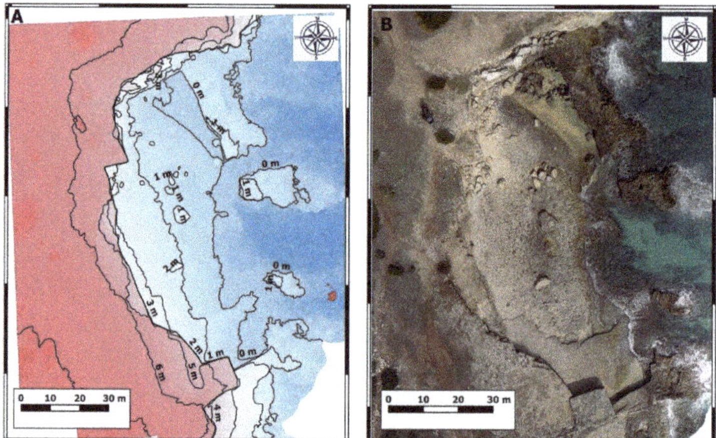

Figure 5. (**A**) High-resolution digital elevation model (2 cm grid cell). (**B**) Orthophotographs (resolution 1 cm/pxl) reconstructed by UAV photogrammetric survey.

These two products were integrated into the general Geo Data Base developed in 2009 and composed by (i) orthophotos and DEM obtained by TLS (2009) and UAV surveys (2015–2018); (ii) orthophotos provided by Regione Sicilia (2007, 2008, 2013); (iii) RTK-GPS data of boulder positions and displacements (2009–2018); and (iv) a 3D model of all boulders dislocated inside the quarry since 2009. Morpho-topographic surveys were performed every year and interpreted in the GIS environment (QGIS) through the digitalization of all the main morphological features such as boulders, fractures, detached part of coastline, score marks and sediment accumulations. These features have been compared with those extracted from the orthophotographs and DEM obtained in the previous year to identify boulder movements or other changes in coastal landscape. As it is possible to observe in Figure 6, several boulders were displaced in the studied area for effects of the impact of Medicane Zorbas; some of these were already in the area and moved from their original position, while others were detached from the infralittoral zone and deposited on the coastline.

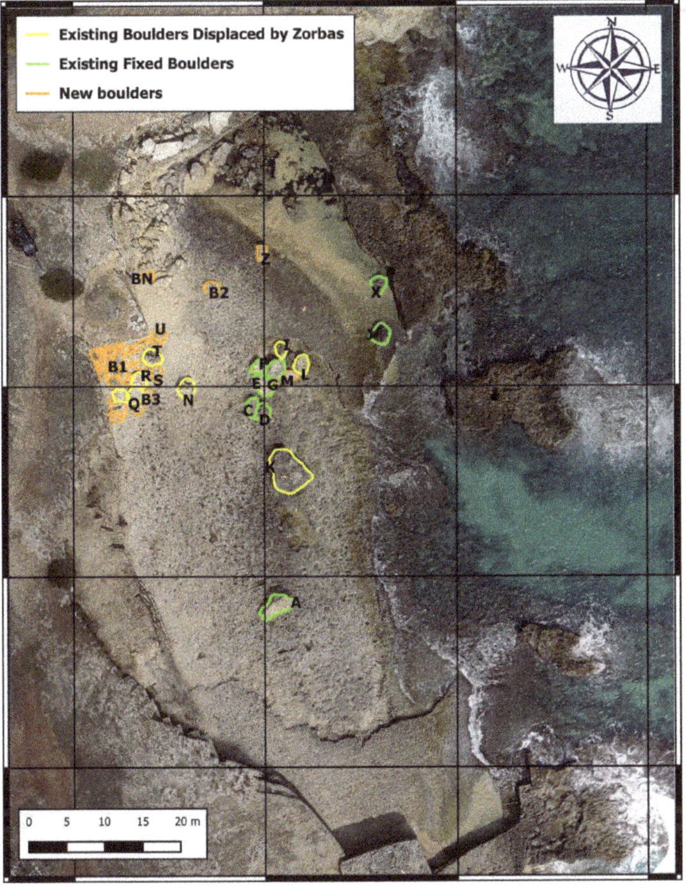

Figure 6. Boulders edited in Quantum Geographic Information System (QGIS) from the Orthophoto reconstructed after the impact of Medicane Zorbas; the boulders already present on the coastline and displaced by Zorbas are marked in yellow, the boulders detached from the coastline and dispersed on the shore-platform by Zorbas are marked in orange, and the boulders not displaced by Zorbas are marked in green.

3.2. Boulders Survey

Once identified in the most recent orthophoto, the new boulders dispersed on the coastline by the impact of Medicane Zorbas have been surveyed with photogrammetry techniques to accurately calculate volume, dimension, and organic encrustations. Pictures were collected with an High Definition camera (resolution 24Mpx, lens 16 mm, f3,5-5,6). Around each boulder, 38 benchmarks were positioned on the boulders and surveyed with a total station for georeferencing the reconstructions of the blocks. Pictures and Benchmarks were processed in Agisoft Photoscan Professional software version 1.4.0 (Figure 7A) to obtain high-resolution 3D surface of the boulders and then post processed in Rhino 6 (Rhino Software 6, Mc Neel Europe, Barcellona, Spain) to close holes and missed surface and generate the final 3d model (Figure 7B). Determination of bulk density was carried out on rock samples collected from the surveyed boulders. The volume of rock samples was determined through the Instantaneous Water Immersion Method [75]. The mass of each boulder was estimated as the product of boulder's density and volume. Density was assigned to each boulder on the basis of results of bulk density measurements as well as considering its thickness and lithology (Table 1).

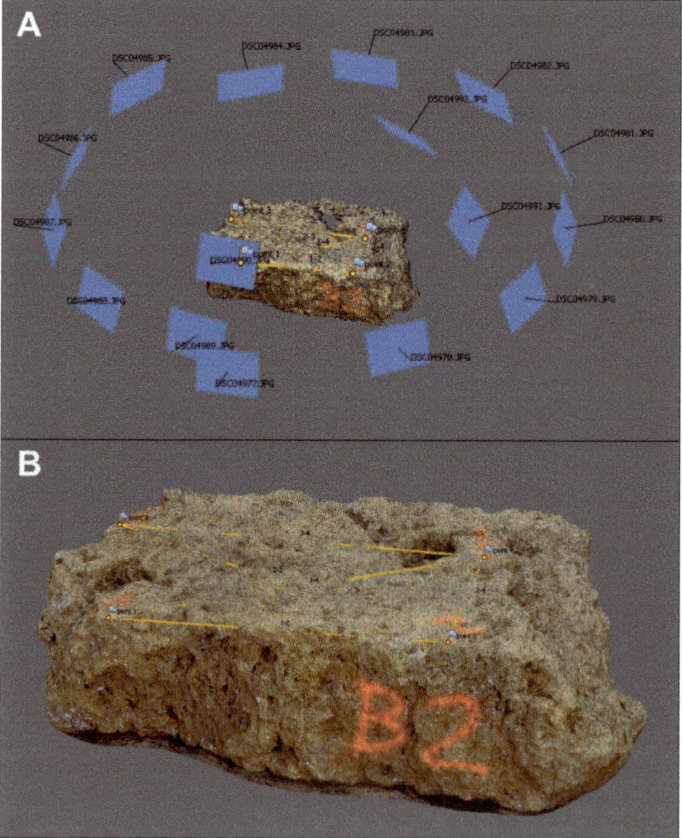

Figure 7. (**A**) Pictures (blue rectangles) and benchmarks (light blue flags) processed in Agisoft Photoscan Professional software version 1.4.0. (**B**) Three-dimensional model of boulder B2 reconstructed with Rhino 6.

Table 1. Boulder features detected through photogrammetry surveys and laboratory analysis.

ID	Volume (m³)	Density (kg/m³)	Masses (kg)	a-Axis (m)	b-Axis (m)	c-Axis (m)	a-Axis Direction
BN	0.367	2080	763.36	1.8	0.95	0.3	N110E
B2	0.705	2080	1466.4	1.5	1.17	0.42	N120E
B3	0.828	2040	1689.12	1.9	0.96	0.45	N210E
B4	0.569	2070	1177.83	1.57	0.97	0.4	N190E
D	2.283	2040	4657.32	2.4	1.95	1	
F	1.461	2040	2980.44	2.1	1.9	0.6	N354E
H	3.377	2040	6889.08	2.6	2.4	0.9	
L	0.89	2040	1815.6	2.5	1.6	0.5	N338E
N	4.031	2040	8223.24	3.2	2.2	1.15	N281E
J	3.367	2040	6868.68	2.2	2.1	1.25	
X	1.61	2040	3284.4	2.2	1.95	0.7	
K	18.782	2200	41320.4	5.8	5.18	1.3	N159E

3.3. Video Editing

More than 7 h of consecutive video recorded on 28 September 2019 were analyzed, detecting 28 distinct boulder movements occurred in the Greek quarry (Table 2). In particular, we focused on five boulders, for which a reliable identification in field and in the orthophoto was possible. Boulders B2, B3, B4, and BN were not present on the coastline before the impact of Zorbas. Boulder K is a very large block, about 41 ton in weight, first displaced by a storm in 2009 and moved again by the two medicanes (2014 and 2018). Videos highlighted the wave impacts on the Greek quarry and showed the boulder movement that occurred in subaerial/submerged scenario. In the video frames, the moment of boulder movements is visible, which occurred with a clear wet surface caused by a continuous wave flow impacting. At the moment of boulder movement, wave flow assumed a turbulent motion with main directions from SE and E, with subsequent backwash flow detected in different directions.

Table 2. Results of video analyses of boulder displacements detected in the Maddalena Peninsula.

No.	Video Time Frame	Boulder	Movement Dynamics	Found in Field	Flow u (m/s)
1	14:21:37.221–14:21:50.945	B1	Sliding	NO	
2	14:29:32.769–14:29:44.885	B1	Sliding/overturning	NO	2.012
3	14:31:08.964–14:31:13.264	B1	Overturning	NO	1.71
4	14:33:11.857–14:33:16.157	B1	Saltation	NO	1.29
5	14:35:32.948–14:35:36.648	B1_F	Overturning	NO	1.588
6	14:41:13.447–14:41:15.147	B1	Overturning	NO	2.113
7	15:01:18.471	B1	Saltation	NO	
8	15:02:17.148–15:02:18.948	B2	Overturning	NO	2.101
9	15:14:57.718-15:15:04.118	B2	Overturning	NO	1.687
10	15:51:53.464	B1_F	Saltation	NO	1.783
11	15:52:46.083	B2	Saltation	YES	2.101
12	15:57:57.546–15:58:02.145	B3	Sliding	YES	
13	15:59:05.542–15:59:07.342	B3	Sliding	YES	2.58
14	15:59:16.666	B3	Sliding	YES	
15	16:05:17.440–16:05:24.040	B3	Overturning	YES	1.663
16	16:08:04.024–16:08:34.941	B3	Overturning/Sliding	YES	2.331
17	16:16:17.351–16:16:23.504	B3	Overturning	YES	1.305
18	16:16:57.984–16:17:05.536	BN	Overturning	YES	1.981
19	16:18:08.880–16:18:13.080	B3	Overturning/Sliding	YES	
20	16:26:28.971	B3	Overturning/Sliding	YES	
21	16:28:04.317–16:28:14.854	BT	Overturning	NO	1.345
22	16:28:30.688–16:28:44.506	BT	Overturning	NO	2.269
23	16:31:27.882	B3	Overturning	YES	
24	16:31:30	K	Sliding	YES	4
25	17:41:35.526	B4	Saltation	YES	
26	17:42:45.483–17:42:52.613	B4	Saltation	YES	
27	17:49:26.616–17:49:36.252	B4	Overturning	YES	
28	18:16:30	B4	Sliding	YES	2.525

Video analysis has been performed by means of Tracker software (Doug Brown, Cabrillo College, CA, USA) using as metric reference the measures of the quarry borders detected by TLS data and some other specific features recognizable on the bedrock of the quarry (Figure 8). Metric references were selected very close to the boulders displaced in the video in order to avoid issues related to perspective distortion that could induce significant errors in the measurements of flow velocity. Videos were acquired with a frequency of 25 fps; detecting a fixed point on the boulders in each frame, it was possible to assess the wave flow velocity, wave period, wave height, velocity, and acceleration of boulders when they start to move. For boulders B2, B3, B4, and BN, we selected the movement showing highest values of flow velocity as reported in Table 3. Boulder K moved just one time, but, unfortunately, the presence of strong turbulence at that time did not permit us to evaluate the flow for each frame of the video. In this case, velocity was calculated considering distance and time between two clearly visible frames.

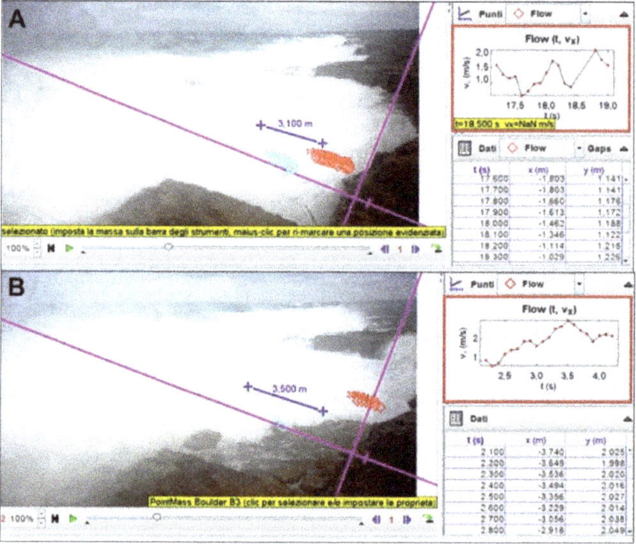

Figure 8. Editing analysis of video frames during boulder movements through Tracker software. (**A**) Rolling of boulder B2 at hour 15:02:00 UTC; (**B**) movement of boulder B3 at hour 16:16:00 UTC. In the red box diagrams are showed the flow velocity estimations marked for each frame of the video.

Table 3. Boulder detected by video editing, selected for flow velocity comparison with hydrodynamic models.

Boulder	Instant Time (Hour UTC)	Maximum Flow Observed in the Videos (m/s)
B2	15:01:59	2.1 ± 0.42
B3	16:26:13	2.33 ± 0.5
BN	18:16:27	1.98 ± 0.11
B4	16:17:49	2.53 ± 1.34
K*	16:30:08	4 ± 2.25

3.4. Hydrodynamic Models

In the last few years, several hydrodynamic models have been proposed by different authors to describe boulders displacement in coastal areas as result of the propagation of an extreme wave [28,29,31,32,36,76]. Nandasena et al. [32], in particular, elaborated a model to calculate the flow velocity needed to start the movement of a boulder for various dynamics of movement (sliding, rolling/overturning, saltation/lifting) and different pre-settings conditions of the boulder (the joint-bounded-JB, subaerial/submerged-SB). At the present, the main limit for applying Nandasena et al. [32] to field evidence has been that the dynamics of the movements were not certain but reconstructed comparing the final position of a boulder with the previous position.

As exposed in the videos editing, the final position of the boulders, displaced during the Medicane Zorbas, is never related to a single wave impact, but it is the result of several different movements that occurred in different directions and with different amplitudes. In this case, we overcome this problem because, through the recorded video, it was possible to extract, for each movement, certain information about the original position and pre-setting condition of the boulder and the dynamics of the movement. Moreover, from video editing, we obtained values of flow velocity at the time of movement of the boulders that could be directly compared with values estimated by Nandasena et al. [32] model. This model has been applied to the five considered boulders in relation to five specific movements recorded in the video (Table 3), in subaerial/submerged scenario:

sliding

$$u^2 = \frac{2(\rho_s/\rho_w - 1)gc(\mu_s\cos\theta + \sin\theta)}{C_d(c/b) + \mu_s C_l} \tag{1}$$

rolling/overturning

$$u^2 = \frac{2(\rho_s/\rho_w - 1)gc(\cos\theta + (c/b)\sin\theta)}{C_d(c^2/b^2) + C_l} \tag{2}$$

saltation/lifting

$$u^2 = \frac{2(\rho_s/\rho_w - 1)gc(\cos\theta)}{C_l} \tag{3}$$

in joint-bounded scenario
saltation/lifting

$$u^2 = \frac{2(\rho_s/\rho_w - 1)gc(\cos\theta + \mu_s\sin\theta)}{C_l} \tag{4}$$

where

u is the flow velocity needed to start the boulder movement;
ρ_s is the density of the boulder;
ρ_w is the density of the water (equal to 1025 kg/m^3);
g is the gravity acceleration;
μ_s is the coefficient of static friction;
θ is the angle of the bed slope at the pre-transport location in degrees;
C_d is the drag coefficient (equal to 1.5);
$a - b - c$ are the axis of the boulder, with the convention a > b > c;
C_l is the lift coefficient (equal to 0.178).

Although some authors attributed to the wave flow the same density value for the seawater [28,29,31,32], recent work [77] has demonstrated that flow density, resulting from a mix of seawater and sediments, should be estimated as follow:

$$\rho_m = (f_s*\rho_s) + (f_w + \rho_w) \tag{5}$$

where ρ_m is the average density of the seawater and sediment mix, f_w is the seawater fraction by volume between 0 and 1, f_s is the sediment fraction by volume between 0 and 1, ρ_w is the density of clear seawater (1.025 g/cm^3), ρ_s is the sediment density, and $f_w + f_s = 1$.

4. Results

Since 2009, the coast of Maddalena Peninsula has been impacted by several storms, three of which induced boulder displacements inside a Greek quarry and two, in particular, were Medicane events

(Qendresa 2014; Zorbas 2018). on 13 January 2009, an intense storm displaced the biggest boulder (K, about 41 ton in weight) about 9 m inland, which was also moved during Medicanes Qendresa and Zorbas (Figure 9). Analyses of orthophotos provided by *Regione Sicilia* for (2007, 2008) revealed that the boulder K was broken from the coastline between August 2017 and August 2018. Field survey suggested that it occurred as a result of erosional process (Figure 9B) and not in response to the direct impact of a wave. Although the storm in 2009 was able to induce the biggest displacement to boulder K, if compared with movements during Medicane (Figure 9C), it did not dislocate many other boulders (only four more small boulders were moved during the storm). In contrast, Qendresa and Zorbas displaced a larger number of small boulders but were able to move boulder K only for short distance (about 1 m for each event).

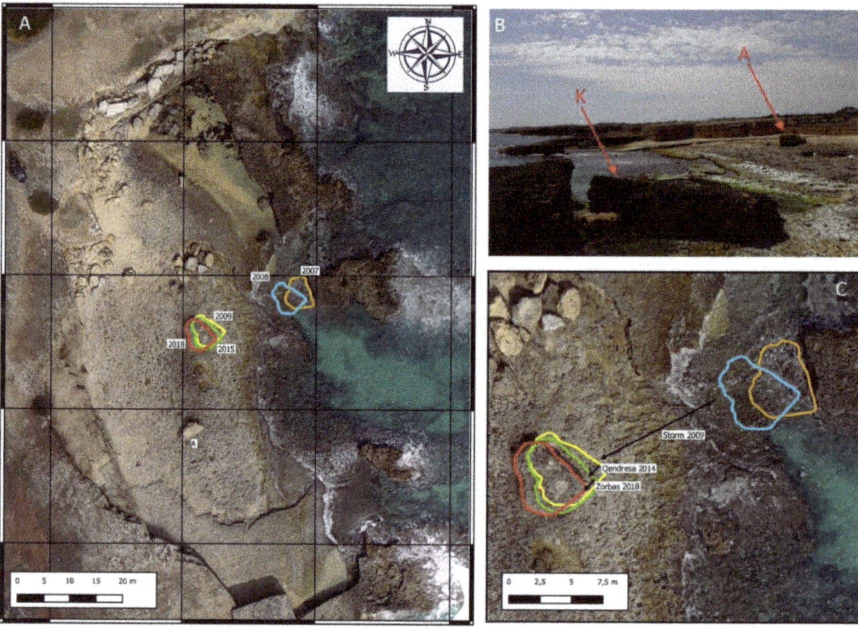

Figure 9. Boulders displacements during storm events; (**A**) Position of Boulders K and A from 2007 to 2018. (**B**) Position of Boulders K and A on 2008. (**C**) Displacements of Boulder K for the impact of the storms.

The two medicanes transported inland dozens of small blocks, ranging in weight between 1 ton and 2 ton, eroded from the external edge of the coastline (Figure 10) and dispersed on the coastal area up to the external walls of the quarry located about 33 m landward. The most relevant boulder deposit was detected in the NW corner of the quarry and it is mainly composed by imbricated boulders lying on a coarse sand of about 40 cm thick. Analyses of the video showed that this berm was the result of boulder accumulation by the action of wave impacts, but mostly due to a turbulent flow, running along the base of the quarry, that moved the boulders along a path parallel to the coastline. The video highlighted that this flow, reaching flow velocity values of up to 4 m/s, was nourished by the strongest waves, hitting the wall of the quarry, and triggered by the topography gently sloping toward the north (about 0.58°).

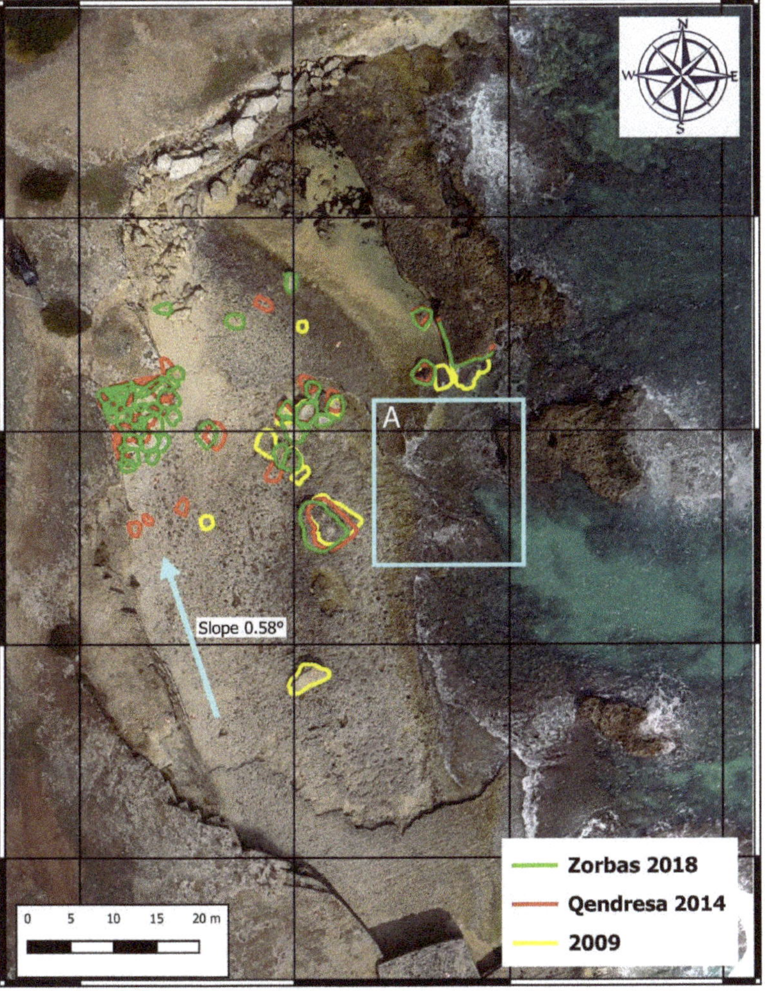

Figure 10. Boulders displaced by the impact of medicanes Qendresa and Zorbas; the rectangle marks the area where boulders have been detached from the coastline, and the arrow indicates the direction of the flow generating boulder deposit on the edge of the quarry.

By video analysis, it was possible to observe the movement of the boulders in relation to the wave flow impacting, highlighting that a surface wetted by a significant number of waves allowed the movement of the boulders. Both sliding, rolling/overturning, and saltation lifting movements were observed in video records, and the main displacements were observed during great wave flow impacting on the shore platform, with Eastern and Southeastern directions, characterized by a minimum flow of 1.2 m/s and a maximum flow of 4 m/s. Considering the relationships of Nandasena et al. [32], the wave flow needed for boulders movements has been assessed, obtaining wave flow values for subaerial/submerged scenario span from 1.96–4.78 m/s in sliding, 3.75–9.73 m/s in rolling/overturning, 5.64–12.63 m/s in saltation lifting and wave flow values in joint bounded scenario span from 5.8–12.76 m/s (Figure 11). In particular, focus was given on boulders BN, B2, B3, B4, K (Figure 12) in submerged scenario, selecting the five most clearly visible movements (one video for each boulder, see Table 3 and Supplementary Materials Video S1: Boulder B2; Video S2: Boulder B3;

Video S3: Boulder B4; Video S4: Boulder BN; Video S5: Boulder K) in order to estimate the wave flow in combination with type of boulder movement. These boulders present volume values spanning from 0.37 m^3 to 0.83 m^3 with masses estimated between 763 kg and 1689 kg. In each one, sliding, rolling/overturning, saltation/lifting movements were detected by video records in combination with fragmentation as well. For the boulder K, about 41 ton in weight, a movement with a displacement of 1 m between 16:31–16:32 UTC was detected (Figure 13). Although the great turbulence at the time of displacement did not permit us to calculate flow velocity in each single frame of the video, we estimated the flow considering the time that the wave spent to run across the distance existing between the edge of the quarry and the boulder K.

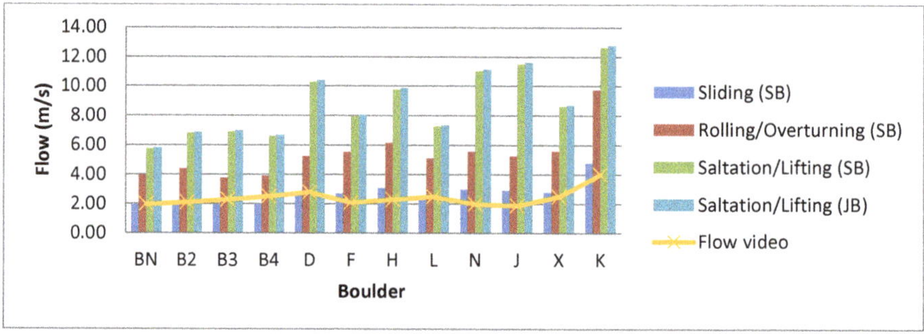

Figure 11. Flow needed to move the boulders assessed with Nandasena et al. [32] relationships and flow obtained by video editing.

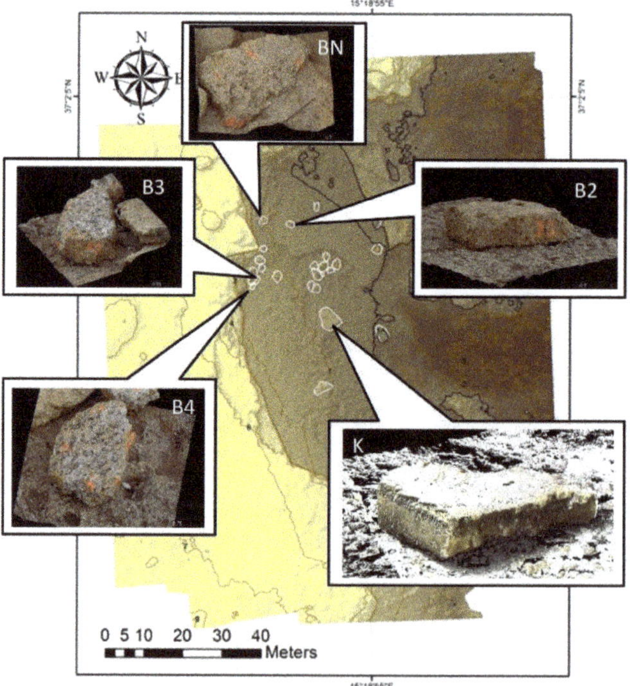

Figure 12. Morpho-topography of the shore-platform showing the boulders features assessed through photogrammetric (B2, B3, B4, BN) and Terrestrial Laser Scanner (TLS) (K) surveys.

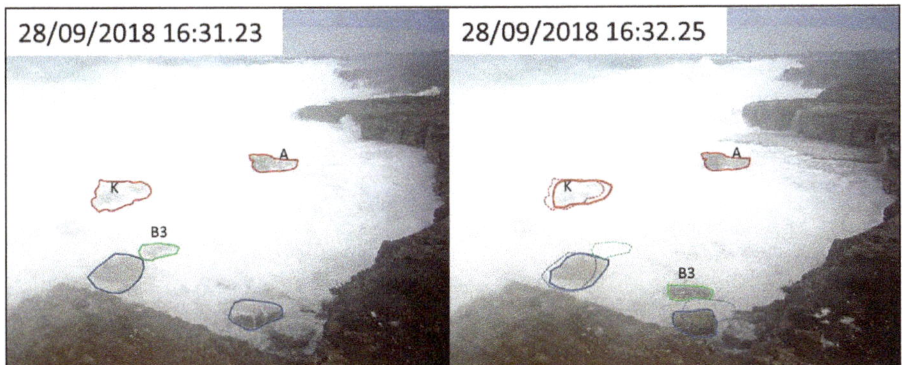

Figure 13. Boulder displacements after a strong wave flow; in green, boulders with volume less than 1 m^3 in blue boulders with volume between 1-5 m^3, in red boulders with volume greater than 5 m^3.

5. Discussions

The Medicane Zorbas caused a very strong storm, characterized by high values of surges and wave height, able to dissipate a lot of flow energy on the coastal zones. The energy dissipated on Maddalena Peninsula was sufficient to move coastal boulders of relevant weight, up to 41 tons. Up to the present, the studies of boulders movements were carried out through a definition of pre and post impact scenarios after the event (e.g., references [19–22,30,75]), but in this work, for the first time, it was possible not only to directly observe how the boulders movements occur, but also to accurately measure some important parameter, as the flow velocity at time of boulder displacement, useful to compare the hydrodynamic model with natural process. From a general analysis of the videos, although all the displaced boulders were already present on the coastline, 90% of the movement occurred only when the boulders were completely submerged by water flows. Single wave impact is rarely responsible of boulder displacements (three cases out of 28), highlighting that it is not correct to connect unique wave height and boulder movements. According to Cox et al. [34], it appears more reasonable that boulder displacements along the coastline occur due to the effects of multiple waves. In our videos, it is evident that multiple waves set up a continuous and turbulent water flow on the coastal area, causing flooding able to submerge the boulders that start the movement.

During the storm on Maddalena Peninsula, flooding generated by multiple waves impact was amplified by a temporary storm surge inducted by the effects of Medicane. The towns of Marzamemi and Santa Croce Camerina, located about 40 km south of the Maddalena Peninsula, registered severe flooding inside the harbor with a sea level increased up to 1 m. In addition, on 28 September 2018, the tide gauge located inside the harbor of Catania recorded anomalous values of sea level—up to 15 cm above normal (Figure 2C). This effect was estimated also in Maddalena Peninsula through a comparison between pre-impact immersive scenario, reconstructed from TLS and UAV photogrammetric data, and images extracted from the videos. This analysis let us reconstruct an increase of water column in Maddalena Peninsula of about 50 cm (Figure 14).

In any case, the main contribution to the flooding that submerged the boulders in the Greek quarry sited in the Maddalena Peninsula is related to the impact of multiple waves, up to nine associated with the displacement of the biggest boulders, which sometimes reached the highest level of the quarry's wall (4 m height). Furthermore, for the boulder K, which is the biggest displaced by the storm, the movement occurred only when 9 multiple waves (wave heights span 0.22–1.3 m in video records) generated a great flow impacted the shore. For boulder K, the Engel and May [36] relationship provides wave height at breaking point needed to start the movement, equal to 11.56 m, much greater than what recorded both by satellite data in off-shore (Hs of about 4.1 m) (source AVISO satellite altimetry, credits CLS/CNES [71]) and by video analysis on the shore-platform.

Figure 14. Flooding generated by Zorbas on the coastal area of the quarry evaluated through the video frames in condition of minimum turbulences.

In the videos, we detected 28 distinct boulders movements; for 10 of these, it was not possible to calculate flow velocity because, at the time of their occurrence, turbulence was too intense to detect certain points in the video images. Moreover, boulders B1, B1_F, and BT were not recognized in field, so we cannot apply the Nandasena et al. [32] model to these boulders. For these reasons we focused our analyses on four little boulders (B2, B3, B4, BN), for which movements occurred through dynamics of sliding, rolling/overturning and saltation/lifting, and one big boulder of 41 tons (boulder K), displaced by a turbulent wave flow. Knowledge of the boulder features, obtained by TLS (K) and photogrammetric (B2, B3, B4, BN) surveys, permitted us to assess the flow needed to displace the boulders for each three types of movements by means of the hydrodynamic equations of Nandasena et al. [32], while the video records allowed us to determine the wave flow velocity at the moment of starting movement with software Tracker. Flow velocity assessed through software Tracker is subjected to relative error dependent on perspective distortion and frames sampling. This error ranges in value between 15% in case of clearly visible flows and 50% in case of turbulent flows that do not permit to identify reference points in the frame of the video. Although the use of a flow meter located inside the quarry could contribute to better evaluate flow velocity, the absence of solid points to anchor the instruments does not permit us to use this technique. For this reason, video editing was the most suitable tool for estimating the wave flow velocity in this case of study.

We selected for each boulder the movements with the highest value of flow velocity measured in the video and compared these values with those calculated with the Nandasena et al. [32] model. Comparison shows that the models provide flow values greater than those measured on the video at the starting movements. The discrepancy is probably related to the fact that the Nandasena et al. [32] model considers equal the flow velocity required to move a boulder in a subaerial and submerged scenario, while evidence highlighted from the analyses of the videos suggested that it is easiest for multiple waves to move a submerged boulder. This could be explained considering that a wave flow able to move a boulder owns a density that is different than normal seawater because it is made by a mix of seawater and sediments. In the literature, the standard density value of seawater is attributed at the wave flow, while the recent work of Terry and Malik [77] has demonstrated that the correct density value is due to the mix of seawater and sediments.

We made different evaluations of the flow velocity based on the Nandasena et al. [32] model considering the different ρ_m values (Table 4) in function of sands fraction. Although assessment of model flow, considering a 20% sand fraction with 80% of seawater fraction, shows a better fitting between calculated and observed flow velocity values in hydrodynamic relationships, and they still show an important disagreement (Figure 15).

Table 4. The Terry and Malik [77] relationship between seawater fraction (f_w) and sand fraction (f_s) connected to the average density of the seawater and sediment mix (ρ_m).

f_w	f_s	$\rho_m (kg/m^3)$
1	0	1025
0.9	0.1	1082.5
0.8	0.2	1140
0.7	0.3	1197.5
0.6	0.4	1255
0.5	0.5	1312.5
0.4	0.6	1370
0.3	0.7	1427.5
0.2	0.8	1485
0.1	0.9	1542.5

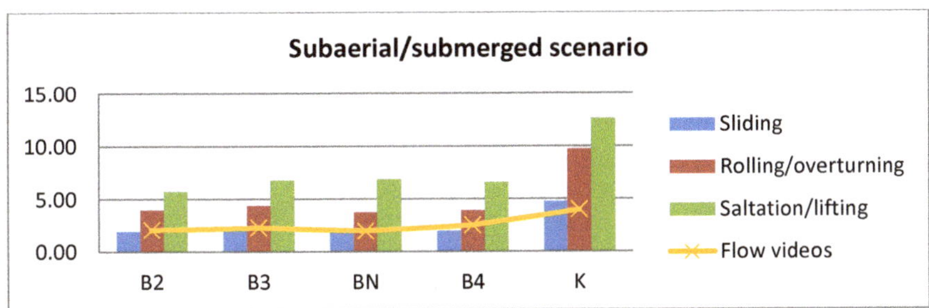

Figure 15. Flow needed to move the boulders, evaluated using Nandasena et al. [32] relationships, included correction for values of water flow density, compared with flow derived by video analysis.

Another parameter that has probably to be redefined is the lift coefficient. The friction forces encountered by the wave flow are conditioned by this coefficient that in literature is equal to 0.178. Recent work by Rovere et al. [37] and Cox et al. [34] show that the value of lift coefficient used in literature was underestimated. We used Nandasena et al. [32] equations to estimate the value of lift coefficient necessary to fill the gap existing between flow velocities calculated, included of density correction, with those observed in the video. In the specific case of the Maddalena Peninsula, we estimated a corrected lift coefficient equal to 1.4.

Another consideration on Boulder A (Figure 9A,B) located in the Southern area of the quarry. This boulder of about 19 ton in weight was interpreted in Scicchitano et al. [16,54] as the result of the impact of a past tsunami event. Although it is much smaller than Boulder K (41 ton in weight), it was not displaced during the three storms that occurred in last 10 years. Boulder A is resulted in the same position (measured with centimetric accuracy trough RTK-GPS and TLS techniques) since the beginning of monitoring in 2007. In the same period Boulder K, that is double in weight, was displaced at least two times (Figure 9). Analyses of the videos, together with comparison of the 3D model of the boulders, suggested that the axis orientation and the contact surface geometry between boulders and bedrock play an important role in the displacements of boulder. Boulder K exposed its longer axis to the main wave impact direction (from East), while Boulder A face has been showed the same direction with the shorter axis. Moreover, the face of the Boulder A, directly in contact with the surface of the quarry, is flat and very regular (it was probably part of the quarry detached and transported inland by an extreme event), resulting in high value of friction. In contrast, the surface of Boulder K is heavily karstified and exposes several erosional polls up to 1 m of diameter and 40 cm depth. These characteristics led to the water flow, induced by the impact of multiple waves, to generate turbulences below Boulder K that are minimized for Boulder A. These considerations support the hypothesis that a tsunami could be reasonably responsible of the displacement of Boulder A [15,16].

Last but not least, analyses of the three storms that impacted Maddalena Peninsula in 2009, 2014 and 2018, suggest that, although medicanes generated wave heights compatible with the annual storm regime of the area (Figure 2A), their impacts produced major effects. Qendresa and Zorbas displaced dozens of blocks up to 2 ton in weight in the quarry of Maddalena Peninsula; most of these were eroded from the coastline and transported landward. The 2009 storm dislocated only one boulder that, although is the biggest located inside the quarry, at that time, it was lying very close to sea in a very unstable position (Figure 9). Major effects produced by medicanes could be explained by the storm surges (up to 1 m in the case of Zorbas) that they induced on coastal area as registered by the tide gauge of Catania and observed in the video recorded on Maddalena Peninsula. As a consequence, coastal areas are probably more vulnerable to the impact of medicanes than to common storms, and this appears more significant when taking into account the theories claiming that, in the future, medicanes will increase in strength in response to climate changes [12,13,78].

6. Conclusions

The dynamics of boulder displacements, in response to the impact of a storm event, are always a difficult aspect to study due to the lack of evidence about the modality of the movements. Although surveys pre and post events allow analysis the entity of boulders displacements, video records represent the best solution for a deeper comprehension of the dynamics correlated to this natural process. In this study, we present the first known video records of boulder displacements, which occurred inside an ancient costal quarry located in the Maddalena Peninsula (Siracusa). Using different techniques of survey and remote sensing, we reconstructed pre and post immersive virtual scenario that were used to geometrically analyze the video and to calculate flow velocity and wave heights at the time of boulder displacements. General observations, together with comparison between data measured in the video and calculated with hydrodynamic model, suggested us the following conclusions:

(1)   Boulder displacements occur mainly in submerged scenario, as consequence of flooding generated by the impact of a series of waves. Although some hydrodynamic models equal the flow velocity necessary to move a boulder in sub-aerial and submerged scenario, our evidence suggest they should probably be described with different specific equations.

(2)   Movements occur for the impact of multiple small waves rather than of a singular big one; in any case, the possibility that a singular big wave could displace a boulder is not excluded at all, but it is probably more attributable to tsunami events. Multiple waves generate a continuous flow that nullifies friction forces, triggering the boulder displacements. Modelling through the Engel and May [36] approach results in wave height values (11.56 m) much higher than those recorded by satellite data in off-shore (Hs of about 4.1 m) and by video analysis on the shore platform (0.22–1.3 m). This confirms that single impact models provide values of wave heights strongly overestimated, respect to the natural process.

(3)   Analyses performed for five displacements, in particular, let us compare values of flow velocity estimated from videos with those calculated through Nandasena et al. [32] model for the same boulders. Comparison shows a strong overestimation of the model, enforcing the thesis that sub-aerial and submerged scenarios have to be treated with different equations and suggesting

that the values of flow density and lift coefficient used in literature are underestimated. For the case of the boulders movements occurred in the Maddalena Peninsula, we estimated a value of flow density of about 1140 kg/m^3, according to Terry and Malik [77] and a different value of lift coefficient according to Rovere et al. [37] and Cox et al. [34].

(4)   Considerations on a big boulder not displaced by the impact of the storm in 2009 or by the impact of the two medicanes—Qendresa (2014) and Zorbas (2018)—suggest that a tsunami could be reasonably responsible for the deposition of the boulders [15,16]. This seems to confirm that, although some authors considered overestimated the number of tsunami reconstructed for Mediterranean Sea [25], it is possible to define field evidence and methodological analyses to discern tsunami and storm events [26].

Analyses of the monitored data suggest that medicanes produce greater effects than common storms; this is probably due to the flooding they induce on the coastal areas and represents an important aspect to investigate to properly assess the vulnerability of the coasts.

Supplementary Materials: The following are available online at http://www.mdpi.com/2073-4441/12/5/1497/s1, Video S1: Boulder B2; Video S2: Boulder B3; Video S3: Boulder B4; Video S4: Boulder BN; Video S5: Boulder K.

Author Contributions: Conceptualization: G.S. (Giovanni Scicchitano), G.S. (Giovanni Scardino), G.M. (Giuseppe Mastronuzzi), A.P., M.M., and C.M.; Data curation: G.S. (Giovanni Scicchitano), G.S. (Giovanni Scardino), G.B., and G.L.; Formal analysis: G.S. (Giovanni Scicchitano), G.S. (Giovanni Scardino), and A.P.; Funding acquisition: G.M. (Giuseppe Mastronuzzi) and G.S. (Giovanni Scicchitano); Investigation: G.S. (Giovanni Scicchitano), S.T., G.M. (Gianfranco Mazza), G.S. (Giovanni Scardino), G.M. (Giuseppe Mastronuzzi), A.P., and M.M.; Methodology: G.S. (Giovanni Scicchitano), G.S. (Giovanni Scardino), G.M. (Giuseppe Mastronuzzi), and A.P.; Project administration: G.S. (Giovanni Scicchitano) and G.M. (Giuseppe Mastronuzzi); Resources: G.M. (Giuseppe Mastronuzzi), G.S. (Giovanni Scicchitano), A.P., M.M., and G.M. (Gianfranco Mazza); Software: G.B. and G.L.; Visualization: G.S. (Giovanni Scicchitano) and G.S. (Giovanni Scardino); Writing—original draft: G.S. (Giovanni Scicchitano), G.S. (Giovanni Scardino), and G.M. (Giuseppe Mastronuzzi); Writing—review and editing: G.S. (Giovanni Scicchitano), G.S. (Giovanni Scardino), G.M. (Giuseppe Mastronuzzi), M.M., C.M., and G.M. (Gianfranco Mazza). All authors have read and agreed to the published version of the manuscript.

Funding: Field operations were supported by Studio Geologi Associati T.S.T. (Catania, Italy) that provided survey instruments (UAV, GPS RTK, Total Station) and technical support.

Acknowledgments: Many thanks are due to Marine Protected Area of Plemmirio (SR, Italy) for providing the video recorded from surveillance camera and for its logistic support during the survey operation with the UAV. Thanks are also due to Dott.ssa Antonia Scicchitano of Sialab s.r.l. (Avola) for the analyses of boulders density. This study has motivated by discussion at the workshops of MOPP-MEDFLOOD (INQUA project 1603P) held in Siracusa in September 2018.

Conflicts of Interest: The authors declare no conflict of interest. The funders had no role in the design of the study; in the collection, analyses, or interpretation of data; in the writing of the manuscript, or in the decision to publish the results.

References

1. Ernst, J.A.; Matson, M. A Mediterranean Tropical Storm? *Weather* **1983**, *38*, 332–337. [CrossRef]
2. Reale, O.; Atlas, R. Tropical Cyclone—Like Vortices in the Extratropics: Observational Evidence and Synoptic Analysis. *Weather Forecast.* **2001**, *16*, 7–34. [CrossRef]
3. Jansà, A. Miniciclons a la Mediterrània. *IX Jornades Meteorol. Eduard Fontserè Assoc. Catalana Meteorol. (ACAM) Barc.* **2003**, *9*, 75–85.
4. Fita, L.; Romero, R.; Luque, A.; Emanuel, K.; Ramis, C. Analysis of the environments of seven Mediterranean tropical-like storms using an axisymmetric, nonhydrostatic, cloud resolving model. *Nat. Hazards Earth Syst. Sci.* **2007**, *7*, 41–56. [CrossRef]
5. Miglietta, M.M.; Laviola, S.; Malvaldi, A.; Conte, D.; Levizzani, V.; Price, C. Analysis of tropical-like cyclones over the Mediterranean Sea through a combined modeling and satellite approach. *Geophys. Res. Lett.* **2013**, *40*, 2400–2405. [CrossRef]
6. Cavicchia, L.; von Storch, H.; Gualdi, S. Mediterranean Tropical-Like Cyclones in Present and Future Climate. *J. Clim.* **2014**, *27*, 7493–7501. [CrossRef]

7.   Nastos, P.T.; Karavana-Papadimou, K.; Matsangouras, I.T. Tropical-like cyclones in the Mediterranean: Impacts and composite daily means and anomalies of synoptic conditions. In Proceedings of the 14th International Conference on Environmental Science and Technology, Rhodes, Greece, 3–5 September 2015.

8.   Cioni, G.; Malguzzi, P.; Buzzi, A. Thermal structure and dynamical precursor of a Mediterranean tropical-like cyclone. *Q. J. R. Meteorol. Soc.* **2016**, *142*, 1757–1766. [CrossRef]

9.   Moscatello, A.; Miglietta, M.M.; Rotunno, R. Observational analysis of a Mediterranean "hurricane" over south-eastern Italy. *Weather* **2008**, *63*, 306–311. [CrossRef]

10.   Pytharoulis, I.; Craig, G.C.; Ballard, S.P. The hurricane-like Mediterranean cyclone of January 1995. *Meteorol. Appl.* **2000**, *7*, 261–279. [CrossRef]

11.   Bakkensen, L.A. Mediterranean Hurricanes and Associated Damage Estimates. *J. Extrem. Events* **2017**, *4*, 1750008. [CrossRef]

12.   Lionello, P.; Giorgi, F. Winter precipitation and cyclones in the Mediterranean region: Future climate scenarios in a regional simulation. *Adv. Geosci.* **2007**, *12*, 153–158. [CrossRef]

13.   Lionello, P.; Conte, D.; Marzo, L.; Scarascia, L. The contrasting effect of increasing mean sea level and decreasing storminess on the maximum water level during storms along the coast of the Mediterranean Sea in the mid 21st century. *Glob. Planet. Chang.* **2017**, *151*, 80–91. [CrossRef]

14.   Portmann, R.; González-Alemán, J.J.; Sprenger, M.; Wernli, H. Medicane Zorbas: Origin and impact of an uncertain potential vorticity streamer. *Weather Clim. Dyn. Discuss.* **2019**, *2019*, 1–30.

15.   Scicchitano, G.; Monaco, C.; Tortorici, L. Large boulder deposits by tsunami waves along the Ionian coast of south-eastern Sicily (Italy). *Mar. Geol.* **2007**, *238*, 75–91. [CrossRef]

16.   Scicchitano, G.; Pignatelli, C.; Spampinato, C.R.; Piscitelli, A.; Milella, M.; Monaco, C.; Mastronuzzi, G. Terrestrial Laser Scanner techniques in the assessment of tsunami impact on the Maddalena peninsula (south-eastern Sicily, Italy). *Earthplanets Space* **2012**, *64*, 889–903. [CrossRef]

17.   Vacchi, M.; Rovere, A.; Zouros, N.; Firpo, M. Assessing enigmatic boulder deposits in NE Aegean Sea: Importance of historical sources as tool to support hydrodynamic equations. *Nat. Hazards Earth Syst. Sci.* **2012**, *12*, 1109–1118. [CrossRef]

18.   Shah-Hosseini, M.; Morhange, C.; De Marco, A.; Wante, J.; Anthony, E.J.; Sabatier, F.; Mastronuzzi, G.; Pignatelli, C.; Piscitelli, A. Coastal boulders in Martigues, French Mediterranean: Evidence for extreme storm waves during the Little Ice Age. *Z. Geomorphol.* **2013**, *57*, 181–199. [CrossRef]

19.   Biolchi, S.; Furlani, S.; Antonioli, F.; Baldassini, N.; Causon Deguara, J.; Devoto, S.; Di Stefano, A.; Evans, J.; Gambin, T.; Gauci, R.; et al. Boulder accumulations related to extreme wave events on the eastern coast of Malta. *Nat. Hazards Earth Syst. Sci.* **2016**, *16*, 737–756. [CrossRef]

20.   Biolchi, S.; Furlani, S.; Devoto, S.; Scicchitano, G.; Korbar, T.; Vilibic, I.; Sepic, J. The origin and dynamics of coastal boulders in a semi-enclosed shallow basin: A northern Adriatic case study. *Mar. Geol.* **2019**, *411*, 62–77. [CrossRef]

21.   Biolchi, S.; Denamiel, C.; Devoto, S.; Korbar, T.; Macovaz, V.; Scicchitano, G.; Vilibić, I.; Furlani, S. Impact of the October 2018 Storm Vaia on Coastal Boulders in the Northern Adriatic Sea. *Water* **2019**, *11*, 2229. [CrossRef]

22.   Piscitelli, A.; Milella, M.; Hippolyte, J.-C.; Shah-Hosseini, M.; Morhange, C.; Mastronuzzi, G. Numerical approach to the study of coastal boulders: The case of Martigues, Marseille, France. *Quat. Int.* **2017**, *439*, 52–64. [CrossRef]

23.   Mastronuzzi, G.; Sansò, P. Large boulder accumulations by extreme waves along the Adriatic coast of southern Apulia (Italy). *Quat. Int.* **2004**, *120*, 173–184. [CrossRef]

24.   Barbano, M.S.; Pirrotta, C.; Gerardi, F. Large boulders along the south-eastern Ionian coast of Sicily: Storm or tsunami deposits? *Mar. Geol.* **2010**, *275*, 140–154. [CrossRef]

25.   Marriner, N.; Kaniewski, D.; Morhange, C.; Flaux, C.; Giaime, M.; Vacchi, M.; Goff, J. Tsunamis in the geological record: Making waves with a cautionary tale from the Mediterranean. *Sci. Adv.* **2017**, *3*, e1700485. [CrossRef]

26.   Vött, A.; Bruins, H.J.; Gawehn, M.; Goodman-Tchernov, B.N.; De Martini, P.M.; Kelletat, D.; Mastronuzzi, G.; Reicherter, K.; Röbke, B.R.; Scheffers, A.; et al. Publicity waves based on manipulated geoscientific data suggesting climatic trigger for majority of tsunami findings in the mediterranean—Response to 'tsunamis in the geological record: Making waves with a cautionary tale from the mediterranean' by marriner et al. (2017). *Z. Geomorphol.* **2019**, *62*, 7–45.

27. Mastronuzzi, G.; Pignatelli, C.; Sansò, P. Boulder fields: A valuable morphological indicator of palaeotsunami in the Mediterranean sea. *Z. Geomorphol. NF* **2006**, *146*, 173–194.

28. Nott, J. Tsunami or Storm Waves? Determining the Origin of a Spectacular Field of Wave Emplaced Boulders Using Numerical Storm Surge and Wave Models and Hydrodynamic Transport Equations. *J. Coast. Res.* **2003**, *19*, 348–356.

29. Nott, J. Waves, coastal boulder deposits and the importance of the pre-transport setting. *Earth Planet. Sci. Lett.* **2003**, *210*, 269–276. [CrossRef]

30. Pignatelli, C.; Sansò, P.; Mastronuzzi, G. Evaluation of tsunami flooding using geomorphologic evidence. *Mar. Geol.* **2009**, *260*, 6–18. [CrossRef]

31. Nandasena, N.A.K.; Paris, R.; Tanaka, N. Numerical assessment of boulder transport by the 2004 Indian ocean tsunami in Lhok Nga, West Banda Aceh (Sumatra, Indonesia). *Comput. Geosci.* **2011**, *37*, 1391–1399. [CrossRef]

32. Nandasena, N.A.K.; Paris, R.; Tanaka, N. Reassessment of hydrodynamic equations: Minimum flow velocity to initiate boulder transport by high energy events (storms, tsunamis). *Mar. Geol.* **2011**, *281*, 70–84. [CrossRef]

33. Milella, M.; Scardino, G.; Piscitelli, A.; De Giosa, F.; Locuratolo, G.; Barracane, G. Experimental determination of the friction coefficient for estimating sea storm induced megaboulders movements. *Alp. Mediterr. Quat.* **2018**, *31*, 27–30.

34. Cox, R.; Ardhuin, F.; Dias, F.; Autret, R.; Beisiegel, N.; Earlie, C.S.; Herterich, J.G.; Kennedy, A.; Paris, R.; Raby, A.; et al. Systematic Review Shows That Work Done by Storm Waves Can Be Misinterpreted as Tsunami-Related Because Commonly Used Hydrodynamic Equations Are Flawed. *Front. Mar. Sci.* **2020**, *7*, 7. [CrossRef]

35. Mastronuzzi, G.; Pignatelli, C. The boulder berm of Punta Saguerra (Taranto, Italy): A morphological imprint of the Rossano Calabro tsunami of April 24, 1836? *Earthplanets Space* **2012**, *64*, 829–842. [CrossRef]

36. Engel, M.; May, S.M. Bonaire's boulder fields revisited: Evidence for Holocene tsunami impact on the Leeward Antilles. *Quat. Sci. Rev.* **2012**, *54*, 126–141. [CrossRef]

37. Rovere, A.; Casella, E.; Harris, D.L.; Lorscheid, T.; Nandasena, N.A.K.; Dyer, B.; Sandstrom, M.R.; Stocchi, P.; D'Andrea, W.J.; Raymo, M.E. Giant boulders and Last Interglacial storm intensity in the North Atlantic. *Proc. Natl. Acad. Sci. USA* **2017**, *114*, 12144–12149. [CrossRef]

38. Grasso, M.; Lentini, F. Sedimentary and tectonic evolution of the eastern Hyblean Plateau (southeastern Sicily) during late Cretaceous to Quaternary time. *Palaeogeogr. Palaeoclimatol. Palaeoecol.* **1982**, *39*, 261–280. [CrossRef]

39. Cultrera, F.; Barreca, G.; Scarfi, L.; Monaco, C. Fault reactivation by stress pattern reorganization in the Hyblean foreland domain of SE Sicily (Italy) and seismotectonic implications. *Tectonophysics* **2015**, *661*, 215–228. [CrossRef]

40. Bianca, M.; Monaco, C.; Tortorici, L.; Cernobori, L. Quaternary normal faulting in southeastern Sicily (Italy): A seismic source for the 1693 large earthquake. *Geophys. J. Int.* **1999**, *139*, 370–394. [CrossRef]

41. Scandone, P.; Patacca, E.; Radoicic, R.; Ryan, W.B.F.; Cita, M.B.; Rawson, M.; Chezar, H.; Miller, E.; McKenzie, J.; Rossi, S. Mesozoic and Cenozoic Rocks from Malta Escarpment (Central Mediterranean). *AAPG Bull.* **1981**, *65*, 1299–1319.

42. Makris, J.; Nicolich, R.; Weigel, W. A seismic study in the Western Ionian Sea. *Ann. Geophys.* **1986**, *4*, 665–678.

43. Catalano, R.; Doglioni, C.; Merlini, S. On the Mesozoic Ionian Basin. *Geophys. J. Int.* **2001**, *144*, 49–64. [CrossRef]

44. Baratta, M. *I Terremoti d'Italia: Saggio di Storia, Geografia e Bibliografia Sismica Italiana*; Fratelli Bocca: Torino, Italy, 1901; p. 984.

45. Postpischl, D.C. *Catalogo dei Terremoti Italiani Dall'Anno 1000 al 1980*; Postpischl, D., Ed.; Consiglio Nazionale Delle Ricerche, Progetto Finalizzato Geodinamica, Sottoprogetto Rischio Sismico e Ingegneria Sismica: Bologna, Italy, 1985; p. 239.

46. Boschi, E.; Guidoboni, E.; Ferrari, G.; Valensise, G.; Gasperini, P. *Catalogo dei forti terremoti in Italia dal 461 a.C. al 1990*; ING-SGA: Bologna, Italy, 1997; p. 644.

47. Tinti, S.; Maramai, A.; Graziani, L. The New Catalogue of Italian Tsunamis. *Nat. Hazards* **2004**, *33*, 439–465. [CrossRef]

48. Scicchitano, G.; Costa, B.; Di Stefano, A.; Longhitano, S.G.; Monaco, C. Tsunami and storm deposits preserved within a ria-type rocky coastal setting (Siracusa, SE Sicily). *Z. Geomorphol. Suppl. Issues* **2010**, *54*, 51–77. [CrossRef]

49. Gerardi, F.; Smedile, A.; Pirrotta, C.; Barbano, M.S.; Martini, P.M.D.; Pinzi, S.; Gueli, A.M.; Ristuccia, G.M.; Stella, G.; Troja, S.O. Geological record of tsunami inundations in Pantano Morghella (south-eastern Sicily) both from near and far-field sources. *Nat. Hazards Earth Syst. Sci.* **2012**, *12*, 1185–1200. [CrossRef]

50. Di Grande, A.; Raimondo, W. Linee di costa pliopleistoceniche e schema litostratigrafico del Quaternario siracusano. *Geol. Romana* **1982**, *21*, 279–309.

51. Ferranti, L.; Antonioli, F.; Mauz, B.; Amorosi, A.; Dai Pra, G.; Mastronuzzi, G.; Monaco, C.; Orrù, P.; Pappalardo, M.; Radtke, U.; et al. Markers of the last interglacial sea-level high stand along the coast of Italy: Tectonic implications. *Quat. Int.* **2006**, *145*, 30–54. [CrossRef]

52. Ferranti, L.; Antonioli, F.; Anzidei, M.; Monaco, C.; Stocchi, P. The timescale and spatial extent of vertical tectonic motions in Italy: Insights from relative sea-level changes studies. *J. Virtual Explor.* **2010**, *36*, 36. [CrossRef]

53. Dutton, A.; Scicchitano, G.; Monaco, C.; Desmarchelier, J.M.; Antonioli, F.; Lambeck, K.; Esat, T.M.; Fifield, L.K.; McCulloch, M.T.; Mortimer, G. Uplift rates defined by U-series and 14C ages of serpulid-encrusted speleothems from submerged caves near Siracusa, Sicily (Italy). *Quat. Geochronol.* **2009**, *4*, 2–10. [CrossRef]

54. Scicchitano, G.; Antonioli, F.; Berlinghieri, E.F.C.; Dutton, A.; Monaco, C. Submerged archaeological sites along the Ionian coast of southeastern Sicily (Italy) and implications for the Holocene relative sea-level change. *Quat. Res.* **2008**, *70*, 26–39. [CrossRef]

55. Scicchitano, G.; Spampinato, C.R.; Antonioli, F.; Anzidei, M.; Presti, V.L.; Monaco, C. Comparing ancient quarries in stable and slowly uplifting coastal area located in Eastern Sicily, Italy. *Geogr. Fis. Din. Quat.* **2018**, *41*, 81–92.

56. Scicchitano, G.; Berlinghieri, E.F.C.; Antonioli, F.; Spampinato, C.R.; Monaco, C. Sacred Landscapes and Changing Sea Levels: New Interdisciplinary Data from the Early Neolithic to the Present in South-Eastern Sicily. In *Under the Sea: Archaeology and Palaeolandscapes of the Continental Shelf*; Bailey, G.N., Harff, J., Sakellariou, D., Eds.; Coastal Research Library, Springer International Publishing: Cham, Switzerland, 2017; pp. 233–253. ISBN 978-3-319-53160-1.

57. Spampinato, C.R.; Costa, B.; Di Stefano, A.; Monaco, C.G.; Scicchitano, G. The contribution of tectonics to relative sea-level change during the Holocene in south-eastern Sicily coastal area: New data from boreholes. *Quat. Int.* **2011**, *232*, 214–227. [CrossRef]

58. Anzidei, M.; Scicchitano, G.; Tarascio, S.; de Guidi, G.; Monaco, C.; Barreca, G.; Mazza, G.; Serpelloni, E.; Vecchio, A. Coastal retreat and marine flooding scenario for 2100: A case study along the coast of Maddalena peninsula (southeastern Sicily). *Geogr. Fis. Din. Quat.* **2018**, *41*, 5–16.

59. Le Cozannet, G.; Nicholls, R.J.; Hinkel, J.; Sweet, W.V.; McInnes, K.L.; Van de Wal, R.S.W.; Slangen, A.B.A.; Lowe, J.A.; White, K.D. Sea Level Change and Coastal Climate Services: The Way Forward. *J. Mar. Sci. Eng.* **2017**, *5*, 49. [CrossRef]

60. *IPCC Climate Change 2014: Synthesis Report. Contribution of Working Groups I, II and III to the Fifth Assessment Report of the Intergovernmental Panel on Climate Change*; Core Writing Team; Pachauri, R.K.; Meyer, L.A. (Eds.) IPCC: Geneva, Switzerland, 2014; p. 151.

61. *IPCC Special Report on the Ocean and Cryosphere in a Changing Climate*; Pörtnerd, H.-O.; Robertsv, C.; Masson-Delmottep, V.; Zhaim, P.; Tignore, M.; Poloczanskak, E.; Mintenbeck, K.; Alegría, A.; Nicolai, M.; Okem, A.; et al. (Eds.) 2019; in press.

62. Vecchio, A.; Anzidei, M.; Serpelloni, E.; Florindo, F. Natural Variability and Vertical Land Motion Contributions in the Mediterranean Sea-Level Records over the Last Two Centuries and Projections for 2100. *Water* **2019**, *11*, 1480. [CrossRef]

63. Smedile, A.; De Martini, P.M.; Pantosti, D.; Bellucci, L.; Del Carlo, P.; Gasperini, L.; Pirrotta, C.; Polonia, A.; Boschi, E. Possible tsunami signatures from an integrated study in the Augusta Bay offshore. *Mar. Geol.* **2011**, *281*, 1–13. [CrossRef]

64. De Martini, P.M.; Barbano, M.S.; Smedile, A.; Gerardi, F.; Pantosti, D.; Del Carlo, P.; Pirrotta, C. A unique 4000 year long geological record of multiple tsunami inundations in the Augusta Bay (eastern Sicily, Italy). *Mar. Geol.* **2010**, *276*, 42–57. [CrossRef]

65. De Martini, P.M.; Barbano, M.S.; Pantosti, D.; Smedile, A.; Pirrotta, C.; Del Carlo, P.; Pinzi, S. Geological evidence for paleotsunamis along eastern Sicily (Italy): An overview. *Nat. Hazards Earth Syst. Sci.* **2012**, *12*, 2569–2580. [CrossRef]

66. Scardino, G.; Piscitelli, A.; Milella, M.; Sansò, P.; Mastronuzzi, G. Tsunami fingerprints along the Mediterranean coasts. *Rend. Lincei* **2020**, 1–17. [CrossRef]

67. Fago, P.; Pignatelli, C.; Piscitelli, A.; Milella, M.; Venerito, M.; Sansò, P.; Mastronuzzi, G. WebGIS for Italian tsunami: A useful tool for coastal planners. *Mar. Geol.* **2014**, *355*, 369–376. [CrossRef]

68. Available online: www.idromare.com (accessed on 1 April 2015).

69. Inglesi, R.; Corsini, S.; Guiducci, F.; Arseni, A. Statistical analysis of extreme waves on the Italian coasts from 1989 to 1999. *Boll. Geofis. Teor. Appl.* **2000**, *41*, 315–337.

70. RON ISPRA. Available online: http://dati.isprambiente.it/dataset/ron-rete-ondametrica-nazionale/ (accessed on 4 May 2020).

71. MSS: Aviso+. Available online: https://www.aviso.altimetry.fr/en/data/products/auxiliary-products/mss.html (accessed on 4 May 2020).

72. RETE MAREOGRAFICA NAZIONALE—HOMEPAGE. Available online: https://mareografico.it/ (accessed on 4 April 2020).

73. Tracker Video Analysis and Modeling Tool for Physics Education. Available online: https://physlets.org/tracker/ (accessed on 4 May 2020).

74. Casella, E.; Rovere, A.; Pedroncini, A.; Stark, C.P.; Casella, M.; Ferrari, M.; Firpo, M. Drones as tools for monitoring beach topography changes in the Ligurian Sea (NW Mediterranean). *Geo-Mar. Lett.* **2016**, *36*, 151–163. [CrossRef]

75. Pepe, F.; Corradino, M.; Parrino, N.; Besio, G.; Presti, V.L.; Renda, P.; Calcagnile, L.; Quarta, G.; Sulli, A.; Antonioli, F. Boulder coastal deposits at Favignana Island rocky coast (Sicily, Italy): Litho-structural and hydrodynamic control. *Geomorphology* **2018**, *303*, 191–209. [CrossRef]

76. Imamura, F.; Goto, K.; Ohkubo, S. A numerical model for the transport of a boulder by tsunami. *J. Geophys. Res. Ocean.* **2008**, *113*, 113. [CrossRef]

77. Terry, J.P.; Malik, S.A. Reconsidering the seawater-density parameter in hydrodynamic flow transport equations for coastal boulders. *New Zealand J. Geol. Geophys.* **2020**, 1–8. [CrossRef]

78. Marcos, M.; Jordà, G.; Gomis, D.; Pérez, B. Changes in storm surges in southern Europe from a regional model under climate change scenarios. *Glob. Planet. Chang.* **2011**, *77*, 116–128. [CrossRef]

Article

Sea-Level Variability in the Gulf of Naples and the *"Acqua Alta"* Episodes in Ischia from Tide-Gauge Observations in the Period 2002–2019

Berardino Buonocore *, Yuri Cotroneo *, Vincenzo Capozzi, Giuseppe Aulicino, Giovanni Zambardino and Giorgio Budillon

Dipartimento di Scienze e Tecnologie, Università degli Studi di Napoli "Parthenope", Ctr Direz Isola C4, 80143 Naples, Italy; vincenzo.capozzi@uniparthenope.it (V.C.); giuseppe.aulicino@uniparthenope.it (G.A.); giovanni.zambardino@uniparthenope.it (G.Z.); giorgio.budillon@uniparthenope.it (G.B.)
* Correspondence: buonocore@uniparthenope.it (B.B.); yuri.cotroneo@uniparthenope.it (Y.C.);
  Tel.: +39-081-547-6572 (B.B.); +39-081-547-6576 (Y.C.)

Received: 28 July 2020; Accepted: 31 August 2020; Published: 2 September 2020

Abstract: This work presents an 18-year-long (2002–2019) tide-gauge dataset collected on the Island of Ischia (Gulf of Naples, Southern Tyrrhenian Sea) that can contribute to the analysis of the basic features of sea-level variability in this region. Analysis of tidal constituents shows that the Gulf of Naples is characterized by the absence of any amphidromic system. In this area, sea-level changes due to the astronomical component of the tide are generally limited to ±20 cm with respect to the mean sea level, but the impact of this variability is enhanced by global sea-level increase and the effect of regional atmospheric perturbations that might also triple sea-level variations. The effects of these events, whose frequency has increased in recent decades, has been dramatic in coastal areas where intense social and economic activity occurs, e.g., in Ischia. On interannual time scales, the results indicate that the relative sea-level rise in Ischia has a magnitude of 3.9 mm/year. Special attention is dedicated to the *"acqua alta"* episodes and to their linkage with long-term sea-level trends and atmospheric forcing.

Keywords: sea level change; tide-gauges; atmospheric variability; Tyrrhenian sea; climate change; interannual variations; acqua alta

---

1. Introduction

Sea level (SL) can change over a wide range of temporal and spatial scales. Even though available in situ observations are characterized by a sparse and uneven spatial distribution [1] that largely limits their use [2], tide-gauge records represent the main source of information about local SL variations from the last two centuries [3–7]. Previously, tide-gauges were mainly used for navigation issues, e.g., to observe and forecast SL variations associated with the tide, but nowadays they are also essential for evaluating oceanographic models, e.g., [8], validating satellite data [9,10], and performing instrumental calibration of satellite missions [11,12], as well as being used in the framework of long-term climatic studies. With growing comprehension of SL variation and its interaction with atmospheric variability, the importance of these records has then enlarged to coastal area management issues. A global SL rising trend has been associated with climate change and human-originated greenhouse effects, e.g., [13–17]. In its fifth assessment report (AR5), the Intergovernmental Panel on Climate Change (IPCC) estimated SL increases of 1.7, 2.0, and 3.2 mm/year in the periods 1901–2010, 1971–2010, and 1993–2010, respectively [18–20], with potential dramatic societal impacts on coastal regions [21]. In particular, AR5 indicates that the Mediterranean Sea is one of the most vulnerable regions in the world in terms of the impacts of climate change and SL rise.

The analysis of the longest tide-gauge series along the Mediterranean Sea coasts indicates a rate of SL rise of 1.1–1.3 mm/year over the 20th century [22]. This trend is not homogeneous, with a period of SL decrease after the 1960s [22] and a fast rise in SL observed since the late 1990s [23,24]. Furthermore, a high-resolution SL time series showed the existence of a seasonal signal superimposed on the long-term trend [19]. Finally, the presence of densely populated areas along the coasts and a high-value economic activity have enhanced these impacts. The Campania Region coast in the Southern Tyrrhenian Sea has been identified as one of the Italian coastal areas that is most susceptible to inundation risk [25–27].

Although estimations of SL trends have been retrieved on the basis of the longest tide-gauges records available, the temporal and spatial distributions of these instruments strongly affect the results. SL can significantly vary from one region to another, and a variety of processes cause local trends to deviate from the global mean rate of change [7,28]. Therefore, it is not straightforward to infer the rate of global ocean volume change from these data. It is necessary to study SL changes on a local or regional basis to determine both the long-term trend and the shorter wavelengths of the signals, as well as to understand the different geophysical processes associated with such changes [13]. These efforts could then help in distinguishing between natural and anthropogenic changes occurring in the Earth's climate system and affecting the livelihoods of millions of people [29]. In this context, the availability of additional tide-gauge data in the Mediterranean Sea and detailed study of the Tyrrhenian sub-basins [30–32] is important to provide more accurate information on basin circulation and on the local SL variability and its trends.

In the Gulf of Naples (GON) in the Southern Tyrrhenian Sea (Figure 1), the University of Naples "Parthenope" manages a large weather and oceanographic monitoring network that includes weather stations, a weather radar, wave-buoys, oceanographic moorings, and a coastal high-frequency radar network. In this framework, since 2002, a tide-gauge has operated on the island of Ischia (IS). In this area, SL variability due to the purely theoretical astronomical component is ±20 cm with respect to the mean SL, but the impact of this variability is enhanced by the global SL increase and the effect of regional atmospheric perturbations that can even triple SL variations. A second tide-gauge is located in the GON at Castellammare di Stabia (CS), but it has only operated since 2011. The locations of both the instruments were planned according to the different geological histories and characteristics of the sites. Ischia is characterized by intense volcanic activity and a complex tectonic history [33], while CS is located next to the Monti Lattari carbonate reliefs that are characterized by greater stability [34].

Together with the Volturno coastal plain, Ischia and Castellammare are characterized by high economic and ecological value, due to the presence of tourist structures, farm activities, and wetland protected zones [27] that increase their vulnerability to sea level rise and storm surge.

In this study, the 18-year IS tide-gauge dataset was carefully analyzed to identify the basic features of SL variability and the interactions with the atmospheric variability and to describe the possible causes of the remarkable *"acqua alta"* ("high water", hereafter AA) episodes in Ischia. To reach this aim, the IS SL data interannual variability was analyzed, and a comparison between IS tidal constants with nearby tide-gauges, including CS, was conducted. The frequency and amplitude of the AA events are discussed. The "Materials and Methods" section describes the tide-gauges, their locations, and the collected observations, as well as the air pressure data and the large scale atmospheric indices included in the analyses. Finally, future perspectives for the analysis of SL variability in the study area and improvements in the tide-gauge network are also described.

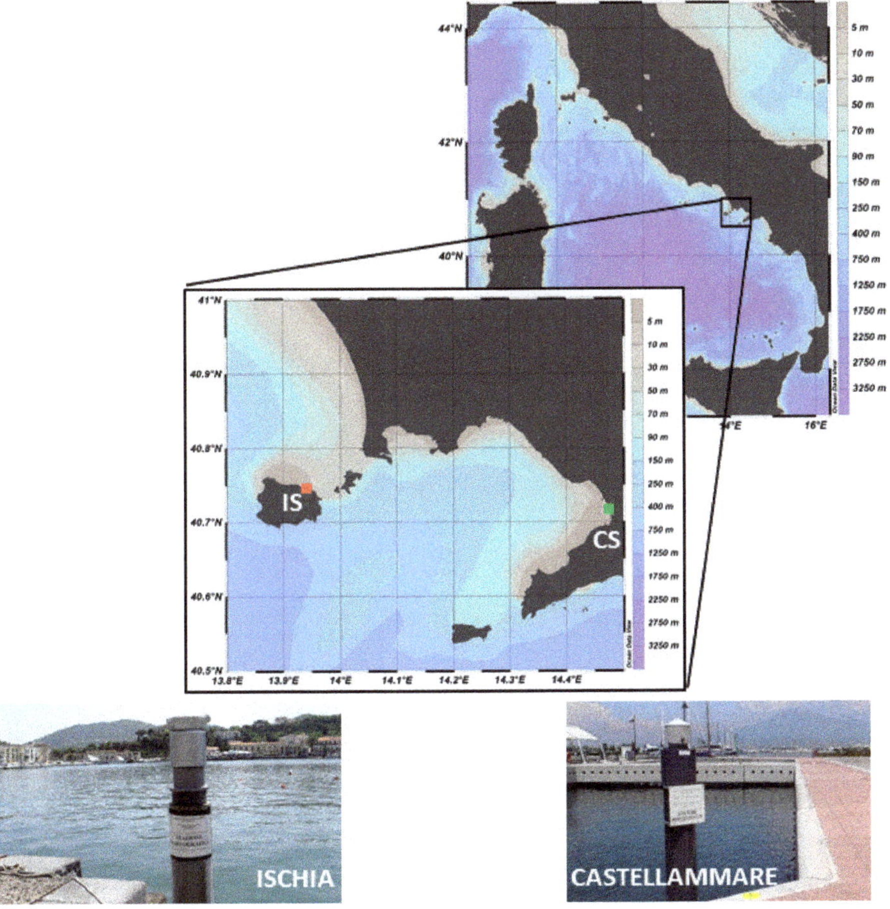

Figure 1. Study area (**upper** panels) and locations of the Ischia (IS—red square) and Castellammare di Stabia (CS—green square) tide-gauges. Bathymetry is indicated through the color scale. The **lower** (**left** and **right**) panels show the IS and CS sites respectively.

2. Materials and Methods

2.1. Tide-Gauge Data

A "Thalimedes" tide-gauge by OTT HydroMet (IS in Figure 1) was installed in 2002 on the island of Ischia (in the GON) by the Department of Science and Technology of the University of Naples "Parthenope." The instrument is located and operates in the northeastern part of the island, inside its main harbor (Lat: 40.745833° N Lon: 13.940556° E) whose entrance faces winds coming from the North. The "Tahlimedes" is a float-operated tide-gauge suitable for continuous water level measurement in both ground and surface water (more details at www.ott.com/). The instrument is placed in a stilling well and SL measurements are made by referring to benchmarks placed on the dock that are not connected to the national leveling network. It is noteworthy that the location of the tide-gauge did not change throughout the entire study period (2002–2019). A regular maintenance plan has been applied on a 6-month basis and is still active.

It is worth noting that the IS tide-gauge is located on the side of the piers used by the Corps of the Port Captaincies from the Coast Guard of Ischia. This pier is 34 cm higher than the old piers located on the right side of the harbor where AA episodes are observed. For this reason, tide-gauge data collected at the Coast Guard pier can efficiently register the AA events.

As stated above, a second "Thalimedes" tide-gauge operated by the University "Parthenope" in the GON was installed more recently (2011) in Castellammare di Stabia (CS in Figure 1). This instrument is located in the touristic harbor of Marina di Stabia (Lat: 40.716111° N Lon: 14.474722° E) and data were used to complement the harmonic analysis of tide constituents derived from IS observations.

Data analysis and quality control procedures were performed in accordance with the guidelines of the Intergovernmental Oceanographic Commission (IOC) [35–37] and the Italian Istituto Superiore per la Protezione e la Ricerca Ambientale (ISPRA) [38,39]. After the removal of episodic spikes and errors, the Pugh filter [40] was used to retrieve the hourly data, and the Doodson filter [41] was used to obtain the daily values. According to the ISPRA guidelines, the *"length"*, *"continuity"*, and *"completeness"* of the time series were estimated.

The *"Length"* (L) of the time series is simply defined as the difference between the year of the last record and the year of the first record plus 1. The *"Continuity"* (*Cy*) parameter gives an indication of the length of the periods characterized by good data with respect to the length of the time series and is expressed as the maximum amount of collectible data. It is defined as

$$Cy = 1 - 2 \times \frac{number\ of\ data\ missing\ intervals}{maximum\ number\ of\ collectable\ data}$$

Cy values near 1 are then associated with time series containing all valid data without any gaps in the records.

The *"Completeness"* (*Co*) gives an indication of the general quality of the collected data, and its value is defined by the equation

$$Co = \frac{number\ of\ valid\ data}{maximum\ number\ of\ collectable\ data}$$

The variability in the surface air pressure is one of the most important elements of meteorological forcing of SL variability over the open ocean as well as near the coast [42]. Due to the inverse barometer (IB) effect, an increase in air pressure of 1 mbar results in a decrease in SL of approximately 1 cm [43]. Furthermore, coastal SL variability on daily to monthly timescales departs significantly from the IB response to air pressure change due to the effect of winds on the coastal areas [42]. In order to remove the atmospheric pressure signal from the SL data, the following equation was used:

$$\Delta h = -\frac{\Delta p}{\rho g}$$

where Δh is the variation in SL due to the IB effect, Δp is the atmospheric pressure variation respect to the long-term local mean (1015.1 mbar) calculated on air pressure data collected in the framework of this work (see Section 2.2), ρ is the mean seawater density for the study area, and g is the gravity.

2.2. Atmospheric Pressure Data and Large Scale Atmospheric Variability Indices

The atmospheric pressure data used in this work for the period 2002–2019 were provided by the University of Naples "Parthenope." Data, with a temporal resolution of 10 min, were collected at the Via Acton weather station for the period from January 2002 to September 2011 and in the new headquarters of the University in Centro Direzionale di Napoli for the period from October 2011 to December 2019. The distance between the two sites is about 3 km. Both weather stations are located in the metropolitan area of Naples and lie at a few meters above SL. The air pressure time series was analyzed and quality controlled according to standard procedures. Within the quality control phases, the air pressure values were related to the SL and compared with hourly data collected at the Napoli

Capodichino Airport, located 6 km away from the Acton and 3 km from Centro Direzionale weather stations, which were provided by the Italian Air Force Weather Service.

Atmospheric data were complemented with wind direction data and speed observations from January to March 2010 collected by ISPRA through its weather station in Naples (Lat: 40.841389° N; Lon: 14.269167° E). These data are available through the webpage https://www.mareografico.it.

The atmospheric variability and its link with SL variation was also investigated through two available climate indices: the North Atlantic Oscillation Index (NAO) and the Mediterranean Oscillation Index (MOI).

The NAO index can be defined by the difference in surface sea-level pressure between the Subtropical High (at Ponta Delgada, Azores) and the Subpolar Low (at Akureyri, Iceland) [44,45]. The literature offers significant evidence of the relationships between NAO phases and the meteorological dynamics of the Mediterranean domain, e.g., [46–50]. The positive phase of the NAO reflects below-normal heights and pressures across the high latitudes of the North Atlantic and above-normal heights and pressure over the central North Atlantic, the eastern United States, and western Europe. The negative phase reflects an opposite pattern of height and pressure anomalies over these regions [51] and is synonymous with an increase in atmospheric transient activity over the western and central Mediterranean basins.

The NAO daily time series used in this study was provided by the Climate Prediction Centre of the National Oceanic and Atmospheric Administration (NOAA). The procedure used to calculate this daily NAO teleconnection index is indeed based on the rotated principal component analysis used by Barnston and Livezey [52]. This procedure isolates the primary teleconnection patterns for all months in order to allow the reconstruction of the time-series patterns. The technique was applied to monthly standardized 500 mbar height anomalies in the analysis region (20–90° N) between January 1950 and December 2000. The monthly teleconnection patterns were linearly interpolated to the day in question and therefore account for the seasonality inherent in the NAO patterns. The daily values were calculated using the same approach, thus representing the combination of teleconnection patterns that accounts for the most spatial variance in the observed anomaly map on any given day (https://www.cpc.ncep.noaa.gov/).

Since the NAO index is based on pressure and geopotential height values west of the European coast, a strong influence of this large-scale pattern on SL via changes in wind stress and sea-level pressure was expected [53]. In particular, NAO variability has been anticorrelated with the southern European mean SL variability through the IB effect and large-scale wind patterns [54] over the coastal areas of Europe and in the Mediterranean Sea, e.g., [54–65].

As highlighted by Brunetti et al. [66], the NAO index explains only a fraction of the central Mediterranean atmospheric variability. Therefore, in our analysis, we also considered the MOI [67–70], which is defined as the normalized pressure difference between Algiers (36.4° N, 3.1° E) and Cairo (30.1° N, 31.4° E). A second version of the index can be calculated from Gibraltar's Northern Frontier (36.1° N, 5.3° W) and Lod Airport in Israel (32.0° N, 34.5° E) [71]. Here, data are presented for both variants, respectively named MOI1 and MOI2, using pressure interpolated (16-point Bessel) from the NCEP/NCAR reanalysis.

The negative phase of MOI is generally associated with an atmospheric pattern characterized by a low-pressure area over the central and western Mediterranean basins; therefore, it describes a general circulation supporting rainy events across a large part of the Italian peninsula. On the contrary, the atmospheric scenario associated with the positive phase of this index has very different features. In fact, a dipole anomaly pattern can be observed, i.e., a negative pressure anomaly on the eastern side of the Mediterranean region and a positive anomaly in the northwestern area. Because of this pattern, dry conditions are generally observed on the northern and western sides of the Italian peninsula during positive MOI events.

2.3. Trend Estimation Methods

SL variability occurs on a large set of temporal scales ranging from hours (minutes and seconds in the case of waves) to years. Several procedures have been developed to distinguish between random fluctuations and more persistent temporal changes. Tests for the detection of significant trends in long time series can be classified as parametric and non-parametric methods. Parametric trend tests require data to be independent and normally distributed, while non-parametric methods only require that the data are independent and can accommodate seasonal variation and serial dependence in a time series. The use of non-parametric methods is widely diffuse in environmental, e.g., [72], meteorological, e.g., [73], and oceanographic, e.g., [74] data.

In this study, the non-parametric Mann–Kendall test was used in order to detect the existence of long-term temporal trends in the IS time series, while the trend magnitude was evaluated by the non-parametric Sen's method.

The basic principle of using Mann–Kendall tests to analyze trends is to examine the signs of all pairwise differences in observed values [75,76]. Through the calculation of the total sum of SL variations for each time step and the analysis of its variance, the Mann–Kendall tests provide an estimation of the so-called "Z" parameter. A positive (negative) value of Z indicates an upward (downward) trend. Z values are compared to a value of 1.96, which should be obtained for a confidence interval (upper and lower) of 95%, referring to a normal distribution.

In cases where a significant trend was detected, Sen's non-parametric method [77] was used to estimate the true slope of the existing trend (as change per time step). To get the slope estimate "Q", the slopes of all data value pairs were first computed, and then the median of the Q values was calculated. For this methodology, missing values are allowed, and the data need not conform to any particular distribution. Furthermore, Sen's method is not greatly affected by single data errors or outliers.

3. Results and Discussion

3.1. Tide-Gauge Data Preliminary Analysis and Tidal Constituents

The IS tide-gauge began operating on 26 March 2002. After quality control procedures were completed, its observations were analyzed in order to compute the *"length"*, *"completeness"*, and *"continuity"* of the 18 year time series (Figure 2). IS tide-gauge records were characterized by values of $L = 18$; $Cy = 0.967$, and $Co = 0.983$. This preliminary analysis was also carried out on the shorter (8 years) CS time series starting on 9 February 2011. The CS dataset was characterized by values of $L = 8$, $Cy = 1$, and $Co = 1$.

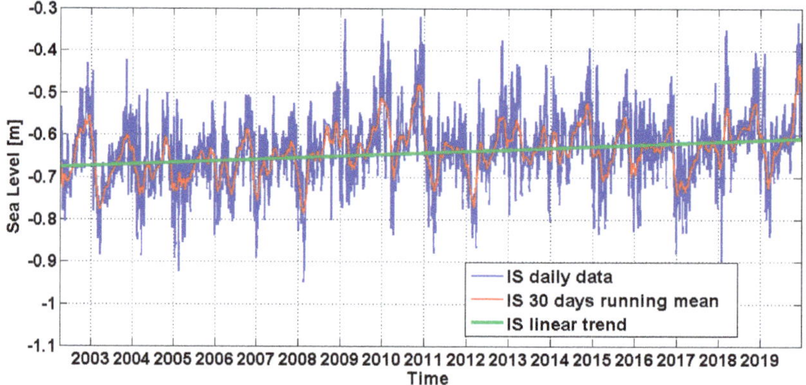

Figure 2. Daily tide-gauge data collected in Ischia from 2002 to 2019 (blue line). A 30-day running mean (red line) and a linear trend of the data (green line) are also shown.

The IS tide-gauge data were used to calculate the tidal constants and estimate the power spectral density (PSD) of the IS timeseries. The calculation of tidal constants is particularly useful to analyze the tide at regional and local scales and to gain insight into their relationships with the dynamics of the basin. Furthermore, as satellite altimetry data are continuously improving their ability to investigate coastal areas, the tidal constants, and tide-gauge-data in IS could provide an important reference for remotely sensed data analysis and instrumental calibration [9–12,78].

In this work, the harmonic analysis of tides was carried out with the MATLAB *"t-tide"* tool [79]. Only tidal components with a signal to noise ratio of larger than 1 are reported (Tables 1 and 2). Thanks to the 18-year-long timeseries of data, the IS tide-gauge makes an important contribution to the determination of tidal constants in the GON and offers well-established average values for the SL. Such a long time series—154,350 records, corresponding to about 6490 days of continuous data collection—allowed the calculation of the first 53 components of the astronomic tide at this site (Table 1).

Table 1. Frequency (Hz), amplitude (m), phase (°), and corresponding error of the 53 tidal constituents where the signal to noise ratio was larger than 1 for the Ischia (IS) timeseries. Standard acronyms are used for tidal components according to the International Hydrographic Organization.

Tidal Component	Frequency	Amplitude	Amplitude Error	Phase	Phase Error
	[Hz]	[m]	[m]	[°]	[°]
SA	0.00011	0.04223	0.00753	275.83488	10.56418
SSA	0.00023	0.01385	0.00815	96.39392	33.77839
2Q1	0.03571	0.00055	0.00039	329.54163	36.07178
SIG1	0.03591	0.00071	0.00037	320.96412	33.24574
Q1	0.03722	0.00199	0.00039	26.05282	9.87697
O1	0.03873	0.00937	0.00038	114.90251	2.13150
TAU1	0.03896	0.00050	0.00034	264.91656	38.48748
NO1	0.04027	0.00083	0.00025	196.13602	17.54705
PI1	0.04144	0.00128	0.00035	129.89058	17.91288
P1	0.04155	0.00817	0.00033	200.46610	2.58984
S1	0.04167	0.00334	0.00056	265.17196	9.52056
K1	0.04178	0.02813	0.00035	219.95858	0.64616
PSI1	0.04189	0.00130	0.00036	279.94776	16.88132
PHI1	0.04201	0.00073	0.00042	199.78408	29.65075
THE1	0.04309	0.00035	0.00032	256.30201	61.34690
J1	0.04329	0.00148	0.00036	248.91639	13.98859
OO1	0.04483	0.00043	0.00027	293.61793	35.72354
OQ2	0.07597	0.00033	0.00023	202.53649	43.24022
EPS2	0.07618	0.00087	0.00027	206.44114	16.35544
2N2	0.07749	0.00320	0.00022	228.53227	4.52386
MU2	0.07769	0.00391	0.00023	224.93708	3.57347
N2	0.07900	0.02378	0.00024	249.21649	0.55513
NU2	0.07920	0.00450	0.00028	253.24170	2.96859

Table 1. *Cont.*

Tidal Component	Frequency	Amplitude	Amplitude Error	Phase	Phase Error
H1	0.08040	0.00036	0.00026	354.22491	40.51804
M2	0.08051	0.11605	0.00024	263.34442	0.12050
MKS2	0.08074	0.00029	0.00025	308.89712	46.55347
LDA2	0.08182	0.00062	0.00022	252.08874	22.79655
L2	0.08202	0.00352	0.00031	267.77840	5.86598
T2	0.08322	0.00275	0.00026	273.87879	5.04700
S2	0.08333	0.04322	0.00026	282.47517	0.32211
R2	0.08345	0.00041	0.00020	298.33811	28.63368
K2	0.08356	0.01147	0.00021	296.83166	1.26689
MSN2	0.08485	0.00028	0.00022	358.10997	58.63309
ETA2	0.08507	0.00056	0.00023	321.46052	24.15966
MO3	0.11924	0.00192	0.00023	77.29497	6.57455
M3	0.12077	0.00407	0.00022	40.07398	3.54530
SO3	0.12206	0.00068	0.00025	128.92401	21.17344
SK3	0.12511	0.00204	0.00026	354.97139	6.63384
MN4	0.15951	0.00141	0.00007	162.39051	2.64839
M4	0.16102	0.00342	0.00007	203.79874	1.04128
SN4	0.16233	0.00034	0.00007	223.72687	11.49480
MS4	0.16384	0.00208	0.00006	264.19105	1.80286
MK4	0.16407	0.00047	0.00007	283.39939	8.08110
S4	0.16667	0.00036	0.00006	166.82859	11.87666
SK4	0.16689	0.00007	0.00006	129.61752	56.47201
2MK5	0.20280	0.00007	0.00004	270.67088	24.37170
2SK5	0.20845	0.00005	0.00003	207.91530	37.88104
2MN6	0.24002	0.00005	0.00004	291.48814	39.60927
M6	0.24153	0.00011	0.00003	321.15579	16.53225
2MS6	0.24436	0.00014	0.00003	2.15540	14.94986
2MK6	0.24458	0.00004	0.00003	26.47919	47.66250
2SM6	0.24718	0.00004	0.00004	344.13630	56.25650
MSK6	0.24741	0.00003	0.00003	21.26731	58.82638

Table 2. Frequency (Hz), amplitude (m), phase (°), and corresponding errors for the 40 tidal constituents with a signal to noise ratio of greater than 1 for the Castellammare di Stabia (CS) timeseries. Standard acronyms are used for tidal components according to the International Hydrographic Organization.

Tidal Component	Frequency	Amplitude	Amplitude Error	Phase	Phase Error
	[Hz]	[m]	[m]	[°]	[°]
SA	0.00011	0.03998	0.01052	268.45971	16.91032
SSA	0.00023	0.01753	0.01082	77.24666	39.59678
2Q1	0.03571	0.00075	0.00056	338.31146	58.67968
SIG1	0.03591	0.00072	0.00061	329.26622	43.17487
Q1	0.03722	0.00209	0.00070	41.78789	19.04889
O1	0.03873	0.00996	0.00059	126.92249	3.31510
NO1	0.04027	0.00108	0.00045	168.58926	25.51024
PI1	0.04144	0.00130	0.00053	141.36931	22.67050
P1	0.04155	0.00736	0.00054	198.66490	4.32768
S1	0.04167	0.00512	0.00070	282.29873	7.91398
K1	0.04178	0.03181	0.00054	211.00035	1.13173
PSI1	0.04189	0.00178	0.00055	308.38762	17.99524
J1	0.04329	0.00188	0.00055	229.76715	22.95559
EPS2	0.07618	0.00088	0.00032	215.45979	23.58849
2N2	0.07749	0.00334	0.00029	237.35181	5.56670
MU2	0.07769	0.00398	0.00029	225.58832	4.31571
N2	0.07900	0.02430	0.00029	247.28484	0.76937
NU2	0.07920	0.00456	0.00031	249.83144	3.56994
H1	0.08040	0.00101	0.00028	279.77635	19.41432
M2	0.08051	0.11791	0.00034	261.62021	0.14881
H2	0.08063	0.00102	0.00029	225.91666	17.48227
MKS2	0.08074	0.00086	0.00039	263.24797	30.76287
LDA2	0.08182	0.00073	0.00028	260.72057	22.89653
L2	0.08202	0.00307	0.00038	273.90925	7.24979
T2	0.08322	0.00291	0.00033	270.56248	5.84740
S2	0.08333	0.04388	0.00033	283.20792	0.39815
K2	0.08356	0.01271	0.00039	281.64550	1.70247
ETA2	0.08507	0.00079	0.00058	270.84843	36.62518
MO3	0.11924	0.00212	0.00047	89.51545	10.54339
M3	0.12077	0.00397	0.00032	36.33811	5.04156
SO3	0.12206	0.00089	0.00041	135.13022	28.36057
SK3	0.12511	0.00232	0.00041	342.76597	9.93294
MN4	0.15951	0.00152	0.00008	158.16051	3.75535
M4	0.16102	0.00362	0.00009	197.55199	1.47829
SN4	0.16233	0.00041	0.00009	219.92647	13.01316
MS4	0.16384	0.00214	0.00010	257.02772	2.34888
MK4	0.16407	0.00051	0.00014	256.71967	12.79714
S4	0.16667	0.00034	0.00009	184.94696	17.02859
SK4	0.16689	0.00017	0.00014	132.65692	46.82863
2SM6	0.24718	0.00010	0.00006	340.98222	42.08318

To complement these results, a harmonic analysis of tide constituents was also carried out on the CS shorter tide-gauge dataset. The results from this site allowed us to estimate the tidal constants for 40 components to determine the astronomical tidal variability (Table 2).

Both the IS and CS values are in agreement with the estimations of the tidal constants based on data collected in Naples since 1986 by the Italian national tidal observation network (RMN—Rete Mareografica Nazionale) [80]. These results show that GON is characterized by the absence of any amphidromic system, with synchronous SL oscillation over the entire gulf. The regional dynamics are included in the larger spatial dynamics of the Tyrrhenian Sea that are characterized by synchronous variation in the SL over the entire basin [81]. According to the general Mediterranean circulation, the Tyrrhenian Sea is indeed characterized by a general cyclonic circulation along the boundaries and the presence of several quasi-permanent and intermittent cyclonic and anticyclonic structures across the basin, e.g., the strong and recurrent anticyclonic circulation that persists offshore of the GON [30]. The southern sector is also influenced by the recirculation processes occurring at its meridional openings, which are mainly driven by steric level gradients existing between this region and the adjacent basins.

The PSD analysis on IS data (Figure 3) shows, as expected, energy peaks corresponding to 12 and 24 h tidal cycles. Nevertheless, the length of the IS timeseries allows the determination of energy peaks at lower frequencies. The longest period observed in the IS data is 8×10^{-6} cycles per hour, corresponding to about 14.2 years. Similar results were achieved for CS (not shown), but the limited duration of this time series discouraged additional in-depth analysis of the CS dataset.

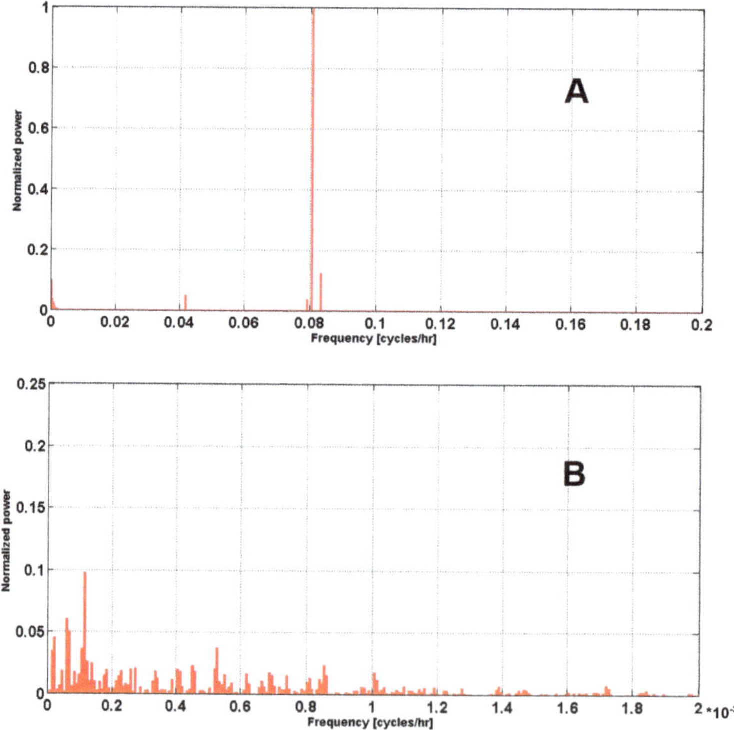

Figure 3. Normalized power spectrum of tide-gauge measurements in Ischia (**A**). The frequency is expressed in cycles per hour. Panel (**B**) offers a zoom-in view of the *x*-axis interval from 0 to 0.002 cycles/hr.

A PSD analysis on the mean sea level pressure data was also performed to evaluate the contribution of the meteorological variability to the computed power spectrum. The results show the presence of an energy peak at about 0.08 cycles per hour (Figure 4), which is probably due to the effects of the breeze regime on the tide gauge-site. However, most of the energy was found across longer time scales and was possibly associated with seasonal and interannual variations.

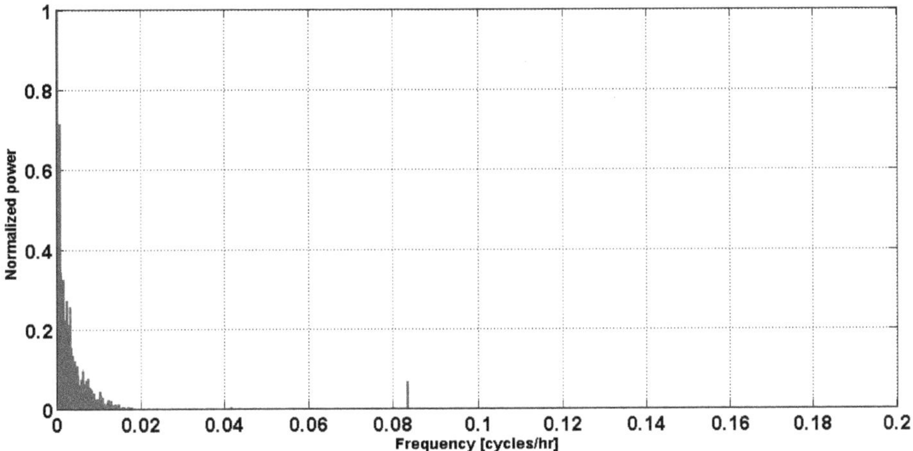

Figure 4. Normalized power spectrum of atmospheric pressure measurements in Napoli. The frequency is expressed in cycles per hour.

3.2. Trend Analysis

The IS area is characterized by seismic and volcanic phenomena affecting the ground elevation that cannot be ignored. In particular, low-intensity shallow seismicity, bradyseism activity, and intense hydrothermalism are active phenomena on the island [82]. The in situ measurements made along the geodetic network of the island of Ischia clearly show persistent and significant subsidence with a larger magnitude in the central-southern and north-western sectors of the island with a movement of up to 10 mm/year [83,84]. Additionally, the island shows large heterogeneity in the vertical displacement values [85]. Based on differential synthetic aperture radar interferometry data, Manzo et al. [83] demonstrated that the vertical displacements do not have any significant uplift but only have stable and subsiding areas. Among these, the Ischia Porto area, where the tide-gauge is located, is characterized by a vertical displacement trend ranging from +1 to −1 mm per year in the period 1992–2003, just before the tide-gauge installation. Successively, an analysis of the GPS survey data collected from 1998 to 2010 showed the existence of an elevation trend estimated to be 2.3 ± 0.9 mm/year in the data of the GPS station next to the Ischia tide-gauge [86], refining a previous estimation of −2.1 ± 1.0 mm/year obtained by De Martino et al. [87]. These values are in agreement with the satellite estimation (−2.3 mm/year) for the period 2002–2011 provided by the Italian National Science Council, Institute for the electromagnetic monitoring of the environment (IREA-CNR) through the webpage http://webgis.irea.cnr.it/webgis.html. A possible contribution of bradyseism to the SL records of Ischia was already hypothesized in the early 1900s basing on tide-gauge data collected by Giulio Glabrovitz, director of the local Casamicciola Geodynamical Observatory. In particular, he observed a relative SL rise of 3.5 mm/year for the period 1890–1910 in the same location as the modern tide-gauge [88].

Even though tide-gauges measure the height of the sea surface relative to crustal reference points that may move with tectonic activity or local subsidence, they can provide venerable information on long-term regional and global relative SL variations. To achieve this aim, IB correction was applied to IS measurements, and relative SL trend values were estimated. Results achieved through the

Mann–Kendall and Sen methods report a linear trend of +3.93 mm per year (99.9% significance level) over the 18-year study period, as shown in Figure 2. This value is in agreement with the global mean SL rise estimate of 3.20 (2.80 to 3.60) mm per year found by different altimetry data processing groups for the period 1993–2012 [20], even though it was not possible to remove any eventual signals associated with vertical ground movement from the IS tide gauge data. Trend values calculated from relative SL data in this work are also consistent with trend values from 1993 to 2012 estimated by Bonaduce et al. [89] over the Mediterranean Sea on the basis of tide-gauge (3.50 ± 0.7 mm per year) and satellite data (2.64 ± 0.6 mm per year). Their analysis suggested the importance of events from 2010 to 2011 on the SL trend results.

3.3. Relationship with Large-Scale Atmospheric Patterns

Air pressure variability has a direct effect on the SL variability on a wide range of time scales, from hourly to seasonal and multi-annual. For this reason, a detailed analysis of air pressure variability was carried out.

Figure 5 presents the behavior of atmospheric pressure data (black line) registered in Via Acton (January 2002–September 2011) and Centro Direzionale sites (October 2011–December 2019) over time and gives a comparison with NAO (red line), MOI1 (cyan line), and MOI2 (green line) indexes. A 30-day running mean was applied to the four signals in order to remove the very-high-frequency variability. According to the evidence presented in Figure 5, the period of interest (2002–2019) can be segmented into four different time intervals:

- the 2002–2008 period, in which the air pressure is generally greater than its mean value;
- the late-2008 to mid-2015 period, which is characterized by relevant oscillations of the atmospheric pressure signal; this interval includes two sub-periods in which negative air pressure anomalies prevailed: 2009–2011 and the segment encompassing late 2012 and early 2014;
- the late 2015–2017 period, in which the air pressure is generally above its mean value;
- 2018–2019, which is characterized by negative pressure anomalies.

As expected, the NAO and MOI indexes explain a large part of the observed variability, especially in the winter season. A monthly-based correlation analysis revealed that the first version of the MOI index (MOI1) is better suited to our study area than the second one (MOI2). More specifically, a clear linkage between MOI1 index and air pressure observed in Naples was found between December and March, with correlation values ranging from 0.5 to 0.7. The NAO index effectively modulates the meteorological features of the study area only in late autumn and early winter (November and December). The selected teleconnection patterns determined from sea-level pressure anomalies (see Section 2.2) are not adequate to capture the summer variability, e.g., [90,91]. The meteorological dynamics of this season, in fact, are generally driven by synoptic transients that are well structured only in the mid and upper tropospheric levels (i.e., cut-off lows and upper-level lows).

For each year, the annual means and standard deviations of the atmospheric pressure data were computed. The results are presented in Table 3. The percentages of samples with pressure levels higher than 1013 and 1026 mbar as well as those lower than 1013 and 1000 mbar are also reported. The value of 1013 mbar was chosen as the globally accepted boundary value between low- and high-pressure conditions, while two additional threshold values were chosen to represent the anomalous high (1026 mbar) and low (1000 mbar) pressure periods for the study area.

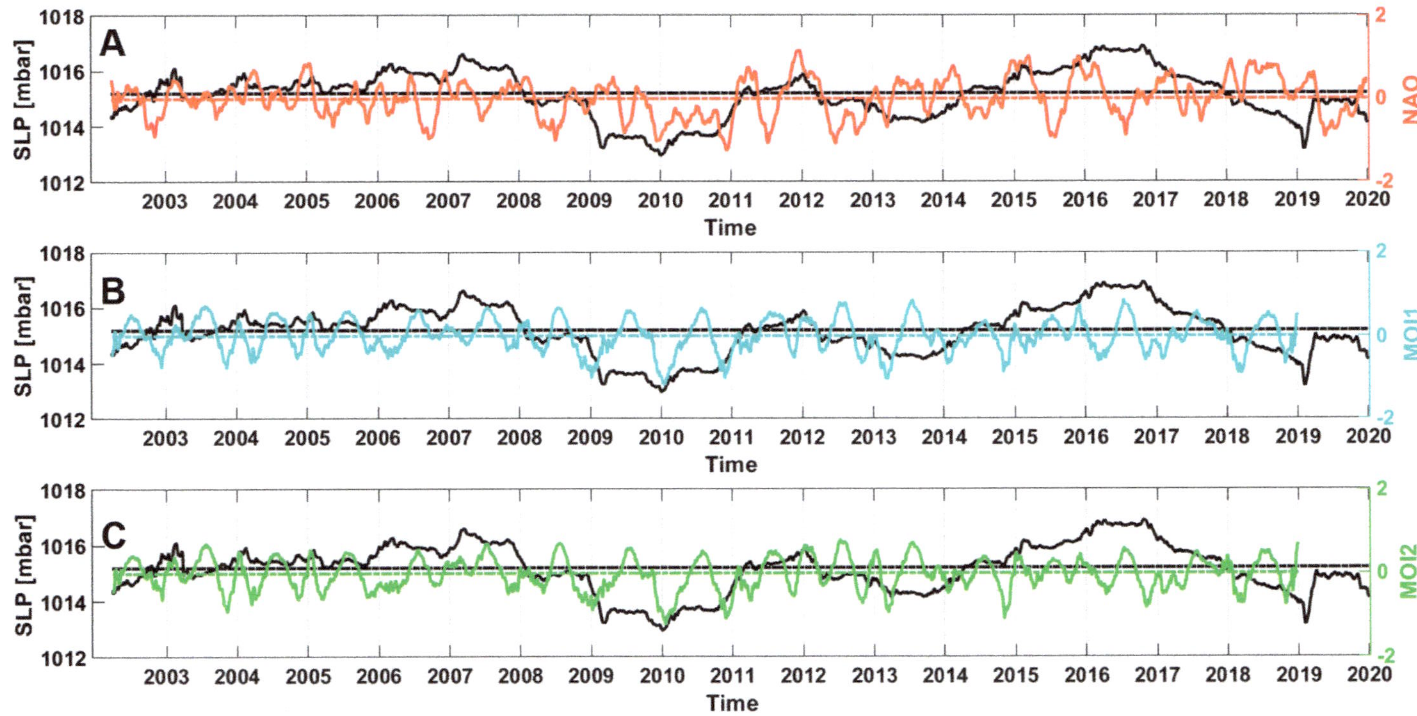

Figure 5. Time series of air pressure data (black line), (**A**) North Atlantic Oscillation Index (NAO) (red line), (**B**) Mediterranean Oscillation Index (MOI)1 (cyan line), and (**C**) MOI2 (green lines) indexes for the period January 2002 to December 2019. Data are presented in terms of the 30-day running mean. Horizontal lines represent the mean value of each signal in the considered time interval. It should be noted that atmospheric pressure data have been registered by Naples Via Acton weather station for the period from January 2002 to September 2011 and by the Naples Centro Direzionale station for October 2011 to December 2019.

Table 3. Annual means of atmospheric pressure at the Via Acton (January 2002 to September 2011) and Centro Direzionale Sites (October 2011 to December 2019). For each year, the amount of data, annual mean, standard deviation value, percentage of samples with pressure higher than 1013 mbar, percentage of samples with pressure higher than 1026 mbar, percentage of samples with pressure lower than 1013 mbar, and percentage of samples with pressure lower than 1000 mbar are presented.

Year	Samples	Mean Pres	Std Dev	Samples > 1013 mbar	Samples > 1026 mbar	Samples < 1013 mbar	Samples < 1000 mbar
	#	[mbar]	[mbar]	[%]	[%]	[%]	[%]
2002	6697	1013.7	5.46	57.17	0.24	42.83	1.91
2003	8761	1015.6	5.99	71.16	3.23	28.84	0.99
2004	8785	1015.4	7	66.01	6.56	33.99	2.25
2005	8761	1015.7	7.34	66.21	7.73	33.79	2.41
2006	8762	1016.3	6.66	71.68	6.46	28.32	1.62
2007	8761	1015.5	6.25	68.02	5.33	31.98	1.61
2008	8785	1015.6	6.92	64.56	8.8	35.44	1.81
2009	8761	1013.3	6.7	58.77	0.98	41.23	4.38
2010	8761	1012.6	6.38	52.6	1.55	44.3	4.37
2011	8761	1016.5	5.47	75.84	5.22	23.84	0.46
2012	8785	1015.2	6.26	69.58	4.04	29.6	2.12
2013	8761	1014.3	7.31	65.53	4.14	33.58	4.51
2014	8761	1014.6	5.79	63.47	2.13	36.23	1.68
2015	8761	1017.1	7.81	74.01	13.65	25.97	2.97
2016	8785	1016	6.82	72.03	8.78	27.97	1.45
2017	8761	1016.3	5.71	75.03	4.23	24.97	0.99
2018	8761	1013.9	6.21	61.53	2.83	38.47	3.14
2019	8737	1014.2	6.39	62.21	2.31	37.79	2.44

A close inspection of Table 3 highlights that in 2009, 2010, and 2013, the number of high-pressure samples was lower than in the other years and that the very-low-pressure records ($\approx$4.5% < 1000 mbar) occurred about three times more often than in other years. On the other hand, during 2015, a relevant number of high-pressure events has been detected (13.7% > 1026 mbar). The footprint of large-scale atmospheric patterns on the strong variability and oscillations found between 2009 and 2015 is evident from Figure 5. This result was confirmed by Wavelet coherence (WC) analysis, a very useful tool that allows time-localized oscillations in non-stationary signals, such as geophysical signals, to be detected [92]. The WC shows that the NAO and MOI1 indexes exhibit common features between 2009 and 2015 with relatively low frequencies (between 2 and 4 years). According to the direction of the black arrows, the two signals are in phase, although NAO lags MOI by a phase angle of about 45° (Figure 6). Moreover, the WC spectrum highlights small areas of strong covariance in the high-frequency region (between 0.25 and 0.5 years) in 2005–2006, 2011–2013, and 2017–2018.

The WC analysis suggests that from 2009 to early 2011, as well as between late 2012 and early 2013, the persistence of both NAO and the MOI1 negative phase caused frequent low-pressure conditions on the central Mediterranean basin [93]. These anomalies in atmospheric pressure patterns may have a strong influence on SL variability, as is discussed in the next section.

To evaluate any correlation with the atmospheric variability as described by large-scale climate indices (i.e., NAO, MOI1, and MOI2), the meteorological tide time series for the IS tide-gauge was obtained by removing the astronomic theoretical tide from the signal. Hourly tidal constants obtained directly by harmonic analysis based on seven standard main constituents were used to obtain a new time series that still includes the IB component of the SL variations.

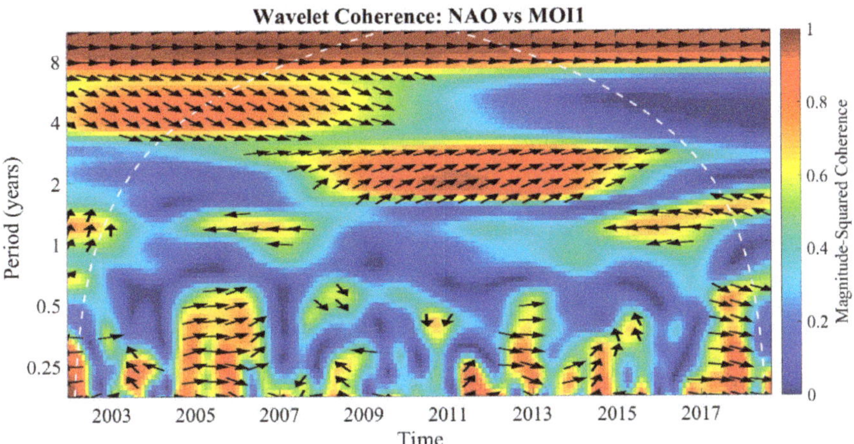

Figure 6. Wavelet coherence spectrum between monthly NAO and MOI1 timeseries for the period 2002–2018. The relative phase relationship between the two indexes is shown by the directions of the black arrows (see [92] for details). The white dashed line indicates the cone of influence, where the edge effects might distort the coherence data. The color bar gives a measure of correlation between the two analyzed signals in the time–frequency plane.

To examine the connections between the IS meteorological tide series and the teleconnection patterns, we focused on the period from November to March, which can be defined as an extended winter season. This choice is justified by the relevant linkages between the meteorological features of the study area and the synoptic variability only in late autumn, winter, and early spring, as previously shown. The results show that a moderate to strong anticorrelation exists on the long timescale between IS tide-gauge data and climate indices during the 2002–2019 study period, with the linear correlation coefficients being R = −0.5, R = −0.7, and R = −0.5 for NAO, MOI1, and MOI2, respectively.

The relationship between the two teleconnection patterns selected in this study, NAO and MOI, was further analyzed through a windowed correlation analysis. In particular, we estimated the correlation between the 30-day running mean of SL and the three teleconnection patterns for each extended winter season (November to March) for the period 2002–2019 (Figure 7).

An anti-phase relationship exists between MOIs and SL variability (Figure 7), as shown by the linear correlation coefficient that generally ranges from −0.4 to −0.8. However, there are some exceptions, i.e., during winter in 2003/2004, 2005/2006, and 2015/2016. In these years, a very weak correlation was found, especially when MOI1 index was considered. The Mediterranean atmospheric circulation had a very strong impact on SL in the following winter seasons: 2002/2003, 2006/2007, 2007/2008, 2009/2010, 2010/2011, 2014/2015, 2016/2017, and 2017/2018. It is not a mere coincidence that, in these cases, the NAO index (Figure 7, upper panel) also exhibited a moderate to strong anticorrelation with SL variability. A deeper examination of these seasons revealed some interesting linkages between atmospheric and mareographic variability, as discussed below.

162

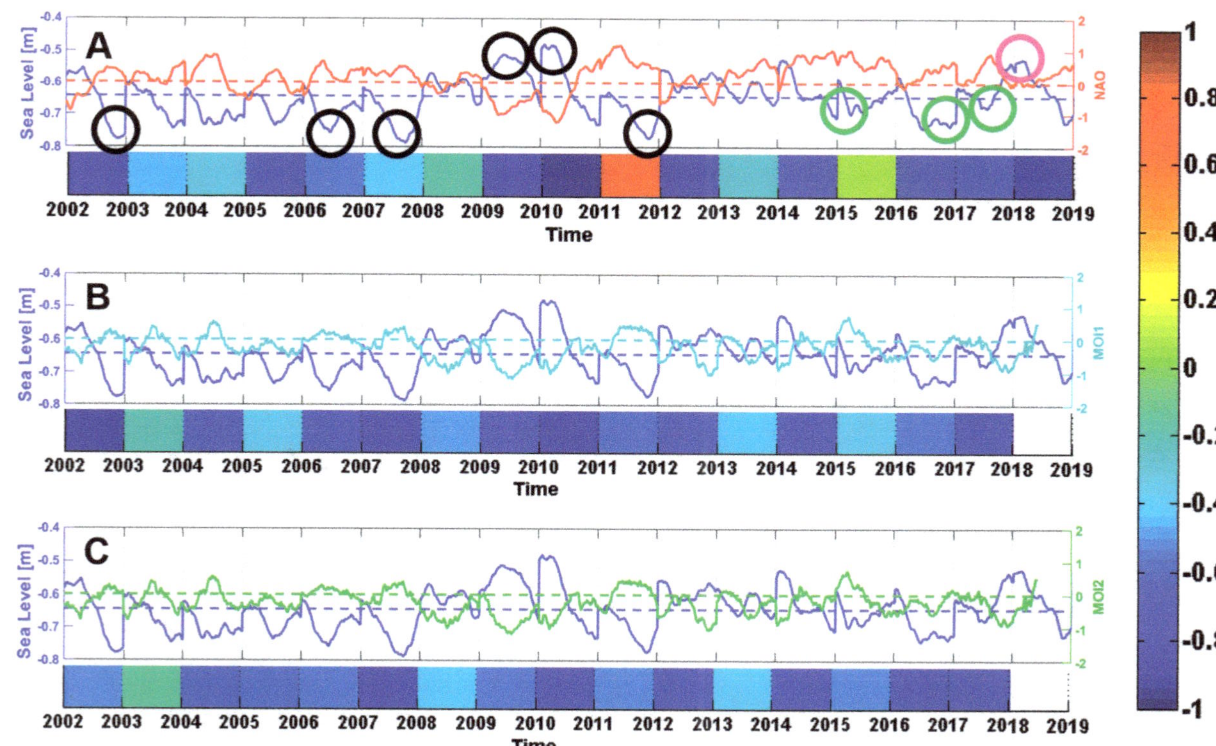

Figure 7. Windowed correlation (color bar) between the IS tide-gauge (blue) and (**A**) NAO (red), (**B**) MOI1 (cyan), and (**C**) MOI2 (green) daily data. All time series were filtered with a 30-day running mean. Dotted lines show the mean and zero values, respectively, for SL and atmospheric indices.

During the first five winter seasons (2002/2003, 2006/2007, 2007/2008, 2009/2010, and 2010/2011) relevant anomalies of SL were observed (i.e., negative in 2002/2003, 2006/2007, and 2007/2008, positive in 2009/2010 and 2010/2011). Such events, marked as a black circle on the upper panel of Figure 7, were forced by a coupling between teleconnection patterns synthetized by the NAO and MOI. During these seasons, the indexes had quite similar behavior, i.e., a shift from a negative to a positive phase in 2002/2003, 2006/2007, and 2010/2011, persistence of the positive phase in 2007/2008, and a shift to a negative phase in 2009/2010. In 2002/2003, 2006/2007, and 2007/2008, the positive phase of both NAO and MOI resulted in a strong anticyclonic pattern over the central and western Mediterranean basins, which led to conditions favorable for a lowering of SL. Conversely, in the 2009/2010 and 2010/2011 seasons, the opposite large-scale circulation anomalies prevailed and the conditions were in favor of an increase in the SL in Ischia. Analogous results were achieved by Landerer and Volkov [93] who focused on recent SL variability in the Mediterranean Sea and found an increase in the mean SL of 10 cm during the boreal winter months of 2009/2010 and 2010/2011. Our findings reinforce their observations, including the assessment of a prominent role of the NAO on the observed SL fluctuations.

During the 2014/2015, 2016/2017, and 2017/2018 seasons, the NAO index was generally in its positive phase, while MOI1 and MOI 2 were neutral or negative. Despite the positive NAO phase having a magnitude comparable with that observed in the last part of 2002/2003, in the 2006/2007 and 2007/2008 seasons, its impact on Ischia SL was less relevant (Figure 7, green circles) due to a missing coupling with the Mediterranean atmospheric variability.

A strong lowering of SL, again marked by a black circle, also occurred in the extended winter season of 2011/2012. It is reasonable to assume that the positive phases of NAO and MOI strongly contributed to the relevant magnitude of the negative SL anomaly. Although in this season, the NAO index exhibited a negative trend, it remained in the positive phase, contributing, together with the positive phase of MOI, to a progressive regression of SL, culminating in January 2012. However, in this case, a positive linear correlation was found between NAO and SL. Available information cannot ensure that this result is ascribed to a mathematical artifact rather than a physical mechanism.

Finally, an interesting but isolated case is the rise of SL observed at the beginning of the 2018/2019 season (see magenta circle on the upper panel of Figure 7). Differently from the ones that occurred in 2009/2010 and in 2010/2011, this event was associated with a neutral NAO phase and a slightly negative MOI.

3.4. "Acqua Alta" Episodes in Ischia

SL variability is the result of several forces that act at different spatial and temporal scales and can be associated with changes in sea water properties, the so-called steric component, or not. In the Mediterranean Sea, at time scales longer than the seasonal scale, SL changes are due to both steric and non-steric forcing [42], while on a shorter timescale, the variability is mainly due to non-steric factors [93,94]. Among these, the combined effects of the wind air pressure and tides can result in the rapid submersion of coastal areas, leading to catastrophic flooding and danger to life and economic activities. This phenomenon is known as *"acqua alta"*, and it affects some Mediterranean areas, such as the city of Venice in the Northern Adriatic Sea and the island of Lipari in the Southern Tyrrhenian Sea.

In Ischia, AA events cause water ingression over the piers in the Ischia Porto area (Figure 8) and over the roads in the Ischia Ponte neighborhood, where intense socio-economic activities, mainly linked to tourism, exist. This phenomenon has gained the attention of the wider public in recent years, and its effects have been reported in a number of national and local newspapers.

In this work, the number, amplitude, and seasonal distribution of the AA events in Ischia Porto were studied (Table 4) through SL data collected by the IS tide-gauge since 2002. As described in Section 2, the IS tide-gauge is installed on the edge of the Coast Guard piers in the harbor that are 34 cm higher than the ancient piers limiting the *"riva destra"* area, where AA is more frequent. For this analysis, the edge of the Coast Guard pier was chosen as a reference. Hourly mean SL values overpassing the height of the old pier edge were considered to be single AA episodes. This analysis was performed on the original quality controlled timeseries recorded by the tide-gauge, in order to analyze the combined effects of tide and meteorological variability on the AA occurrences. Hourly means were used in order to minimize the effect of waves induced by ship movements or wind gusts as well as transient values that could influence the tide-gauge data during these events. For each year from 2002 to 2019, the total number of AA events is expressed as the percentage over the total number of hourly records. Table 4 also shows the number of events overpassing heights of 5, 10, 20, and 30 cm over the pier edge thresholds. After 2009, the AA episodes seem to have been more frequent, with only the years 2011 and 2017 having a percentage of single events lower than 1%. A higher number of AA episodes was registered in 2010 (8.7%), corresponding to about 762 submersion hours. During that year, the higher SL height over the pier was registered (32 cm) on October 1 at 09:00.

Figure 8. "*Acqua alta*" events at the "*riva destra*" in the Ischia Porto area: (**A**) 2 March 2018, photo from "il Dispari"; (**B**) 30 November 2019, photo from Repubblica, Napoli by Peppe Trani.

A similar analysis was carried out on a monthly basis in order to highlight the presence of a seasonal component in the AA occurrences. Figure 9 shows the monthly distribution of AA events from 2002 to 2019. The late fall and winter months from October to January were found to be characterized by an higher number of AA episodes, while during the spring and summer months, AA barely reached 30 episodes (June 2010).

Even though the estimated long-term SL rise has an effect on the frequency of the AA episodes through the expected increase in the mean SL, it cannot explain this phenomenon and its seasonal variability. Instead, the change in the frequency of AA events after 2009 may have been strongly influenced by large-scale atmospheric patterns. A detailed comparison between the results presented in Table 4 and Figure 9 and those reported in Table 1 and Figure 7 suggests that a relationship exists between the variability and fluctuations in atmospheric and AA events during the 2008–2015 interval.

A significant contribution could then be ascribed to the combined effect of low-pressure systems and wind forcing on the sea surface that generates so-called storm surges. To confirm this hypothesis, IS tide-gauge measurements and Via Acton air pressure data were complemented with wind direction and intensity data from 1 January 2010 to 31 March 2010 collected at the ISPRA weather station site in Naples. The analysis of the atmospheric variability during these three months shows that the large scale and Mediterranean circulation were conditioned by negative phases for both NAO and MOI1, especially in the first days of January and in the first twenty days of February (Figure 10a,b). The precondition for a storm surge event, i.e., a coupling between the two atmospheric teleconnections involved in our study, is recognizable in mid-February, when the atmospheric asset designed by negative phases of NAO and MOI caused negative pressure anomalies over the GON (Figure 10a), creating favorable conditions for an increase in SL (Figure 10b). Additionally, during those days, the wind regime over the GON was characterized by high-frequency winds from the N/NE that were able to directly impact on the Ischia harbor whose entrance is oriented toward North (Figure 10c). The effects of this atmospheric variability, associated with tidal SL variations, generated a significant storm surge event, as registered on 19 February (Figure 10d).

Therefore, the performed analyses suggest that the occurrence of AA events is the result of different degrees of forcing on the SL variability of Ischia, more so than the effects of tides or global SL rise. While SL rise can obviously contribute to setting a long-term situation that supports AA events, in situ tide-gauge data reveal that neither the number of AA events, nor the marked seasonal signals observed in their occurrence, are directly correlated with the SL trend. On the other hand, an increase in Ischia AA events is clearly associated with the NAO negative phases in conjunction with the low atmospheric pressure systems and wind patterns in the GON and over the Southern Tyrrhenian Sea. The combination of these forces should be considered as one leading factor fostering AA episodes.

Table 4. Maximum sea-level height over the pier (cm), total number of *"acqua alta"* events (%), and number of *"acqua alta"* events higher than 5, 10, 20, and 30 cm over the pier edge from 2002 to 2019.

YEAR	H max [cm]	AA %	AA% > 5 cm	AA% > 10 cm	AA% > 20 cm	AA% > 30 cm
2003	12.0	0.82	0.74	0.02	0.00	0.00
2004	11.0	0.52	0.44	0.02	0.00	0.00
2005	5.5	0.38	0.27	0.00	0.00	0.00
2006	5.8	0.58	0.48	0.00	0.00	0.00
2007	6.5	0.35	0.32	0.00	0.00	0.00
2008	13.0	0.91	0.75	0.06	0.00	0.00
2009	26.0	4.50	4.20	0.91	0.08	0.00
2010	32.0	8.70	8.20	1.70	0.13	0.02
2011	13.0	0.41	0.35	0.01	0.00	0.00
2012	20.0	2.00	1.80	0.30	0.00	0.00
2013	18.0	3.00	2.70	0.17	0.00	0.00
2014	13.0	3.50	3.10	0.14	0.00	0.00
2015	12.0	2.10	2.00	0.06	0.00	0.00
2016	9.1	1.20	0.99	0.00	0.00	0.00
2017	7.2	0.47	0.39	0.00	0.00	0.00
2018	24.0	4.40	4.00	0.83	0.07	0.00
2019	24.0	4.30	4.10	1.20	0.14	0.00

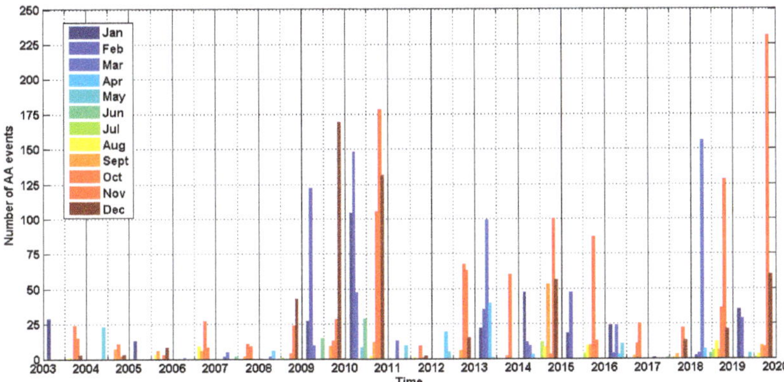

Figure 9. Number of *"acqua alta"* events from 2002 to 2019. Each event is defined as a single hourly mean sea level overpassing the *"riva destra"* pier edge.

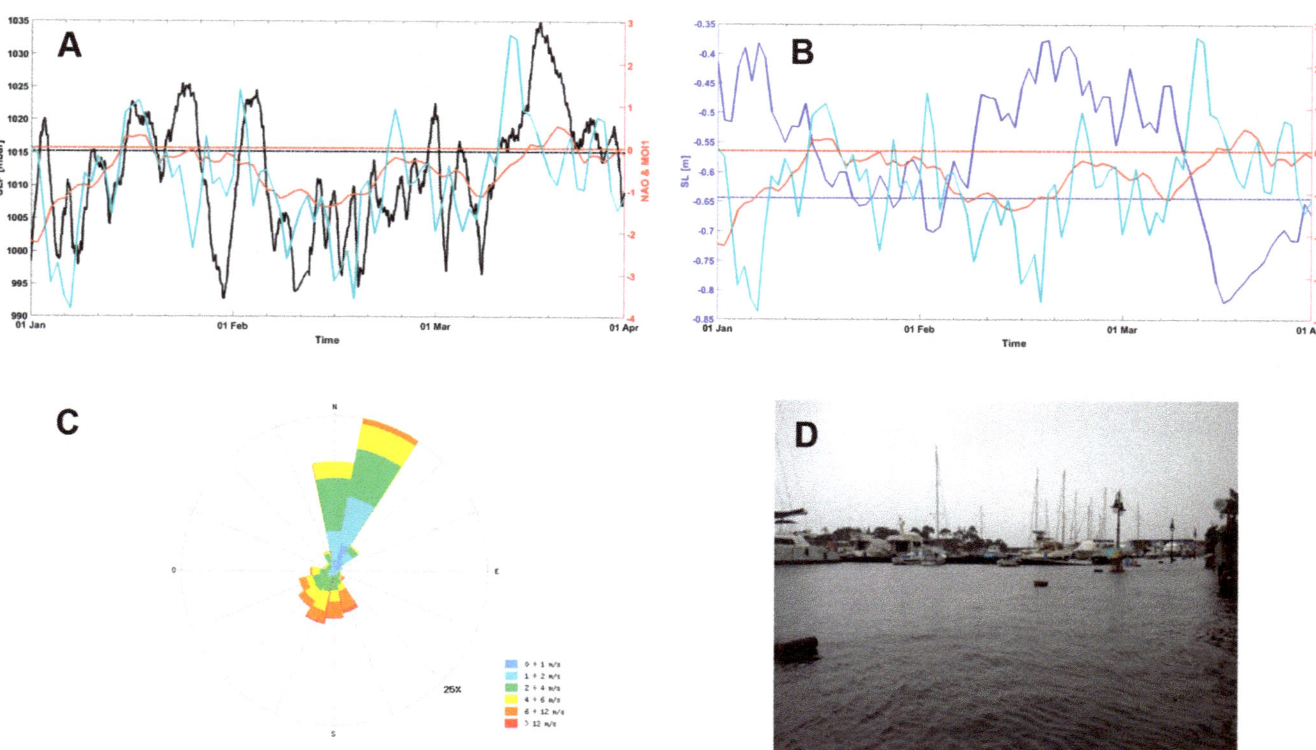

Figure 10. *"Acqua alta"* event on February 2010: (**A**) Daily (black line) and mean air pressure (black dotted line) at the Via Acton weather station, NAO (red line), and MOI1 (cyan line) with their neutral phases (red dotted line) are also indicated; (**B**) Daily (blue line) and mean sea level (blue dotted line) in IS, NAO (red line), and MOI1 (cyan line) with their neutral phases (red dotted line); (**C**) increase in wind in the ISPRA weather station in Naples (courtesy of ISPRA); (**D**) 2 February 2010 in the Ischia Porto area—during this event, the sea level was about 35 cm higher than the pier edge.

4. Conclusions and Future Perspectives

The exploitation of IS tide-gauge data allowed the analysis of several aspects associated with SL variability in the Campania coastal area. The IS tidal constituents, corroborated by the analysis of the shorter CS time series, confirmed that the GON is characterized by non-amphidromic properties. Long-term trend analysis on the IS dataset showed a relative SL rise of 3.9 mm/year, in agreement with previous studies that focused on the Mediterranean Sea and the global mean SL rise.

The tidal variability and the global SL rise can explain a minor part of the SL variability measured by the IS tide-gauge, while the registered increasing occurrence of extreme events over the island, i.e., *"acqua alta"*, can be better described by a combination of factors. Firstly, an antiphase relationship was found between SL and two teleconnections that effectively describe the large-scale atmospheric variability and its anomalies, i.e., NAO and MOI. The latter modulates the tidal variability of the study area through the well-known IB effect in late fall, winter, and early spring, the months where most relevant SL anomalies have been detected. According to the results of our analysis, the main positive departures from the average SL, such as those observed in the 2009/2010 and 2010/2011 winter seasons, occurred when both NAO and MOI were simultaneously in their negative phase, which implies that there were low-pressure conditions in the central Mediterranean basin. At the same time, the strong SL lowering registered in the 2002/2003, 2006/2007, and 2007/2008 seasons can be explained by the coupling of NAO and MOI positive phases, which implies that there were anticyclonic conditions in the central and western Mediterranean area.

As demonstrated by the discussion of the AA event of February 2010, the storm surge dynamics are characterized by the combination of atmospheric forcing with the shape and morphology of Ischia harbor, which is most affected by the northern wind regime.

Of course, these results might improve as the tide-gauge time series becomes longer, so the accurate and regular maintenance of both IS and CS will be continued by the University "Parthenope" through planning and realizing the necessary operations during the forthcoming years.

Future developments are also expected, and these include the possibility of installing GPS and weather stations corresponding to the IS and CS tide-gauges, as recommended by international organizations and programs to discriminate between the effects of ground vertical movement and storms on SL variations. Additional analyses on long-term time series from different tide-gauges located in the Campania region and the Southern Tyrrhenian Sea are also advisable, as well as the opportunity to implement efficient procedures for comparing these data with available (and forthcoming) SL satellite observations.

Author Contributions: Conceptualization, B.B. and Y.C.; methodology, B.B., Y.C., V.C.; software, B.B. and G.Z.; validation, formal analysis, B.B., G.Z., V.C. and Y.C.; investigation, B.B., Y.C., V.C., G.A. and G.B.; writing—original draft preparation, B.B., Y.C., V.C., G.A.; writing—review and editing, G.A., G.B.; funding acquisition, G.B. All authors have read and agreed to the published version of the manuscript.

Funding: This research received no external funding.

Acknowledgments: The authors want to thank the Italian Air Force Weather Service for providing the air pressure data. We also thank the technicians working at the University of Naples "Parthenope" for their support on the design and maintenance of the tidal and meteorological network. The NAO daily timeseries used in this study was provided by the Climate Prediction Centre of the National Oceanic and Atmospheric Administration (NOAA). The MOI1 and MOI2 daily time series used in this study were provided by the University of East Anglia Climatic Research Unit. The authors thank the Marina di Stabia harbor management and the Corps of the Port Captaincies—Coast Guard of Ischia. We thank ISPRA for the wind data in Naples provided through the webpage https://www.mareografico.it/. Finally, the authors want to thank two anonymous reviewers for their work and the suggestions received. GA work was realized in the framework of the PON R&I 2014–2020 "AIM—Attraction and International Mobility" at Università degli Studi di Napoli Parthenope.

Conflicts of Interest: The authors declare no conflict of interest.

References

1. Woodworth, P.L.; White, N.J.; Jevrejeva, S.; Holgate, S.J.; Church, J.A.; Gehrels, W.R. Evidence for the accelerations of sea level on multi-decade and century timescales. *Int. J. Climatol.* **2009**, *29*, 777–789. [CrossRef]
2. Frederikse, T.; Riva, R.; Slobbe, C.; Broerse, T.; Verlaan, M. Estimating decadal variability in sea level from tide gauge records: An application to the North Sea. *J. Geophys. Res. Oceans* **2016**, *121*, 1529–1545. [CrossRef]
3. Zerbini, S.; Raicich, F.; Prati, C.M.; Bruni, S.; Del Conte, S.; Errico, M.; Santi, E. Sea-level change in the Northern Mediterranean Sea from long-period tide-gauge time series. *Earth Sci. Rev.* **2017**, *167*, 72–87. [CrossRef]
4. Ekman, M. The world's longest continuous series of sea level observations. *Pure Appl. Geophys.* **1988**, *127*, 73–77. [CrossRef]
5. Van Veen, J. Bestaat er een geologische bodemdaling te Amsterdam sedert 1700? *Tijdschrift Koninklijk Nederlandsch Aardrijkskundig Genootschap.* 1945. Available online: http://www.kwaad.net/VanVeen_1945_PeilAmsterdam_1700AD.pdf (accessed on 2 September 2020).
6. Gornitz, V.; Lebedeff, S.; Hansen, J. Global sea level trend in the past century. *Science* **1982**, *215*, 1611–1614. [CrossRef] [PubMed]
7. Douglas, B.C. Global sea level rise. *J. Geophys. Res.* **1991**, *96*, 6981–6992. [CrossRef]
8. Higginson, S.; Thompson, K.R.; Woodworth, P.L.; Hughes, C.W. The tilt of mean sea level along the east coast of North America. *Geophys. Res. Lett.* **2015**, *42*, 1471–1479. [CrossRef]
9. Passaro, M.; Rose, S.K.; Andersen, O.B.; Boergens, E.; Calafat, F.M.; Dettmering, D.; Benveniste, J. ALES+: Adapting a homogenous ocean retracker for satellite altimetry to sea ice leads, coastal and inland waters. *Remote Sens. Environ.* **2018**, *211*, 456–471. [CrossRef]
10. Volkov, D.L.; Pujol, M.I. Quality assessment of a satellite altimetry data product in the Nordic, Barents, and Kara seas. *J. Geophys. Res. Oceans* **2012**, *117*, C03025. [CrossRef]
11. Chambers, D.P.; Ries, J.C.; Shum, C.K.; Tapley, B.D. On the use of tide gauges to determine altimeter drift. *J. Geophys. Res. Oceans* **1998**, *103*, 12885–12890. [CrossRef]
12. Wunsch, C. Calibrating an altimeter: How many tide gauges is enough? *J. Atmos. Ocean. Technol.* **1986**, *3*, 746–754. [CrossRef]
13. Ding, X.; Zheng, D.; Chen, Y.; Chao, J.; Li, Z. Sea level change in Hong Kong from tide gauge measurements of 1954–1999. *J. Geod.* **2001**, *74*, 683–689. [CrossRef]
14. Houghton, L.G.; Meira Filho, L.G.; Callander, B.A. (Eds.) *Climate Change 1995: The Science of Climate Change*; Cambridge University Press: Cambridge, UK, 1996.
15. Warrick, R.A.; Oerlemans, J.; Woodworth, P.L.; Meier, M.F.; le Provost, C. Changes in sea level. In *Climate Change 1995: The Science of Climate Change*; Houghton, J.T., Meira Filho, L.G., Callander, B.A., Eds.; Cambridge University Press: Cambridge, UK, 1996; pp. 359–405.
16. Church, J.A.; Clark, P.U.; Cazenave, A.; Gregory, J.M.; Jevrejeva, S.; Levermann, A.; Merrifield, M.A.; Milne, G.A.; Nerem, R.S.; Nunn, P.D.; et al. Sea Level Change. In *Climate Change 2013: The Physical Science Basis*; Contribution of Working Group I to the Fifth Assessment Report of the Intergovernmental Panel on Climate Change; Stocker, T.F., Qin, D., Plattner, G.-K., Tignor, M., Allen, S.K., Boschung, J., Nauels, A., Xia, Y., Bex, V., Midgley, P.M., Eds.; Cambridge University Press: Cambridge, UK; New York, NY, USA, 2013; pp. 1137–1216.
17. Nerem, R.S.; Beckley, B.D.; Fasullo, J.T.; Hamlington, B.D.; Masters, D.; Mitchum, G.T. Climate-change-driven accelerated sea-level rise detected in the altimeter era. *Proc. Natl. Acad. Sci. USA* **2018**, *115*, 2022–2025. [CrossRef] [PubMed]
18. Mahdi, H.; Hebib, T. Mediterranean Sea level trends from long-period tide gauge time series. *Acta Oceanol. Sin.* **2020**, *39*, 157–165. [CrossRef]
19. Intergovernmental Panel on Climate Change (IPCC). Climate Change 2013: The Physical Science Basis; 2013. Available online: http://www.climatechange2013.org (accessed on 17 July 2018).
20. Intergovernmental Panel on Climate Change (IPCC). *Climate Change 2014: Synthesis Report. Contribution of Working Groups I, II and III to the Fifth Assessment Report of the Intergovernmental Panel on Climate Change*; Core Writing Team, Pachauri, R.K., Meyer, L.A., Eds.; IPCC: Geneva, Switzerland, 2014; p. 151.

21. Bruni, S.; Zerbini, S.; Raicich, F.; Errico, M. Rescue of the 1873–1922 high and low waters of the Porto Corsini/Marina di Ravenna (northern Adriatic, Italy) tide gauge. *J. Geod.* **2019**, *93*, 1227–1244. [CrossRef]
22. Tsimplis, M.N.; Baker, T.F. Sea level drop in the Mediterranean Sea: An indicator of deep water salinity and temperature changes? *Geophys. Res. Lett.* **2000**, *27*, 1731–1734. [CrossRef]
23. Cazenave, A.; Cabanes, C.; Dominh, K.; Mangiarotti, S. Recent sea level changes in the Mediterranean Sea revealed by TOPEX/POSEIDON satellite altimetry. *Geophys. Res. Lett.* **2001**, *28*, 1607–1610. [CrossRef]
24. Fenoglio-Marc, L. Analysis and representation of regional sealevel variability from altimetry and atmospheric-oceanic data. *Geophys. J. Int.* **2001**, *145*, 1–18. [CrossRef]
25. Antonioli, F.; Silenzi, S. Variazioni Relative del Livello del Mare e Vulnerabilita Delle Pianure Costiere Italiane. *Quaderno della Soc. Geol. Ital.* **2007**, *2*, 1–29.
26. Lambeck, K.; Antonioli, F.; Anzidei, M.; Ferranti, L.; Leoni, G.; Scicchitano, G.; Silenzi, S. Sea level change along the Italian coast during the Holocene and projections for the future. *Quatern. Int.* **2011**, *232*, 250–257. [CrossRef]
27. Aucelli, P.C.; Di Paola, G.; Incontri, P.; Rizzo, A.; Vilardo, G.; Benassai, G.; Buonocore, B.; Pappone, G. Coastal inundation risk assessment due to subsidence and sea level rise in a Mediterranean alluvial plain (Volturno coastal plain–southern Italy). *Estuar. Coast. Shelf Sci.* **2016**. [CrossRef]
28. Thompson, P.R.; Hamlington, B.D.; Landerer, F.W.; Adhikari, S. Are long tide gauge records in the wrong place to measure global mean sea level rise? *Geophys. Res. Lett.* **2016**, *43*, 10403–10411. [CrossRef]
29. Dangendorf, S.; Hay, C.C.; Calafat, F.M.; Marcos, M.; Berk, K.; Jensen, J. Persistent acceleration in global sea-level rise since the 1970s. *Nat. Clim. Chang.* **2019**, *9*, 705–710. [CrossRef]
30. Aulicino, G.; Cotroneo, Y.; Lacava, T.; Sileo, G.; Fusco, G.; Carlon, R.; Satriano, V.; Pergola, N.; Tramutoli, V.; Budillon, G. Results of the first wave glider experiment in the southern Tyrrhenian Sea. *Adv. Oceanogr. Limnol.* **2016**, *7*, 16–35. [CrossRef]
31. Durante, S.; Schroeder, K.; Mazzei, L.; Pierini, S.; Borghini, M.; Sparnocchia, S. Permanent thermohaline staircases in the Tyrrhenian Sea. *Geophys. Res. Lett.* **2019**, *46*, 1562–1570. [CrossRef]
32. Castagno, P.; de Ruggiero, P.; Pierini, S.; Zambianchi, E.; De Alteris, A.; De Stefanoet, M.; Budillon, G. Hydrographic and dynamical characterization of the Bagnoli-Coroglio Bay (Gulf of Naples, Tyrrhenian Sea). *Chem. Ecol.* **2020**. [CrossRef]
33. ISPRA-CARG. Geological Map of Italy 1:50,000 Scale, Sheet 464 "Ischia". 2009. Available online: https://www.isprambiente.gov.it/Media/carg/464_ISOLA_DISCHIA/Foglio.html (accessed on 5 August 2020).
34. ISPRA-CARG. Geological Map of Italy 1:50,000 Scale, Sheet 466 "Castellammare". 2009. Available online: https://www.isprambiente.gov.it/Media/carg/466_485_SORRENTO_TERMINI/Foglio.html (accessed on 5 August 2020).
35. IOC. *Manual on Sea Level Measurement and Interpretation: Volume I—Basic Procedures*; IOC Manuals and Guides 14; UNESCO: Paris, France, 1985; p. 75.
36. IOC. *Manual on Sea Level Measurement and Interpretation: Volume II—Emerging Technologies*; IOC Manuals and Guides 14; UNESCO: Paris, France, 1994; p. 52.
37. IOC. *Manual on Sea Level Measurement and Interpretation: Volume IV—An Update to 2006*; IOC Manuals and Guides 14; UNESCO: Paris, France, 2006; p. 80.
38. ISPRA. *Manuale di Mareografia e Linee Guida per i Processi di Validazione dei Dati Mareografici*; Manuali e Linee Guida; ISPRA: Rome, Italy, 2012; ISBN 978-88-448-0532-6.
39. ISPRA. *Linee Guida pe L'analisi e L'elaborazione Statistica di Base Delle Serie Storiche di Dati Idrologici*; ISPRA: Rome, Italy, 2013.
40. Pugh, D.T. *Tides, Surges and Mean Sea-Level*; John Wiley & Sons: New York, NY, USA, 1987.
41. Doodson, A.T. The analysis of tidal observations. *Philos. Trans. R. Soc. Lond. Ser. A Contain. Pap. Math. Phys. Character* **1928**, *227*, 223–279.
42. Woodworth, P.; Melet, A.; Marcos, M.; Ray, R.D.; Wöppelmann, G.; Sasaki, Y.N.; Cirano, M.; Hibbert, A.; Huthnance, J.M.; Monserrat, S.; et al. Forcing Factors Affecting Sea Level Changes at the Coast. *Surv. Geophys.* **2019**, *40*, 1351–1397. [CrossRef]
43. Roden, G.I.; Rossby, H.T. Early Swedish contribution to oceanography: Nils Gissler (1715–1771) and the inverted barometer effect. *Bull. Am. Meteorol. Soc.* **1999**, *80*, 675–682. [CrossRef]
44. Rogers, J.C. The association between the North Atlantic Oscillation and the Southern Oscillation in the Northern Hemisphere. *Mon. Weather Rev.* **1984**, *112*, 1999–2015. [CrossRef]

45. Rogers, J.C. Atmospheric circulation changes associated with the warming over the northern North Atlantic in the 1920s. *J. Clim. Appl. Meteorol.* **1985**, *24*, 1303–1310. [CrossRef]

46. Hurrel, J.W. Decadal Trends in the North Atlantic Oscillation: Regional Temperatures and Precipitation. *Science* **1995**, *269*, 676–679. [CrossRef] [PubMed]

47. Rodwell, M.J.; Rowell, D.P.; Folland, C.K. Oceanic forcing of the wintertime North Atlantic Oscillation and European climate. *Nature* **1999**, *398*, 320–323. [CrossRef]

48. Eshel, G.; Cane, M.A.; Farrell, B.F. Forecasting eastern Mediterranean droughts. *Mon. Weather Rev.* **2000**, *128*, 3618–3630. [CrossRef]

49. Capozzi, V.; Budillon, G. Detection of heat and cold waves in Montevergine time series (1884–2015). *Adv. Geosci.* **2017**, *44*, 35–51. [CrossRef]

50. Rezaeian, M.; Mohebalhojeh, A.R.; Ahmadi-Givi, F.; Nasr-Esfahany, M. A wave-activity view of the relation between the Mediterranean storm track and the North Atlantic Oscillation in winter. *Q. J. R. Meteorol. Soc.* **2016**, *142*, 1662–1671. [CrossRef]

51. Hastenrath, S.; Greischar, L. The North Atlantic Oscillation in the NCEP-NCAR reanalysis. *J. Clim.* **2001**, *14*, 2404–2413. [CrossRef]

52. Barnston, A.G.; Livezey, R.E. Classification, seasonality and persistence of low-frequency atmospheric circulation patterns. *Mon. Weather Rev.* **1987**, *115*, 1083–1126. [CrossRef]

53. Han, W.; Stammer, D.; Thompson, P.; Ezer, T.; Palanisamy, H.; Zhang, X.; Domingues, C.; Zhang, L.; Yuan, D. Impacts of basin-scale climate modes on coastal sea level: A review. *Surv. Geophys.* **2019**. [CrossRef] [PubMed]

54. Tsimplis, M.N.; Shaw, A.G.P. The forcing of mean sea level variability around Europe. *Glob. Planet. Chang.* **2008**, *63*, 196–202. [CrossRef]

55. Wakelin, S.L.; Woodworth, P.L.; Flather, R.A.; Williams, J.A. Sea-level dependence on the NAO over the NW European Continental Shelf. *Geophys. Res. Lett.* **2003**, *30*, 1403. [CrossRef]

56. Woolf, D.K.; Shaw, A.G.P.; Tsimplis, M.N. The influence of the North Atlantic Oscillation on sea-level variability in the North Atlantic region. *J. Atmos. Ocean. Sci.* **2003**, *9*, 145–167. [CrossRef]

57. Yan, Z.W.; Tsimplis, M.N.; Woolf, D. Analysis of the relationship between the North Atlantic oscillation and sea-level changes in northwest Europe. *Int. J. Climatol.* **2004**, *24*, 743–758. [CrossRef]

58. Hughes, C.W.; Meredith, C.P. Coherent sea-level fluctuations along the global continental slope. *Philos. Trans. R. Soc. A* **2006**, *364*, 885–901. [CrossRef]

59. Tsimplis, M.N.; Shaw, A.G.P.; Flather, R.A.; Woolf, D.K. The influence of the North Atlantic Oscillation on the sea-level around the northern European coasts reconsidered: The thermosteric effects. *Philos. Trans. R. Soc. A* **2006**, *364*, 845–856. [CrossRef]

60. Miller, L.; Douglas, B.C. Gyre-scale atmospheric pressure variations and their relation to 19th and 20th century sea level rise. *Geophys. Res. Lett.* **2007**, *34*, L16602. [CrossRef]

61. Gomis, D.; Ruiz, S.; Sotillo, M.G.; Alvarez-Fanjul, E.; Terradas, J. Low frequency Mediterranean Sea level variability: The contribution of atmospheric pressure and wind. *Glob. Planet. Chang.* **2008**, *63*, 215–229. [CrossRef]

62. Calafat, F.M.; Chambers, D.P.; Tsimplis, M.N. Mechanisms of decadal sea level variability in the eastern North Atlantic and the Mediterranean Sea. *J. Geophys. Res. Oceans* **2012**, *117*, C09022. [CrossRef]

63. Tsimplis, M.N.; Calafat, F.M.; Marcos, M.; Jorda, G.; Gomis, D.; Fenoglio-Marc, L.; Struglia, M.V.; Josey, S.A.; Chambers, D.P. The effect of the NAO on sea level and on mass changes in the Mediterranean Sea. *J. Geophys. Res. Oceans* **2013**, *118*, 944–952. [CrossRef]

64. Dangendorf, S.; Calafat, F.M.; Arns, A.; Wahl, T.; Haigh, I.D.; Jensen, J. Mean sea level variability in the North Sea: Processes and implications. *J. Geophys. Res. Oceans* **2014**, *119*, 6820–6841. [CrossRef]

65. Ezer, T.; Haigh, I.D.; Woodworth, P.L. Nonlinear sea-level trends and long-term variability on Western European coasts. *J. Coast. Res.* **2016**, *32*, 744–755. [CrossRef]

66. Brunetti, M.; Maugeri, M.; Nanni, T. Atmospheric circulation and precipitation in Italy for the last 50 years. *Int. J. Climatol.* **2002**, *22*, 1455–1471. [CrossRef]

67. Conte, M.; Giuffrida, A.; Tedesco, S. *The Mediterranean Oscillation. Impact on Precipitation and Hydrology in Italy Climate Water*; Publications of the Academy of Finland: Helsinki, Finland, 1989.

68. Colacino, M.; Conte, M. Greenhouse effect and pressure patterns in the Mediterranean Basin. *Il Nuovo Cimento C* **1993**, *16*, 67–76. [CrossRef]

69. Palutikof, J.P.; Conte, M.; Casimiro Mendes, J.; Goodess, C.M.; Espirito Santo, F. Climate and climate change. In *Mediterranean Desertification and Land Use*; Brandt, C.J., Thornes, J.B., Eds.; John Wiley and Sons: London, UK, 1996.

70. Piervitali, E.; Colacino, M.; Conte, M. Rainfall over the central–western Mediterranean basin in the period 1951–1995. Part II: Precipitation scenarios. *Il Nuovo Cimento C* **1999**, *22*, 649–661.

71. Palutikof, J.P. Analysis of Mediterranean climate data: Measured and modelled. In *Mediterranean Climate: Variability and Trends*; Bolle, H.J., Ed.; Springer: Berlin, Germany, 2003.

72. Dietz, E.J.; Killeen, T.J. A Nonparametric Multivariate Test for Monotone Trend with Pharmaceutical Applications. *J. Am. Stat. Assoc.* **1981**, *76*, 169–174. [CrossRef]

73. Gocic, M.; Trajkovic, S. Analysis of changes in meteorological variables using Mann–Kendall and Sen's slop e estimator statistical tests in Serbia. *Glob. Planet. Chang.* **2013**, *100*, 172–182. [CrossRef]

74. Fusco, G.; Artale, V.; Cotroneo, Y.; Sannino, G. Thermohaline variability of Mediterranean Water in the Gulf of Cádiz, 1948–1999. *Deep-Sea Res. I* **2008**, *55*, 1624–1638. [CrossRef]

75. Mann, H.B. Nonparametric tests against trend. *Econometrica* **1945**, *13*, 245–259. [CrossRef]

76. Kendall, M.G. *Rank Correlation Methods*; Griffin: London, UK, 1975.

77. Sen, P.K. Estimates of the regression coefficient based on Kendall's tau. *J. Am. Stat. Assoc.* **1968**, *63*, 1379–1389. [CrossRef]

78. Piccioni, G.; Dettmering, D.; Bosch, W.; Seitz, F. TICON: TIdal CONstants based on GESLA sea-level records from globally located tide gauges. *Geosci. Data J.* **2019**, *6*, 97–104. [CrossRef]

79. Pawlowicz, R.; Beardsley, R.; Lentz, S. Classical tidal harmonic analysis including error estimates in MATLAB using T_TIDE. *Comput. Geosci.* **2002**, *28*, 929–937. [CrossRef]

80. Lama, R.; Corsini, S. *Analisi dei dati Storici della Rete Mareografica Italiana*; Ed. Poligrafico dello Stato: Rome, Italy, 2003.

81. Mosetti, F.; Purga, N. Le maree del mar Tirreno. *Boll. Oceanol. Teor. Appl.* **1985**, *3*, 83–102.

82. Rapolla, A.; Paoletti, V.; Secomandi, M. Seismically-induced landslide susceptibility evaluation: Application of a new procedure to the island of Ischia, Campania Region, Southern Italy. *Eng. Geol.* **2010**, *114*, 10–25. [CrossRef]

83. Manzo, M.; Ricciardi, G.P.; Casu, F.; Ventura, G.; Zeni, G.; Borgstrom, S.; Berardino, P.; Del Gaudio, C.; Lanari, R. Surface deformation analysis in the Ischia Island (Italy) based on spaceborne radar interferometry. *J. Volcanol. Geotherm. Res.* **2006**, *151*, 399–416. [CrossRef]

84. Sepe, V.; Atzori, S.; Ventura, G. Subsidence due to crack closure and depressurization of hydrothermal systems: A case study from Mt Epomeo (Ischia Island, Italy). *Terra Nova* **2007**, *19*, 127–132. [CrossRef]

85. Ricco, C.; Petrosino, S.; Aquino, I.; Del Gaudio, C.; Falanga, M. Some Investigations on a Possible Relationship between Ground Deformation and Seismic Activity at Campi Flegrei and Ischia Volcanic Areas (Southern Italy). *Geosciences* **2019**, *9*, 222. [CrossRef]

86. Cubellis, E.; Luongo, G.; Obrizzo, F.; Sepe, V.; Tammaro, U. Contribution to knowledge regarding the sources of earthquakes on the island of Ischia (Southern Italy). *Nat. Hazards* **2020**, *100*, 955–994. [CrossRef]

87. De Martino, P.; Tammaro, U.; Obrizzo, F.; Sepe, V.; Brandi, G.; D'Alessandro, A.; Dolce, M.; Pingue, F. La rete GPS dell'isola d'Ischia: Deformazioni del suolo in un'area vulcanica attiva (1998–2010). In *Quaderni di Geofisica INGV Roma*; INGV Istituto Nazionale di Geofisica e Vulcanologia: Rome, Italy, 2011; ISSN 1590-2595-95:4-23; Available online: http://istituto.ingv.it/images/collane-editoriali/quaderni-di-geofisica/quaderni-di-geofisica-2011/quaderno95.pdf (accessed on 25 June 2020).

88. Grablovitz, G. Il mareografo d'Ischia in relazione ai bradisismi. *Boll. Della Soc. Sismol. Ital.* **1911**, *15*, 144–153.

89. Bonaduce, A.; Pinardi, N.; Oddo, P.; Spada, G.; Larnicol, G. Sea-level variability in the Mediterranean Sea from altimetry and tide gauges. *Clim. Dyn.* **2016**, 1–16. [CrossRef]

90. Milošević, D.D.; Savić, S.M.; Pantelić, M.; Stankov, U.; Žiberna, I.; Dolinaj, D.; Leščešen, I. Variability of seasonal and annual precipitation in Slovenia and its correlation with large-scale atmospheric circulation. *Open Geosci.* **2016**, *8*, 593–605. [CrossRef]

91. Folland, C.K.; Knight, J.; Linderholm, H.W.; Fereday, D.; Ineson, S.; Hurrell, J.W. The summer North Atlantic Oscillation: Past, present, and future. *J. Clim.* **2009**, *22*, 1082–1103. [CrossRef]

92. Grinsted, A.; Moore, J.C.; Jevrejeva, S. Application of the cross wavelet transform and wavelet coherence to geophysical time series. *Nonlin. Process. Geophys.* **2004**, *11*, 561–566. [CrossRef]

93. Landerer, F.W.; Volkov, D.L. The anatomy of recent large sea level fluctuations in the Mediterranean Sea. *Geophys. Res. Lett.* **2013**, *40*. [CrossRef]

94. Fukumori, I.; Menemenlis, D.; Lee, T. A near-uniform basinwide sea level fluctuation of the Mediterranean Sea. *J. Phys. Oceanogr.* **2007**, *37*, 338–358. [CrossRef]

Article

Relative Sea-Level Rise and Potential Submersion Risk for 2100 on 16 Coastal Plains of the Mediterranean Sea

Fabrizio Antonioli [1,2,*], Giovanni De Falco [3], Valeria Lo Presti [4], Lorenzo Moretti [2], Giovanni Scardino [5], Marco Anzidei [6], Davide Bonaldo [7], Sandro Carniel [8], Gabriele Leoni [9], Stefano Furlani [10], Antonella Marsico [5], Marcello Petitta [2], Giovanni Randazzo [11], Giovanni Scicchitano [12] and Giuseppe Mastronuzzi [5]

[1]   Istituto Nazionale di Geofisica e Vulcanologia, 00143 Rome, Italy
[2]   ENEA, 40129 Bologna, 00123 Rome, Italy; lorenzo.moretti@enea.it (L.M.); marcello.petitta@enea.it (M.P.)
[3]   Institute of Anthropic Impacts and Sustainability in Marine Environment—IAS CNR, 09170 Oristano, Italy; giovanni.defalco@cnr.it
[4]   Studio Tecnico, 90142 Palermo, Italy; valeria.lopresti@gmail.com
[5]   Department of Earth and Geoenvironmental Sciences, University of Bari, 70125 Bari, Italy; giovanni.scardino@uniba.it (G.S.); antonella.marsico@uniba.it (A.M.); giuseppe.mastronuzzi@uniba.it (G.M.)
[6]   National Institute of Geopysics and Volcanomology, 00143 Rome, Italy; marco.anzidei@ingv.it
[7]   National Research Council, Institute of Marine Sciences (CNR-ISMAR), 30122 Venice, Italy; davide.bonaldo@ve.ismar.cnr.it
[8]   National Research Council, Institute of Polar Science (CNR-ISP), 30122 Venice, Italy; sandro.carniel70@gmail.com
[9]   ISPRA, 00144 Roma, Italy; gabriele.leoni@isprambiente.it
[10]   Department of Mathematics and Geosciences, University of Trieste, 34125 Trieste, Italy; sfurlani@units.it
[11]   Department of Mathematics, Pysics and Geosciences, MIFT, University of Messina, 98166 Messina, Italy; grandazzo@unime.it
[12]   Studio Geologi Associati TST, 95129 Catania, Italy; scicchitano@studiogeologitst.com
*   Correspondence: fabrizioantonioli2@gmail.com

Received: 5 June 2020; Accepted: 29 July 2020; Published: 1 August 2020

Abstract: The coasts of the Mediterranean Sea are dynamic habitats in which human activities have been conducted for centuries and which feature micro-tidal environments with about 0.40 m of range. For this reason, human settlements are still concentrated along a narrow coastline strip, where any change in the sea level and coastal dynamics may impact anthropic activities. In the frame of the RITMARE and the Copernicus Projects, we analyzed light detection and ranging (LiDAR) and Copernicus Earth Observation data to provide estimates of potential marine submersion for 2100 for 16 small-sized coastal plains located in the Italian peninsula and four Mediterranean countries (France, Spain, Tunisia, Cyprus) all characterized by different geological, tectonic and morphological features. The objective of this multidisciplinary study is to provide the first maps of sea-level rise scenarios for 2100 for the IPCC RCP 8.5 and Rahmstorf (2007) projections for the above affected coastal zones, which are the locations of touristic resorts, railways, airports and heritage sites. On the basis of our model (eustatic projection for 2100, glaciohydrostasy values and tectonic vertical movement), we provide 16 high-definition submersion maps. We estimated a potential loss of land for the above areas of between about 148 km^2 (IPCC-RCP8.5 scenario) and 192 km^2 (Rahmstorf scenario), along a coastline length of about 400 km.

Keywords: Mediterranean Sea; coastal plains; relative sea-level rise; 2100; marine submersion

1. Introduction

Sea-level rise is one of the major consequences of climate change which affects global communities living along coasts. This phenomenon is due to planetary-scale processes that are contributing to the current sea-level trend. The main responsible factors are global warming, which is driving the melting of ice and the thermal expansion of the oceans and finally the geological vertical movements of the land along the coastal zones, which may accelerate (or decrease) the submersion of low-elevated coasts.

Instrumental and observational data show that over the past two centuries, the global sea level has risen at faster rates than in the last two millennia [1–3], with values up to 3.2 mm/year over the last decades [4–7]. Recently, Toimil et al. [8] delineated the requirements for a scientific approach for future projections of shoreline change, which should take into account changes in the mean sea level, storm surges, coastal erosion and other additional factors, with the aim of reducing uncertainty in shoreline change estimates.

This rise is exposing many coastal areas around the world to coastal hazards and marine flooding. Sea-level rise and storm surge has been the indirect cause of dramatic effects in conjunction with extreme storm surge events, such as in 1953 for the coast of Netherlands (1953) or more recently in New Jersey (2012, Hurricane Sandy), Louisiana (2015, Hurricane Katrina), Florida (2016, Hurricane Matthew), Louisiana and Texas (2017, Hurricane Harvey).

In the Mediterranean basin, the sea level is rising at 1.8 mm/year, as inferred from the analysis of tidal gauge data for the last century. In 2100, the sea level could be 500–1400 mm higher than today due to the melting of continental ice and thermal expansion/steric effects as a consequence of global warming.

In this scenario, the coastal low lands will be more prone to marine flooding during extreme sea events, threating many highly populated areas of the Mediterranean coasts.

In the last two decades, many studies have focused on sea-level rise and its effects along the continental (i.e., Gornitz et al. [9] for New York; Walsh et al. [10] for Australia; Miller et al. [11], Ezer and Atkinson [12] for USA; Rehman et al. [13] for India) and insular coasts (i.e., Wadey et al. [14] for Maldives) highlighting the relevant exposure to coastal hazard for lowlands and small islands. These studies are relevant in the Mediterranean, where about 75% of the local population lives in coastal areas [15] and is therefore exposed to the multiple effects of sea-level rise. Although several studies have attempted to predict global sea-level rise for 2100 [2,16–20] or even until 2200 [21], only in a few cases have sea-level rise projections been used in combination with high-resolution digital terrain models (DTM) and with geological data to evaluate the geomorphological landscape changes at different time scales, to draw detailed maps of the expected coastal submersion. In the Mediterranean, many coasts are expected to be flooded as a consequence of sea-level rise, storm surge and tsunamis, as inferred from seismic, geodetic, geological and archaeological evidence ([22,23], www.savemedcoasts.eu [24]). In this region, rapid urbanization started after the middle of the 19th century, which led to the uncontrolled expansion of coastal settlements; these settlements are now exposed to increasing coastal hazards. In addition, the coasts, which are often characterized by cultural and natural heritage sites, host important urban and industrial installations and continuously growing tourist activities.

An integrated approach to the analysis of the impacts of climate change in coastal areas is crucial for a thorough projection of coastal hazards and for the identification of response strategies and priorities. To this end, in the frame of the RITMARE project [25], the Adriatic Sea (in the north-eastern Mediterranean, Figure 1) has been considered a test site for a multidisciplinary study on coastal vulnerability to sea-level rise and erosion. This activity, funded by the Italian Ministry of University and Research, provided a proof-of-concept for how different observational and numerical modelling tools can be combined in the characterization of the physical drivers of coastal processes in a changing climate perspective. As emerged from the outcomes of RITMARE, the spatial modulation of interplaying geological/sedimentological and meteo-oceanographic processes in such complex systems puts serious limitations on the applicability of global and regional climate change projections for the assessment of the local features of coastal vulnerability. A typical example is given by the possible impacts of storm surge hazards and erosion in future scenarios. Notwithstanding a general tendency toward a weaker storminess at the Mediterranean scale, the increasing sea level is expected to supersede the decrease of sea state severity [26,27]. In addition, local intensifications of the wave climate are possible in response to the northbound migration of the Mediterranean cyclone tracks, contributing to the further enhancement of the potential coastal hazard in coastal regions [28].

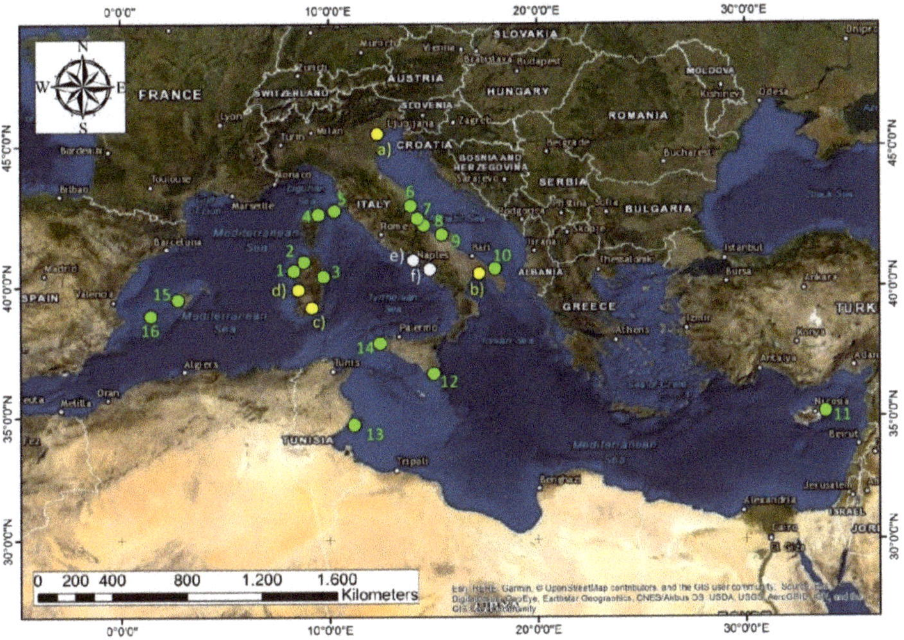

Figure 1. The Mediterranean coastal plains investigated in this paper (green dots) and previous studies by Antonioli et al. [29], Marsico et al. [30] (yellow marks and letters), Aucelli et al. [31,32] and Giordano et al. [33] (white marks and letters).

In the frame of the C3S European Tourism CLIMTOUR project (a sectorial climate service in the field of tourism which engages and interacts with a number of stakeholders across the six sectors of agriculture, tourism, insurance, coasts, infrastructure and health to establish an inventory of existing policy needs and user requirements in terms of climate data and climate impact indicators), researchers have been investigating the potential scenarios of sea-level rise in specific coasts of the Mediterranean European countries. Moreover, in the framework of the "Integrated Sea Storm Management Strategies" project (IWS, I-STORMS Web System; https://iws.seastorms.eu [34,35]) by the National Research Council of Italy Marine Sciences Institute (CNR-ISMAR), the Agency for Prevention, Environment and Energy of Emilia-Romagna, the Hydro-Meteo-Climate Service (Arpae-SIMC), and Apulia Region Civil Protection, some other areas have been investigated along the coasts of Apulia in Southern Italy.

In this study, sea-level projections for 2100 estimated by the two global models IPCC-AR5 [36] and Rahmstorf [17] were used according to the method reported in Antonioli et al. [29] and Marsico et al. [30]. The proposed scenarios for the selected coastal plains include the contribution of vertical land movements (tectonics and isostasy) with rates based on the interpretation of geological data or modeling analysis. Our maps show the maximum sea-level height expected for 2100 for the above reference climatic projections and the corresponding flooded area. We remark that our analysis does not take into account hydrodynamics models, and the contribution of sediment flow from rivers, coastal erosion and possible anthropic defenses may change the estimated extension of the flooded areas proposed in this study. This work provides additional scenarios with respect to previous studies [23,29,31,33,37–46] focusing on new areas which are at risk of marine submersion over the next decades.

2. Material and Method

This study is based on the same method described in [29] and [30], to ensure homogeneity and enable a comparison with previous results, but extending the study toward other Mediterranean countries, besides Italy.

Sea-level change along the Mediterranean coast is the sum of eustatic, glaciohydro-isostatic, and tectonic factors. The first is time dependent while the latter two also vary with location, consists to sum the different components of sea level rise in the following main steps: (a) the IPCC-AR5 projections (RCP-8.5 upper limits scenarios) or Rahmstorf 2007; (b) the long term land vertical movements from geological data; (c) the glacio-hydro-isostatic movement (GIA); (d) by combining eustatic, isostatic and tectonic data projected up to 2100, we provided the expected sea-levels at 2100 for the investigated coastal areas and the expected inland extent of related marine flooding.

The choice of the study areas was decided by several different factors: (i) tectonics (stable areas only); (ii) exposure, (iii) max fetch, (iv) sedimentological material, (v) wave energy flux, (vi) bedrock and (vii) geomorphological features.

For the realization of the 16 maps (Figure 1 for geographic location, Figures 2–9) we preferred to use the global projections released by the IPCC AR5 in 2013, with the aim of comparing the new results with those previously obtained by [29] and [30], using the same parameters and models. Particularly the upper limit of the likely range of IPCC AR-5 RCP 8.5 [36] and Rahmstorf [17] scenarios (Figures 2–9 and Figure S1). The latter are in good agreement with the most recent global projections, once the uncertainties on the estimates are considered in the analysis (Table 1 and references therein).

Those values have been used in our maps to evaluate the uncertainties of the relative sea-level rise.

Most of the studied areas are well covered by EO data that were retrieved from the Internet from different agencies (Table 2) and analyzed through Geographic Information System software to obtain information on surface features and elements exposed to sea-level rise.

For the relative sea level rise case studies, we used the IPCC global mean sea level rise estimation. Indeed, the local data for the Mediterranean Sea are not accurate for several reasons. First of all, most of the global models do not consider the exchange of mass and energy across the semi-enclosed basin of the Mediterranean Sea. Therefore, we decided to not estimate the uncertainties assuming to follow the upper limit of the likely range of the RCP8.5.

Most of the studied areas are well covered by geographic information, as GIS layers, describing topographical features, morphology and exposed elements. This availability led to the generation of a geographical database and the use of GIS tools for all study phases. GIS tools enable the use of original data, thus preventing any loss of detail during the elaboration. In addition, a GIS archive allows the easy update of the resulting maps depending either on the environmental changes or on the refinement of the sea-level rise estimates.

The DTMs used for the 16 coastal areas (Tables 2 and 3) have different resolutions. The realization of the maps of (1) Fertilia, (2) Valledoria, (3) Orosei, (7) Sangro, (8) Pescara, (9) Lesina, (10) Brindisi, (12) Granelli and (14) Stagnone di Marsala were also supported by in situ surveys including low-elevation aerial views through unmanned aerial vehicle (UAV) systems and some geomorphological observations, such as the measurements of dune size, and by checking for drainage and protection systems capable of preventing or reducing the ongoing submersion (Figures 2–9; Figure S1 and Video S1).

Table 1. Global projections (cm) of sea-level rise at 2100 for different authors.

RCP	IPCC 2013 cm	IPCC 2019 cm	Kopp et al., 2016 cm	Mengel et al., 2016 cm	Horton et al., 2014 cm	Rahmstorf 2007 cm	Bamber al., 2019 cm
RCP 8.5	53–97	61–110	52–131	57–131	50–150	50–140	21–163

Table 2. Link to the websites and data (shapefiles).

Link	Project	Map Number
http://www.pcn.minambiente.it/mattm/	RITMARE	5, 6, 7, 8, 9, 10
http://www.sitr.regione.sicilia.it/?page_id=419	RITMARE	14
	CLIMTOUR	12
http://www.sardegnageoportale.it/areetematiche/modellidigitalidielevazione/	CLIMTOUR	1, 2, 3
http://centrodedescargas.cnig.es/CentroDescargas/buscador.do#	CLIMTOUR	15, 16
https://land.copernicus.eu/	CLIMTOUR	4, 11, 13
https://www.geoportail.gouv.fr/carte	CLIMTOUR	4

Table 3. A: site number; B: coastal zone; C: (digital terrain model, DTM) resolution and accuracy; D: epoch of DTM; E: IPCC projection AR5-RCP8.5 for 2100; F: IPCC projection AR5-RCP8.5 for 2050; G: Rahmstorf projection (2007); H: vertical tectonic rate mm/year; I: glacial isostatic adjustment (GIA) rate (Lambeck et al., 2011); J: relative sea-level rise for 2050, IPCC AR5-RCP8.5 scenario; K: relative sea-level rise for 2100, IPCC AR5-RCP8.5 scenario; L: sea-level rise for 2100, Rahmstorf (2007); M: relative sea-level rise for SROCC 2100; N: potential flooded area in 2100, IPCC RCP8.5; O: flooded area for 2100 in km^2, Rahmstorf (2007); P: exposed coastline length (km^2).

A	B	C	D	E	F	G	H	I	J	K	L	M	N	O	P
Site no.	Coastal Zone	DTM Resolution and Vertical Accuracy (m)	Year	SLR Projection 2100 IPCC-8.5 (2013) (mm)	SLR Projection 2050 IPCC-8.5 -2013 (mm)	SLR Projection 2100 Rahmstorf -2007 (mm)	Vertical Tectonic Rate (mm/year)	GIA (mm/year)	Total SLR -IPCC (2013) 2050 (mm)	Total SLR- IPCC 2100 (2013) (mm)	Total SLR 2100 Rahmstorf (2007) (mm)	Total SLR-IPCC SROCC 2100 (mm)	Flooded Area 2100 IPCC-8.5 (km^2)	Flooded area 2100 Rahmstorf (2007) (km^2)	Exposed Coastline Length (km)
1	Fertilia	LiDAR 5 × 5 (±0.2)	2008	970	225	1400	0	0.57	249	1022 ±134	1452 ±134	1152.44 ±134	1.7	2.3	6.2
2	Valledoria	LiDAR 1 × 1 (±0.2)	2008	970	225	1400	0	0.57	249	1022 ±134	1452 ±134	1152.44 ±134	1	2.2	5.6
3	Orosei	LiDAR 1 × 1 (±0.2)	2008	970	225	1400	0	0.62	251	1027 ±134	1457 ±134	1157.04 ±134	1.6	3	7.6
4	Bastia	25 × 25 ±1	2010	970	225	1400	0	0.44	243	1010 ±134	1440 ±134	1139.6 ±134	21	25	29.4
5	Marina di campo	LiDAR 2 × 2 (±0.2)	2013	970	225	1400	0	0.37	239	1002 ±134	1432 ±134	1132.19 ±134	0.1	0.34	4.8
6	Tronto	LiDAR 2 × 2 (±0.2)	2008	970	225	1400	0	0.3	238	998 ±134	1428 ±134	1127.6 ±134	0.06	0.09	9.2
7	Sangro	LiDAR 2 × 2 (±0.2)	2008	970	225	1400	0	0.37	241	1004 ±134	1434 ±134	1134.04 ±134	0.08	0.13	7.2
8	Pescara	LiDAR 2 × 2 (±0.2)	2008	970	225	1400	0	0.385	241	1005 ±134	1435 ±134	1135.42 ±134	0.3	0.5	20.7
9	Lesina	LiDAR 2 × 2 (±0.2) (coast) 1 × 1 (±0.2) (inland)	2008 2014	970	225	1400	See Section 3.2	0.42	243	1009 ±134	1439 ±134	1138.64 ±134	13	13	40
10	Brindisi	LiDAR 2 × 2 (±0.2)	2008	970	225	1400	0	0.44	243	1010 ±134	1440 ±134	1140.48 ±134	0.5	0.9	9.9
11	Larnaka Ciprus	25 × 25 ±1	2010	970	225	1400	0	0.21	227	989 ±134	1419 ±134	1118.9 ±134	10.3	11.6	32.7
12	Granelli	LiDAR 2 × 2 (±0.2)	2007 2008	970	225	1400	0	0.56	248	1022 ±134	1452 ±134	1151.52 ±134	5.4	6.8	14.4
13	Kerkennah	SRTM 30 × 30	2010	970	225	1400	0	0.25	234	999 ±134	1420 ±134	1122.5 ±134	82.3	109	220.1
14	Stagnone e saline di	LiDAR 2 × 2 (±0.2)	2008	970	225	1400	0	0.56	249	1022 ±134	1452 ±134	1151.52 ±134	5.4	6.9	18.5
15	Marsala Mallorca	LiDAR 5 × 5 (±0.2)	2010	970	225	1400	0	0.61	249	1025 ±134	1455 ±134	1154.9 ±134	1.3	3.7	23.1
16	Ibiza	LiDAR 5 × 5 (±0.2)	2010	970	225	1400	0	0.61	249	1025 ±134	1455 ±134	1154.9 ±134	2.9	4.2	51
	Total												148.5	192.4	407.8

2.1. Vertical Land Movements

Because the relative sea-level changes along the coastal zones also depend on the rates of the vertical land movements caused by tectonics and GIA, these geological processes were considered in the analysis, including natural ground compaction and GIA, as in Lambeck et al. [47]. To this end, we used the GIA values (Table 3) from [47] which have rates between 0.2 and 0.65 mm/year and assumed that they will be constant for the next 500 years. The Lambeck et al., 2011, is a model tested in the field of stable Mediterranean areas comparing radiocarbon ages of fossil shells sampled on cores crossing fossil lagoon areas (minimum slr error).

To account for vertical tectonics, we have used as reference the tectonic rates reported in Antonioli et al. [29] which are representative of a calibrated balance between the long-term tectonic signal (i.e., last Interglacial, [48,49]), the short-term (i.e., mid- to late Holocene, [50]; additional material of Lambeck et al. [47]) and instrumental data [22]. In addition to these, we used the upgraded Mediterranean database for the last 125 ka BP of Antonioli et al. [51], (see Section 3.1, Data). Finally, the GIA and tectonic values were included in the analysis and added to the sea-level rise projections to obtain the values of relative sea-level change expected for 2100 at each site. Vertical tectonic rates were calculated using the MIS 5.5 highstand (125 ka BP) altitude, considering the eustatic altitude in the Mediterranean Sea of 6 ± 2 m.

2.2. Digital Terrain Models

To map the sea-level rise scenarios, a data set of high-resolution topography based on light detection and ranging (LiDAR) observations produced by different agencies from 2008 to 2019 was used (Table 2). The extracted Digital Terrain Models (DTM) were obtained at variable spatial resolutions depending on the data set and in the range at about 20 cm of mean vertical resolution [52].

The details of the characteristics of the DTM are described in Table 3 and in the maps available in the online supporting material. The details regarding the link to the website from which we downloaded or requested the digital data are also described in the online supporting material. DTMs were mapped and analyzed by Global Mapper Software® (www.globalmapper.com [53]) (Version 21, Hallowell, ME, USA) to create 3D high-resolution maps of the investigated areas, on which the position of the present-day coastline and its potential position in 2100 as a result of relative sea-level rise are shown by contour lines. The DTMs with contour lines and submerged surfaces were represented using the color shaded option and exported as georeferenced images through GIS composer.

Since high-resolution LiDAR data were not available for all the investigated areas, the DTMs for Larnaka (Cyprus), Bastia (France) and Kerkennah Islands (Tunisia), were extracted from Earth Observation data retrieved at https://land.copernicus.eu/ [54]. The technical specifications of data sources and DTMs are shown in Table 3 and discussed in Section 4.

To link the land surface to the seafloor along the coasts, the bathymetric data were obtained from GEBCO (www.gebco.net) [55] which includes regional low-resolution grids) and the European Marine Observation and Data Network (EMODnet, http://portal.emodnet-bathymetry.eu/ [55]). Marine and terrestrial topographic data were co-registered and georeferenced into the same UTM-WGS84 (Zones 30, 31, 32 and 33, 36) reference frame, and the shoreline position was determined relative to the epoch of the surveys for each area.

The choice of study areas was dictated by several factors. The first factor was the tectonic activity; in fact, we chose only tectonically stable areas to eliminate a variable (with consequent error) in the sea-level change equation. As is known for relative sea-level change, the combined results of eustasy, glacio-hydro-isostasy and vertical tectonic motion exhibit considerable spatial variability; the first of these is time-dependent, while the latter two vary with location. As shown in Table 4, the coastal areas chosen for our research have very different exposures, maximum fetch, sedimentological material, wave energy flux, and kinds of geomorphology; our choice was precisely due to their great heterogeneity. A common characteristic of all the areas chosen was that of not having vertical tectonic movements.

Table 4. A: coastal areas chosen for our research; B, C: region and Country; D: coordinates; E: wind\wave exposure; F: maximum fetch (km); G: geological coastal material; H: annual mean energy flux (kW\m) according with [56]; I: kind of geomorphological outcrop; L: human made structures occupying the area planned to be submerged on 2100.

A	B	C	D	E	F	G	H	I	L
Coastal Site	Region	Country	Latitude and Longitude (Dedimal Degree)	Exposure Direction	Max Fetch (km)	Coastal Material	Wave Energy Flux kW/m	Geomorphology	Human-Made Structures
1 Fertilia	Sardinia	Italy	40.617889° N 8.200782° E Gr	S	505	Sand	7	Pocket beach	Fertilia town
2 Valledoria	Sardinia	Italy	40.961910° 8.840432°	NW	520	Sand ravels	4	Embayed Beach	Agricultural crops
3 Orosei	Sardinia	Italy	40.373995° 9.725289°	SW	574	Sand	1.5	Embayed Beach	Agricultural crops
4 Bastia	Corse	France	42.660470° 9.448431°	W	700	Rock	2	Barrier Lagoon	Bastia Airport
5 Marina di Campo	Island of Elba	Italy	42.748714° 10.238112°	SW	660	Sand	1,5	Pocket beach	Marina di Campo town
6 Tronto	Marche	Italy	42.896730° 13.911675°	WNW	518	Sand	1	River Delta/mainland beach	Agricultural crops
7 Sangro	Abruzzo	Italy	42.241935° 14.517749°	NW	423	Sand	1	River Delta/mainland beach	Agricultural crops
8 Pescara	Abruzzo	Italy	42.467166° 14.225779°	NW	518	Sand	1	River Delta/mainland beach	Pescara town
9 Lesina	Apulia	Italy	41.883915° 15.452643°	N	416	Sand	1.5	Coastal lake and lagoon, tombolo	Agricultural crops
10 Brindisi	Apulia	Italy	40.668235° 17.948512°	NE	688	Rock	1.5	Gently sloping rocky coast	Brindisi Airport
11 Larnaka	District of Larnaka	Cyprus	34.947170° 33.643878°	SW	403	Silt	5	Embayed Beach	Larnaka airport
12 Granelli	Sicily	Italy	36.700054° 15.024145°	SSW	770	Sand	3	Barrier Lagoon	Agricultural crops
13 Kerkennah	Governorate of Sfax	Tunisia	34.740386° 11.221183°	E	2167	Rock	2.5	Low-lying rock platform	Tourist activities
14 Marsala	Sicily	Italy	37.862205° 12.442988°	W	1605	Sand	5	Barrier Lagoon	Saltponds
15 Mallorca	Island of Mallorca	Spain	39.530120° 2.729394°	SW	637	Sand	5	Embayed Beach	Mallorca airport
16 Ibiza	Island of Ibiza	Spain	38.858203° 1.370749°	S	2730	Sand	3	Embayed Beach	Saltponds

3. Results

The 16 maps described in this section (Figures 1–9 and Figure S1, Table 3) show the expected submerged area for 2100 for the selected areas of Italy, Spain, France, Tunisia and Cyprus. For the sites investigated in the CLIMTOUR Project, 324.7 km of coast with a potential submerged area of 176.2 km^2 was estimated, while from the RITMARE Italian National Project and I-STORMS data set, 68.1 lm of at-risk coast with possible submerged areas of 13.8 km^2 and 9.9 km, and 1 km^2 for Brindisi (I-STORMS), were estimated, respectively.

3.1. Geographic Descripion of the Areas at Potential Risk of Submersion

In addition to the values reported in Table 3 for the studied areas (Figure 1), we provide some information on the infrastructures located in the selected areas, such as roads and transport networks, airports, and humid protected areas, which are prone to be flooded in 2100 (Figures 2–9 and Figure S1). In situ inspections were carried out at Fertilia, Valledoria, Orosei, Sangro, Pescara, Lesina, Brindisi, Granelli and Stagnone di Marsala (Video S1).

3.1.1. Fertilia

Map 1: This zone is one of the most important coastal wetlands in Sardinia and is locally called "Il Calich". It faces the bay of Alghero, extending from its neighbors to the town of Fertilia, and is connected to the sea by a channel Figure 2. The coast is characterized by very well-preserved coastal dunes up to 7 m high. The presence of *Posidonia oceanica* in the marine area and the dune–lagoon system could be able to retreat in the case of sea submersion. The sector exposed to submersion risk is presently a farming area, with the exception of a large campsite located between the lagoon and the road.

3.1.2. Valledoria

Map 2: The area includes the mouth of the Coghinas River, which is interesting due to the intense agricultural cultivation of artichokes (50% of Sardinian artichoke production). It is partly located in the fluvial marine arm of the mouth of the Coghinas River, which forms a protected zone parallel to the coastline. The latter is characterized by a 5–6 m high dune system on average, with up to 12 m of elevation. Because of the absence of roads or buildings on the dune area, with the exception of a campsite, the dunes are free to retreat in case of sea-level rise (Figure S1).

3.1.3. Orosei

Map 3: The Cedrino river flows near the Orosei bay, with an estuary mouth which is closed in dry periods by a sandy shoreline which forms a marshy area—particularly rich in fauna—in the area behind it. It is the fifth longest river in Sardinia and has caused dangerous flooding in the past. The construction of a dam 20 km upstream (under Dorgali) and 5 m high embankments in the area just before the mouth prevent the river from overflowing into the plain in the depressed areas shown on the map during the floods. There are human activities and agricultural fields which are protected from the floods of the river. This high-value coastal wetland retains a subtle balance between sediments brought by the river and the rising sea, Figure 3.

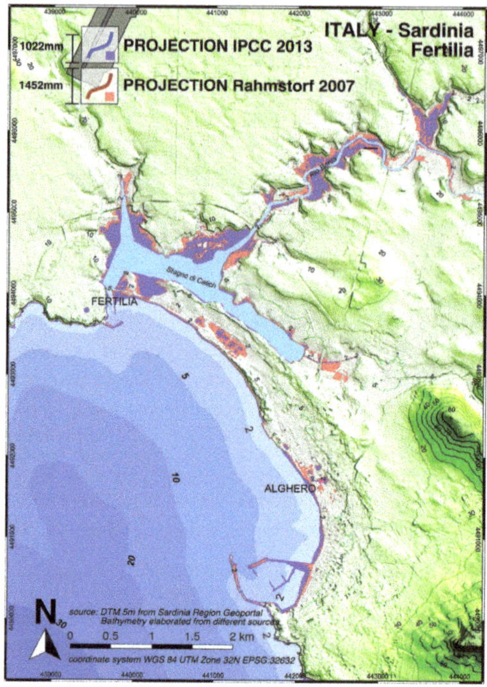

Figure 2. Map 1 Fertilia (Sardinia Italy, see also Figure 1 for location). The potential submersion area, using Rahmstorf 2007 projections are 2.3 km^2.

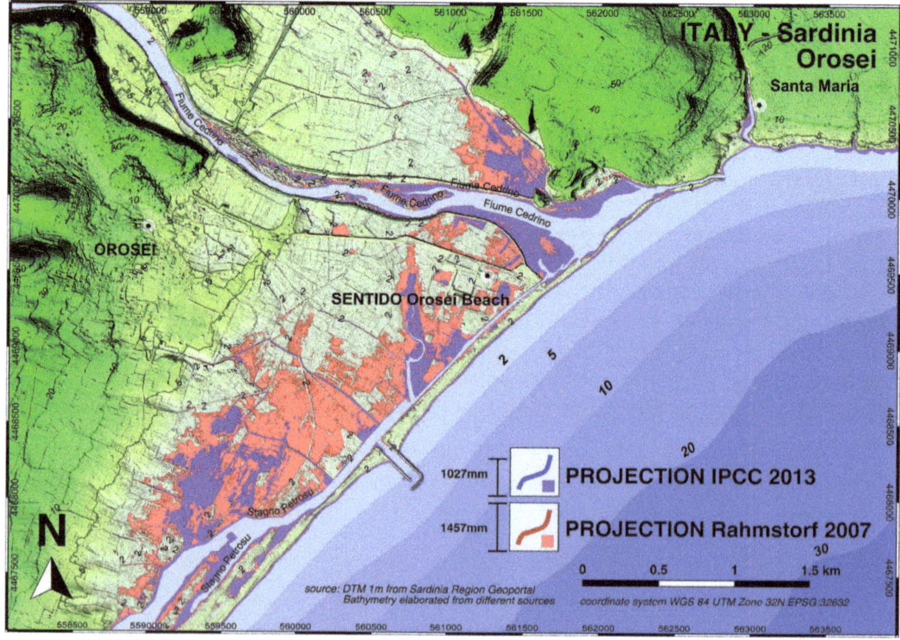

Figure 3. Map 3 Orosei (Sardinia Italy, see also Figure 1 for location). The potential submersion area, using Rahmstorf 2007 projections are 3.1 km^2.

3.1.4. Bastia

Map 4: The Biguglia lagoon is located south of Bastia (France) and is separated from the Tyrrhenian Sea by a sandy dune belt. It is an important wetland zone of interest that includes the largest lake in Corsica. It covers an area of 14.5 km^2, with a length of 11 km and a width of 2.5 km. The lagoon has two Natura 2000 protection zones and is a site of European relevance. Its large extent makes it an area of great ecological importance with high biodiversity in the Mediterranean basin due to the presence of avifauna fauna and aquatic flora. The aquatic fauna allows a large number of birds to nest, live and reproduce on the site. The external dune cord seems fairly well preserved, but there are many houses that will be exposed to sea-level rise if the dune recedes. Some rivers flow into the area and bring numerous sedimentary deposits. The area of potential submersion also falls in the northern portion of the Bastia airport (Figure S1).

3.1.5. Marina di Campo

Map 5: The Marina di Campo beach (Elba Island, Italy) is the largest of the islands (1.4 km). It is made of granite-source sand and surrounded by a thick pine forest. The dune no longer exists, and many houses in the town and the southern part of the airport are exposed to sea-level rise, Figure 4.

3.1.6. Tronto

Map 6: The Tronto River (Adriatic Sea, Italy) has a typical Apennine regime with strong floods in the rainy season (about 1500 m^3/s) and in summer. The coastal dune has been eroded, except for the northern coastal portion of the river-mouth, with depressed swampy areas and farming areas at risk of submersion in 2100. On the southern side is the inhabited center of Martinsicuro, with large beaches and bridles that collect the sediments of the river that feed the beaches. Some local houses will be exposed to marine submersion in 2100.

3.1.7. Sangro

Map 7: In this area, there are small depressed zones behind the beach, mainly devoted to agriculture. Although the railway and the highway are not exposed to marine submersion as they are protected by the continuous deposit of fluvial sediments along the coasts, we note that the dunes have been eroded. The fluvial–marsh geomorphological history of the plain is described in Parlagreco et al. [57] (Figure S1).

3.1.8. Pescara

Map 8: The geomorphological evolution of the largely anthropized coastal area of Pescara is characterized by a slow progradation [58]. About 7000 years ago, the area in which the sports field of Pescara is located was a lagoonal environment. The dune system has been completely eroded. There are small depressed areas within the city, but the flow of the river sediments should counteract any submersion (Figure S1).

3.1.9. Lesina

Map 9: The Lesina Lake (a protected natural area since 1981, now part of Gargano National Park) is a brackish lagoon and an important habitat for bird species, located along the northwestern shore of the Gargano promontory (Apulia, Italy) and connected to the Adriatic Sea through three narrow artificial channels. The depth of the brackish waters does not exceed two meters. Most of the dune belts are preserved, and they do not show evidence of fast retreat, representing an important obstacle to sea-level rise. The low lands bordering the lagoonal area are devoted to agriculture and zoo technique activities. The village of Marina di Lesina may experience local instabilities due to the karst sinking into the gypsum layers, as described in Section 3.2.

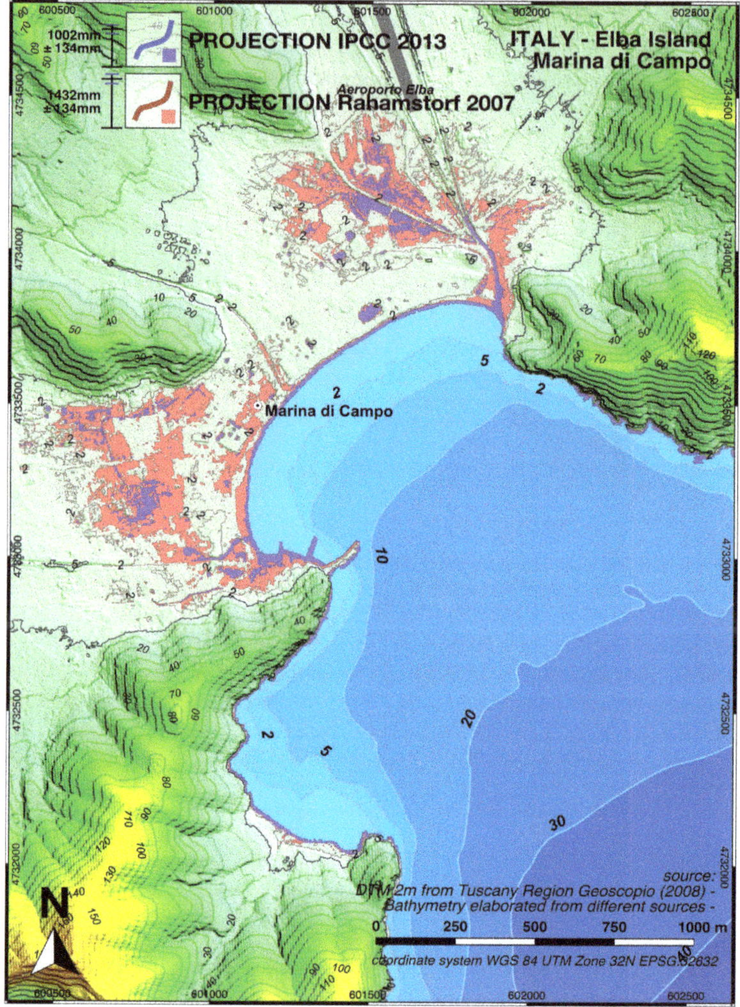

Figure 4. Map 5 Marina di Campo (Elba Island, Italy, see also Figure 1 for location). The potential submersion area, using Rahmstorf 2007 projections are 3.4 km^2.

3.1.10. Brindisi

Map 10: This coastal area is located in a very sensitive rocky area occupied by a peripheral part of Brindisi city, touristic settlements, archaeological and architectural sites, air–industrial facilities and the Casale airport, whose NE portion, at about 1 m above sea level, falls near the touristic harbor and connected shipyards. However, the main runway runs SE–NW at safe altitudes between 7 and 8 m (Figure S1).

3.1.11. Larnaka (Cyprus)

Map 11: The international airport area occupies the wetland area of Larnaka, with a runway placed at about 1 m above sea level. The low resolution of the map prevents any further detailed coastal hazard assessment, Figure 5.

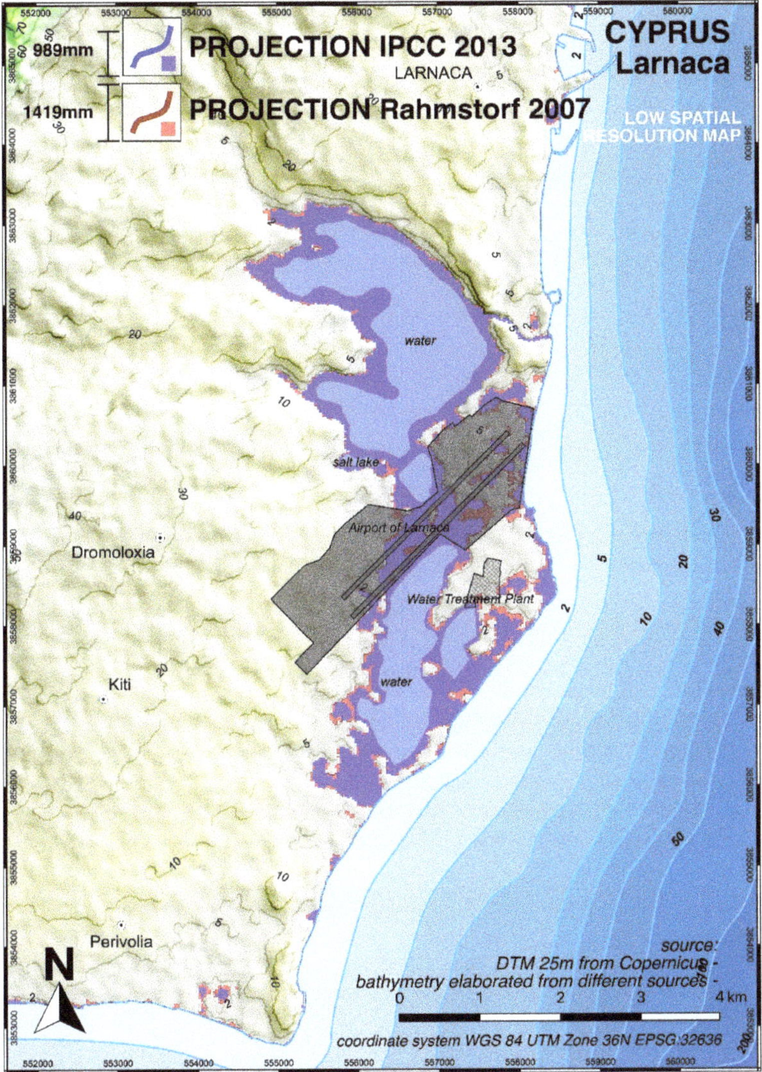

Figure 5. Map 11 Larnaka (Cyprus), see also Figure 1 for location). The potential submersion area, using Rahmstorf 2007 projections are 11.6 km^2.

3.1.12. Granelli (Italy)

Map 12: The Longarini lagoon is a wetland of about 0.02 km^2 and is known as Porto Ulisse due to its frequentation in Greek and Roman times; it is part of the wider Site of Community Importance (European Union (EU) Habitat Directive), which is one of the most important wetlands in the Mediterranean for the protection of biodiversity. It is separated from the Sicily Channel by a dune belt which is strongly urbanized and shows several paths that interrupt its continuity; thus, it does not avoid the sea flooding during extreme events and partial submersion in the next decades due to sea-level rise, Figure 6.

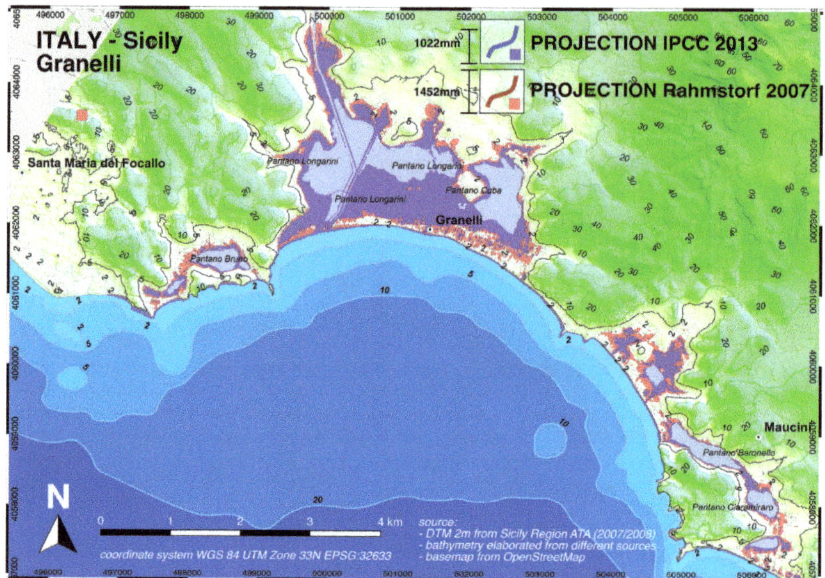

Figure 6. Map 12 Granelli (Sicily, Italy), see also Figure 1 for location). The potential submersion area, using Rahmstorf 2007 projections are 6.8 km^2.

3.1.13. Kerkennah

Map 13: This archipelago is located in the Gulf of Gabes (Tunisia)—a few kilometers from Sfax. It is composed of seven small and poorly inhabited rocky islands placed between 0 and 4–5 m above the sea level. In this area, the available data do not allow us to create a high-resolution DTM, preventing any reliable description of a submersion scenario. Being located close to the sea level in a stable tectonic environment, the lowest elevated coastal zones of this area are prone to be flooded by 2100, Figure 7.

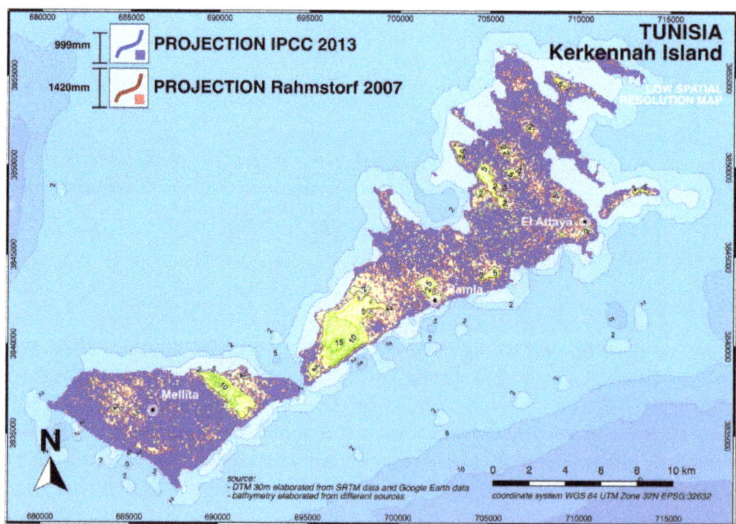

Figure 7. Map 13 Kerkennah (Tunisia), see also Figure 1 for location. The potential submersion area, using Rahmstorf 2007 projections are 109 km^2.

3.1.14. Stagnone e Saline di Marsala

Map 14: The Stagnone of Marsala is considered a semi-enclosed lagoon that was partially flooded by about 1 m during the Punic Age (2.3 ka BP) [41]. This area is occupied by the airport of Trapani Birgi and the nearby salt pans, with the latter being at risk of marine submersion and consequent economic loss. In particular, the runway of Birgi airport is placed between 7 and 2 m above sea level and is exposed to marine submersion during storm surges (Figure S1).

3.1.15. Mallorca

Map 15: The map shows an extremely anthropized low-elevation rocky coast, without dunes and with depressed marshy areas inside. Additionally, the north runway of the international airport is prone to marine submersion due to its low elevation above sea level, Figure 8.

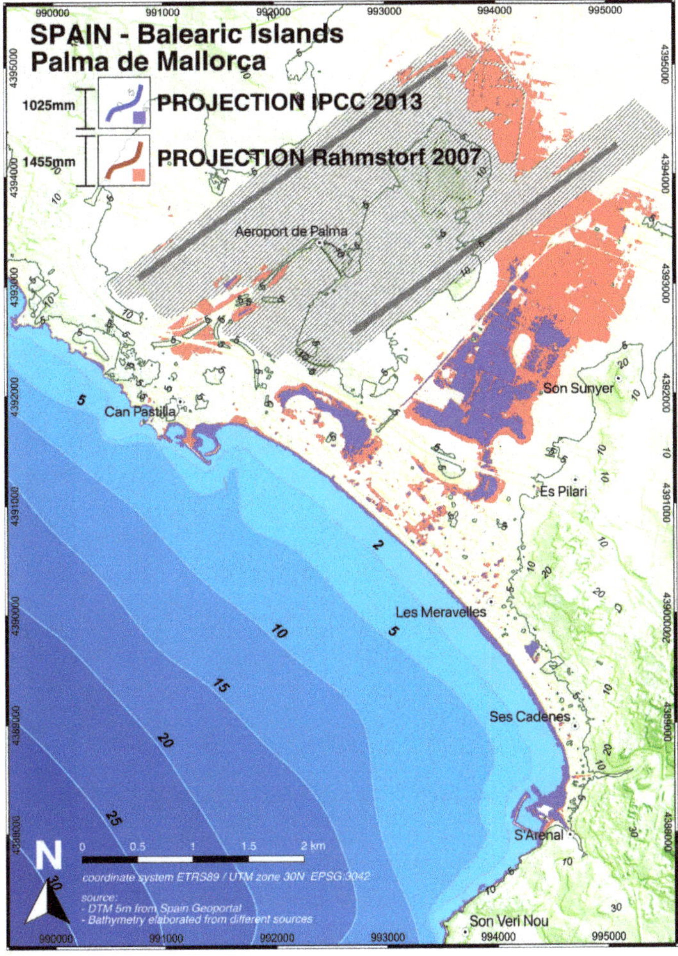

Figure 8. Map 15 Palma de Mallorca (Spain), see also Figure 1 for location). The potential submersion area, using Rahmstorf 2007 projections are 3.7 km^2.

3.1.16. Ibiza

Map 16: Our map highlights some depressed zones around the city of Ibiza, belonging to marshy areas. The international airport is located between 5 and 6 m of elevation above sea level and presently is not at risk of marine submersion. Conversely, some docks and parking areas of the Ibiza harbor, placed at around 1 m above sea level, are prone to sea-level rise in 2100, as well as the salt pans located east of the city (Ses Salines Natural Park). The dune that currently protects the salt marshes is active and has no obstacles to its movement; however, it is cut by a road in its northern sector, which will facilitate marine ingression over the next decades, Figure 9.

Figure 9. Map 16 Ibiza (Spain), see also Figure 1 for location). The potential submersion area, using Rahmstorf 2007 projections are 4.2 km².

3.2. New Tectonic Data

In this section, we show the vertical tectonic trends for the studied areas, which are considered tectonically stable (see Introduction and references therein). Radiocarbon data from lagoon fossil shells sampled in cores and the predicted sea-level rise for the Holocene [47] are also included in the analysis.

A recent paper for the Fertilia area [59] (for a site located 9 km from Fertilia) dated the *Lithopyllum* reef sampled into the fossil tidal notch, confirming it belongs to MIS 5.5, 125 ka BP). The tectonic stability considered for Marina di Campo (island of Elba) derives from a deposit containing *Strombus bubonius* (now *Persitrombus latus*) found in the island of Pianosa, which is 26 km distant and at an altitude of 5 m. However, a new paper [60] was able to discuss our considerations; the radiocarbon ages of marine and brackish shells extracted from some cores, if compared with the local sea-level curve from Lambeck et al. [47], show a clear subsidence of 2–3 mm/year. If this is not due to compaction or in situ sediments, the areas of the maps highlighted by us as being at risk of possible sea submersion could be wider. There are no fossil deposits of MIS 5.5 found on the island. For the coastal area at the mouth of the Sangro river, it was possible to examine some works such as Aucelli et al. [32] and Parlagreco et al. [57] that confirmed the tectonic stability of the area. For the small depressed areas mapped around the city of Pescara, data were reported in Aucelli et al. [32] and Parlagreco et al. [58]. In this work, ENEA carried out a core in the city for the study of the lagoon's evolution down to 15 m. Apart from the first meters of soil compaction, the radiocarbon age of shells from the lagoonal environment, when compared with the sea-level curve from Lambeck et al. [47], confirm the remarkable tectonic stability.

The Lesina area is placed eastward of the mouth of the Fortore River. It is dominated by a marine terrace with pebbly deposits with the inner margin at about 25 m above sea level, attributed on a morphostratigraphic basis to a generic last Interglacial. When attributed to the MIS 5.5 last interglacial, as suggested by Mastronuzzi et al. [61], we can speculate a slow tectonic uplift. A high cliff separates it from the plain of the Fortore River; its sediments are distributed from NNW to ESE by the drift of a long shore that over time has supplied a littoral spit that has its root at Punta delle Pietre Nere and which limits Lesina Lake [62]. This area is very complex from a neotectonic point of view. A few kilometers north of the lagoon, a core carried out near to the Biferno river mouth and radiocarbon analysis on lagoon shells [63], when compared against Lambeck et al. [47], indicated a slight uplift. During the Holocene, the whole area of the plain of Fortore River underwent repeated events of slow subsidence and uplift [64,65]; these were identified thanks to the presence of the uplifted *Cladocora caespitosa* "coralligenous-like" bioconstruction of Punta di Pietre Nere (today, the top is at about 1.5 m above sea level) and by the recognized bio concretions of *Dendropoma* spp; when compared with Lambeck et al. [47], this indicates a general uplift of about 2.49 ± 0.54 mm/year [50]. In this area, the uplifting can be related to the combination of the dynamics imposed by the activity of the gypsum structure of Punta delle Pietre Nere and the regional tectonics conditioned by the structures of the Fortore River, Tremiti Islands and Gargano Promontory [66–68]. On the other hand, in the town of Lesina Marina, Caporale et al. [69] highlighted the localized subsidence of the area with rates of up to 5 mm/year due to the karst activity that produces more sinkholes in the gypsum units, while, on the basis of radiocarbon age determinations on samples derived by cores sampled near Lesina—Punta delle Pietre Nere performed by Longhitano et al. [70], a low rate of uplift may be supposed using the model of Lambeck et al. [47]. In the presence of these neotectonic differences, it was not easy to attribute tectonic rates to the map (Table 2); thus, we considered it more appropriate to make a mean of the values because there is no regional study on these local phenomena nor instrumental geodetic data. In this case, we have assumed zero tectonics, and for some small areas of the map the values could, therefore, be either in excess or too low. A few kilometers further south, just beyond the Gargano promontory, Caldara and Simone [71] published the results of a core carried out in Palude Frattarolo where the radiocarbon ages of the lagoon *Cerastoderma*, when compared with Lambeck et al. [47], indicate substantial stability [50].

The investigated area of Brindisi is characterized by outcrops of Middle–Late Pleistocene units on which the sea shaped a gently sloping rocky coast. In general, the area preserves archaeological and geomorphological evidence that indicates a very low rate of subsidence over the long term and a substantial stability during the Holocene [72–75].

The Island of Cyprus is located in a tectonically quasi-stable area (0–0.15 tectonic uplift). Galili et al. [76] reports the elevation of MIS 5.5 across the island, falling at Larnaca airport at 7 ± 0.5 m, indicating the stability of this coastal area (sea level during MIS 5.5 was 7 m higher than today in Mediterranean Sea [50]).

Regarding the coastal area of Stagnone and Saline di Marsala, there are some data [48,77] related to MIS 5.5 which indicate tectonic stability. Basso et al. [78] carried out several surveys in the Stagnone di Marsala and performed radiocarbon aging on brackish organisms such as the *Cerithium*. Findings evidenced depths of 50–70 cm and ages up to 3 and 6 ka cal BP, as if there had been considerable uplift (when compared with the curve from Lambeck et al. [47]); after extensive discussions based on the ecology of the place (personal communication with Prof. Renato Chemello), there are high chances that these environments have undergone major changes in the last 6 ka cal BP. The confirmation that this is the well-known Punic-era road (2300 year BP) which is now up to 1 m below sea level shows a paleo level of the sea of at least 1.8 m.

As regards the Mallorca and Ibiza islands, Muhs et al. [79] and the references therein testify the tectonic stability of this islands on the basis of speleothem and coral U/Th dating and amino acid analyses.

4. Discussion

The 16 maps represent an update for new areas of the flooding scenarios reported by Lambeck et al. [47] and Antonioli et al. [29]. Because some areas are not covered by high-resolution LiDAR data, three of the generated maps are affected by a lower resolution which prevented the realization of detailed submersion scenarios for the IPCC-AR5 RCP8.5 [36] and Rahmstorf [17] projections for 2100 (Table 3). Although our maps' detail shows the expected scenarios and include the contribution of the vertical land movements (due to GIA and vertical tectonics) that may accelerate the submerging process, they have been realized without taking into account the effects of the drainage systems that could keep the investigated areas dry.

Therefore, we can only speculate on the extension of the expected marine submersion in 2100 with potential related impacts on coastal morphology and infrastructures, given the chosen climatic scenarios and the estimated trend of vertical land movements.

Table 3 shows the uncertainties of the DTM obtained by LiDAR and Copernicus data (Table 2). For the latter, the RMSE (Root Mean Square Error) in terms of vertical accuracy is about ±1 m for lowlands (less than 10% slope) as reported in the DTM v1.0 statistical validation document (https://land.copernicus.eu/user-corner/technical-library/eu-dem-v1.0 [54]). In particular, the RMSE in Cyprus is 1.39 m, which is still too high to build up a high resolution DTM suitable to analyze in detail the sea level rise scenario for 2100. Similarly, also for Bastia (France) and Kerkennah (Tunisia) the low resolution of the spatial data prevented to realize flooding scenarios.

In particular, the relative sea level rise falls between just over a meter and the average error of Copernicus data is 1.6 m on the Z-axis, preventing the significance of the three maps of Larnaka Bastia and Kerkennah. To this issue, we compared the elevation in the DTMs from Lidar and the altitudes from Google Earth, finding a high correspondence. This comparison was also carried out on the three maps built by DTMs using Copernicus data (25 × 25 m), and also in this case the values are similar to the Copernicus DTM. Scientific papers concerning the vertical error of Google Earth [80–84] report that the use of new satellite since 2016, improved the quality of the data, including the Z-axis between 2.5 m and less of 1 m, with an average error of 1.6 m and even better values in flat areas.

For the Bastia area, the1:5000 scale map (Figure 10a,b, Table 2). provides good details which are in excellent agreement with the Copernicus data (25 × 25 m) on the Z-axis.

The DTMs from Copernicus and SRTM (Shuttle Radar Topography Mission) compared for the vertical with Google Earth, show the uncertainty regarding the low resolution of this data: using resolution of 25 × 25 or 30 × 30 m, the slight changes in the lowlands such as dunes, bars, embankments and other features are smoothed in the pixels, thus losing information about those morphological features (natural and anthropic) that could combat the sea-level rise. Furthermore, regarding this issue, we cannot exactly assess the sea-submersion extent on these areas, but we certainly assert that they will suffer marine submersion effects in the future (on the maps of Bastia, Larnaka and Kerkennah, have been annotated with the label "LOW SPATIAL-RESOLUTION MAP").

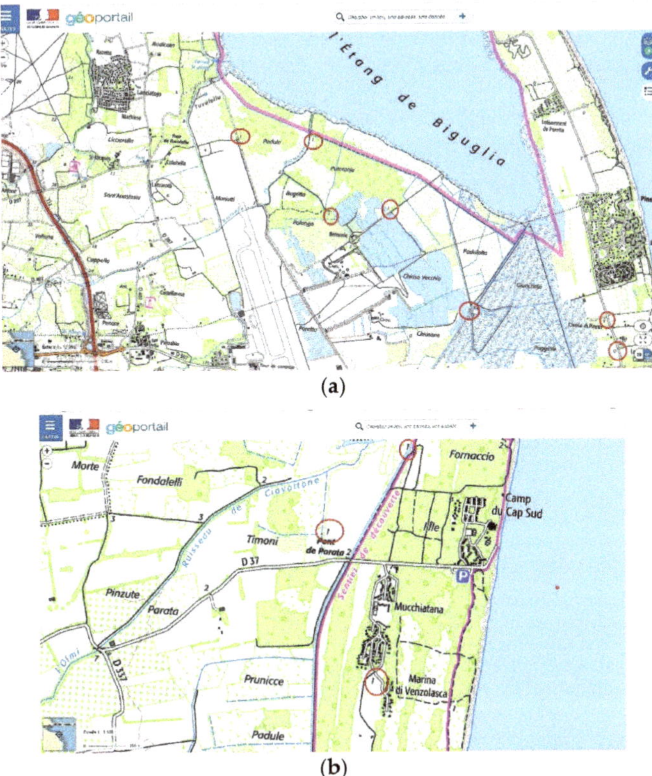

Figure 10. (**a**) and (**b**). Bastia 1:5000 scale map, the red circle underline the altitude showing a excellent agreement with the Map 4 Bastia.

5. Conclusions

In this work, we have shown a methodology to create maps with potentially expected submersion scenarios for 2100 in 16 different selected coastal zones of the Mediterranean basin which are prone to marine submersion under the effects of relative sea-level rise. For these areas, were produced thematic maps that were based on climatic scenarios, tectonics, local geological behavior and the best available digital topography. Maps have been produced for five small coastal areas of the Adriatic Sea (Italy) and 11 areas for eight islands of the Mediterranean Sea (Sardinia, Sicily, Elba Island in Italy; Corsica in France; Cyprus; Kerkennah in Tunisia; Majorca and Ibiza in Spain). All these areas are exposed to coastal hazard due to the their low elevation above the present mean sea level, the rates of vertical land movements and their morphological features. The different characteristics affect the expected scenarios that were built for both gently sloping rocky coasts or wave-dominated coasts. The latter are often characterized by sandy mobile systems marked by the presence of extended back dune areas with swamp, lagoon or coastal lake that are particularly sensitive to sea level rise. Some areas guest natural high value sites belonging to protected areas or National Parks; some are deeply urbanized by residential or touristic settlements; others are characterized by the presence of cultural heritage, and infrastructure such as communication routes, harbors or airports.

All maps were realized with the same methodology, but only 13 out of 16 were elaborated using high-resolution data derived by LiDAR surveys. The remaining three maps (Larnaka in Cyprus; Bastia in France and Kerkennah in Tunisia) were elaborated through data freely released by the Copernicus program with a resolution that did not allow us to map detailed scenarios. However, the land elevation of these areas was a critical factor for the expected sea-level rise for 2100.

From our analysis we estimated a potential loss of land for the above areas between about 148 km^2 for the IPCC-RCP8.5 scenario and 192 km^2 for the Rahmstorf scenario, impacting a coastline length of about 400 km, where are often located densely inhabited settlements and infrastructures. The expected scenario and the exposition of the investigated areas to coastal hazard should be considered for a cognizant management of the coastal zone.

Supplementary Materials: The following are available online at http://www.mdpi.com/2073-4441/12/8/2173/s1, Figure S1: Eight maps of possible submerging coastal areas in the Mediterranean Sea, Video S1: Short drone flight on some coastal plains in Italy (mp4 short film).

Author Contributions: Conceptualization: F.A., M.A., G.M., G.D.F.; Data curation: G.D.F., L.M., V.L.P., G.S. (Giovanni Scardino); Formal analysis: L.M., V.L.P., G.D.F., M.A., D.B., S.C., G.L., S.F., A.M., M.P., G.S. (Giovanni Scicchitano), G.R.; Funding acquisition: S.C., F.A., G.M., M.P.; Investigation: F.A., V.L.P., G.S. (Giovanni Scicchitano), M.A., S.C., G.L., S.F., Methodology: F.A., M.A., G.M., G.D.F., G.S. (Giovanni Scardino) Project administration: S.C., M.P. Resources: S.C., M.P., F.A., G.M. Visualization: G.R., L.M., V.L.P., G.D.F., M.A., D.B., S.C., G.L., S.F., A.M., M.P., G.S. (Giovanni Scicchitano); Writing—original draft: F.A., M.A., G.D.F., G.L., D.B., M.P.; Writing—review and editing: M.A., F.A., S.F., G.M., G.R., G.D.F., G.S. (Giovanni Scardino). All authors have read and agreed to the published version of the manuscript.

Funding: This study has been carried out in the frame of the RITMARE Project The Italian Research for the Sea, coordinated by the Italian National Research Council and funded by the Italian Ministry of Education, University and Research (Resp. F. Antonioli, S. Carniel), the CLIMTOUR Projects (Resp. M. Petitta) and I-STORMS Project from DISTEGEO—UNIBA and Regione Puglia Civil Protection (resp. G. Mastronuzzi).

Acknowledgments: We are thankful to Egidio Trainito for the Surveys in Sardinia. Additional maps by Antonioli et al. [29] are available online at the ISPRA website at the following link: http://portalesgi.isprambiente.it/it/news/news/scenari-di-innalzamento-del-livello-del-mare-su-alcune-aree-costiere-italiane [85]. We are thankful to the Italian Geological Service (ISPRA), which will continue to update this database. This work was also carried out under the umbrella of the IGCP Project n. 639 "Sea-level change from minutes to millennia" (Project Leaders: S. Engelhart, G. Hoffmann, F. Yu and A. Rosentau), the savemedcoasts (www.savemedcoasts.eu) and savemedcoasts-2 (www.savemedcoasts2.eu) projects.

Conflicts of Interest: The authors declare no conflict of interest. The funders had no role in the design of the study; in the collection, analyses, or interpretation of data; in the writing of the manuscript; or in the decision to publish the results.

Abbreviations

The following abbreviations are used in the manuscript:

EO	Earth Observation
CET	Copernicus European Tourism
CLIMTOUR	Copernicus Climate Change Service: European Tourism
DTM	Digital Terrain Model
GEBCO	General Bathymetric Chart of the Oceans
GIS	Geographic Information System
MIS	Marine Isotope Stage
LiDAR	Light Detection and Ranging
IPCC AR5	Intergovernmental Panel on climate change, 2013
RCP	Representative Concentration Pathways
sRMSE	Root Mean Square Error
SRTM	Shuttle Radar Topography Mission
RITMARE	is one of the Flag Projects of the Italian Research Program funded by the Ministry of University and Research

References

1. Vermeer, M.; Rahmstorf, S. Global sea level linked to global temperature. *Proc. Natl. Acad. Sci. USA* **2009**, *106*, 21527–21532. [CrossRef] [PubMed]

2. Church, J.A.; Woodworth, P.L.; Aarup, T.; Wilson, W.S. *Understanding Sea-Level Rise and Variability*; Wiley-Blackwell: Hoboken, NJ, USA, 2010; ISBN 978-1-4443-3452-4.

3. Kemp, A.C.; Horton, B.P.; Donnelly, J.P.; Mann, M.E.; Vermeer, M.; Rahmstorf, S. Climate related sea-level variations over the past two millennia. *Proc. Natl. Acad. Sci. USA* **2011**, *108*, 11017–11022. [CrossRef] [PubMed]

4. Meyssignac, B.; Cazenave, A. Sea level: A review of present-day and recent-past changes and variability. *J. Geodyn.* **2012**, *58*, 96–109. [CrossRef]

5. Mitchum, G.T.; Nerem, R.S.; Merrifield, M.A.; Gehrels, W.R. Modern sea level changes estimates. In *Understanding Sea Level Rise and Variability*; Church, J.A., Woodworth, P.L., Aarup, T., Wilson, W.S., Eds.; Wiley: Chichester, UK, 2010; pp. 122–138.

6. Jevrejeva, S.; Moore, J.C.; Grinsted, A.; Matthews, A.P.; Spada, G. Trends and acceleration in global and regional sea levels since 1807. *Glob. Planet. Chang.* **2014**, *113*, 11–22. [CrossRef]

7. Wöppelmann, G.; Marcos, M. Coastal sea level rise in southern Europe and the nonclimate contribution of vertical land motion. *J. Geophys. Res. Ocean.* **2012**, *117*. [CrossRef]

8. Toimil, A.; Camus, P.; Losada, I.J.; Le Cozannet, G.; Nicholls, R.J.; Idier, D.; Maspataud, A. Climate change-driven coastal erosion modelling in temperate sandy beaches: Methods and uncertainty treatment. *Earth-Sci. Rev.* **2020**, *202*, 103110. [CrossRef]

9. Gornitz, V.; Couch, S.; Hartig, E.K. Impacts of sea level rise in the New York City metropolitan area. *Glob. Planet. Chang.* **2001**, *32*, 61–88. [CrossRef]

10. Walsh, K.J.E.; Betts, H.; Church, J.; Pittock, A.B.; McInnes, K.L.; Jackett, D.R.; McDougall, T.J. Using Sea Level Rise Projections for Urban Planning in Australia. *J. Coast. Res.* **2004**, *20*, 586–598. [CrossRef]

11. Miller, K.G.; Kopp, R.E.; Horton, B.P.; Browning, J.V.; Kemp, A.C. A geological perspective on sea-level rise and its impacts along the U.S. mid-Atlantic coast. *Earth's Future* **2013**, *1*, 3–18. [CrossRef]

12. Ezer, T.; Atkinson, L.P. Accelerated flooding along the U.S. East Coast: On the impact of sea-level rise, tides, storms, the Gulf Stream, and the North Atlantic Oscillations. *Earth's Future* **2014**, *2*, 362–382. [CrossRef]

13. Rehman, S.; Sahana, M.; Kumar, P.; Ahmed, R.; Sajjad, H. Assessing hazards induced vulnerability in coastal districts of India using site-specific indicators: An integrated approach. *GeoJournal* **2020**. [CrossRef]

14. Wadey, M.; Brown, S.; Nicholls, R.J.; Haigh, I. Coastal flooding in the Maldives: An assessment of historic events and their implications. *Nat. Hazards* **2017**, *89*, 131–159. [CrossRef]

15. Carreau, P.R.; Gallego, F.J. *EU25 Coastal Zone Population Estimates from the Disaggregated Population Density Data 2001*; European Commission, DG Joint Research Centre: Brussels, Belgium, 2006.

16. Church, J.A.; Clark, P.U.; Cazenave, A.; Gregory, J.M.; Jevrejeva, S.; Levermann, A.; Merrifield, M.A.; Milne, G.A.; Nerem, R.S.; Nunn, P.D.; et al. Sea-Level Rise by 2100. *Science* **2013**, *342*, 1445. [CrossRef] [PubMed]

17. Rahmstorf, S. A Semi-Empirical Approach to Projecting Future Sea-Level Rise. *Science* **2007**, *315*, 368–370. [CrossRef] [PubMed]

18. Galassi, G.; Spada, G. Sea-level rise in the Mediterranean Sea by 2050: Roles of terrestrial ice melt, steric effects and glacial isostatic adjustment. *Glob. Planet. Chang.* **2014**, *123*, 55–66. [CrossRef]

19. Kopp, R.E.; Kemp, A.C.; Bittermann, K.; Horton, B.P.; Donnelly, J.P.; Gehrels, W.R.; Hay, C.C.; Mitrovica, J.X.; Morrow, E.D.; Rahmstorf, S. Temperature-driven global sea-level variability in the Common Era. *Proc. Natl. Acad. Sci. USA* **2016**, *113*, E1434–E1441. [CrossRef]

20. Bamber, J.L.; Oppenheimer, M.; Kopp, R.E.; Aspinall, W.P.; Cooke, R.M. Ice sheet contributions to future sea-level rise from structured expert judgment. *Proc. Natl. Acad. Sci. USA* **2019**, *116*, 11195–11200. [CrossRef]

21. Zecca, A.; Chiari, L. Lower bounds to future sea-level rise. *Glob. Planet. Chang.* **2012**, *98–99*, 1–5. [CrossRef]

22. Anzidei, M.; Lambeck, K.; Antonioli, F.; Furlani, S.; Mastronuzzi, G.; Serpelloni, E.; Vannucci, G. Coastal structure, sea-level changes and vertical motion of the land in the Mediterranean. *Geol. Soc. Lond. Spec. Publ.* **2014**, *388*, 453–479. [CrossRef]

23. Reimann, L.; Vafeidis, A.T.; Brown, S.; Hinkel, J.; Tol, R.S.J. Mediterranean UNESCO World Heritage at risk from coastal flooding and erosion due to sea-level rise. *Nat. Commun.* **2018**, *9*, 1–11. [CrossRef]

24. Savemedcoasts. Available online: http://www.savemedcoasts.eu/ (accessed on 4 April 2020).

25. Bonaldo, D.; Antonioli, F.; Archetti, R.; Bezzi, A.; Correggiari, A.; Davolio, S.; De Falco, G.; Fantini, M.; Fontolan, G.; Furlani, S.; et al. Integrating multidisciplinary instruments for assessing coastal vulnerability to erosion and sea level rise: Lessons and challenges from the Adriatic Sea, Italy. *J. Coast. Conserv.* **2019**, *23*, 19–37. [CrossRef]

26. Marcos, M.; Jordà, G.; Gomis, D.; Pérez, B. Changes in storm surges in southern Europe from a regional model under climate change scenarios. *Glob. Planet. Chang.* **2011**, *77*, 116–128. [CrossRef]

27. Lionello, P.; Conte, D.; Marzo, L.; Scarascia, L. The contrasting effect of increasing mean sea level and decreasing storminess on the maximum water level during storms along the coast of the Mediterranean Sea in the mid 21st century. *Glob. Planet. Chang.* **2017**, *151*, 80–91. [CrossRef]

28. Bonaldo, D.; Bucchignani, E.; Pomaro, A.; Ricchi, A.; Sclavo, M.; Carniel, S. Wind waves in the Adriatic Sea under a severe climate change scenario and implications for the coasts. *Int. J. Climatol.* **2020**. [CrossRef]

29. Antonioli, F.; Anzidei, M.; Amorosi, A.; Lo Presti, V.; Mastronuzzi, G.; Deiana, G.; De Falco, G.; Fontana, A.; Fontolan, G.; Lisco, S.; et al. Sea-level rise and potential drowning of the Italian coastal plains: Flooding risk scenarios for 2100. *Quat. Sci. Rev.* **2017**, *158*, 29–43. [CrossRef]

30. Marsico, A.; Lisco, S.; Presti, V.L.; Antonioli, F.; Amorosi, A.; Anzidei, M.; Deiana, G.; Falco, G.D.; Fontana, A.; Fontolan, G.; et al. Flooding scenario for four Italian coastal plains using three relative sea level rise models. *J. Maps* **2017**, *13*, 961–967. [CrossRef]

31. Aucelli, P.P.C.; Di Paola, G.; Incontri, P.; Rizzo, A.; Vilardo, G.; Benassai, G.; Buonocore, B.; Pappone, G. Coastal inundation risk assessment due to subsidence and sea level rise in a Mediterranean alluvial plain (Volturno coastal plain—Southern Italy). *Estuar. Coast. Shelf Sci.* **2017**, *198*, 597–609. [CrossRef]

32. Aucelli, P.P.C.; Di Paola, G.; Rizzo, A.; Rosskopf, C.M. Present day and future scenarios of coastal erosion and flooding processes along the Italian Adriatic coast: The case of Molise region. *Environ. Earth Sci.* **2018**, *77*, 371. [CrossRef]

33. Giordano, L.; Alberico, I.; Ferraro, L.; Marsella, E.; Lirer, F.; Di Fiore, V. A new tool to promote sustainability of coastal zones. The case of Sele plain, southern Italy. *Rend. Fis. Acc. Lincei* **2013**, *24*, 113–126. [CrossRef]

34. Welcome! - IWS I-STORMS Web System. Available online: https://iws.seastorms.eu/ (accessed on 2 May 2020).

35. Ferrarin, C.; Valentini, A.; Vodopivec, M.; Klaric, D.; Massaro, G.; Bajo, M.; Pascalis, F.D.; Fadini, A.; Ghezzo, M.; Menegon, S.; et al. Integrated sea storm management strategy: The 29 October 2018 event in the Adriatic Sea. *Nat. Hazards Earth Syst. Sci.* **2020**, *20*, 73–93. [CrossRef]

36. IPCC. *Climate Change 2013: The Physical Science Basis.Contribution of Working Group I to the Fifth Assessment Report of the Intergovernmental Panel on Climate Change*; Stocker, T.F., Qin, D., Plattner, G.-K., Tignor, M., Allen, S.K., Boschung, J., Nauels, A., Xia, Y., Bex, V., Midgley, P.M., Eds.; Cambridge University Press: Cambridge, UK; New York, NY, USA, 2013; p. 1535.

37. Aucelli, P.; Cinque, A.; Mattei, G.; Pappone, G.; Rizzo, A. Studying relative sea level change and correlative adaptation of coastal structures on submerged Roman time ruins nearby Naples (southern Italy). *Quat. Int.* **2019**, *501*, 328–348. [CrossRef]

38. Anzidei, M.; Bosman, A.; Carluccio, R.; Casalbore, D.; Caracciolo, F.D.; Esposito, A.; Nicolosi, I.; Pietrantonio, G.; Vecchio, A.; Carmisciano, C.; et al. Flooding scenarios due to land subsidence and sea-level rise: A case study for Lipari Island (Italy). *Terra Nova* **2017**, *29*, 44–51. [CrossRef]

39. Anzidei, M.; Scicchitano, G.; Tarascio, S.; de Guidi, G.; Monaco, C.; Barreca, G.; Mazza, G.; Serpelloni, E.; Vecchio, A. Coastal retreat and marine flooding scenario for 2100: A case study along the coast of Maddalena Peninsula (southeastern Sicily). *Geogr. Fis. Din. Quat.* **2019**, *41*, 5–16. [CrossRef]

40. Anzidei, M.; Doumaz, F.; Vecchio, A.; Serpelloni, E.; Pizzimenti, L.; Civico, R.; Greco, M.; Martino, G.; Enei, F. Sea Level Rise Scenario for 2100 A.D. in the Heritage Site of Pyrgi (Santa Severa, Italy). *J. Mar. Sci. Eng.* **2020**, *8*, 64. [CrossRef]

41. Ravanelli, R.; Riguzzi, F.; Anzidei, M.; Vecchio, A.; Nigro, L.; Spagnoli, F.; Crespi, M. Sea level rise scenario for 2100 A.D. for the archaeological site of Motya. *Rend. Fis. Accad. Lincei* **2019**, *30*, 747–757. [CrossRef]

42. Perini, L.; Calabrese, L.; Luciani, P.; Olivieri, M.; Galassi, G.; Spada, G. Sea-level rise along the Emilia-Romagna coast (Northern Italy) in 2100: Scenarios and impacts. *Nat. Hazards Earth Syst. Sci.* **2017**, *17*, 2271–2287. [CrossRef]

43. Rovere, A.; Furlani, S.; Benjamin, J.; Fontana, A.; Antonioli, F. MEDFLOOD project: Mediterranean Sea-level change and projection for future FLOODing. *Alp. Mediterr. Quat.* **2012**, *25*, 3–5.

44. Snoussi, M.; Ouchani, T.; Khouakhi, A.; Niang-Diop, I. Impacts of sea-level rise on the Moroccan coastal zone: Quantifying coastal erosion and flooding in the Tangier Bay. *Geomorphology* **2009**, *107*, 32–40. [CrossRef]

45. Azidane, H.; Benmohammadi, A.; Hakkou, M.; Magrane, B.; Haddout, S. A Geospatial approach for assessing the impacts of sea-level rise and flooding on the ! Kenitra coast (Morocco). *J. Mater. Environ. Sci.* **2018**, *9*, 1480–1488. [CrossRef]

46. Scardino, G.; Sabatier, F.; Scicchitano, G.; Piscitelli, A.; Milella, M.; Vecchio, A.; Anzidei, M.; Mastronuzzi, G. Sea-Level Rise and Shoreline Changes Along an Open Sandy Coast: Case Study of Gulf of Taranto, Italy. *Water* **2020**, *12*, 1414. [CrossRef]

47. Lambeck, K.; Antonioli, F.; Anzidei, M.; Ferranti, L.; Leoni, G.; Scicchitano, G.; Silenzi, S. Sea level change along the Italian coast during the Holocene and projections for the future. *Quat. Int.* **2011**, *232*, 250–257. [CrossRef]

48. Ferranti, L.; Antonioli, F.; Mauz, B.; Amorosi, A.; Dai Pra, G.; Mastronuzzi, G.; Monaco, C.; Orrù, P.; Pappalardo, M.; Radtke, U.; et al. Markers of the last interglacial sea-level high stand along the coast of Italy: Tectonic implications. *Quat. Int.* **2006**, *145–146*, 30–54. [CrossRef]

49. Ferranti, L.; Antonioli, F.; Anzidei, M.; Monaco, C.; Stocchi, P. The timescale and spatial extent of vertical tectonic motions in Italy: Insights from relative sea-level changes studies. *J. Virtual Explor.* **2010**, *36*. [CrossRef]

50. Antonioli, F.; Ferranti, L.; Fontana, A.; Amorosi, A.; Bondesan, A.; Braitenberg, C.; Dutton, A.; Fontolan, G.; Furlani, S.; Lambeck, K.; et al. Holocene relative sea-level changes and vertical movements along the Italian and Istrian coastlines. *Quat. Int.* **2009**, *206*, 102–133. [CrossRef]

51. Antonioli, F.; Ferranti, L.; Stocchi, P.; Deiana, G.; Lo Presti, V.; Furlani, S.; Marino, C.; Orru, P.; Scicchitano, G.; Trainito, E.; et al. Morphometry and elevation of the last interglacial tidal notches in tectonically stable coasts of the Mediterranean Sea. *Earth-Sci. Rev.* **2018**, *185*, 600–623. [CrossRef]

52. Cobby, D.M.; Mason, D.C.; Davenport, I.J. Image processing of airborne scanning laser altimetry data for improved river flood modelling. *ISPRS J. Photogramm. Remote Sens.* **2001**, *56*, 121–138. [CrossRef]

53. Global Mapper—All-in-One GIS Software. Available online: https://www.bluemarblegeo.com/products/global-mapper.php (accessed on 15 April 2020).

54. Copernicus Land Monitoring Service. Available online: https://land.copernicus.eu/ (accessed on 15 April 2020).

55. EMODnet Bathymetry Viewing and Download Service. Available online: https://portal.emodnet-bathymetry.eu/ (accessed on 15 April 2020).

56. Soukissian, T.; Denaxa, D.; Karathanasi, F.; Prospathopoulos, A.; Sarantakos, K.; Iona, A.; Georgantas, K.; Mavrakos, S. Marine Renewable Energy in the Mediterranean Sea: Status and Perspectives. *Energies* **2017**, *10*, 1512. [CrossRef]

57. Parlagreco, L.; Miccadei, E.; Mascioli, F.; Devoti, S.; Silenzi, S.; Antonioli, F.; Di Palo, C. Application of relative sea level rise scenarios to coastal management policy: The case of the Abruzzo Region. In Proceedings of the Geoitalia 2009, Rimini, Italy, 9–11 September 2009; p. 1.

58. Parlagreco, L.; Mascioli, F.; Miccadei, E.; Antonioli, F.; Gianolla, D.; Devoti, S.; Leoni, G.; Silenzi, S. New data on Holocene relative sea level along the Abruzzo coast (central Adriatic, Italy). *Quat. Int.* **2011**, *232*, 179–186. [CrossRef]

59. Sechi, D.; Andreucci, S.; Stevens, T.; Pascucci, V. Age and significance of late Pleistocene Lithophyllum byssoides intertidal algal ridge, NW Sardinia, Italy. *Sediment. Geol.* **2020**, *400*, 105618. [CrossRef]

60. D'Orefice, M.; Graciotti, R.; Bertini, A.; Fedi, M.; Foresi, L.M.; Ricci, M.; Toti, F. Latest Pleistocene to Holocene environmental changes in the Northern Tyrrhenian area (central Mediterranean). A case study from southern Elba Island. *Alp. Mediterr. Quat.* **2020**, *33*, 1–25.

61. Mastronuzzi, G.; Palmentola, G.; Ricchetti, G. Aspetti dell'evoluzione olocenica della costa pugliese. *Mem. Soc. Geol. Ital.* **1989**, *42*, 287–300.

62. Mastronuzzi, G.; Aringoli, D.; Aucelli, P.P.C.; Baldassarre, M.A.; Bellotti, P.; Bini, M.; Biolchi, S.; Bontempi, S.; Brandolini, P.; Chelli, A.; et al. Geomorphological map of the Italian Coast: From a descriptive to a morphodynamic approach. *Geogr. Fis. Din. Quat.* **2017**, *40*, 161–195. [CrossRef]

63. Amorosi, A.; Bracone, V.; Campo, B.; D'Amico, C.; Rossi, V.; Rosskopf, C.M. A late Quaternary multiple paleovalley system from the Adriatic coastal plain (Biferno River, Southern Italy). *Geomorphology* **2016**, *254*, 146–159. [CrossRef]

64. Mastronuzzi, G.; Sansò, P. Holocene uplift rates and historical rapid sea-level changes at the Gargano promontory, Italy. *J. Quat. Sci.* **2002**, *17*, 593–606. [CrossRef]

65. Mastronuzzi, G.; Sansò, P. The role of strong earthquakes and tsunami in the Late Holocene evolution of the Fortore River coastal plain (Apulia, Italy): A synthesis. *Geomorphology* **2012**, *138*, 89–99. [CrossRef]

66. Refice, A.; Pasquariello, G.; Bovenga, F.; Festa, V.; Acquafredda, P.; Spilotro, G. Investigating uplift in Lesina Marina (Southern Italy) with the aid of persistent scatterer SAR interferometry and in situ measurements. *Environ. Earth Sci.* **2016**, *75*, 1–13. [CrossRef]

67. Teofilo, G.; Festa, V.; Sabato, L.; Spalluto, L.; Tropeano, M. 3D modelling of the Tremiti salt diapir in the Gargano offshore (Adriatic Sea, southern Italy): Constraints on the Tremiti Structure development. *Ital. J. Geosci.* **2016**, *135*, 474–485. [CrossRef]

68. Festa, V.; Fregola, R.A.; Acquafredda, P.; De Giosa, F.; Monno, A.; Ventruti, G. The enigmatic ascent of Ca-sulphate rocks from a deep dense source layer: Evidences of hydration diapirism in the Lesina Marina area (Apulia, southern Italy). *Int. J. Earth Sci. (Geol. Rundsch.)* **2019**, *108*, 1897–1912. [CrossRef]

69. Caporale, F.; De Venuto, G.; Leandro, G.; Spilotro, G. Interventi di mitigazione del rischio da sinkholes nell'area di Lesina marina (Provincia di Foggia, Italia). *Mem. Descr. Carta Geol. D' Ital.* **2013**, *93*, 121–142.

70. Longhitano, S.G.; Della Luna, R.; Milone, A.L.; Cilumbriello, A.; Caffau, M.; Spilotro, G. The 20,000-years-long sedimentary record of the Lesina coastal system (southern Italy): From alluvial, to tidal, to wave process regime change. *Holocene* **2016**, *26*, 678–698. [CrossRef]

71. Caldara, M.; Simone, O. Coastal changes in the eastern Tavoliere Plain (Apulia, Italy) during the Late Holocene: Natural or anthropic? *Quat. Sci. Rev.* **2005**, *24*, 2137–2145. [CrossRef]

72. Mastronuzzi, G.; Sansò, P. Pleistocene sea-level changes, sapping processes and development of valley networks in the Apulia region (southern Italy). *Geomorphology* **2002**, *46*, 19–34. [CrossRef]

73. Mastronuzzi, G.; Caputo, R.; Di Bucci, D.; Fracassi, U.; Iurilli, V.; Milella, M.; Pignatelli, C.; Sansò, P.; Selleri, G. Middle-Late Pleistocene evolution of the Adriatic coastline of Southern Apulia (Italy) in response to relative sea-level changes. *Geogr. Fis. Din. Quat.* **2011**, *34*, 207–221. [CrossRef]

74. Mastronuzzi, G.; Antonioli, F.; Anzidei, M.; Auriemma, R.; Alfonso, C.; Scarano, T. Evidence of relative sea level rise along the coasts of central Apulia (Italy) during the late Holocene via maritime archaeological indicators. *Quat. Int.* **2017**, *439*, 65–78. [CrossRef]

75. Mastronuzzi, G.; Milella, M.; Piscitelli, A.; Simone, O.; Quarta, G.; Scarano, T.; Calcagnile, L.; Spada, I. Landscape analysis in Torre Guaceto area (BRindisi) aimed at the reconstruction of the late Holocene sea level curve. *Geogr. Fis. Din. Quat.* **2018**, *41*, 65–79. [CrossRef]

76. Galili, E.; Sevketoglu, M.; Salamon, A.; Zviely, D.; Mienis, H.K.; Rosen, B.; Moshkovitz, S. Late Quaternary beach deposits and archaeological relicts on the coasts of Cyprus, and the possible implications of sea-level changes and tectonics on the early populations. *Geol. Soc. Lond. Spec. Publ.* **2015**, *411*, 179–218. [CrossRef]

77. Antonioli, F.; Kershaw, S.; Renda, P.; Rust, D.; Belluomini, G.; Cerasoli, M.; Radtke, U.; Silenzi, S. Elevation of the last interglacial highstand in Sicily (Italy): A benchmark of coastal tectonics. *Quat. Int.* **2006**, *145–146*, 3–18. [CrossRef]

78. Basso, D.; Bernasconi, M.P.E.; Robba, E.M.S. Environmental evolution of the Marsala sound, Sicily, during the last 6000 years. *J. Coast. Res.* **2008**, *24*, 177–197. [CrossRef]

79. Muhs, D.R.; Simmons, K.R.; Porat, N. Uranium-series ages of fossil corals from Mallorca, Spain: The "Neotyrrhenian" high stand of the Mediterranean Sea revisited. *Palaeoclimatology* **2015**, *438*, 408–424. [CrossRef]

80. Benker, S.C.; Langford, R.P.; Pavlis, T.L. Positional accuracy of the Google Earth terrain model derived from stratigraphic unconformities in the Big Bend region, Texas, USA. *Geocarto Int.* **2011**, *26*, 291–303. [CrossRef]

81. Satge, F.; Denezine, M.; Pillco, R.; Timouk, F.; Pinel, S.; Molina, J.; Garnier, J.; Seyler, F.; Bonnet, M.-P. Absolute and relative height-pixel accuracy of SRTM-GL1 over the South American Andean Plateau. *ISPRS J. Photogramm. Remote Sens.* **2016**, *121*, 157–166. [CrossRef]

82. Mulu, Y.A.; Derib, S.D. Positional Accuracy Evaluation of Google Earth in Addis Ababa, Ethiopia. *Artif. Satell.* **2019**, *54*, 43–56. [CrossRef]

83. Ragheb, A.E.; Ragab, A.F. Enhancement of Google Earth Positional Accuracy. *Int. J. Eng. Res. Technol.* **2015**, *4*, 627–630.

84. Hachani, M.A.; Ziadi, B.; Langar, H.; Sami, D.A.; Turki, S.; Aleya, L. The mapping of the *Posidonia oceanica* (L.) Delile barrier reef meadow in the southeastern Gulf of Tunis (Tunisia). *J. Afr. Earth Sci.* **2016**, *121*, 358–364. [CrossRef]

85. Scenari di Innalzamento del Livello del Mare su Alcune Aree Costiere Italiane. Available online: http://portalesgi.isprambiente.it/it/news/news/scenari-di-innalzamento-del-livello-del-mare-su-alcune-aree-costiere-italiane (accessed on 15 April 2020).

Article

Sea-Level Rise and Shoreline Changes Along an Open Sandy Coast: Case Study of Gulf of Taranto, Italy

Giovanni Scardino [1], François Sabatier [2], Giovanni Scicchitano [3], Arcangelo Piscitelli [4], Maurilio Milella [4], Antonio Vecchio [5,6], Marco Anzidei [7] and Giuseppe Mastronuzzi [1,*]

[1] Dipartimento di Scienze della Terra e Geoambientali, Università degli Studi di Bari Aldo Moro, 70121 Bari, Italy; giovanni.scardino@uniba.it

[2] Aix-Marseille Univ, CNRS, IRD, Coll de France, CEREGE, 13545 Aix en Provence, France; sabatier@cerege.fr

[3] Studio Geologi Associati TST, 95045 Misterbianco Catania, Italy; scicchitano@studiogeologitst.com

[4] Environmental Surveys s.r.l. Spin Off c/o Università degli Studi di Bari, 74121 Taranto, Italy; arcangelopiscitelli@ensu.it (A.P.); mauriliomilella@ensu.it (M.M.)

[5] Radboud Radio Lab, Department of Astrophysics/IMAPP, Radboud University, 6500GL Nijmegen, The Netherlands; A.Vecchio@astro.ru.nl

[6] Lesia Observatoire de Paris, PSL Research University, Université de Paris, 92195 Meudon, France

[7] Istituto Nazionale di Geofisica e Vulcanologia, 00143 Rome, Italy; marco.anzidei@ingv.it

* Correspondence: giuseppe.mastronuzzi@uniba.it; Tel.: +39-080-544-2634

Received: 5 April 2020; Accepted: 13 May 2020; Published: 15 May 2020

Abstract: The dynamics of the sandy coast between Castellaneta and Taranto (Southern Italy) has been influenced by many natural and anthropogenic factors, resulting in significant changes in the coastal system over the last century. The interactions between vertical components of sea-level changes and horizontal components of the sedimentary budget, in combination with anthropogenic impact, have resulted in different erosion and accretion phases in the past years. Local isostatic, eustatic, and vertical tectonic movements, together with sedimentary budget changes, must be considered in order to predict the shoreline evolution and future marine submersion. In this study, all morpho-topographic data available for the Gulf of Taranto, in combination with Vertical Land Movements and sea-level rise trends, were considered by assessing the local evolution of the coastal trend as well as the future marine submersion. Based on the predicted spatial and temporal coastal changes, a new predictive model of submersion was developed to support coastal management in sea-level rise conditions over the next decades. After that, a multi-temporal mathematical model of coastal submersion was implemented in a Matlab environment. Finally, the effects of the relative sea-level rise on the coastal surface prone to submersion, according to the Intergovernmental Panel on Climate Change Assessment Reports (AR) 5 Representative Concentration Pathways (RCP) 2.6 and RCP 8.5 scenarios, were evaluated up to 2100.

Keywords: sea-level rise; coastal dynamics; erosion; accretion; submersion

---

1. Introduction

In recent years, the behavior of the Mediterranean coasts in relation to shoreline migration, as a consequence of the changes in the local sedimentary budget (e.g., [1–5]), and to the sea-level rise (e.g., [6–10]), has been of great interest. Currently, the sea-level change observed along the coasts depends on the sum of eustatic, steric, isostatic, and tectonic factors [11,12]. These factors, which prevail along the vertical direction, have been used to evaluate the submersion surfaces (e.g., [2,13–20]) for different sea-level rise scenarios for 2100 linked to climate change [21–27].

On a global scale, satellite altimetry revealed the absolute sea-level variations relative to the reference ellipsoid, showing a sea-level rise of 3.3 mm/year (NASA Goddard Space Flight Center data).

In the Mediterranean basin, however, tide gauge records revealed the sea-level height to the land upon which benchmarks are grounded, and a sea-level rise of 1.8 mm/year in the last century [28,29]. Sea level is expected to be 500–1400 mm higher than present by 2100 AD [21,22,27,30]. According to Church et al. [23,24] and the Intergovernmental Panel on Climate Change (IPCC) scenarios [25–27] during the 21st century, it is very likely that the global mean sea-level rise will exceed values observed in the last three decades, increasing up to 8–16 mm/year during 2081–2100.

1.1. Local Sea-Level Changes

Sea-level records in the Mediterranean over the last century show a sea-level variability depending on the location of the individual stations [28,29,31,32]. Rates vary in accordance with the length of the records, the local Vertical Land Movements (VLM), as well as the changes in atmospheric dynamics. In Antonioli et al. [33], the tide gauge data analyzed for the time span 2000–2013 at given localities along the Italian coasts, show sea-level rates ranging from 3.9 ± 0.2 mm/year (Ancona, Adriatic Sea) to 6.8 ± 0.1 mm/year (Cagliari, Tyrrhenian Sea), up to as much as 10.7 mm/year (Venice, Adriatic Sea). This is due to the decadal sea-level variability. Vecchio et al. [31] analyzed a set of long sea-level recordings (>60 years) for the northern Adriatic Sea, and estimated rates of around 1.2 mm/yr. However, at the subsiding tidal station of Venice Punta della Salute (Venice), rates reached up to 2.4 mm/year. In the Italian peninsula, as on other coasts of the Mediterranean and beyond, sea-level data show the occurrence of a continuous sea-level rise over the last decades compared to the previous centuries. Based on these rates, together with the contribution of VLM and the climatic scenarios reported in the IPCC Assessment Reports (AR) 5, and Representative Concentration Pathway (RCP) 8.5, a relative sea-level rise of approximately 0.8 m is expected by 2100 in the northern Adriatic Sea, and about 0.6 m for the Gulfs of Cagliari and Taranto. However, several authors [21,22,30,34] predict a more severe global sea-level rise of up to about 2 m by 2100, applying regional climate models [32] and a variety of statistical approaches in combination with VLM [31]. The latter value would cause a dramatic submersion of global coastal areas, and would affect human settlements, industrial and commercial facilities, archaeological and cultural heritage sites, and natural areas. Sea-level rise warnings and coastal hazards have been issued globally [35], as well as in Italy [2,16–20,33,36]. In addition, human activities may produce land subsidence due to compaction or extraction of fluids from the alluvial sequences, determining local subsidence up to several mm/yr, thereby increasing local sea-level rise [2,37–40].

1.2. Coastal Behavior in Function of Sea-Level Rise

While many studies have focused on the sea-level rise, few have described its effects along the coasts. This may be due to the difficult approach to sedimentary coastal budget and its dynamics, which only the common Bruun rule model [41] can help overcome. Currently, the Italian coast is influenced by a decrease in the sedimentary input owing to natural and anthropogenic conditions, e.g., dams and hydrological adjustments, changes in vegetation cover and climate, erosional/depositional capacity of the rivers, defense interventions along the coastal stretches. These conditions are making the coastal system changes, caused by atmosphere dynamics and rising sea level [31,33,38], difficult to assess. Thus, only a fallacious answer may explain how coastal landscapes respond to the landform re-organization during an accelerated sea-level rise. This is especially the case for those tracts, such as sedimentary coasts, that may undergo rapid morphological changes in river drainage basins in order to attain new equilibrium profiles [9,10,42,43]. Predicting a future scenario for every coast set within Valentin's milestone diagram [44] (Figure 1) is not an easy task. Hence, many areas along the Mediterranean coasts will be susceptible to submersion by future sea-level rise and severe flooding [20,29,31].

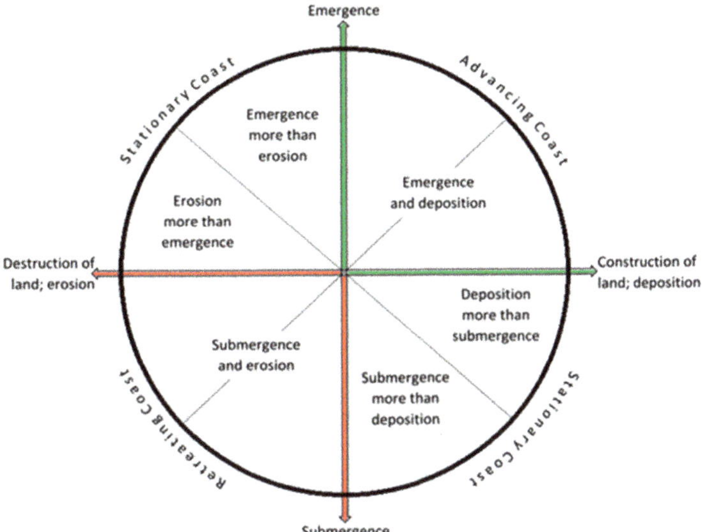

Figure 1. Valentin Diagram [44] which defines the changes in the shoreline in combination with relative sea-level rise.

1.3. Objective

Projected sea-level rises could determine significant future surface loss on the low-lying coastal regions in the Mediterranean basin.

Furthermore, sea-level rise could cause permanent surface submersions in many Mediterranean coastal areas. Thus, it is important to assess the surfaces that could be submerged at any given moment of the future [2,19,33,45].

This paper proposes a new mathematical model for predicting possible submersion scenarios as the one developed for the area of the northern coast of the Gulf of Taranto in Southern Italy. It will consider the changes in both the vertical sea-level components and the horizontal shoreline components, beginning from the past articulation and behavior of the mobile coastal systems. To highlight the coastal response in function of sea-level rise, knowledge of coastal changes and decadal sea-level changes will be reviewed, while, to assess the coastal response linked to the sea-level rise, various existing models will be considered e.g., [9,41,43,46,47].

The aim of this paper is to develop a submersion model of the relative sea-level rise scenario estimated for the coastal plain on the Gulf of Taranto by 2100.

2. Geomorphological Setting

The Metaponto coastal plain covers most of the northern coast of the Gulf of Taranto. The plain takes its name from the ancient Magna Grecia colony of Metapontum, whose remains are still standing today. It stretches NNE to SSW from the Taranto area, and runs along the foredeep of the Bradanic Trough, up to the area of Roseto Capo Spulico, at the eastern border of the Apennine Chain (Figure 2a).

The coastal plain is the result of the sedimentary body accumulated during the last 7 kyear (10^3 years) on the local basement. It was shaped by marine processes which produced a ravinement surface at the base of an old cliff. This coastal plain was conditioned by the interplay between the eustatic sea-level rise during the post Last Glacial Maximum (LGM) transgression and the regional uplift [48,49]. The coastal plain has been heavily supplied by siliciclastic sediments discharged by the main rivers crossing the Apennine chain (Agri, Sinni, Cavone, Bradano, and Basento Rivers). It is composed of quartz, polymineralic lithics and bioclasts [48,49] which are distributed SSW to NNE by the longshore drift.

The landscape of the coastal area adapted to the ice-sheet melting after the Last Glacial Maximum (LGM) of 18 kyear (e.g., [7,49–54]).

In particular, these main phases can be recognized:

- Marine Isotope Stage (MIS) 2—Last Glacial Maximum (LGM): sea-level stand at about 120 m below the present sea level and deep engravings on the coastal wedge up to the shelf break
- 20-6/7 kyear: sea-level transgression with a rate of 8 mm/year and re-shaping of incised valleys along the Gulf of Taranto
- from 6/7 kyear to the second half of twentieth century: sea-level rise allowing the growth of the coastal plain and beach-dune system progradation conditioned by NE longshore drift.

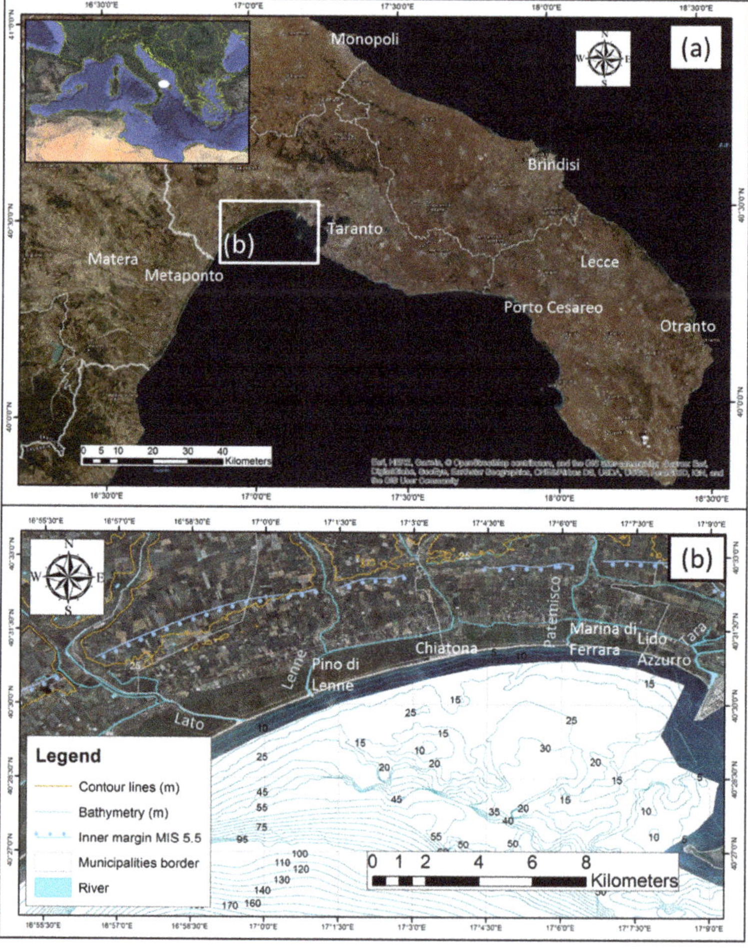

Figure 2. The study area of the Apulia region (Southern Italy): (**a**) the coastal plain of the Gulf of Taranto, Northeastern to Southwestern trending, highlighted by the white box and (**b**) the coastal stretch extending between Castellaneta and Taranto.

The Metaponto plain is bordered landward by a progradated mobile coastal system [55,56] consisting of a series of polyphasic, quasi-parallels dune belts, extending more than 1000 m inland

and characterized by an altitude ranging between 8 and 17 m [56,57] (Figure 2b). The dune belts are characterized by an accentuated lateral continuity, and are largely colonized by the "Macchia Mediterranea", e.g., vegetation assemblages stabilizing the innermost dunes. The back-dune areas are characterized by low-lands that, in the past, hosted a lagoon, reclaimed at the beginning of the twentieth century. A primary dune belt is bordered and supplied by a medium/fine sandy beach with a gently sloping profile, currently undergoing strong erosion.

The entire coastal area was in progradation up to the late 1950s. In many areas, dunes show strong evidence of significant shoreline retreat, highlighted by cliffs shaped in the dune deposits (Figure 3). This retreat has been induced by a negative sedimentary input as a result of: (i) the anthropogenic modification of the river basin catchments with the realization of large dams and hydrological works during the second half of the twentieth century and, (ii) more recently, the conditioning of the longshore drifting due to the building of some touristic harbors and coastal defenses ([58] and references therein).

This mobile coastal system is characterized by an intermediate surf zone domain between reflective and dissipative hydrodynamic regimes, producing marked changes of nearshore sea-floor features [4,49]. Originally, the studied mobile coastal system was characterized by an intermediate morpho-hydrodynamic state, with rhythmic bars located where wave interactions cancel net bottom stress.

At present, these bars act as natural breakwaters, reducing wave energy in the proximity of the shoreline. This influences the currents that allow sediment movement along the entire littoral zone up to a seaward annual closure depth equal to 7.3 m [59].

The study area corresponds to the easternmost part of the plain, stretching between Castellaneta and Taranto. As in the rest of the plain, the mobile coastal system is characterized by well-defined beach-dune belt-back dune areas; several orders of parallel dune belts with elevations of up to about 15 m above sea level (a.s.l.), crowned by a backshore with widths ranging from 25 to 40 m and a foreshore width of up to 15 m. Four short rivers, the Lato, Lenne, Patemisco, and Tara, are mainly supplied by karst aquifers from East to West across the coastal plain without significant sediment supply to the mobile coastal system.

Figure 3. Dune erosion observed in the Gulf of Taranto: (**a**) The Lido Azzurro coast was flooded by storm events in the winter of 2018. Note the erosion of the dune belt; (**b**) a detail of the Lido Azzurro coast with the eroded dune and the exposed plant roots; (**c**) flooding and erosion of a secondary dune in the proximity of the Lato river mouth; and (**d**) dune belt erosion of the Chiatona beach.

3. Methodology

3.1. Data

3.1.1. Tectonics and Isostasy

The Taranto area is characterized by a weak vertical tectonic uplift of the Late Pleistocene and the Holocene. This can be inferred by the elevation of the inner margin of the Marine Isotope Stage (MIS) 5.5 deposits in the Taranto area at 23 m a.s.l. and up to 40 m between Metaponto and Policoro (e.g., [52–54,60–62]).

The different altitudes of the MIS 5.5 terrace indicate decreasing values of long-term uplift rates, from West to East, along the Gulf of Taranto, from 0.39 mm/year close to the Apennine Chain, to 0.26 mm/year in the Taranto area [40,52].

To evaluate the isostatic components, the Lambeck et al. models [11,13] were applied. They consider three-layers, an elastic lithospheric thickness of 110 km, an upper mantle viscosity of 3×10^{20} Pa·s and a lower mantle viscosity of 3×10^{22} Pa·s, providing a rate of the Vertical Land Movement at about 0.45 mm/year in the Gulf of Taranto.

An Interferometric Synthetic Aperture Radar (InSAR) analysis [63] did not show evidence of anthropogenic-induced displacement in the Taranto area. This reveals the general tectonic stability in the NE end of the gulf and a very slight subsidence trend in the northern part of the City of Taranto. To assess the current rates of VLM along the coastal zones of the Mediterranean region, a continuous GPS (cGPS) dataset, analyzed in the frame of the SAVEMEDCOASTS Project (www. savemedcoasts.eu [64]), was used. This study focuses on the uplift velocity estimated at the cGPS station of MMET (Matera METaponto) which falls within the investigated area located in Metaponto (Matera, Southern Italy). This station is managed by the Italian Space Agency (ASI/CGS) and is integrated in the Rete Integrata Nazionale GNSS (RING) network (http://ring.gm.ingv.it/ [65]). Vertical GPS velocity has been determined in the IGb08 realization of the global ITRF08 absolute geocentric reference frame [31,64,66] providing an uplift of 0.29 ± 0.14 mm/year.

3.1.2. Tide and Sea-Level Trend

The current sea-level trend between Castellaneta and Taranto was obtained using sea-level data recorded by the Istituto Superiore per la Protezione e la Ricerca Ambientale (ISPRA), tidal station located at the Molo Sant'Eligio, in the Taranto harbor. Data were continuously recorded with a sampling rate of 10 min using an ultrasound transducer SIAP ± MICROS ID0710 (from 1999 to 2010, ISPRA, Taranto, Italy), and a radar sensor SIAP + MICROS TLR (from 2010 to 2020, ISPRA, Taranto, Italy). During the analysis, gaps and outliers were recognized and corrected to avoid artificial signals in the time series. The analysis highlighted a semidiurnal tidal range of 0.4 ± 0.1 m and a mean sea-level rise of 1.04 ± 0.5 mm/year in the time span 1999–2020 (20 years) (Figure 4). The tide gauges located along the Apulian coasts, managed by ISPRA and the Autorità di Bacino della Puglia (AdBP), have shown a continuous sea-level rise in the last two decades, with similar trends for Taranto (Figure 5).

3.1.3. Orthophoto and Satellite Images

Aerial orthophotos were obtained from the Istituto Geografico Militare (IGM), the Apulian Servizio d'Informazione Territoriale (SIT) of the Apulia Region, and the Ministero dell'Ambiente. Images were collected during the past 70 years (about 1947–2018). Satellite images were derived from WorldView2/3 (DigitalGlobe, Westminster, U.S.), RapidEye (Satellite Imaging Corporation, Canada, U.S.), and Declass1 satellites (USGS, U.S.) for the time span period 1961–2018 (Table S1, supplementary material).

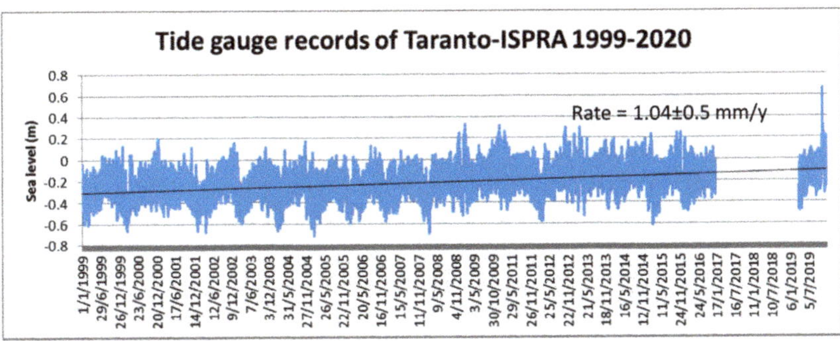

Figure 4. Sea-level time series for the Taranto station in the time span 1999–2020, recorded at a 10-minute sampling rate. The black line is the linear fit of the data that provide a sea-level rate of 1.04 ± 0.5 mm/year.

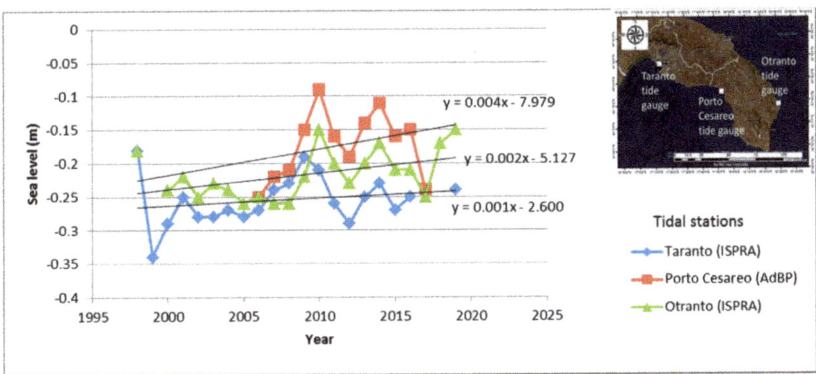

Figure 5. Annual mean sea-level records at the Taranto, Otranto, and Porto Cesareo tidal stations (retrieved from www.mareografico.it [67] and www.adb.puglia.it [68], ISPRA—Istituto Superiore per la Protezione e la Ricerca Ambientale and AdBP—Autorità di Bacino della Puglia).

From each image, the shoreline was extracted, the weather conditions at the time of acquisition were considered, and the barometric effects reduced, using the Theiler et al. method [69].

3.1.4. LIDAR and TLS

Laser Imaging Detection and Ranging (LIDAR) data were derived from the 2008–2009 surveys performed by the Ministero dell'Ambiente, at a resolution 2 × 2 m. In addition, Terrestrial Laser Scanner (TLS) surveys were performed at different times during the winter of 2018 by means of the Faro Focus X130 (Polo Scientifico Tecnologico Magna Grecia, Taranto, Italy), at a resolution of 2 cm. LIDAR and TLS data and Digital Elevation Models (DEMs) were combined to extract the shoreline configuration and the shore morpho-topography (Figure 6). For the considered coastal stretches, DEMs reveal a mean shoreface slope between 2 and 5 degrees which, in response to the tide amplitude, determines a daily horizontal shoreline displacement of 12.5 ± 3 m under a low hydrodynamic condition. A reference value to evaluate the shoreline rate changes was obtained by integrating long-term changes, detected by orthophotos and satellite images, with short-term changes detected by TLS data.

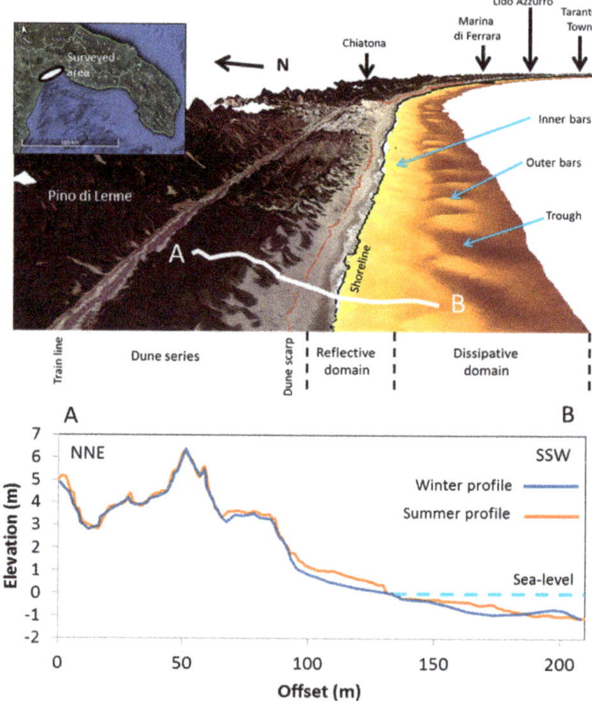

Figure 6. Morpho-topographic and morpho-bathymetric digital models obtained by direct surveys performed on the sandy coast stretching from the Castellaneta beach to the City of Taranto.

3.2. Data Processing

An analysis of the spatial and temporal changes in the Ionic littoral zone between Castellaneta and Taranto was performed by digitalizing all the shorelines via aerial photographs, orthophotos, satellite images, and LIDAR data available for that stretch of coast. They were collected for a medium-term analysis and integrated with TLS data acquired at different hours on different days. From all the collected images, the shorelines were digitalized in ArcGIS for each year. Digital Shoreline Analysis System (DSAS) tools were also used [69] in order to obtain the shoreline 1947 to 2018 changes. During the processing, uncertainty about the position of the shoreline had to be defined. The following factors were considered:

- Line drawing of the operator on the swash zone with an error range of ±3 m
- Weather conditions (tide excursion, atmospheric pressure, temperature, sea level, wind) at time of image acquisition with an error range of ±0.4 m
- Instrumental accuracy with an error range of 0.31–0.5 m
- Computer errors with an error range of ±0.1 m

For the entire stretch of the sandy coast, 345 transects (50 m spacing) were extracted perpendicularly to a reference baseline placed 50 m seaward. The distance between the baseline and each shoreline intersection point made it possible to obtain metric measurements and all shoreline change rates, expressed as meters of change along the transects per year. To determine the different behaviors of the sandy coastal stretches, these estimates were obtained for three different portions of the sandy coast, each in lateral continuity from West to East (Figure 7). For each time span, the minimum, maximum, weighted average, and error range were evaluated. The choice of the weighted average is related to the density data for a given time range (Figures 8 and 9).

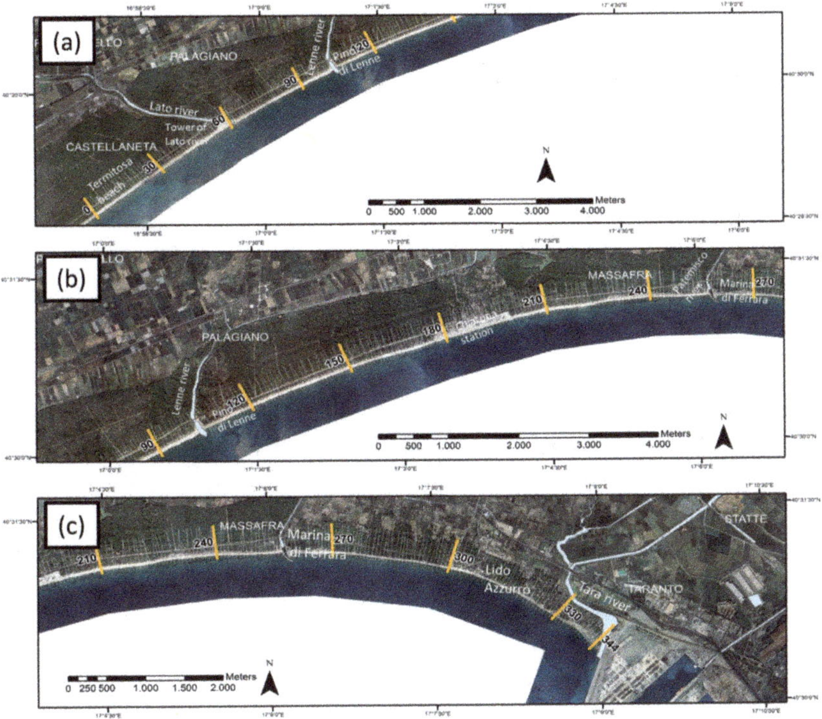

Figure 7. Coastal stretches analyzed in this study. Each transect is reported in the figures; (**a**) CS1 from the Pino di Lenne beach to the Castellaneta beach; (**b**) CS2 from the Marina di Ferrara beach to the Pino di Lenne beach; and (**c**) CS3 from the Tara River mouth to the Lido Azzurro beach.

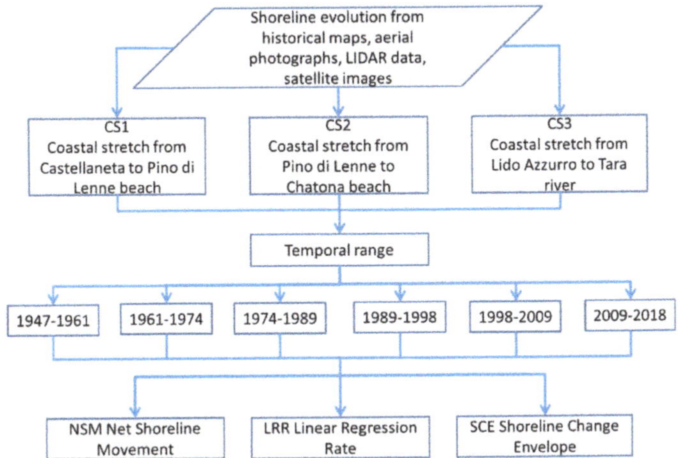

Figure 8. Flow chart of shoreline change analysis performed over different time spans using Digital Shoreline Analysis System (DSAS).

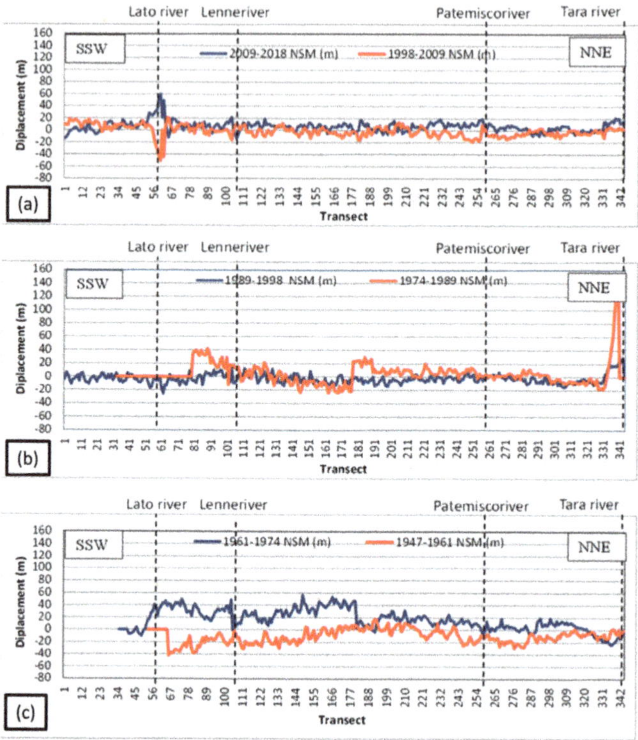

Figure 9. Net Shoreline Movement (NSM) for the study area: (**a**) NSM during the 2009–2018 and 1998–2009 time spans; (**b**) NSM during the 1989–1998 and 1974–1989 time spans; and (**c**) NSM during the 1947–1961 and 1961–1974 time spans.

3.3. Predictive Model of Submersion Surfaces Connected to Sea-Level Rise for 2100

Predicted coastal changes in response to sea-level rise may reflect a different behavior in function of the geomorphological elements of a sandy coast. These may determine three different scenarios: (i) a barrier erosion with dispersed sediments on the shoreface just above the wave base, as described by Bruun [41]; (ii) a rolling-over of the barrier, migrating onshore through mass relocation [43]; and (iii) an overstepping or in-place drowning in response to a fast sea-level rise with a complete submersion of the barrier.

A mathematical submersion model was implemented according to the concept that the 3D geometric shoreline displacement and subsequent migration of the intertidal zone are caused by a sea-level rise and horizontal shoreline movements. The development of a submersion model along a sandy coast requires knowledge of the following parameters: (i) sea-level trend; (ii) VLM rates, including isostatic adjustments; and (iii) shoreline erosion/accretion. The model was implemented mathematically in a Matlab environment by considering the components conditioning both the vertical and horizontal coastal displacements for the sea level and shoreline changes:

$$\Delta z = \left(v_{RSLR} \times 10^{-3} \right) \times \Delta t \pm \Delta z_{\text{tide}} \tag{1}$$

$$\Delta x = v_{LRR} \times \Delta t \times \cos \beta \tag{2}$$

$$\Delta y = v_{LRR} \times \Delta t \times \sin \beta \tag{3}$$

where:

Δz—relative sea-level rise (m)

v_{RSLR}—relative sea-level rate (mm/year)

Δt—prediction time span (year)

Δz_{tide}—tide amplitude (m)

Δx—longitude shoreline displacement (m)

Δy—latitude shoreline displacement (m)

v_{LRR}—shoreline rate changes (m/year)

β—normal shoreline angle (degrees).

The output provides all the points corresponding to the submersion surface predicted up to 2100, presented in a GIS-layer format with the xyz coordinates in world metric reference.

4. Results

4.1. Shoreline Changes

The multi-temporal analysis of the shoreline changes highlighted a marked erosion between 1947 and 1967, followed by a significant shoreline accretion up to 1998 (already shown by [48,49]) and, finally, another erosion period up to 2018. Important changes in the entire sandy coast were revealed, together with river mouth changes (Figure 10).

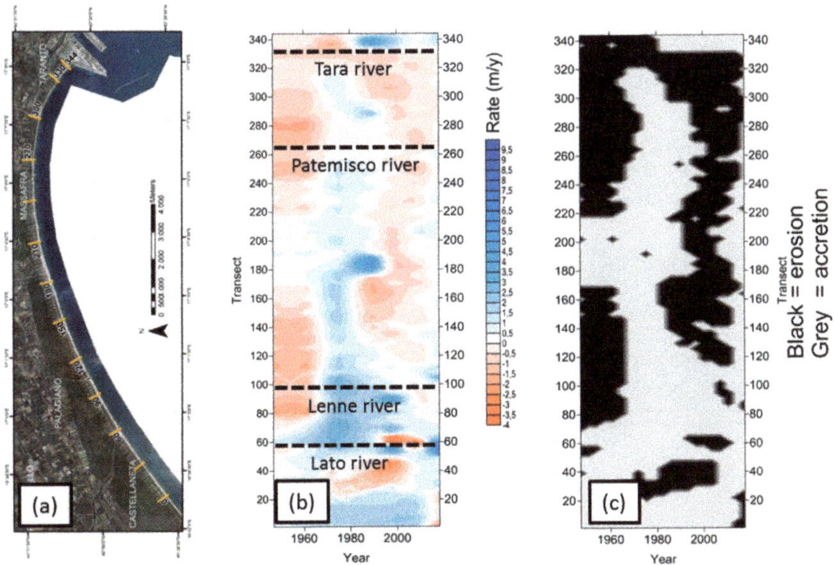

Figure 10. Shoreline evolution of the investigated sandy coast: (**a**) Map with transects; (**b**) shoreline change rates (red indicates the erosion, while blue, the accretion); and (**c**) shoreline changes: grey indicates the advancements, while black the retreats.

The first and most important event is connected to the shoreline accretion close to the Tara River. The 120 m accretion which occurred in 1987 was significant, and can be attributed to the construction of the Taranto Harbor (Figure 11).

The Patemisco River mouth showed a substantial modification in width during 1960, subsequent to the armored anthropogenic construction devised to deviate the longshore drift. In this area, a marked past change on the shoreline and dune (Figure 12) can be noted. In particular, the nearshore bar movements caused by wave diffraction may be observed (Figure 12e).

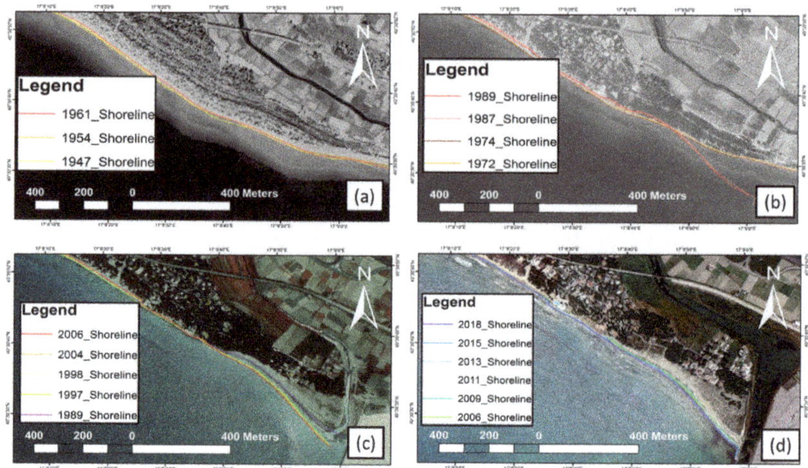

Figure 11. Shoreline changes close to the Tara River mouth: (**a**) 1954 basemap with 1947–1961 super-imposed shorelines; (**b**) 1974 basemap with 1972–1989 super-imposed shorelines; (**c**) 1989 basemap with 1989–2006 super-imposed shorelines; and (**d**) 2013 basemap with 2006–2018 super-imposed shorelines.

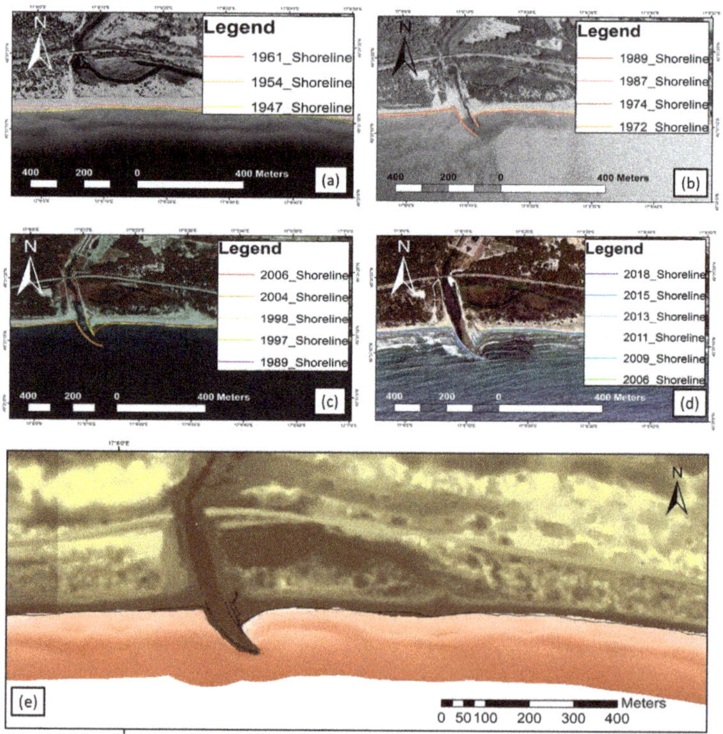

Figure 12. Shoreline changes close to the Patemisco River mouth: (**a**) 1954 basemap with 1947–1961 super-imposed shorelines; (**b**) 1974 basemap with 1972–1989 super-imposed shorelines; (**c**) 1989 basemap with 1989–2006 super-imposed shorelines; (**d**) 2013 basemap with 2006–2018 super-imposed shorelines; and (**e**) Laser Imaging Detection and Ranging of 2009 nearshore topography and morpho-bathymetry.

The Lenne River revealed different full-mouth events, marked by the building of dykes. The latter hinder transversal sediment transports and amplify sand accumulation near the river mouth during storm events (Figure 13). Compared to other rivers, the Lato River is less affected by anthropogenic contributions, and shows little accretion or variable mouth shape, connected to the sea-state regime and longshore currents (Figure 14).

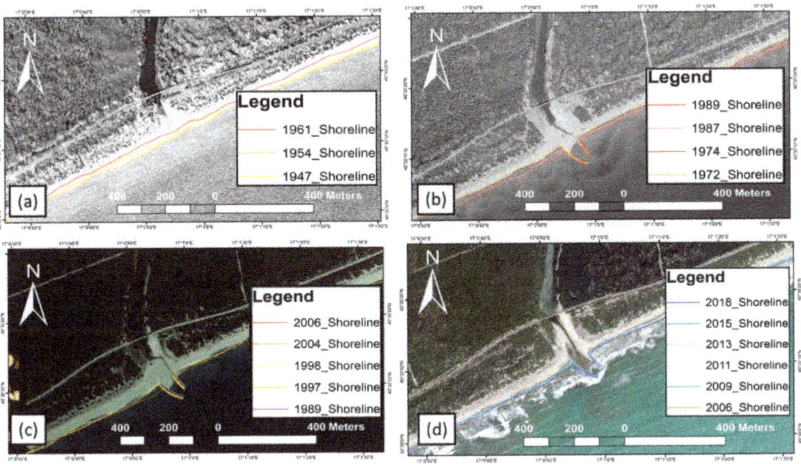

Figure 13. Shoreline changes close to the Lenne River mouth: (**a**) 1954 basemap with 1947–1961 super-imposed shorelines; (**b**) 1974 basemap with 1972–1989 super-imposed shorelines; (**c**) 1989 basemap with 1989–2006 super-imposed shorelines; and (**d**) 2013 basemap with 2006–2018 super-imposed shorelines.

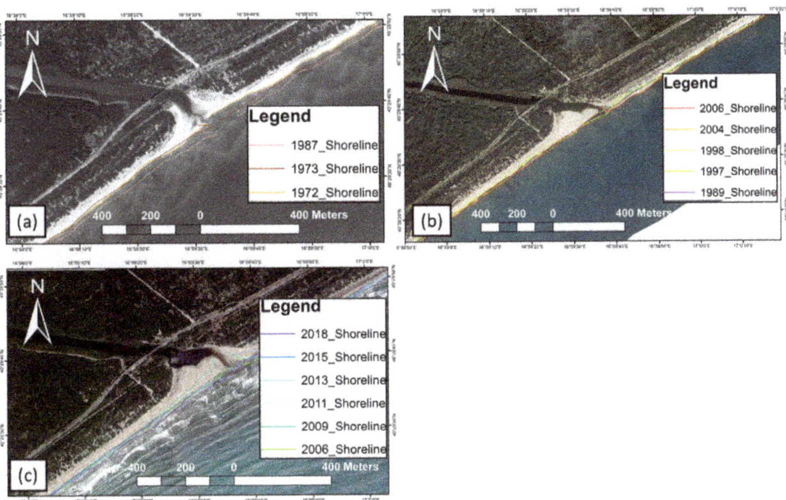

Figure 14. Shoreline changes close to the Lato River mouth; (**a**) 1972 basemap with 1972–1987 super-imposed shorelines; (**b**) 2006 basemap with 1989–2006 super-imposed shorelines; (**c**) 2013 basemap with 2006–2018 super-imposed shorelines.

A variable dynamic is recognizable along the Castellaneta beach, close to the Lato River. This is directly connected to the sea-state influence together with the behavior of the Lato River mouth in the past. Indeed, the hot spot values of the shoreline migration for the transects, corresponding to the Lato River mouth (Figure 10), may be identified. During 1947 and up to 1961, a shoreline erosion trend of −0.38 ± 0.17 m/year occurred, with a maximum net movement of about 7.49 m. This was followed, in the subsequent years (1961–1987), by a shoreline accretion having the same maximum net movement.

From the Lenne River mouth, crossing the Pino di Lenne beach and Chiatona beach (Figures 2–6), a significant shoreline erosion rate of −0.42 ± 0.19 m/year occurred up to the 1970s. This was followed by a general accretion rate of 1.26 ± 0.66 m/year, occurring, in particular, in proximity to the Chiatona railway station, and to the numerous holiday resorts built along the coast. Subsequent sediment redistribution took place between the 1970s and the 2010s. Then, during the period 1990–2010, an erosion rate of −0.36 ± 0.18 m/year occurred.

During the mid-twentieth century (1950–1970), the shoreline change rate between the Marina di Ferrara and the Lido Azzurro highlighted the retreat of some portions of the dune systems, at a rate of −0.81 ± 0.36 m/year. Furthermore, with the construction of the Lido Azzurro holiday resort, an acceleration of this retreat occurred, showing that the current position of the shoreline corresponds to the position of the dune scarp in the 1970s. As of 1989, there has been an erosion rate of −0.23 ± 0.11 m/year, and, as of 1998, an erosion rate of −0.40 ± 0.13 m/year. On the contrary, the building of the breakwater structures in the Taranto harbor, together with the supply trend of the Tara River, changed the shoreline configuration dramatically during the 1970s–1990s.

4.2. Geomorphological Scenario for the Future

An analysis of the dune-beach system topography indicates geomorphological scenarios similar to those described by Davidson-Arnott [47], e.g., a sea-level rise corresponds to a shoreline retreat and beach/dune erosion as a result of the reduction in the sedimentary budget connected to a river transport decrease. A conceptual geomorphological model for the sea-level rise along the coast stretching between Castellaneta and Taranto was deduced by analyzing a sequence of aerial photos from 1947 to present, as well as by using direct surveys (Figure 15).

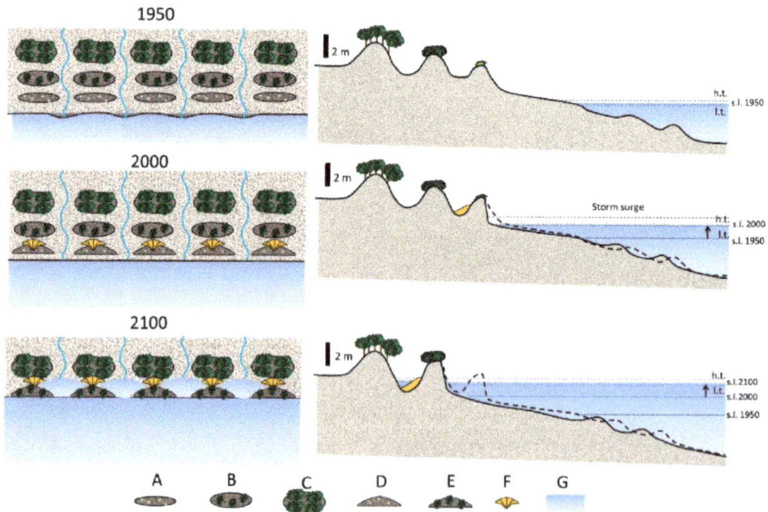

Figure 15. The conceptual model for coastal submersion in function of the mean sea-level rise, and changes in foreshore between high tide (h.t.) and low tide (l.t.). A—Primary Dune Ridge; B—Secondary Dune Ridge; C—Tertiary Dune Ridge; D—Eroding Primary Dune Ridge; E—Eroding Secondary Dune Ridge; F—Aeolian Sand Deposits; and G—Sea.

In the first half of the twentieth Century, a progradation of the mobile coastal system was observed [48,49,56,70] with a well-extended dune system up to 1 km landward stabilized by vegetation cover.

Between the 1950s and 2000s, two phases of coastal erosion were observed throughout the entire littoral area: sea-level rise and increase in the number of storm events, and a decrease in the amount of natural sedimentary nourishment coming from the land. A foredune erosion was observed, and the foredunes along the coastal stretches near the Lato River were completely destroyed. Furthermore, erosion of the secondary dunes was also observed. The building of holiday resorts in the area of the Chiatona and Lido Azzurro coastal stretches caused foredune leveling and increased vulnerability to effects of the sea-level rise.

Within the framework of an increase in the speed of the sea-level rise up to 2100 [27,31], together with an increase in storm intensity [32], the coasts around the Gulf of Taranto could be subjected to a shoreline regression, coupled with a partial foredune erosion and subsequent marine ingressions through river mouths, not to mention submersion of low-lands behind the dune ridge.

5. Discussion

5.1. Simulating the Impact of a Rising Sea Level—A Difficult Methodology

The sandy coast of the Gulf of Taranto is characterized by a complex dynamics system, making the exact shoreline position very difficult to assess. First of all, it can be assumed that if the coast is considered to be in a steady-state, with no significant sediment movements, the shoreline migration should be subjected to only vertical displacements of the relative sea level. By considering the sea-level rise recorded at the Taranto-ISPRA station, different horizontal displacement rates, in function of the coastal slopes (Figure 16), can be derived. Second of all, geometric horizontal shoreline movements due to sea-level rise were subtracted from the observed shoreline movements in order to obtain effective shoreline change (Table 1) and to apply a correct value of horizontal displacement for the three coastal slopes considered in Figure 7.

The observed shoreline movements show very different results from the geometric horizontal migrations induced by a sea-level rise (Table 1 and Figure 10). According to this methodology, the sea-level contribution to the shoreline retreat is lower than 10%. Factors contributing to the shoreline retreat, other than the sea-level rise, such as the gradient of longshore sediment transport, negative sediment input due to dam and hydrological adjustments or construction of holiday resorts along the coast, were all considered. Albeit the investigated area is part of a bay, the beach dynamics have most probably been wave-dominated up to now, but increasing sea levels could modify this behavior.

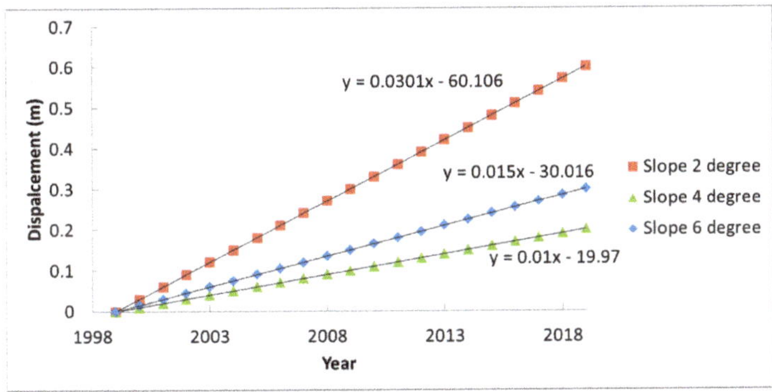

Figure 16. Horizontal displacements obtained following a geometric steady-state approach: sea-level rise determines a landward shoreline movement without accretion, where the shoreline rate change is dependent only on the coastal slope.

Table 1. Effective shoreline rate changes between 1999 and 2020 in the Gulf of Taranto, corrected with geometric horizontal migration rates.

ID	Coastal Slope (degrees)	Observed Shoreline Changes (m/year)	Geometric Horizontal Migration (m/year)	Effective Shoreline Changes (m/year)
CS1	6	−0.40	−0.01	−0.39
CS2	2	−0.36	−0.03	−0.33
CS3	4	−0.38	−0.02	−0.36

5.2. Sea-Level Rise and Predicted Shoreline Positions for Two IPCC Scenarios

In order to estimate both the sea level for 2050 and 2100 AD and the subsequent sea-level rise scenario along the coast of Taranto, two regional sea-level (SL) projections, discussed in the Fifth Assessment Report of the IPCC-AR5 [23,24] and made available by the Integrated Climate data Center-ICDC of the University of Hamburg, rescaled for the Mediterranean Sea (http://icdc.cen. uni-hamburg.de/1/daten/ocean/ar5-slr.html [71]), were used. These data consist of mean values at the upper 95% and lower 5% confidence bounds of the SL, obtained by adding the contributions from geophysical sources driving long-term sea-level changes. The SL projections are based on two different Representative Concentration Pathways (RCP 2.6 and RCP 8.5), and estimated by including the thermosteric/dynamic contribution, obtained from the 21 Coupled Model Intercomparison Project Phase 5 (CMIP5). In addition, the atmosphere–ocean general circulation models (AOGCMs), the surface mass balance and dynamic ice-sheet contributions from Greenland and Antarctica, the glacier and land water storage contributions, the glacial isostatic adjustment, and the inverse barometer effect [23–25,31] were also employed. The predicted sea-level rise along the coast of Taranto was finally estimated as the sum of the above contributions and the tectonic uplift inferred from the elevation of the marine terraces of MIS 5.5 (placed at an elevation of 43 m), and the nearest Global Navigation Satellite System (GNSS) MMET stations, following the method previously applied in other areas of the Mediterranean [17,19,27,31,33,64]. The results are reported in Table 2 and Figure 17. It can be observed that both long-term and instrumental VLM rates are in agreement and provide similar mean uplift velocity of about 0.3 mm/year (MMET is uplifting at 0.29 ± 0.14 mm/year [29]; www.savemedcoasts.eu [64]), while the MIS 5.5 marine terrace provides values between 0.39 and 0.26 mm/year [40,52], supporting detailed relative sea-level (RSL) projections for 2050 and 2100.

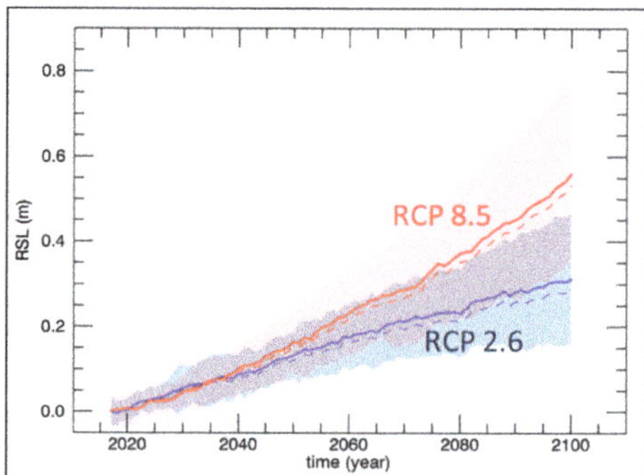

Figure 17. Graph of the sea-level rise projection estimated at the upper 95% bounds of the regional Intergovernmental Panel on Climate Change (IPCC) sea-level projections, integrated with the contribution of the mean Vertical Land Movements (VLM) rate derived from the cGPS data and long-term geological elevation of the MIS 5.5 marine terrace, relative to 2019 AD. Upper and lower curves refer to the RCP 8.5 and RCP 2.6 climate scenarios, respectively. The small scale sea-level variations are related to the modeled sea-level variability due to the ocean component contribution accounting for the effects of dynamic Sea Surface Height (SSH), the global thermosteric SSH anomaly, and the inverse barometer effects ([23,24]; http://icdc.cen.unihamburg.de/ [71]).

Table 2. Sea-level projections in the Assessment Reports 5 Representative Concentration Pathways (RCP) 2.6 and RCP 8.5 scenarios estimated for the Taranto coastal zone.

	Relative Sea-Level Rise (m)	
	2050	2100
RCP 2.6	0.133 ± 0.055	0.313 ± 0.153
RCP 8.5	0.166 ± 0.068	0.559 ± 0.209

Over the last decades, the sea-level rise in the Gulf of Taranto has been causing a generalized landward shoreline displacement and the foredune erosion, while secondary dune belts have only been partially eroded owing to the stability provided by the vegetation cover (Figure 15). With a future rise in the sea-level, the sandy coast could become extremely vulnerable due to the submersion of the entire shoreface and the marine ingression through the inlets (Figure 18).

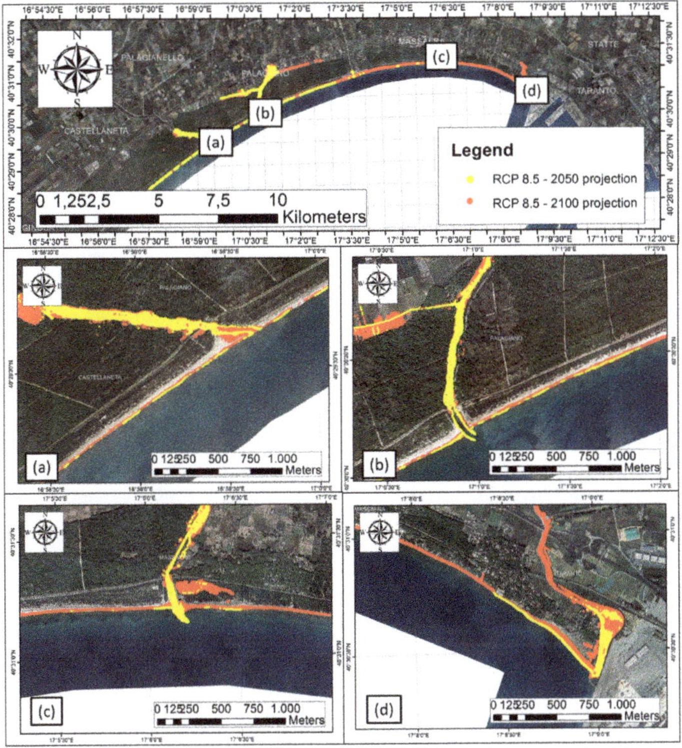

Figure 18. Submersion predictions for the coast around the Gulf of Taranto. Scenarios are relative to the AR 5 RCP 2.6 and RCP 8.5 projections of sea-level rise for 2050 and 2100 together with the VLM and horizontal displacement caused by shoreline change and sea-level rise: (**a**) the Lato River mouth; (**b**) the Lenne River mouth; (**c**) the Patemisco River mouth coastal stretch; (**d**) the Tara River mouth.

In view of a sea-level rise for 2100, the model developed in the present study describes a framework where inlets, constituted by rivers, determine preferential pathways to marine ingression. According to the Brunel and Sabatier model [9], shoreline retreats determine the destruction of the dunes during storms as some of the sand is projected behind the dunes (e.g., washover). Therefore, a shoreline retreat is accompanied by an increase in the elevation of the land behind the dunes which conditions the future position of the shoreline, and compensates, at least partially, for the sea-level rise [72]. To date, no methodologies which include this process have been proposed. Hence, it is necessary to interpret the present results attentively. In any case, river flow regimes will certainly be affected by the increase in the sea-level and, consequently, their ability to carry sediment onto beaches will be altered.

6. Conclusions

Over the past years, the Apulian coastal systems have been subjected to relevant erosion, in part, caused by meteo-marine processes and climate changes, and, in part, by anthropogenic activities. This coastal erosion, together with sea-level rise and storm flooding, is increasing the vulnerability of the coasts.

The authors' analysis, based on available data and predictive models, enabled the reconstruction of the past dynamics of the coastal area between Castellaneta and Taranto in function of the present-day trend, as well as the development of a submersion model for future sea-level rise scenarios.

Since the mid-twentieth century, the coast has been undergoing various changes in response to sea-level rise:

- First half of the twentieth Century—progradation of the coastal system stretching from Castellaneta to Taranto, plantations of Pino d'Aleppo along the entire coast stabilized the dune system;
- 1950s–1970s—shoreline erosion highlighted along the entire littoral stretch, the building of holiday resorts in proximity to railway stations (Castellaneta Marina, Chiatona, Lido Azzurro);
- 1970s—construction of the Taranto Harbor outside the Mar Grande inlet with a substantial modification of the shoreline, and significant changes caused by the Tara, Patemisco, Lato and Lenne river mouths;
- 1970s–1990s—shoreline accretion for most of the littoral area, mainly for the beaches in proximity to the Tara river mouth, except for the Lido Azzurro and Marina di Ferrara coastal stretches, where a significant shoreline erosion and dune retreat were observed;
- 1999–2016—sea-level rise, recorded at the tide gauge in Taranto, at a rate of 1.04 ± 0.5 mm/year, and shoreline erosion recorded throughout the entire littoral area, with the exception of the area near the Tara mouth where accretion had continued;
- 2016–2018—shoreline stabilization and slow accretion in some coastal stretches, such as the Pino di Lenne beach, Chiatona beach, and Tara river mouth and an exception was made for Castellaneta, Marina di Ferrara and Lido Azzurro where a slow erosion was observed.

Considering the RCP 8.5 of sea-level projections for 2050 and 2100, together with VLM, the model output shows how open beaches will be subjected to a significant shoreline retreat with a foredune erosion and subsequent marine ingressions (Figure 18). This behavior will accelerate locally in coastal stretches already undergoing erosion, such as at Lido Azzurro and Metaponto. This will cause a complete foredune destruction with part of the sediments transported nearshore. On the other hand, adjacent to the coastal areas of Chiatona and Pino di Lenne, the model shows a significant foreshore submersion up to the dune scarp and marine ingressions through inlets of the river mouths.

Supplementary Materials: The following are available online at http://www.mdpi.com/2073-4441/12/5/1414/s1, Table S1: Images and LIDAR/TLS data used to extract the shoreline position.

Author Contributions: Conceptualization: G.S. (Giovanni Scardino), F.S., and G.M.; Data curation: G.S. (Giovanni Scardino), G.S. (Giovanni Scicchitano), F.S., M.A., A.V., and M.M.; Formal analysis: G.S. (Giovanni Scardino) and A.V.; Funding acquisition: G.M. and M.M.; Investigation: G.S. (Giovanni Scardino), F.S.; M.M., and A.P.; Methodology: G.S. (Giovanni Scardino), G.S. (Giovanni Scicchitano), F.S., M.M., A.P., and G.M.; Project administration: M.M. and A.P.; Resources: G.M., F.S., and M.M.; Software: G.S. (Giovanni Scicchitano); Visualization: M.A., F.S., and A.P.; Writing—original draft: G.S. (Giovanni Scardino), F.S., G.S. (Giovanni Scicchitano), and G.M.; Writing—review & editing: G.S. (Giovanni Scicchitano), F.S., M.A., and G.M. All authors have read and agreed to the published version of the manuscript.

Funding: This work was carried out within the Ph.D. Project of the "Programma Operativo Nazionale Ricerca e Inno-vazione 2014—2020 (CCI 2014IT16M2OP005)", Fondo Sociale Europeo, Azione I.1 "Dottorati Innovativi con caratterizzazione Industriale", Università degli Studi di Bari Aldo Moro and Integrated Sea sTORM management Stategies, Interreg ADRION European Regional Development Fund- Instrument for Pre Accession II-Fund.

Acknowledgments: We thank all those who collaborated with their logistic and technical support throughout every phase of this study. This work was carried out under the umbrella of the IGCP Project n. 639 "Sea-level change from minutes to millennia" (Project Leaders: S. Engelhart, G. Hoffmann, F. Yu and A. Rosentau) and SAVEMEDCOASTS (www.savemedcoasts.eu), funded by the European Union (DG-ECHO) (Project Coordinator Marco Anzidei). Our appreciation also goes to Victoria Sportelli, a native English speaker and lecturer, for the English language revision of the paper. We are also grateful to the reviewers for their revisions that have allowed us to improve the quality of the paper.

Conflicts of Interest: The authors declare no conflict of interest. The funders had no role in the design of the study; in the collection, analyses, or interpretation of data; in the writing of the manuscript, or in the decision to publish the results.

References

1. Aucelli, P.P.C.; Iannantuono, E.; Rosskopf, C.M. Evoluzione recente e rischio di erosione della costa molisana (*Italia meridionale*). *Ital. J. Geosci.* **2009**, *128*, 759–771. [CrossRef]
2. Aucelli, P.P.C.; Di Paola, G.; Incontri, P.; Rizzo, A.; Vilardo, G.; Benassai, G.; Buonocore, B.; Pappone, G. Coastal inundation risk assessment due to subsidence and sea level rise in a Mediterranean alluvial plain (Volturno coastal plain—Southern Italy). *Estuar. Coast. Shelf Sci.* **2017**, *198*, 597–609. [CrossRef]
3. Sabatier, F.; Samat, O.; Brunel, C.; Heurtefeux, H.; Delanghe-Sabatier, D. Determination of set-back lines on eroding coasts. Example of the beaches of the Gulf of Lions (French Mediterranean Coast). *J. Coast. Conserv.* **2009**, *13*, 57. [CrossRef]
4. Sabato, L.; Longhitano, S.G.; Gioia, D.; Cilumbriello, A.; Spalluto, L. Sedimentological and morpho-evolution maps of the 'Bosco Pantano di Policoro' coastal system (Gulf of Taranto, Southern Italy). *J. Maps* **2012**, *8*, 304–311. [CrossRef]
5. Ayadi, K.; Boutiba, M.; Sabatier, F.; Guettouche, M.S. Detection and analysis of historical variations in the shoreline, using digital aerial photos, satellite images, and topographic surveys DGPS: Case of the Bejaia bay (East Algeria). *Arab. J. Geosci.* **2015**, *9*, 26. [CrossRef]
6. Paskoff, R.P. Potential Implications of Sea-Level Rise for France. *J. Coast. Res.* **2004**, *20*, 424–434. [CrossRef]
7. Antonioli, F.; Ferranti, L.; Fontana, A.; Amorosi, A.; Bondesan, A.; Braitengerg, C.; Dutton, A.; Fontolan, G.; Furlani, S.; Lambeck, K.; et al. Holocene relative sea-level changes and vertical movements along the Italian and Istrian coastlines. *Quat. Int.* **2009**, *206*, 102–133. [CrossRef]
8. Mastronuzzi, G.; Antonioli, F.; Anzidei, M.; Auriemma, R.; Alfonso, C.; Scarano, T. Evidence of relative sea level rise along the coasts of central Apulia (Italy) during the late Holocene via maritime archaeological indicators. *Quat. Int.* **2017**, *439*, 65–78. [CrossRef]
9. Brunel, C.; Sabatier, F. Potential influence of sea-level rise in controlling shoreline position on the French Mediterranean Coast. *Geomorphology* **2009**, *107*, 47–57. [CrossRef]
10. Le Cozannet, G.; Garcin, M.; Yates, M.; Idier, D.; Meyssignac, B. Approaches to evaluate the recent impacts of sea-level rise on shoreline changes. *Earth-Sci. Rev.* **2014**, *138*, 47–60. [CrossRef]
11. Lambeck, K.; Anzidei, M.; Antonioli, F.; Benini, A.; Esposito, A. Sea level in Roman time in the Central Mediterranean and implications for recent change. *Earth Planet. Sci. Lett.* **2004**, *224*, 563–575. [CrossRef]
12. Rovere, A.; Stocchi, P.; Vacchi, M. Eustatic and Relative Sea Level Changes. *Curr. Clim. Chang. Rep.* **2016**, *2*, 221–231. [CrossRef]
13. Lambeck, K.; Antonioli, F.; Anzidei, M.; Ferranti, L.; Leoni, G.; Scicchitano, G.; Silenzi, S. Sea level change along the Italian coast during the Holocene and projections for the future. *Quat. Int.* **2011**, *232*, 250–257. [CrossRef]
14. Spampinato, C.R.; Costa, B.; Di Stefano, A.; Monaco, C.; Scicchitano, G. The contribution of tectonics to relative sea-level change during the Holocene in coastal south-eastern Sicily: New data from boreholes. *Quat. Int.* **2011**, *232*, 214–227. [CrossRef]
15. Spampinato, C.R.; Scicchitano, G.; Ferranti, L.; Monaco, C. Raised Holocene paleo-shorelines along the Capo Schisò coast, Taormina: New evidence of recent co-seismic deformation in northeastern Sicily (Italy). *J. Geodyn.* **2012**, *55*, 18–31. [CrossRef]
16. Anzidei, M.; Bosman, A.; Carluccio, R.; Casalbore, D.; Caracciolo, F.D.; Esposito, A.; Nicolosi, I.; Pietrantonio, G.; Vecchio, A.; Carmisciano, C.; et al. Flooding scenarios due to land subsidence and sea-level rise: A case study for Lipari Island (Italy). *Terra Nova* **2017**, *29*, 44–51. [CrossRef]

17. Anzidei, M.; Scicchitano, G.; Tarascio, S.; De Guidi, G.; Monaco, C.; Barreca, G.; Mazza, G.; Serpelloni, E.; Vecchio, A. Coastal retreat and marine flooding scenario for 2100: A case study along the coast of Maddalena peninsula (Southeastern Sicily). *Geogr. Fis. Din. Quat.* **2018**, *41*, 5–16. [CrossRef]

18. Anzidei, M.; Doumaz, F.; Vecchio, A.; Serpelloni, E.; Pizzimenti, L.; Civico, R.; Greco, M.; Martino, G.; Enei, F. Sea Level Rise Scenario for 2100 A.D. in the Heritage Site of Pyrgi (Santa Severa, Italy). *J. Mar. Sci. Eng.* **2020**, *8*, 64. [CrossRef]

19. Marsico, A.; Lisco, S.; Presti, V.L.; Antonioli, F.; Amorosi, A.; Anzidei, M.; Deiana, G.; Falco, G.D.; Fontana, A.; Fontolan, G.; et al. Flooding scenario for four Italian coastal plains using three relative sea level rise models. *J. Maps* **2017**, *13*, 961–967. [CrossRef]

20. Bonaldo, D.; Antonioli, F.; Archetti, R.; Bezzi, A.; Correggiari, A.; Davolio, S.; De Falco, G.; Fantini, M.; Fontolan, G.; Furlani, S.; et al. Integrating multidisciplinary instruments for assessing coastal vulnerability to erosion and sea level rise: Lessons and challenges from the Adriatic Sea, Italy. *J. Coast. Conserv.* **2019**, *23*, 19–37. [CrossRef]

21. Rahmstorf, S. A Semi-Empirical Approach to Projecting Future Sea-Level Rise. *Science* **2007**, *315*, 368–370. [CrossRef] [PubMed]

22. Rahmstorf, S.; Perrette, M.; Vermeer, M. Testing the robustness of semi-empirical sea level projections. *Clim. Dyn.* **2012**, *39*, 861–875. [CrossRef]

23. Church, J.A.; Clark, P.U.; Cazenave, A.; Gregory, J.M.; Jevrejeva, S.; Levermann, A.; Merrifield, M.A.; Milne, G.A.; Nerem, R.S.; Nunn, P.D.; et al. Sea-Level Rise by 2100. *Science* **2013**, *342*, 1445. [CrossRef] [PubMed]

24. Church, J.A.; Clark, P.U.; Cazenave, A.; Gregory, J.M.; Jevrejeva, S.; Levermann, A.; Merrifield, M.A.; Milne, G.A.; Nerem, R.S.; Nunn, P.D.; et al. Chapter 13: Sea-level change. In *Climate Change 2013: The Physical Science Basis*; Contribution of Working Group I to the Fifth Assessment Report of the Intergovernmental Panel on Climate Change; Stocker, T.F., Qin, D., Plattner, G.-K., Tignor, M., Allen, S.K., Boschung, J., Nauels, A., Xia, Y., Bex, V., Midgley, P.M., Eds.; Cambridge University Press: Cambridge, UK; New York, NY, USA, 2013; pp. 1137–1216.

25. Stocker, T.F.; Qin, D.; Plattner, G.-K.; Tignor, M.; Allen, S.K.; Boschung, J.; Nauels, A.; Xia, Y.; Bex, V.; Midgley, P.M. (Eds.) *IPCC Climate Change 2013: The Physical Science Basis*; Contribution of Working Group I to the Fifth Assessment Report of the Intergovernmental Panel on Climate Change; Cambridge University Press: Cambridge, UK; New York, NY, USA, 2013; p. 1535.

26. Pachauri, R.K.; Meyer, L.A. (Eds.) *IPCC Climate Change 2014: Synthesis Report*; Contribution of Working Groups I, II and III to the Fifth Assessment Report of the Intergovernmental Panel on Climate Change; Core Writing Team; IPCC: Geneva, Switzerland, 2014; p. 151.

27. *IPCC Special Report on the Ocean and Cryosphere in a Changing Climate*, Pörtner, H.-O.; Roberts, D.C.; Masson-Delmotte, V.; Zhai, P.; Tignor, M.; Poloczanska, E.; Mintenbeck, K.; Alegría, A.; Nicolai, M.; Okem, A.; et al. (Eds.) 2019; in press.

28. Wöppelmann, G.; Marcos, M. Coastal sea level rise in southern Europe and the nonclimate contribution of vertical land motion. *J. Geophys. Res. Ocean.* **2012**, *117*. [CrossRef]

29. Anzidei, M.; Lambeck, K.; Antonioli, F.; Furlani, S.; Mastronuzzi, G.; Serpelloni, E.; Vannucci, G. Coastal structure, sea-level changes and vertical motion of the land in the Mediterranean. *Geol. Soc. Lond. Spec. Publ.* **2014**, *388*, 453–479. [CrossRef]

30. Vermeer, M.; Rahmstorf, S. Global sea level linked to global temperature. *Proc. Natl. Acad. Sci. USA* **2009**, *106*, 21527–21532. [CrossRef]

31. Vecchio, A.; Anzidei, M.; Serpelloni, E.; Florindo, F. Natural Variability and Vertical Land Motion Contributions in the Mediterranean Sea-Level Records over the Last Two Centuries and Projections for 2100. *Water* **2019**, *11*, 1480. [CrossRef]

32. Lionello, P.; Conte, D.; Marzo, L.; Scarascia, L. The contrasting effect of increasing mean sea level and decreasing storminess on the maximum water level during storms along the coast of the Mediterranean Sea in the mid 21st century. *Glob. Planet. Chang.* **2017**, *151*, 80–91. [CrossRef]

33. Antonioli, F.; Anzidei, M.; Amorosi, A.; Lo Presti, V.; Mastronuzzi, G.; Deiana, G.; De Falco, G.; Fontana, A.; Fontolan, G.; Lisco, S.; et al. Sea-level rise and potential drowning of the Italian coastal plains: Flooding risk scenarios for 2100. *Quat. Sci. Rev.* **2017**, *158*, 29–43. [CrossRef]

34. Bamber, J.L.; Oppenheimer, M.; Kopp, R.E.; Aspinall, W.P.; Cooke, R.M. Ice sheet contributions to future sea-level rise from structured expert judgment. *Proc. Natl. Acad. Sci. USA* **2019**, *116*, 11195–11200. [CrossRef]

35. Kulp, S.A.; Strauss, B.H. New elevation data triple estimates of global vulnerability to sea-level rise and coastal flooding. *Nat. Commun.* **2019**, *10*, 1–12. [CrossRef]

36. Mastronuzzi, G.; Aringoli, D.; Aucelli, P.P.C.; Baldassarre, M.A.; Bellotti, P.; Bini, M.; Biolchi, S.; Bontempi, S.; Brandolini, P.; Chelli, A.; et al. Geomorphological map of the Italian Coast: From a descriptive to a morphodynamic approach. *Geogr. Fis. Din. Quat.* **2017**, *40*, 161–195. [CrossRef]

37. Carbognin, L.; Teatini, P.; Tosi, L. Eustacy and land subsidence in the Venice Lagoon at the beginning of the new millennium. *J. Mar. Syst.* **2004**, *51*, 345–353. [CrossRef]

38. Caldara, M.; Capolongo, D.; Damato, B.; Pennetta, L. Can the ground laser scanning technology be useful for coastal defenses monitoring? *Ital. J. Eng. Geol. Environ.* **2006**, *1*, 35–49.

39. Syvitski, J.P.M.; Kettner, A.J.; Overeem, I.; Hutton, E.W.H.; Hannon, M.T.; Brakenridge, G.R.; Day, J.; Vörösmarty, C.; Saito, Y.; Giosan, L.; et al. Sinking deltas due to human activities. *Nat. Geosci.* **2009**, *2*, 681–686. [CrossRef]

40. De Santis, V.; Caldara, M.; Torres, T.; Ortiz, J.E.; Sánchez-Palencia, Y. A review of MIS 7 and MIS 5 terrace deposits along the Gulf of Taranto based on new stratigraphic and chronological data. *Ital. J. Geosci.* **2018**, *137*, 349–368. [CrossRef]

41. Bruun, P. Sea-Level Rise as a Cause of Shore Erosion. *J. Waterw. Harb. Div.* **1962**, *88*, 117–132.

42. Dubois, R.N. How does a Barrier Shoreface Respond to a Sea-Level Rise? *J. Coast. Res.* **2002**, *18*, 3–5.

43. Lorenzo-Trueba, J.; Ashton, A.D. Rollover, drowning, and discontinuous retreat: Distinct modes of barrier response to sea-level rise arising from a simple morphodynamic model. *J. Geophys. Res. Earth Surf.* **2014**, *119*, 779–801. [CrossRef]

44. Valentin, H. Die Küsten der Erde. *Petermanns Geogr. Mitt. Ergänzsungheft* **1952**, *246*, 118.

45. Wong, T.E.; Bakker, A.M.; Keller, K. Impacts of Antarctic fast dynamics on sea level projections and coastal flood defense. *Clim. Change* **2017**, *144*, 347–364. [CrossRef]

46. Bruun, P. *Coast Erosion and the Development of Beach Profiles*; U.S. Beach Erosion Board: Washington, DC, USA, 1954.

47. Davidson-Arnott, R.G.D. Conceptual Model of the Effects of Sea Level Rise on Sandy Coasts. *J. Coast. Res.* **2005**, *21*, 1166–1172. [CrossRef]

48. Cilumbriello, A.; Sabato, L.; Tropeano, M.; Gallicchio, S.; Grippa, A.; Maiorano, P.; Mateu-Vicens, G.; Rossi, C.A.; Spilotro, G.; Calcagnile, L.; et al. Sedimentology, stratigraphic architecture and preliminary hydrostratigraphy of the Metaponto coastal-plain subsurface (Southern Italy). *Mem. Descr. Carta Geol. d'Ital.* **2010**, *XC*, 67–84.

49. Tropeano, M.; Cilumbriello, A.; Sabato, L.; Gallicchio, S.; Grippa, A.; Longhitano, S.G.; Bianca, M.; Gallipoli, M.R.; Mucciarelli, M.; Spilotro, G. Surface and subsurface of the Metaponto Coastal Plain (Gulf of Taranto—Southern Italy): Present-day- vs. LGM-landscape. *Geomorphology* **2013**, *203*, 115–131. [CrossRef]

50. Nakada, M.; Lambeck, K. The melting history of the late Pleistocene Antarctic ice sheet. *Nature* **1988**, *333*, 36–40. [CrossRef]

51. Mastronuzzi, G.; Sansò, P. Morfologia e genesi delle Isole Chéradi e del Mar Grande (Taranto, Puglia, Italia). *Geogr. Fis. Din. Quat.* **1998**, *21*, 131–138.

52. Mastronuzzi, G.; Sansò, P. Quaternary coastal morphology and sea level changes. In *Proceedings of the Puglia 2003, Final Conference—Project IGCP 437 UNESCO—IUGS, Otranto/Taranto, Puglia, Italy, 22–28 September 2003*; GIS Coast Coast—Gruppo Informale di Studi Costieri, Research Publication: Taranto, Italy, 2003; Volume 5, p. 184.

53. Valenzano, E.; Scardino, G.; Cipriano, G.; Fago, P.; Capolongo, D.; De Giosa, F.; Lisco, S.; Mele, D.; Moretti, M.; Mastronuzzi, G. Holocene Morpho-Sedimentary Evolution of the Mar Piccolo Basin (Taranto, Southern Italy). *Geogr. Fis. Din. Quat.* **2018**, *41*, 119–135. [CrossRef]

54. Zander, A.; Fulling, A.; Brückner, H.; Mastronuzzi, G. OSL dating of upper pleistocene littoral sediments: A contribution to the chronostratigraphy of raised marine terraces bordering the Gulf of Taranto, South Italy. *Geogr. Fis. Din. Quat.* **2006**, *29*, 33–50.

55. Carter, R.W.G. *Coastal Evolution: Late Quaternary Shoreline Morphodynamics*; Reprint Edizione; Cambridge University Press: Cambridge, UK, 1997; ISBN 978-0-521-59890-3.

56. Mastronuzzi, G.; Sansò, P. Pleistocene sea-level changes, sapping processes and development of valley networks in the Apulia region (Southern Italy). *Geomorphology* **2002**, *46*, 19–34. [CrossRef]

57. Mastronuzzi, G.; Romaniello, L. Holocene aeolian morphogenetic phases in Southern Italy: Problems in 14C age determinations using terrestrial gastropods. *Quat. Int.* **2008**, *183*, 123–134. [CrossRef]

58. Caldara, M.; Centenaro, E.; Mastronuzzi, G.; Sansò, P.; Sergio, A. Features and present evolution of Apulian Coast (Southern Italy). *J. Coast. Res.* **1998**, *SI*, 55–64.

59. Lisi, I.; Bruschi, A.; Del Gizzo, M.; Archina, M.; Barbano, A.; Corsini, S. Le unità fisiografiche e le profondità di chiusura della costa italiana. *L'ACQUA* **2010**, *2*, 35–52.

60. Di Bucci, D.; Caputo, R.; Mastronuzzi, G.; Fracassi, U.; Selleri, G.; Sansò, P. Quantitative analysis of extensional joints in the Southern Adriatic foreland (Italy), and the active tectonics of the Apulia region. *J. Geodyn.* **2011**, *51*, 141–155. [CrossRef]

61. Amorosi, A.; Antonioli, F.; Bertini, A.; Marabini, S.; Mastronuzzi, G.; Montagna, P.; Negri, A.; Rossi, V.; Scarponi, D.; Taviani, M.; et al. The Middle–Upper Pleistocene Fronte Section (Taranto, Italy): An exceptionally preserved marine record of the Last Interglacial. *Glob. Planet. Chang.* **2014**, *119*, 23–38. [CrossRef]

62. Lisco, S.; Corselli, C.; De Giosa, F.; Mastronuzzi, G.; Moretti, M.; Siniscalchi, A.; Marchese, F.; Bracchi, V.; Tessarolo, C.; Tursi, A. Geology of Mar Piccolo, Taranto (Southern Italy): The physical basis for remediation of a polluted marine area. *J. Maps* **2016**, *12*, 173–180. [CrossRef]

63. Polcari, M.; Albano, M.; Montuori, A.; Bignami, C.; Tolomei, C.; Pezzo, G.; Falcone, S.; La Piana, C.; Doumaz, F.; Salvi, S.; et al. InSAR Monitoring of Italian Coastline Revealing Natural and Anthropogenic Ground Deformation Phenomena and Future Perspectives. *Sustainability* **2018**, *10*, 3152. [CrossRef]

64. Savemedcoasts. Available online: http://www.savemedcoasts.eu/ (accessed on 4 April 2020).

65. RING—Rete Integrata Nazionale GPS. Available online: http://ring.gm.ingv.it/ (accessed on 25 April 2020).

66. Altamimi, Z.; Collilieux, X.; Métivier, L. ITRF2008: An improved solution of the international terrestrial reference frame. *J. Geod.* **2011**, *85*, 457–473. [CrossRef]

67. Rete Mareografica Nazionale—Homepage. Available online: https://mareografico.it/ (accessed on 4 April 2020).

68. Attivazione Del Nuovo Sistema Informativo Meteo Oceanografico Delle Coste Pugliesi (Simop). Available online: https://www.adb.puglia.it/public/news.php?extend.282 (accessed on 4 April 2020).

69. Theiler, E.; Himmelstoss, E.; Zichichi, J.; Ergul, A. *Digital Shoreline Analysis System (DSAS) Version 4.0-An ArcGIS Extension for Calculating Shoreline Change (Ver. 4.4, July 2017)*; U.S. Geological Survey Open-File Report; USGS, U.S.: Reston, VA, USA, 2017; p. 1278.

70. Bonora, N.; Immordino, F.; Schiavi, C.; Simeoni, U.; Valpreda, E. Interaction between Catchment Basin Management and Coastal Evolution (Southern Italy). *J. Coast. Res.* **2002**, 81–88. [CrossRef]

71. AR5 Sea Level Rise. Available online: http://icdc.cen.uni-hamburg.de/1/daten/ocean/ar5-slr.html (accessed on 25 April 2020).

72. Sabatier, F.; Provansal, M.; Fleury, T.-J. Discussion of: PASKOFF, R., 2004. Potential Implications of Sea-Level Rise for France. *J. Coast. Res.* **2004**, *20*, 424–434. [CrossRef]

Article

The Costs of Sea-Level Rise: Coastal Adaptation Investments vs. Inaction in Iberian Coastal Cities

Luis Maria Abadie [1] , **Elisa Sainz de Murieta** [1,2,*] and **Ibon Galarraga** [1]

[1]   BC3 Basque Centre for Climate Change, Sede Building 1, 1st floor, Scientific Campus of the University of the Basque Country, 48940 Leioa, Spain; lm.abadie@bc3research.org (L.M.A.); ibon.galarraga@bc3research.org (I.G.)

[2]   Grantham Research Institute, London School of Economics and Political Science (LSE), London WC2A 2AZ, UK

*   Correspondence: elisa.sainzdemurieta@bc3research.org; Tel.: +34-944-014-690

Received: 3 March 2020; Accepted: 21 April 2020; Published: 24 April 2020

Abstract: Iberian coastal cities are subject to significant risks in the next decades due to climate change-induced sea-level rise. These risks are quite uncertain depending on several factors. In this article, we estimate potential economic damage in 62 Iberian coastal cities from 2020 to 2100 using regional relative sea-level rise data under three representative concentration pathways (RCP 8.5, RCP 4.5 and RCP 2.6). We analyze the expected accumulated damage costs if no adaptation actions take place and compare this scenario to the investment cost of some adaptation strategies being implemented. The results show that some adaptation strategies are less costly than the potential damage under inaction. In other words, it is economically rational to invest in adaptation even in a context of high uncertainty. These calculations are very relevant to inform climate change adaptation decisions and to better manage the risk posed by sea-level rise. Moreover, our findings show the importance of a good understanding of the shape of the sea-level rise and damage cost distributions to calculate the expected damage. We show that using the 50th percentile for these calculations is not adequate as it leads to a serious underestimation of expected damage and coastal risk.

Keywords: climate change; adaptation costs; investment decision; Spain and Portugal coastal cities; uncertainty; stochastic model

1. Introduction

Sea-level rise is one of the main consequences of anthropogenic climate change [1]. Sea level worldwide was stable from approximately 7000 years before present until the end of the 19th century, but since then the geological and tide gauge record presents signs of acceleration [2]. During the 20th century, a significant acceleration of sea-level rise has occurred, quantified at about 1.2 mm per year from 1901 to 1990 [3,4]. The same analysis applied to the period 1993–2010 revealed a much larger acceleration, of about three millimetres per year [4]. Other calculations [5] show an increasing sea-level rise since 2010 of up to 4.4 ± 0.5 mm yr^{-1}, so since the 1880s, the global sea level has increased more than 20 cm and continues rising at more than 4 mm yr^{-1} [6]. Nonetheless, sea-level rise can vary significantly in magnitude and rate of change regionally and these differences can be of up to four times larger in some areas [7].

The global sea-level rise projections by the Intergovernmental Panel for Climate Change (IPCC) [1], as well as the regionalized assessments for each emission scenario (representative concentration pathways, RCP), foresee an acceleration in sea-level rise over the course of the present century [8–11]. Sea-level rise will accelerate in the future and over the next centuries even if emissions stabilize [12]. The magnitude of change, however, remains uncertain because it depends on the ambition of mitigation efforts, as well as the location and the time considered. Let us illustrate this point: if the global average temperature is stabilized at 2 °C above preindustrial levels, by 2100 global sea-level rise could reach 26 cm at the median of the distribution and 81 cm at the 95th percentile. In a high emission scenario in which global temperatures could reach +5 °C, sea-level rise could reach 51 cm and 178 cm at the median and 95th percentile, respectively. Moreover, sea level could even exceed 2 m in 2100 (95th percentile) if thermal expansion and the contribution of major ice sheets are accounted for [13].

Sea-level rise will also lead to more frequent coastal flooding. The latest global estimations show that, by 2100, extreme sea level events that have occurred once every century are expected to take place annually under all emission scenarios considered [14]. The combination of sea-level rise and extreme events represents a growing risk to coastal areas around the world. In Europe, one-third of the population resides in a 50 km strip of the coast, where many ecosystems, assets and infrastructures are located [15]. Therefore, even if during the 20th century, socioeconomic factors have been the main cause of increased exposure and vulnerability in coastal areas, climate change is expected to aggravate this situation in the future [16].

These changes in sea level and the frequency and magnitude of extreme events may cause severe economic impacts. With no additional investments in adaptation, global damage due to coastal flooding could reach between 93,000 and 961,000 million euros, and the number of people affected would rise to 1.5–3.65 million by 2100, considering three climatic and socio-economic scenario combinations (RCP 4.5-SSP1, RCP 8.5-SSP3 and RCP 8.5-SSP5) [17]. In Europe, coastal flooding and erosion by 2050 could have a cost of between 6500 and 40,000 million euros per year for RCP4.5 and RCP 8.5, respectively, combined with the SSP5 socio-economic development scenario [15]. The same study estimated for the same scenarios that the number of people affected could vary between 460,000 and 740,000 annually. Nonetheless, the differences between RCPs is small until mid-century, when sea-level rise scenarios start to diverge. This situation could lead to a current underestimation of coastal risk and a delay in the definition and implementation of adaptation measures [18].

Just as for future sea-level rise, economic impacts (and adaptation investment costs) depend on the emission scenario, the time considered and location [19]. The use of mean values or probable ranges in the assessments of the impacts of climate change is very frequent, but due to the existing uncertainty, several authors have insisted on the importance of considering the full distribution of probabilities and especially paying attention to the upper tail of the distribution, i.e., the most negative impacts [20–22]. These types of risk approaches that analyse situations of low probability but great impact are not new in disciplines such as financial or energy economics and allow for more informed risk management as well as more risk-averse approaches.

In recent years, these approaches have proved very useful for the analysis of climate change impacts. For example, financial risk measures, despite their complex modelling, can support decision-making with regards to climate change adaptation as they offer a more comprehensive picture of the risks faced. These risk measures can be used to assess the risk of low probability situations whose impacts may have catastrophic consequences [23], to assess the suitability of an investment plan [24] or to perform stress testing for urban planning, adaptation measures, large investments or infrastructure plans [25].

In this study, we aim at determining the costs of adaptation inaction in 62 Iberian coastal cities in the context of sea-level rise and increased climate risks. We define the cost of inaction as the damage due to climate change in the lack of additional adaptation measures. In order to do so, we compare adaptation investment costs with expected damage due to sea-level rise under three scenarios (RCP8.5, RCP 4.5 and RCP 2.6). These estimations are first obtained annually, and in a second step, we calculate accumulated costs in 2030, 2050 and 2100. These accumulated costs of adaptation inaction are then compared to the costs of adaptation investment under different adaptation strategies. This work contributes to the literature by, first, calculating the full distribution of sea-level rise and, second, combining the percentiles of sea-level rise distributions [9] with deterministic damage curves for a large number of European coastal cities [19]. This allows the full distribution of economic damage (or costs) to be estimated.

The rest of the paper is organised as follows: Section 2 presents a summary of the foreseen impacts of sea-level rise in the Iberian coastal areas. Section 3 describes the methods used; the results are presented in Section 4, and the overall discussion is included in Section 5.

2. The Impacts of and Adaptation to Sea-Level Rise in the Iberian Peninsula

Coastal risks have increased in the Iberian Peninsula during the last decades. The urban development model and the overexploitation of some resources have put enormous pressure on these coastal areas, causing biodiversity loss and environmental degradation, while increasing their exposure and vulnerability to climatic events [26]. In some areas of the Basque Coast, habitat and ecosystem destruction induced by sea-level rise could generate losses of up to 2.3 million euros per hectare [27]. The increase in flood exposure in certain areas of the Spanish Mediterranean coast (Murcia and Alicante) has been directly linked to two urban development peaks, between 1978–1982 and 1997–2007 [28]. In fact, the Spanish Mediterranean coast has high or very high vulnerability and exposure values, compared to other areas of the Mediterranean subject to larger hazards. This situation repeats, however, in other densely populated areas such as the French southwest coast and the north Adriatic coast [29].

Assuming that the trend of rising sea levels continues during the first half of the century, flood-risk could increase by 2040 by 8% in the Atlantic and Cantabrian coasts of Spain and the Alboran Sea, 6% in the Canary Islands and between 2% and 3% in the rest of the Mediterranean coast and the Gulf of Cádiz [30]. In some locations, the frequency and intensity of extreme events are expected to increase in the future. For example, in the city of Bilbao, located on the Cantabrian coast, the intensity, measured as the change in the flood level, could increase from 3.85 m in 2010 to 4 m in 2040, and its frequency will increase from once every 50 years (2010) to once every 15 years in 2040 [30]. In Barcelona, however, changes in intensity are not expected and the increase in frequency is expected to be smaller: the return period could vary from once every 50 to once in 40 years [30,31].

In Portugal, sea-level rise and changes in storminess are expected to cause higher coastal flooding and erosion, among other impacts such as coastal wetland inundation and retreat, and have been identified as one of the most important consequences of climate change [32,33]. The coast north of Lisbon and the Algarve were identified as the areas most at risk [32]. Recent studies have addressed these areas of the Portuguese coast in further detail, identifying the areas of Aveiro [34], south of Porto, and the Algarve as hotspots for coastal inundation and erosion [35,36]. Moreover, it has been estimated that 900 km^2 of the Portuguese coast could be subject to flood risk in 2050, considering 50 year return period extreme events and empirical sea-level rise projections. At the end of the century, the areas at risk of coastal flooding could increase by 27%, where Lisbon, Faro and Aveiro are the districts most at risk [37,38].

With regards to coastal protection and adaptation to climate change, Spain was already among those countries in Europe with a larger expenditure in the period between 1998 and 2015 [39]. Projections to 2040 estimate that the population exposed to permanent flooding on the Cantabrian coast could reach 2–3%. If extreme events are also incorporated, the exposed population could range between 4% and 9%. As for the economic impact, at the end of the century, permanent flood damage in the Bay of Biscay could reach between 1 billion euros (0.1% of regional GDP(2008)) under RCP 4.5 and 8 billion euros (0.6% of regional GDP(2008)) under a high-end sea-level rise scenario. Direct average damage could double if extreme weather events are considered [30].

3. Materials and Methods

3.1. Estimating the Costs of Adaptation Inaction

The methodological approach followed in this study to assess the costs due to sea-level rise if adaptation is not implemented is summarised in Figure 1. The first step is to identify the cities for our assessment. A recent study estimated the economic damage due to coastal flood height for

600 European cities, as well as the adaptation costs, obtained as a function of defence height [19]. This is done in a deterministic way. Of all these cities, we focus on those located in the Iberian Peninsula, our target study zone. Our sample includes 62 main coastal cities in Portugal and Spain, as shown in Figure 2. The second step is estimating future sea-level rise in each city using a stochastic method and considering three scenarios (RCP 2.6, 4.5 and 8.5) as described next in Section 3.1.1. The third step is calculating, also with a stochastic approach, the annual and accumulated expected damage due to sea-level rise in each of our 62 cities. This is explained in Section 3.1.2.

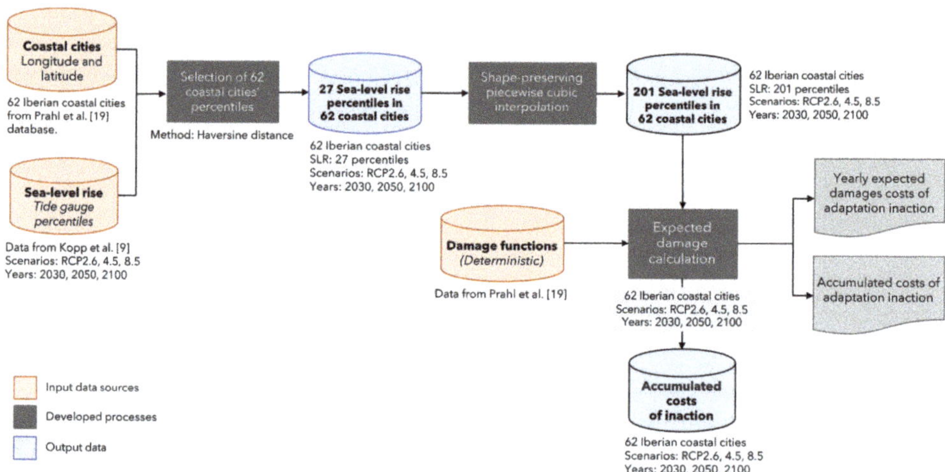

Figure 1. Summary of the methodology followed, including input data and output products.

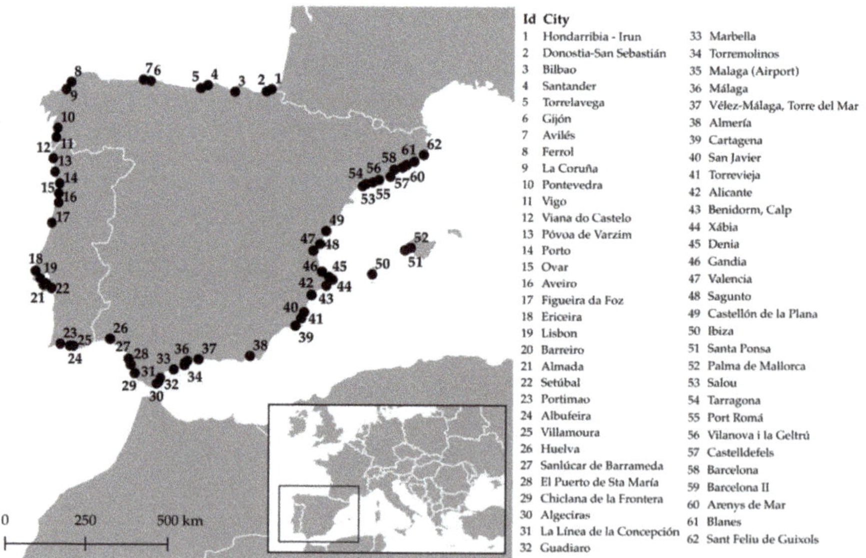

Figure 2. Selected 62 Iberian coastal cities in Spain and Portugal. Source: own elaboration.

3.1.1. Modelling Projected Sea-Level Rise in Each City

First, local sea-level rise data for each city is taken from a database that includes the probabilistic distribution of sea-level rise under three emission scenarios (RCP 2.6, 4.5 and 8.5) in 2030, 2050 and 2100, for more than a thousand tide gauges worldwide [9]. With regard to sea-level rise projections, RCP 8.5 represents the highest emission or the business-as-usual scenario where no climate mitigation policies are adopted; RCP 4.5 describes a middle of the road scenario, and RCP 2.6 is the deep-emission reduction scenario, compatible with a 2 °C warming. The database presents a number of sea-level rise percentiles to a large tide gauge dataset worldwide. For every city in our sample, the closest tide gauge is selected using the Haversine distance between the coastal city and the tide gauges [40]. From the dataset, a total of 27 sea-level rise percentiles with values of xx.0th and xx.5th were selected for each city, scenario and selected year, where xx is an integer between zero and 100. Note, however, that the percentiles are irregularly spaced, while regularly spaced percentiles are needed to apply the methodology proposed. For this reason we calculated an additional 174 percentiles using shape-preserving piecewise cubic interpolation. Thus, 201 percentiles of sea-level rise evenly distributed between 0 and 100 were estimated, spaced by 0.5 (see Figure 1).

As the objective is to estimate the expected (average) damage, the average sea-level rise is calculated using these 201 percentiles. The expected sea-level rise for city i at time j is denoted by E(SLRi,j). This average value is greater than the corresponding 50th percentile, which indicates that the tail of the distribution is long, and thus, it is possible that risks are underestimated if median values are used. This again confirms that knowing the full distribution, additionally to estimating average values, may be of great importance for adequate risk management. This process is followed for each city, each year (2030, 2050 and 2100) and the three aforementioned RCPs.

Note that these percentiles are chosen in order to find the best manner to represent the shape of all distributions. An alternative to this approximation is the so-called parametric approach that consists of calibrating a distribution with two or three parameters [40]. However, this method is not recommended in this case as it may cause important calibration errors that lead to differences between the value of the original percentiles and those obtained from the modelled distribution. Consequently, it will negatively impact the calculation of expected values. We thus argue, in this case, in favour of using the percentiles for calibration as explained above.

As an illustration, Figure 3 shows a standard cumulative distribution function for sea-level rise for nine representative Iberian coastal cities in the year 2100 considering the highest emission scenario, RCP 8.5. The ten cities were selected for being those with the largest population based on data from Eurostat (European Commission, Brussels, Belgium).

Figure 4 shows how cumulative distribution functions may evolve in time, illustrated by the case of Barcelona. One can see that the probabilities of sea-level rise being less than or equal to 50 cm in 2100 are 25% under RCP 8.5, 48.25% under RCP 4.5 and 61% in RCP 2.6.

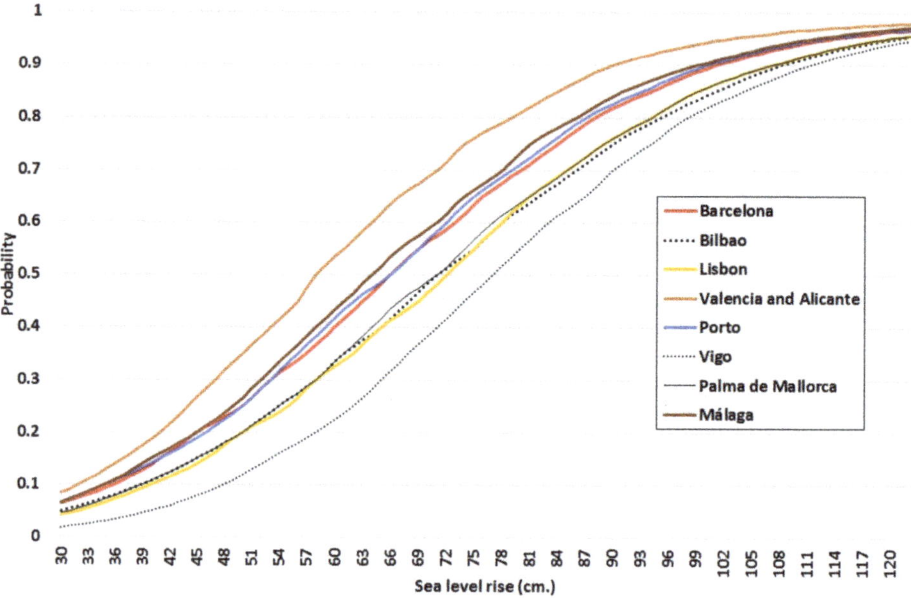

Figure 3. Cumulative distribution function of sea-level rise in 2100 for the nine Iberian cities with largest population under emission scenario RCP 8.5.

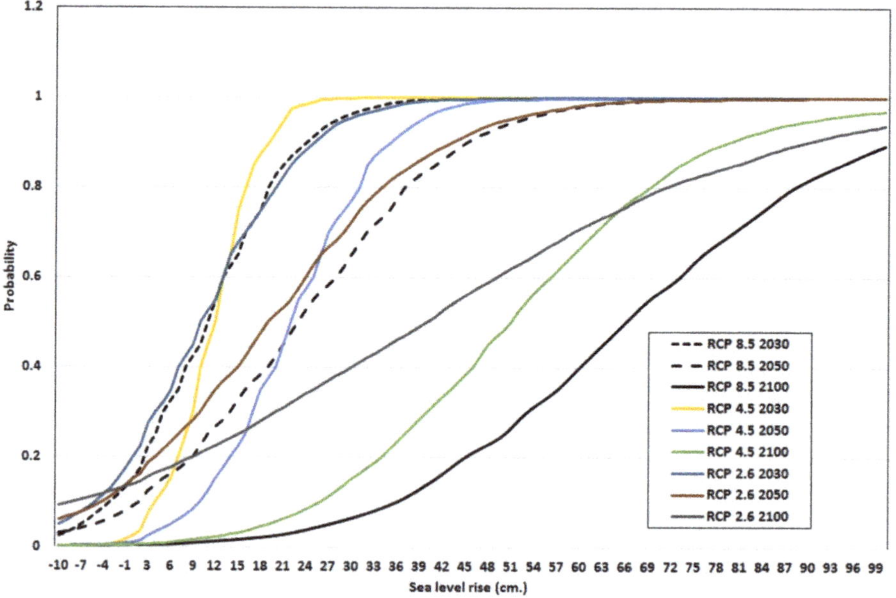

Figure 4. Evolution in time of the cumulative distribution function of sea-level rise in Barcelona under RCP 2.6, 4.5 and 8.5.

3.1.2. Estimation of Economic Damage

A recent study presented the economic damage and adaptation costs as a deterministic function of coastal flood height and height of the defences, respectively, for 600 European coastal cities [19]. The deterministic damage functions used in this work for each of the 62 coastal cities in our sample are taken from the aforementioned study [19]. These functions provide an estimation of damage as a function of sea-level height. The inundation model used to obtain the damage functions used a digital terrain model of 25 m resolution, and hydraulically connected areas were calculated for different coastal flood levels between 0 and 12 m at intervals of half a metre. The damage cost functions [19] were built based on the economic value of the assets exposed to coastal flooding. This exposure was measured considering the economic value of land use, which in turn is based on country level data from a previous study [41], and then adjusting the monetary estimates to 2016 prices by using inflation rates and consumer price index (see [19](p. 5) for further details).

The cost curves [19] are not linear, that is, at higher levels of sea-level rise (SLR), we might obtain, in some cases, higher than proportional costs (Figure 5). Accordingly, the damage distribution is not a linear transformation of the sea-level rise distribution. We denote using E(Di,j) the expected annual damage value for the city i at the time j. Using the sea-level rise percentiles, we calculate 201 damage values for each city at year j. As these values are equally likely, the expected value is calculated as their average value.

Figure 5 illustrates the annual damage depending on local sea-level rise in nine cities of our sample as taken from the damage function database [19]. Note that these deterministic functions are non-linear and not time-dependent, as damage depend only on flood depth. The time component will be incorporated when combined with the sea-level rise scenarios, which vary with time.

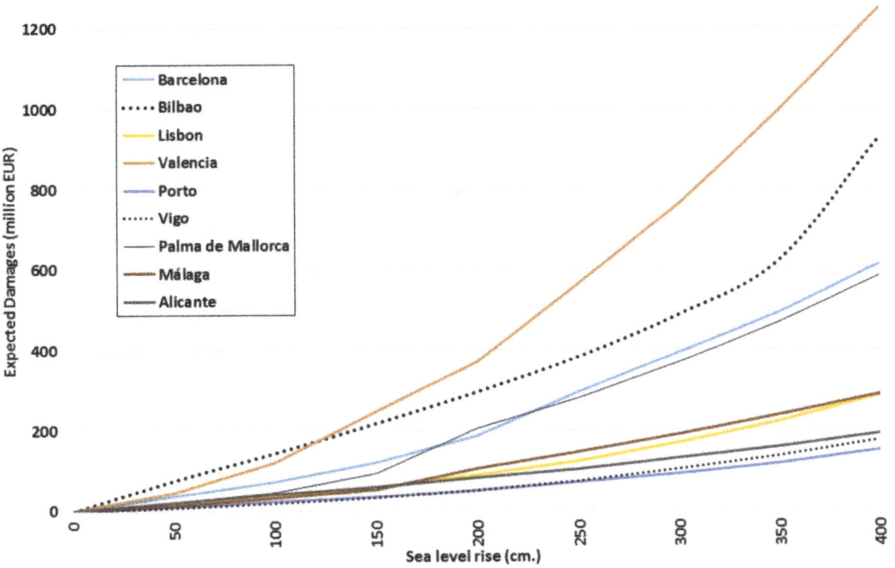

Figure 5. Deterministic damage curves for nine cities based on data by [27].

We define the annual costs of adaptation inaction as the expected damage under three sea-level rise scenarios, which we estimate for the years 2030, 2050 and 2100. Next, we estimate the expected costs for the rest of the years between 2020 and 2100 using shape-preserving piecewise cubic interpolation. Last, we accumulate the calculated damage from 2020 onwards to obtain the accumulated cost of inaction in 2030, 2050 and 2100 [42].

3.2. Measuring the Investment Costs of Adaptation

Adaptation costs for each city were obtained from the study that developed the damage function database [19], which includes the protection needs to be calculated for each city. Adaptation costs are defined based on a theoretical urban protection strategy which consists of a hypothetical defence (adaptation infrastructures, such as dikes or seawalls) that responds to the protection needs in each city. A country-specific dike-construction cost range is used as a proxy for the investment needed for such defences [19]. We acknowledge that this is a rather limited analysis of adaptation options as cities may identify a wide portfolio of both soft and hard adaptation options. However, this is a limitation of the available data but can well illustrate the use of the method proposed in this paper.

A range of adaptation investment costs needed (minimum and maximum cost estimates) are provided in [19]. The average value of this range is also calculated. Based on the available data and for illustrative purposes, two different strategies are assumed to assess adaptation investment needs:

- The first strategy aims at reducing climate change risk, for example by building a defence that can only be overcome in a certain percentage of the cases by a given year under one of the RCP scenarios. In this case we have defined 5% and 0.5% of the cases by the year 2100 under RCP 8.5. In other words, this option requires building an infrastructure to protect from sea-level rise percentiles 95th and 99.5th. We have named this strategy as "risk tailoring".
- The second strategy consists of building defences of standard height (e.g., 2 m or 3 m) and comparing the protection level and costs with the "risk tailoring" strategy.

The costs of adaptation investment for each city and scenario are then compared to the accumulated costs of inaction estimated previously in Section 3.1.2, to determine in which cases the benefits of adaptation (in terms of avoided damage or costs) exceed its costs. Note that for simplicity it is assumed defences protect the city of coastal flood heights smaller or equal to their height, but we acknowledge that waves could act on adjacent areas or circumvent the defences in some cases.

A summary of the methodological steps followed in this section is described in Figure 6 below.

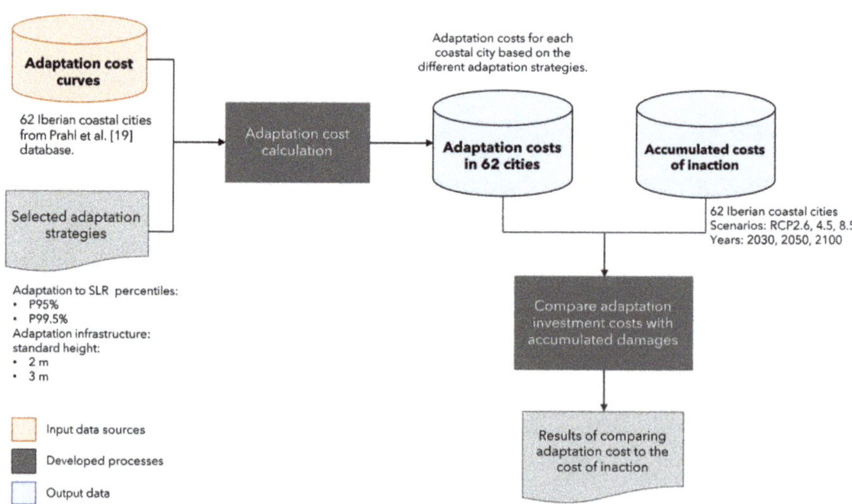

Figure 6. Summary of the methodological approach followed to estimate adaptation investment needs in each city under three sea-level rise scenarios. Investment costs are then compared to the accumulated costs of inaction to assess adaptation benefits.

Discount rates can also be incorporated into the modelling. On this occasion, however, we have decided not to include them for two main reasons. First, setting the appropriate discount factor may become a rather difficult task, especially when such long terms are considered, as fat tails may exist and potential damage (even with very low probability) can be so high [43,44]. There is plenty of literature discussing this issue [45–47] and no clear consensus exists on how to address this (see [48] for a more detailed discussion). Small variations in how the discount rates are applied may have substantial effects on the economic damage and the investment results. Second, in this study we are interested in analysing the decision of inaction and investing in adaptation without the interference of other factors, so we chose to avoid their effect on the calculations. The higher the discount rate the lower the value of the benefit (or costs) will be in the future, but there are other factors that may also substantially affect the results that have not been considered, such as the growth rate in the value of the properties at risk. In fact, one could expect that damage will increase as the value of the properties increase, and therefore, it is the difference between the growth rate and the discount rate that really matters. The impact of discount rates on actual values of investments (or benefits) occurring over long periods of time is always very significant. In any case, with net discunt rates of 1% and 2%, investing in adaptation would still be a good decision in our study (see Table S10 in the Supplementary Materials).

4. Results

4.1. Local Sea-Level Rise and the Costs of Inaction

We calculate three sea-level rise percentiles for each city and the accumulated damage in 2030, 2050 and 2100. Table 1 shows the results in 2100 for the 10 coastal cities with the largest population, under RCP 8.5. The results for all cities and scenarios are provided in the supplementary information. We present the RCP 8.5 in Table 1 because of the greater influence of this RCP on the sea-level rise risk [49].

Table 1. Sea-level rise percentiles and accumulated damage costs in 2100 for those cities with the largest number of inhabitants under RCP 8.5.

City	Year	P50 (cm)	P95 (cm)	P99.5 (cm)	Damage (million EUR)
Barcelona	2030	11	28	39	75.8
Lisbon	2030	13	21	26	34.5
Valencia	2030	10	18	24	79.4
Porto	2030	12	20	24	22.1
Bilbao	2030	14	23	29	177.0
Málaga	2030	11	20	25	28.3
Palma de Mallorca	2030	12	30	41	48.0
Alicante	2030	10	18	24	32.3
Vigo	2030	15	23	27	21.3
Gijón	2030	14	23	30	35.4
Barcelona	2050	23	52	71	343.1
Lisbon	2050	26	42	53	150.0
Valencia	2050	21	36	47	365.8
Porto	2050	24	39	51	98.8
Bilbao	2050	27	45	57	806.4
Málaga	2050	23	38	50	132.8
Palma de Mallorca	2050	25	52	70	201.1
Alicante	2050	21	36	47	159.2
Vigo	2050	30	45	56	89.4
Gijón	2050	27	45	58	155.8

Table 1. *Cont.*

City	Year	P50 (cm)	P95 (cm)	P99.5 (cm)	Damage (million EUR)
Barcelona	2100	66	113	171	2059.5
Lisbon	2100	72	119	181	946.3
Valencia	2100	58	102	161	2504.3
Porto	2100	66	112	174	601.7
Bilbao	2100	71	120	180	4592.3
Málaga	2100	64	111	168	831.9
Palma de Mallorca	2100	71	119	176	1288.4
Alicante	2100	58	102	161	991.2
Vigo	2100	77	123	184	550.9
Gijón	2100	72	122	182	944.9

Note that Bilbao, Valencia and Barcelona are the coastal cities with the largest expected accumulated damage in the case of inaction. Both the sea-level rise percentiles and the damage grow rapidly over time, particularly in the second part of the century, in line with studies for other cities worldwide [49]. An illustration of accumulated expected damage for four cities in the top ranking of damage is shown in Figure 7.

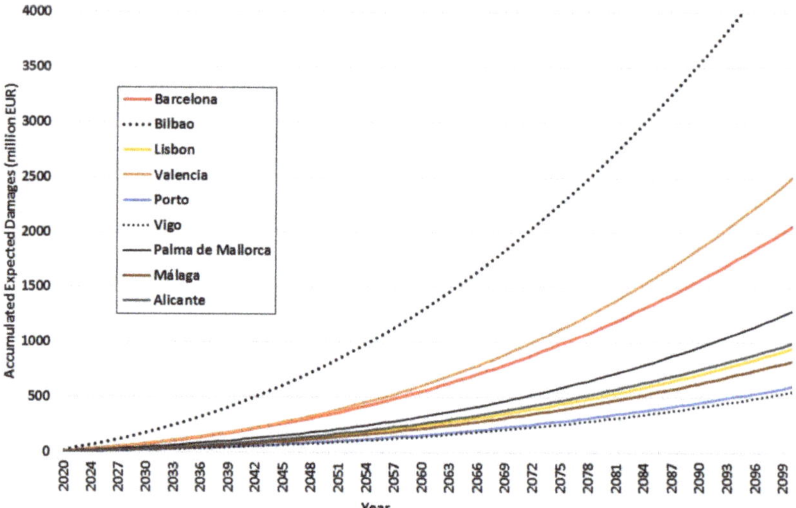

Figure 7. Accumulated expected damage in nine Iberian coastal cities with largest population under RCP 8.5.

To show that averages and median values may differ substantially we present the differences in damage for some cities and years in Table 2. Average damage have been calculated using the full distribution of sea-level rise probabilities, and the damage curves are shown in Figure 5. Damage in Table 2 correspond to the selected years, and they are not accumulated data. As expected, by 2100, differences in mean and average values are very important. This is mainly driven by two effects: (1) the shape of the sea-level rise distribution and (2) the shape of the damage curves. The main and very important conclusion here is that, in the long term, the use and estimation of median values (50th percentile) provide insufficient information and can lead to underestimating coastal risks and consequently to under protection. The methodology proposed in this paper should help overcome this limitation by offering complete information on the risk faced.

In more detail, it can be noted that the shape distribution effect is caused by skewness. Positive skewness is obtained when the distribution has an asymmetric tail extending toward more positive values. In these right-skewed distributions, the mean is greater than the median, and the former is affected by extreme sea-level rise values. The skewness found for Barcelona, Lisbon, Valencia, Porto and Bilbao has also been included in Table 2. Note that skewness increases with time, causing a growing impact of extreme values in the mean values of sea-level rise.

Table 2. Differences in the six largest cities between expected annual damage and damage using the 50th percentile under RCP 8.5. The skewness of the sea-level rise distribution is also presented for each city.

City	Values	2030	2050	2100
Barcelona	Expected damage (A)	8.76	18.00	52.93
	Damage using 50th percentile (B)	8.25	17.24	49.00
	Difference in damage (A−B)	0.52	0.75	3.93
	Skewness	*−0.03*	*0.00*	*3.67*
Lisbon	Expected damage (A)	3.84	7.81	25.36
	Damage using 50th percentile (B)	3.79	7.58	23.23
	Difference in damage (A−B)	0.05	0.23	2.13
	Skewness	*−0.09*	*0.65*	*3.88*
Valencia	Expected damage (A)	9.05	19.94	70.54
	Damage using 50th percentile (B)	9.33	19.58	58.53
	Difference in damage (A−B)	−0.28	0.36	12.01
	Skewness	*−0.06*	*0.62*	*4.41*
Porto	Expected damage (A)	2.53	5.16	15.55
	Damage using 50th percentile (B)	2.54	5.09	14.55
	Difference in damage (A−B)	−0.01	0.07	1.00
	Skewness	*0.01*	*0.73*	*4.05*
Bilbao	Expected damage (A)	20.78	41.82	110.93
	Damage using 50th percentile (B)	21.41	41.28	105.04
	Difference in damage (A−B)	−0.63	0.54	5.89
	Skewness	*−0.18*	*0.38*	*3.47*

Table 3 shows the accumulated damage for a total of 62 Iberian cities and per country for the three scenarios. Damage for RCP 8.5 are very significant, but even in the most favourable scenario (RCP 2.6) important accumulated damage are expected due to sea-level rise.

Table 3. Accumulated damage costs (million EUR) for 64 Iberian coastal cities, aggregated by country.

Country	Scenario	2030	2050	2100
Total Iberian		1764	8100	50,476
Spain	RCP 8.5	1574	7261	45,342
Portugal		189	839	5133
Total Iberian		1718	7588	41,261
Spain	RCP 4.5	1536	6796	37,044
Portugal		183	792	4216
Total Iberian		1732	7474	36,895
Spain	RCP 2.6	1545	6684	33,089
Portugal		188	789	3806

4.2. Adaptation Strategies and Investment Costs

4.2.1. Adapting to the Risk of Sea-Level Rise or "Risk Tailoring"

In this case, we focus on the "risk tailoring" strategy of building protective defences of a height that would only be surpassed in 5% and 0.5% of the cases (under the scenario RCP 8.5 in 2100). These are the heights for the 95th and 99.5th percentiles. Results are shown in Table 4. One can see that inaction is costlier than implementing adaptation in all cases, as expected damage are much greater than the adaptation costs. If defences are to be exceeded only in 5% of the cases, this means that protection height in the top cities with the largest damage would need to be higher than 1 m, and between 1.61 m and 1.80 m to avoid damage in 99.5% of the cases. Table 4 shows an interval (minimum and maximum values) of infrastructure construction costs taken from [19] for each city, depending on its characteristics and defence height. The average is the mean of these two values.

Table 4. Expected accumulated damage and adaptation strategies linked to sea-level rise percentiles (in million EUR) in 2100 for the 10 Iberian coastal cities with the largest populations under the RCP 8.5 scenario.

City	Accumulated Damage	Sea-Level Rise (cm)		Adaptation costs					
				Minimum		Maximum		Average	
		P95	P99.5	P95	P99.5	P95	P99.5	P95	P99.5
Barcelona	2060	113	171	318	595	459	859	388	727
Lisbon	946	119	181	223	425	323	614	273	520
Valencia	2504	102	161	326	645	472	932	399	788
Porto	602	112	174	87	164	126	236	106	200
Bilbao	4592	120	180	424	702	613	1,015	518	858
Málaga	832	111	168	127	221	184	319	155	270
Palma de Mallorca	1288	119	176	240	531	347	767	293	649
Alicante	991	102	161	164	289	237	417	200	353
Vigo	551	123	184	154	283	223	409	189	346
Gijón	945	122	182	211	348	305	503	258	425

4.2.2. Adapting by Fixing a Certain Height

If defence heights were to be decided exogenously, for instance, at 2 m and 3 m by 2100, the average adaptation costs for the top 10 cities of the ranking would range between 282 and 1225 million euros for a protection infrastructure of 2 m (Table 5). In the case of a 3 m defence, average costs could reach 400–2400 million euros, but even in this case, the investment can be well justified to avoid the accumulated damage. Note that if standard height defences of 2 m or 3 m are to be implemented, the probability of damage in the year 2100 is rather small.

Table 5. Expected accumulated damage and adaptation strategies with standard height defences in 2100 (costs shown in million euros) for the 10 largest Iberian coastal cities, under RCP 8.5.

City	Accumulated Damage	Adaptation costs					
		Minimum		Maximum		Average	
		2 m	3 m	2 m	3 m	2 m	3 m
Barcelona	2060	758	1434	1096	2072	927	1753
Lisbon	946	491	947	710	1369	601	1158
Valencia	2504	873	1638	1262	2368	1067	2003
Porto	602	201	404	291	584	246	494
Bilbao	4592	799	1343	1156	1942	978	1642
Málaga	832	279	496	404	717	341	607
Palma de Mallorca	1288	712	1342	1029	1940	870	1641
Alicante	991	378	644	547	931	463	787
Vigo	551	319	614	461	888	390	751
Gijón	945	391	660	565	955	478	808

5. Discussion

5.1. The Economic Damage of Sea-Level Rise

In this paper, we propose a model for the calculation of accumulated expected damage under uncertainty. The model uses an expanded version of the projected sea-level rise percentiles [9] and a deterministic function that depends on flood height (in our case, sea-level height) [19]. The combination of both provides a group of stochastic damage distributions that allow us to calculate annual expected damage and, consequently, also the expected accumulated damage. Accumulated average damage in all 62 cities of our sample exceed 1700 million euros in 2030, and the difference between the results under RCP 4.5 and RCP 8.5 is 45 million euros. However, by the end of the century, accumulated damage reach 41,200 million euros under RCP 4.5 and more than 50,400 million euros in the highest emission scenarios (RCP 8.5). Accumulated damage increase by more than 20 times by 2100 and the difference between RCP 4.5 and RCP 8.5 could reach 10 billion euros (Table S8 and Table S9).

Furthermore, our results show that limiting the assessments to median values (50th percentile) significantly underestimates the potential damage. While differences between median and average values might be small by 2030, they increase considerably by one order of magnitude by the end of the century. For example, in Barcelona the difference between median and average damage is 0.5 million euros in 2030, but increases to almost 4 million euros in 2100 (Table 2). This difference occurs because the shape of the sea-level rise distribution is positively skewed, i.e., the distribution is not symmetrical but has a heavier upper tail The shape of the distribution of the damage is also another cause of important underestimations in the medium term when calculating expected damage. These findings are consistent with previous studies that warned about the larger potential risk of climate-induced impacts.

Damage costs do not account for coastal erosion or ecosystem loss, impacts that are both expected to be relevant in the Iberian Peninsula. Additionally to their intrinsic value, ecosystems provide a number of services to people that can be monetised [50]. The value of services such as storm protection or erosion control provided by salt marshes and mangroves worldwide is estimated to be 194,000 $/ha/year [51], which has not been accounted for in this study. Of course, one should acknowledge that other important factors such as shoreline evolution have to be integrated into a cost-benefit analysis to account for the complexities of coastal defence solutions, as argued earlier by other authors [52]. In this case, we have simplified this part of the analysis to focus on the impact of uncertainty on the economic estimates but by no means do we neglect the need to integrate those factors.

5.2. Comparing the Costs of Inaction to Protection Investment Costs

Two different adaptation options have been explored in this study: the first is based on tailoring the level of risk cities might decide to protect themselves from. This is illustrated through the assumption that coastal or city managers decide to protect each city for 95% or 99.5% of the cases under RCP 8.5 by 2100. In the first case, with protection strategies to face 95% of the cases, investment needs in the largest coastal cities by population would range between 87 (Porto) and 1015 (Bilbao) million euros. Investment costs almost double if a higher protection threshold (99.5%) is considered, and these would vary between 164 (Porto) and 1015 (Bilbao) million euros. The second adaptation option explored the investment costs needed for certain protection heights (2 and 3 m). In the case of 3 m protection infrastructures, construction costs range between 404 and 2368 million euros.

An important finding of this analysis is that adaptation costs in the long term are much smaller than the increasingly expensive costs of inaction, and thus, investing in adaptation is a good decision when comparing costs and benefits of the actions, a finding in line with previous research carried out in this area of study [53].

Of course, one should acknowledge that many other adaptation options that have not been considered in this paper may exist, and that many other criteria other than economic costs and benefits may be worth being taken into account when making such decisions. Another clear finding is that adaptation investment needs will be smaller under more favourable scenarios (RCP 2.6 and RCP 4.5). However, even under the most optimistic scenario (RCP 2.6), expected damage will be very important in the second half of the century. In other words, despite strong emission reduction efforts being really effective, adaptation will be inevitable. This is to highlight the importance of the need to implement both mitigation and adaptation policies.

Our findings show that not adapting to climate change is not, by any means, a good strategy in the medium and long term. Note that, even if many types of adaptation options should be considered when protecting coasts from climate risks, the building of defence infrastructures that require strong investments would be justified based on the accumulated damage by 2100, which are much greater than the investment costs of protection. Moreover, the adaptation strategies considered in this paper are shown to be very cost-effective in every city studied and reduce considerably the probability of experiencing high economic damage by the end of the century.

5.3. Policy Implications

In this paper we propose a method to avoid the underestimation of climate risk and estimate what the size of it may be for the case of 62 cities in the Iberian peninsula. We have used two infrastructure-based adaptation options for illustrative purposes. This has been done because the data needed for the analysis is available in these cases, and not because we argue that these solutions should be the ones implemented. Results show that the underestimation of damage can be very important if we consider median values instead of the average (expected) damage. Therefore, we argue that strong efforts should be made to understand the full probability distributions of sea-level rise and the consequent economic damage, for each city, emission scenario and year. Not accounting for all these may lead to investment decisions (and even maladaptation) that underestimate future risks and cannot respond to them.

The method can also be applied to adequately assess the cost effectiveness of other adaptation options when data becomes available. The purpose of the paper has been to propose a methodology and illustrate the case for sea-level rise and hard adaptation measures for which good data was available. Other pieces of research should help us to better consider alternative adaptation options or even other impacts related to the adaptation options considered. Nonetheless, the main message of this paper remains, not accounting for the full distribution of sea level rise as well as damage costs is a clear and very significant underestimation of climate risks that may lead to inadequate policy decisions.

Supplementary Materials: A dataset with additional results for all 62 cities, scenarios and years is available online at http://www.mdpi.com/2073-4441/12/4/1220/s1.

Author Contributions: Conceptualization, L.M.A.; formal analysis, L.M.A. and E.S.d.M.; methodology, L.M.A.; writing—original draft, L.M.A., E.S.d.M. and I.G. All authors have read and agreed to the published version of the manuscript.

Funding: This research is supported by the Basque Government through the BERC 2018-2021 program and by the Spanish Ministry of Economy and Competitiveness MINECO through BC3 María de Maeztu excellence accreditation MDM-2017-0714. Additionally, Luis M. Abadie is grateful for financial support from The Spanish Ministry of Science and Innovation (ECO2015-68023). E. Sainz de Murieta acknowledges funding from the Basque Government (Postdoctoral Fellowship grant no. POS_2018_2_0027). Ibon Galarraga received financial support from the European Union's Horizon 2020 research and innovation programme under grant agreement No 776479 for the project CO-designing the Assessment of Climate CHange costs.

Conflicts of Interest: The authors declare no conflicts of interest.

References

1. IPCC. *IPCC Climate Change 2013: The Physical Science Basis. Contribution of Working Group I to the Fifth Assessment Report of the Intergovernmental Panel on Climate Change*; Stocker, T.F., Qin, D., Plattner, G.K., Tignor, M., Allen, S.K., Boschung, J., Nauels, A., Xia, Y., Bex, V., Midgley, P.M., Eds.; Cambridge University Press: Cambridge, UK, 2013.

2. Church, J.A.; White, N.J.; Aarup, T.; Wilson, W.S.; Woodworth, P.L.; Domingues, C.M.; Hunter, J.R.; Lambeck, K. Understanding global sea levels: Past, present and future. *Sustain. Sci.* **2008**, *3*, 9–22. [CrossRef]

3. Church, J.A.; White, N.J. A 20th century acceleration in global sea-level rise. *Geophys. Res. Lett.* **2006**, *33*. [CrossRef]

4. Hay, C.C.; Morrow, E.; Kopp, R.E.; Mitrovica, J.X. Probabilistic reanalysis of twentieth-century sea-level rise. *Nature* **2015**, *517*, 481–484. [CrossRef] [PubMed]

5. Yi, S.; Sun, W.; Heki, K.; Qian, A. An increase in the rate of global mean sea level rise since 2010. *Geophys. Res. Lett.* **2015**, *42*, 3998–4006. [CrossRef]

6. Hardy, R.D.; Nuse, B.L. Global sea-level rise: Weighing country responsibility and risk. *Clim. Change* **2016**, *137*, 333–345. [CrossRef]

7. Hamlington, B.D.; Fasullo, J.T.; Nerem, R.S.; Kim, K.-Y.; Landerer, F.W. Uncovering the Pattern of Forced Sea Level Rise in the Satellite Altimeter Record. *Geophys. Res. Lett.* **2019**, *46*, 4844–4853. [CrossRef]

8. Slangen, A.B.A.; Carson, M.; Katsman, C.A.; Wal, R.S.W.; van de Köhl, A.; Vermeersen, L.L.A.; Stammer, D. Projecting twenty-first century regional sea-level changes. *Clim. Chang.* **2014**, *124*, 317–332. [CrossRef]

9. Kopp, R.E.; Horton, R.M.; Little, C.M.; Mitrovica, J.X.; Oppenheimer, M.; Rasmussen, D.J.; Strauss, B.H.; Tebaldi, C. Probabilistic 21st and 22nd century sea-level projections at a global network of tide-gauge sites. *Earths Future* **2014**, *2*, 383–406. [CrossRef]

10. Grinsted, A.; Jevrejeva, S.; Riva, R.; Dahl-Jensen, D. Sea level rise projections for northern Europe under RCP8.5. *Clim. Res.* **2015**, *64*, 15–23. [CrossRef]

11. Jackson, L.P.; Jevrejeva, S. A probabilistic approach to 21st century regional sea-level projections using RCP and High-end scenarios. *Glob. Planet. Chang.* **2016**, *146*, 179–189. [CrossRef]

12. Nauels, A.; Gütschow, J.; Mengel, M.; Meinshausen, M.; Clark, P.U.; Schleussner, C.-F. Attributing long-term sea-level rise to Paris Agreement emission pledges. *Proc. Natl. Acad. Sci. USA* **2019**. [CrossRef] [PubMed]

13. Bamber, J.L.; Oppenheimer, M.; Kopp, R.E.; Aspinall, W.P.; Cooke, R.M. Ice sheet contributions to future sea-level rise from structured expert judgment. *Proc. Natl. Acad. Sci. USA* **2019**, *116*, 11195–11200. [CrossRef] [PubMed]

14. IPCC. *IPCC Special Report on the Ocean and Cryosphere in a Changing Climate*; Pörtner, H.O., Roberts, D.C., Masson-Delmotte, V., Zhai, P., Tignor, M., Mintenbeck, K., Nicolai, M., Okem, A., Petzold, J., Rama, B., et al., Eds.; IPCC: Geneva, Switzerland, 2019; In press.

15. Ciscar, J.C.; Ibarreta, D.; Soria, A.; Dosio, A.; Toreti, A.; Ceglar, A.; Fumagalli, D.; Dentener, F.; Lecerf, R.; Zucchini, A.; et al. *Climate Impacts in Europe: Final Report of the JRC PESETA III Project*; European Commission: Luxembourg, 2018; p. 95.

16. Wong, P.P.; Losada, I.J.; Gattuso, J.P.; Hinkel, J.; Khattabi, A.; McInnes, K.L.; Saito, Y.; Sallenger, A. Coastal systems and low-lying areas. In *Climate Change 2014: Impacts, Adaptation, and Vulnerability. Part A: Global and Sectoral Aspects. Contribution of Working Group II to the Fifth Assessment Report of the Intergovernmental Panel on Climate Change*; Field, C.B., Barros, V.R., Dokken, D.J., Mach, K.J., Mastrandrea, M.D., Bilir, T.E., Chatterjee, M., Ebi, K.L., Estrada, Y.O., Genova, R.C., et al., Eds.; Cambridge University Press: Cambridge, UK, 2014; pp. 361–409.

17. Vousdoukas, M.I.; Mentaschi, L.; Voukouvalas, E.; Bianchi, A.; Dottori, F.; Feyen, L. Climatic and socioeconomic controls of future coastal flood risk in Europe. *Nat. Clim. Chang.* **2018**, *8*, 776–780. [CrossRef]

18. Sánchez-Arcilla, A.; García-León, M.; Gracia, V.; Devoy, R.; Stanica, A.; Gault, J. Managing coastal environments under climate change: Pathways to adaptation. *Sci. Total Environ.* **2016**, *572*, 1336–1352. [CrossRef] [PubMed]

19. Prahl, B.F.; Boettle, M.; Costa, L.; Kropp, J.P.; Rybski, D. Damage and protection cost curves for coastal floods within the 600 largest European cities. *Sci. Data* **2018**, *5*, 180034. [CrossRef]

20. Weitzman, M.L. A precautionary tale of uncertain tail fattening. *Environ. Resour. Econ.* **2013**, *55*, 159–173. [CrossRef]

21. Hinkel, J.; Jaeger, C.; Nicholls, R.J.; Lowe, J.; Renn, O.; Peijun, S. Sea-level rise scenarios and coastal risk management. *Nat. Clim. Chang.* **2015**, *5*, 188–190. [CrossRef]

22. Thiéblemont, R.; Le Cozannet, G.; Toimil, A.; Meyssignac, B.; Losada, I.J. Likely and High-End Impacts of Regional Sea-Level Rise on the Shoreline Change of European Sandy Coasts Under a High Greenhouse Gas Emissions Scenario. *Water* **2019**, *11*, 2607. [CrossRef]

23. Abadie, L.M.; Galarraga, I.; Sainz de Murieta, E. Understanding risks in the light of uncertainty: Low-probability, high-impact coastal events in cities. *Environ. Res. Lett.* **2017**, *12*. [CrossRef]

24. Abadie, L.M.; Sainz de Murieta, E.; Galarraga, I. Investing in adaptation: Flood risk and real option application to Bilbao. *Environ. Model. Softw.* **2017**, *95*, 76–89. [CrossRef]

25. Galarraga, I.; Sainz de Murieta, E.; Markandya, A.; Abadie, L.M. Addendum to Understanding risks in the light of uncertainty: Low-probability, high-impact coastal events in cities. *Environ. Res. Lett.* **2018**, *13*, 029401. [CrossRef]

26. Losada, I.J.; Toimil, A.; Muñoz, A.; Garcia-Fletcher, A.P.; Diaz-Simal, P. A planning strategy for the adaptation of coastal areas to climate change: The Spanish case. *Ocean. Coast. Manag.* **2019**, 104983. [CrossRef]

27. Galarraga, I.; Osés, N.; Markandya, A.; Chiabai, A.; Khatun, K. Aportaciones desde la economía de la adaptación a la toma de decisiones sobre Cambio Climático: Un ejemplo para la Comunidad Autónoma del País Vasco. *Econ. Agrar. Recur. Nat.* **2011**, *11*, 113–142. [CrossRef]

28. Pérez-Morales, A.; Gil-Guirado, S.; Olcina-Cantos, J. Housing bubbles and the increase of flood exposure. Failures in flood risk management on the Spanish south-eastern coast (1975–2013): Housing bubbles and increase of flood risk in Spain. *J. Flood Risk Manag.* **2015**, *11*, S302–S313. [CrossRef]

29. Satta, A.; Puddu, M.; Venturini, S.; Giupponi, C. Assessment of coastal risks to climate change related impacts at the regional scale: The case of the Mediterranean region. *Int. J. Disaster Risk Reduct.* **2017**, *24*, 284–296. [CrossRef]

30. Losada, I.J.; Izaguirre, C.; Díaz, P. *Cambio Climático en la Costa Española*; Oficina Española de Cambio Climático, Ministerio de Agricultura, Alimentación y Medio Ambiente: Madrid, Spain, 2014; p. 133.

31. Mendoza, E.T.; Jimenez, J.A.; Mateo, J. A coastal storms intensity scale for the Catalan sea (NW Mediterranean). *Nat. Hazards Earth Syst. Sci.* **2011**, *11*, 2453–2462. [CrossRef]

32. Freitas, M.C.; Andrade, C.; Cachado, C.; Cardoso, A.C.; Monteiro, J.H.; Brito, P.; Rebelo, L. Coastal land-loss associated with sea-level rise assessed by aerial Videotape-Assisted Vulnerability Analysis—The case of mainland Portugal. *J. Coast. Res.* **2006**, 1310–1315.

33. Gornitz, V. Global coastal hazards from future sea level rise. *Glob. Planet. Change* **1991**, *3*, 379–398. [CrossRef]

34. Pereira, C.; Coelho, C. Mapping erosion risk under different scenarios of climate change for Aveiro coast, Portugal. *Nat. Hazards* **2013**, *69*, 1033–1050. [CrossRef]

35. Ferreira, O.; Viavattene, C.; Jiménez, J.; Bole, A.; Plomaritis, T.; Costas, S.; Smets, S. CRAF Phase 1, a framework to identify coastal hotspots to storm impacts. *E3S Web Conf.* **2016**, *7*, 11008. [CrossRef]

36. Viavattene, C.; Jiménez, J.A.; Ferreira, O.; Priest, S.; Owen, D.; McCall, R. Selecting coastal hotspots to storm impacts at the regional scale: A Coastal Risk Assessment Framework. *Coast. Eng.* **2018**, *134*, 33–47. [CrossRef]

37. Rocha, C.; Antunes, C.; Catita, C. Coastal Vulnerability Assessment Due to Sea Level Rise: The Case Study of the Atlantic Coast of Mainland Portugal. *Water* **2020**, *12*, 360. [CrossRef]

38. Antunes, C. Assessment of Sea Level Rise at West Coast of Portugal Mainland and Its Projection for the 21st Century. *J. Mar. Sci. Eng.* **2019**, *7*, 61. [CrossRef]

39. Semeoshenkova, V.; Newton, A. Overview of erosion and beach quality issues in three Southern European countries: Portugal, Spain and Italy. *Ocean. Coast. Manag.* **2015**, *118*, 12–21. [CrossRef]

40. Abadie, L.M.; Galarraga, I.; Markandya, A.; Sainz de Murieta, E. Risk measures and the distribution of damage curves for 600 European coastal cities. *Environ. Res. Lett.* **2019**, *14*, 064021. [CrossRef]

41. Huizinga, J.; De Moel, H.; Szewczyk, W. *Global Flood Depth-Damage Functions. Methodology and the Database with Guidelines*; European Commission: Luxembourg, 2017; p. 114.

42. Abadie, L.M.; Chamorro, J.M. Income risk of EU coal-fired power plants after Kyoto. *Energy Policy* **2009**, *37*, 5304–5316. [CrossRef]

43. Weitzman, M.L. On modeling and interpreting the economics of catastrophic climate change. *Rev. Econ. Stat.* **2009**, *91*, 1–19. [CrossRef]

44. Stern, N. *The Economics of Climate Change: The Stern Review*; Cambridge University Press: Cambridge, UK, 2007.

45. Gollier, C. *Pricing the Planet's Future: The Economics of Discounting in an Uncertain World*; Princeton University Press: Princeton, NJ, USA, 2012.

46. Groom, B. Discounting. In *Routledge Handbook of the Economics of Climate Change Adaptation*; Markandya, A., Galarraga, I., Sainz de Murieta, E., Eds.; Routledge International Handbooks: Oxon, UK; New York, NY, USA, 2014; pp. 138–168, ISBN 978-0-415-63311-6.

47. Drupp, M.A.; Freeman, M.C.; Groom, B.; Nesje, F. Discounting Disentangled. *Am. Econ. J. Econ. Policy* **2018**, *10*, 109–134. [CrossRef]

48. Markanday, A.; Galarraga, I.; Chiabai, A.; Sainz de Murieta, E.; Lliso, B.; Markandya, A. Determining discount rates for the evaluation of natural assets in land-use planning: An application of the Equivalency Principle. *J. Clean. Prod.* **2019**, *230*, 672–684. [CrossRef]

49. Abadie, L.M. Sea level damage risk with probabilistic weighting of IPCC scenarios: An application to major coastal cities. *J. Clean. Prod.* **2018**, *175*, 582–598. [CrossRef]

50. TEEB. *The Economics of Ecosystems and Biodiversity: Ecological and Economic Foundations*; Kumar, P., Ed.; Earthscan: London, UK, 2010; ISBN 978-1-84971-212-5.

51. Costanza, R.; de Groot, R.; Sutton, P.; van der Ploeg, S.; Anderson, S.J.; Kubiszewski, I.; Farber, S.; Turner, R.K. Changes in the global value of ecosystem services. *Glob. Environ. Chang.* **2014**, *26*, 152–158. [CrossRef]

52. Lima, M.; Coelho, C.; Veloso-Gomes, F.; Roebeling, P. An integrated physical and cost-benefit approach to assess groins as a coastal erosion mitigation strategy. *Coast. Eng.* **2020**, *156*, 103614. [CrossRef]

53. Coelho, C.; Cruz, T.; Roebeling, P. Longitudinal revetments to mitigate overtopping and flooding: Effectiveness, costs and benefits. *Ocean. Coast. Manag.* **2016**, *134*, 93–102. [CrossRef]

Article

Mitigating the Effects of Sea-Level Rise on Estuaries of the Mississippi Delta Plain Using River Diversions

Eric D. White [1,2,*], Ehab Meselhe [2], Denise Reed [3], Alisha Renfro [4], Natalie Peyronnin Snider [5] and Yushi Wang [6]

1   Planning & Research Division, Louisiana Coastal Protection and Restoration Authority, 150 Terrace Avenue, Baton Rouge, LA 70802, USA

2   Department of River-Coastal Science and Engineering, Tulane University, 627 Lindy Boggs Center, 6823 St. Charles Ave., New Orleans, LA 70118, USA; emeselhe@tulane.edu

3   Pontchartrain Institute for Environmental Sciences, University of New Orleans, 2000 Lakeshore Drive, New Orleans, LA 70148, USA; djreed@uno.edu

4   National Wildlife Federation, 3801 Canal Street, Suite 32, New Orleans, LA 70119, USA; renfroa@nwf.org

5   Environmental Defense Fund, 1875 Connecticut Avenue NW, Suite 600, Washington, DC 20009, USA; nsnider@edf.org

6   The Water Institute of the Gulf, 1110 River Road S., Suite 200, Baton Rouge, LA 70802, USA; ywang@thewaterinstitute.org

*   Correspondence: eric.white@la.gov; Tel.: +1-225-342-1286

Received: 7 August 2019; Accepted: 23 September 2019; Published: 28 September 2019

Abstract: Using the Mississippi River as a tool for restoration has been a key element of restoration planning in Louisiana for decades. The results of allowing river water and sediment back into the coastal system are manifested in a number of places in present day Louisiana, with additional plans for large scale sediment and water diversions from the Mississippi River. Many previous numerical modeling studies have focused on sediment delivery to Louisiana estuaries. This study examines the effects of river diversions on salinity gradients in receiving estuarine basins. The Integrated Compartment Model, a planning-level model that simulates multi-decadal change in estuarine hydrodynamics and wetland systems under assumed sea-level rise scenarios, was used to assess the estuarine salinity gradient under potential management regimes. The simulations for current conditions are compared to a future 50-year simulation with additional diversions, as well as cases with a variety of diversion options. This modeling analysis shows that without additional action, 50-years of sea-level rise could result in substantial increases in salinity throughout the Mississippi Delta Plain estuaries. This can be largely offset with additional large river diversions which can maintain variable salinity gradients throughout the estuary basins.

Keywords: Mississippi River Delta; Louisiana; salinity; sediment; wetland loss; estuary; diversion; sea-level rise; environmental planning; coastal restoration

---

1. Introduction

The impacts from projected future sea-level rise are expected to be extensive for the World's River Deltas, with the loss of land and habitat diversity threatening the inhabitants, resources and ecology of these low-lying coastal areas [1,2]. Many deltas have already experienced broad change, resulting in a loss of habitat extent and diversity due to human activities [3]. The Mississippi River Delta is a river-dominated delta that was built in a series of successive lobes over the last 7500 years [4]. However, over the past 80 years, Louisiana has lost 3800 km^2 of land on the Mississippi River's deltaic coastal plain [5] due to a variety of human-induced and natural factors, such as rising sea levels, levee construction along the Mississippi River, sediment compaction due to pressure from overlying sediments and organic decomposition, fluid withdrawal, decreased fluvial sediment discharge, hurricanes, and widespread hydrologic alteration [6,7].

Over time, the interaction between the Mississippi River and the delta plain has been progressively managed [8,9] such that there are presently a few uncontrolled connections upstream of the Birdsfoot Delta [10]. Reconnecting the river with the delta plain is now widely recognized as an essential tool in the future sustainability of the wetland-dominated coastal ecosystem [11–14], and this approach has been central to restoration planning in Louisiana for decades [15]. Two freshwater diversions, Caernarvon and Davis Pond (Table 1) were built in the late 20th century specifically for the management of the estuarine gradient to support conditions for fisheries in the receiving basins. Future managed reconnections are usually in the form of sediment diversions. These are structures with gates built in the existing river levee system that would be operated to allow river water, sediment and nutrients to flow into the wetlands adjacent to the river to build new land and sustain existing wetlands. The results of allowing river water and sediment back into the coastal system are manifest in a number of places including the delta of the Wax Lake Outlet, mining of accumulated sediment in the Bonnet Carre Spillway, crevasse splays in the Birdsfoot Delta and revitalized marshes adjacent to freshwater diversions and siphons [14]. At the same time, adjustments in the estuarine gradient as freshwater is introduced are expected to have some effect on the habitat for important commercial fisheries, including oysters [16], brown shrimp [17,18], and other commercial species [19].

Table 1. Locations of existing or planned connections between the Mississippi River and estuarine basins. Barataria (BA), Breton (BR) and Pontchartrain (PO) identify the receiving basin. ID tracks the locations on Figure 1.

Location	ID	Type	Max. Flowrate Modeled	Operational Regime	Current Conds.	FWOA	Base Run	MP17
					| Model Simulations ||||			
Caernarvon—BR	1	Controlled Diversion	100 cms	Modeled as proportion of Mississippi River (MR) flow	x	x	x	x
Davis Pond—BA	2	Controlled Diversion	180 cms	Modeled as proportion of MR flow	x	x	x	x
Bonnet Carre Spillway—PO	3	Spillway	N/A	Flowrate required to keep MR flow no greater than 35,400 cms	x	x	x	x
Pointe-a-la-Hache—BA	4	Siphon	25 cms	Modeled as proportion of MR flow	x	x	x	x
Mardi Gras Pass—BR	5	Uncontrolled pass	N/A	Modeled as proportion of MR flow			x	x
Bohemia Spillway—BR	6	Spillway	N/A	Modeled as proportion of MR flow when MR > 26,335 cms			x	x
Fort St. Philip—BR	7	Uncontrolled pass	N/A	Modeled as proportion of MR flow	x	x	x	x
Mid-Barataria Sediment Diversion *—BA	8	Controlled Diversion	2123 cms	Min flow of 141 cms; linear increase to capacity from 12,742 to 28,316 cms in MR			x	x
Mid-Breton Sediment Diversion *—BR	9	Controlled Diversion	991 cms	Min flow of 71 cms; linear increase to capacity from 12,742 to 28,316 cms in MR			x	x
E. Maurepas Freshwater Diversion—PO	10	Controlled Diversion	56.6 cms	Constant flow			x	x

Table 1. *Cont.*

Location	ID	Type	Max. Flowrate Modeled	Operational Regime	Model Simulations			
					Current Conds.	FWOA	Base Run	MP17
Bayou Lafourche Diversion—BA	11	Controlled Diversion	28 cms	Constant flow				x
Lower Breton Diversion—BR	12	Controlled Diversion	1415 cms	Closed below 5663 cms in MR; linear increase to 1415 at 28,316 cms in MR; variable above 28,316 cms MR				x
Central Wetlands Diversion—PO	13	Controlled Diversion	141 cms	Constant flow				x
Union Freshwater Diversion—PO	14	Controlled Diversion	708 cms	Closed below 5663 cms or above 16,990 in MR; 708 cms for MR at 11,326 cms; linear from 0 to 708 cms from 5663 cms to 11,326 cms in MR and held constant between 11,326 cms and 16,990 cms in MR.				x
Ama Diversion—BA	15	Controlled diversion	1415 cms	Closed below 5663 cms in MR; linear increase to 1415 at 28,316 cms in MR; variable above 28,316 cms MR				x
Ostrica—BR	16	Uncontrolled pass	N/A	Modeled as proportion of MR flow when MR > 22,653	x	x	x	x

Louisiana has a long history of coastal restoration and protection planning. In response to the past and the growing threat of future coastal land loss due to rising sea levels, subsidence, and altered hydrology, Louisiana has developed a coastwide plan for coastal restoration and storm surge-based risk reduction over the next 50 years. The implementation of this plan would reduce land loss and expected annual damages, compared to a future with no action [20–22]. Through the 2017 Coastal Master Plan process, the restoration projects were analyzed and selected based on their ability to build or sustain land against a potential future landscape that incorporated up to 0.8 m of eustatic sea-level rise over the next 50 years. With the effort and investment envisaged in the 2017 plan, nearly 3000 km^2 of land is projected to be built or sustained that may have otherwise been lost. The expected annual damages from storm surge-based flooding are projected to be reduced by $12.2 billion at year 50 [22]. The 2017 plan selected a variety of restoration project types including river sediment diversions.

A number of sediment diversions from the Mississippi River have been identified to support land building and wetland sustainability [21,22]. The focus on sediment recognizes the value of that resource. Freshwater diversions of various scales have been used for decades in Louisiana to manage estuarine salinity gradients. The 2017 Coastal Master Plan also introduced the concept of using diversions at strategic and suitable locations that operate only under low-moderate river discharges to provide supplemental freshwater to estuarine wetlands, especially freshwater wetlands, during the times of the year when salinities often increase. The ecosystem diversity can be important outcomes of strategically placed riverine diversions, as well as land-building and land maintenance.

Moving water and sediment from the river into the estuarine basins results in ecosystem change relative to current or past conditions. However, the forward-looking analysis that considers future change, especially that associated with sea-level rise is needed.

The analysis described here explores how the use of river freshwater and sediment resources at multiple locations, individually or in combinations, year-round or seasonally, and in different magnitudes can be used to sustain the coastal ecosystem. It uses numerical experiments to consider several strategies for diversion operation and shows how the effects of diversion vary over time as the sea-level rises, between the years as the river fluctuates, and among estuarine basins with different landscape settings.

2. Methods

This analysis utilizes tools developed for the 2017 Coastal Master Plan to further explore the potential of the Mississippi River as a tool to support the sustainability of the Mississippi Delta Plain and the economic activities that depend on it. The Integrated Compartment Model (ICM) is a planning-level model that was developed by integrating into a single modeling platform several models that had previously been used for coastal zone planning and research in Louisiana [23]. The models that are included as subroutines within the ICM framework include a hydrologic and hydraulic model [24], a vegetation dynamics model [25], a wetland morphology/elevation change model [26], as well a barrier island morphology model and several receptor models that summarize the hydrologic and landscape conditions into numerous habitat indices and decision-making metrics. The hydraulic and hydrologic model, ICM-Hydro, simulates on a sub-daily timestep: water level (stage), flow rate, salinity, water temperature, suspended sediment concentration, sediment deposition and resuspension within open water areas, sediment deposition on the marsh surface, and a variety of water quality/nutrient constituents. The seasonal hydrologic conditions and salinity are then utilized by the vegetation dynamics model, ICM-LAVegMod, to predict the relative likelihood of wetland vegetation species being present under such conditions. The patterns of inundation, salinity penetration, sediment deposition and organic matter accretion are all taken into account within the wetland morphology/elevation change model, ICM-Morph, to predict whether wetland areas remain as viable emergent vegetation or collapse into open water due to either persistent inundation stress or acute salinity stress. Finally, as the coastal wetlands change in vegetative cover, and the resultant changes in elevation and land/water composition, the hydraulic network of the estuary is updated prior to the next model year in ICM-Hydro, to simulate the changing hydraulic and hydrologic conditions that occur as a result of collapsing wetland areas.

Prior to any modeling analysis, the ICM was calibrated using observational data from 2010 to 2013, and then validated on observations from 2006 to 2009. These time periods correspond to the development and widespread availability of data from the Coastwide Reference Monitoring System (CRMS), which collects hourly water level and salinity data across coastal Louisiana [27]. The water level predictions were calibrated with a goal that 80% of compartments should have a prediction bias of less than 0.15 m in the daily mean water level prediction. The magnitude of this bias corresponds approximately to the error in the underlying topographic digital elevation model (DEM) used in this analysis, which varies from a RMSE of 0.07 to 0.3 m [28]. The hydraulic link capacity and bed roughness were adjusted if the observed stage signal and amplitude were not in satisfactory agreement with the model. Due to the apparent datum inconsistencies in certain observed stations (as well as clear patterns of hydraulic controls influencing observed water levels), certain observed data were excluded when assessing the overall model fit. These inconsistent datasets were still used to visually compare modeled and observed hydrographs, but they were excluded from any aggregate model performance statistics.

Salinity calibration was also conducted for the same 2010 through 2013, time frame. The model agreement was defined as 80% of the compartments with a bias in salinity prediction of less than 1 ppt. Coastwide, 76% of the model-observation pairs met this goal. In the Mississippi River Delta (Pontchartrain, Breton and Barataria basins, Figure 1), 87% of the model-observation pairs met the bias target of 1 ppt or less (refer to Tables S2 and S3 and Figures S2–S5 in Supplementary Material). To attain this, a model parameter representing a combined dispersion diffusion coefficient, E_{xy}, was adjusted if the observed salinity signal and amplitude were not matched by the model.

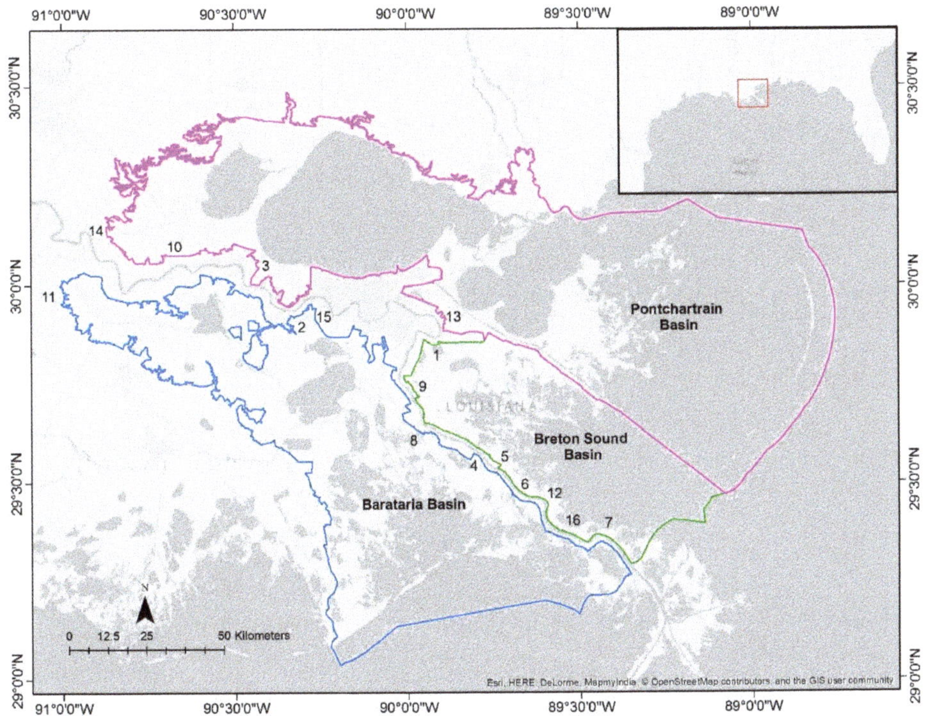

Figure 1. Location map showing the estuarine basins and the location of existing or planned connections between the Mississippi River and estuarine basins. See Table 1 for more information.

The higher E_{xy} values allowed for more exchange between compartments and lower values allowed for less exchange between compartments. In addition to adjusting the dispersion-diffusion coefficient, the salinity predictions were improved by the addition of a term in the hydrodynamic code that replaced the original central-difference method used for salinity convection with a first-order upwinding scheme [29]. This scheme was used for compartments with typical river/canal flows, where upwinding was an appropriate approximation, as compared to slower estuarine/marsh flow regimes where the upwinding scheme was not applied. This addition of the upwinding technique greatly increased the stability of salinity predictions. Further discussion and analysis of the ICM methodology, calibration, validation, model performance and analysis are provided in the Supplementary Materials as well as several peer reviewed articles [24–26,30–34] and technical reports [35–41].

This paper focuses on six sets of analyses. Five of the six simulations were analyzed using the 50-year flow hydrograph used in the 2017 Coastal Master Plan analysis, whereas the current conditions simulation used the observed flow hydrograph for 2006 through 2013. Three distinct Mississippi River flow conditions were identified from the historical river discharge data from 1964 to 2013 and were used in single year model runs for some of the analysis. Figure 2 shows the daily mean flow in the Mississippi River at Tarbert Landing, from 1964 to 2013. Figure 3 shows the annual mean flow anomaly (i.e., deviation from the long-term mean) for the Mississippi River for this 50-year flow hydrograph. A low river year (1988, elapsed year 25), a high river year (1993, elapsed year 30), and an average river year (1992, elapsed year 29) were selected between year 25–35. These river years were used to examine the effects of project configurations under a range of river flow conditions for some of the simulations described here.

Table 1 shows the operation regimes for diversions and other river outlets to estuarine basins. The operational flow rates for most passes, cuts, and diversions on the Mississippi River were based upon the flow in the river at each respective pass/diversion location. Therefore, the operation of any diversion was impacted by all operated diversions located upstream.

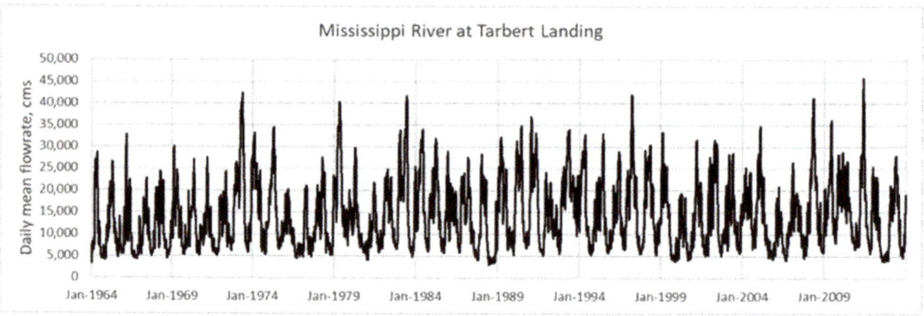

Figure 2. Daily mean flow in the Mississippi River at Tarbert Landing, 1964 to 2013.

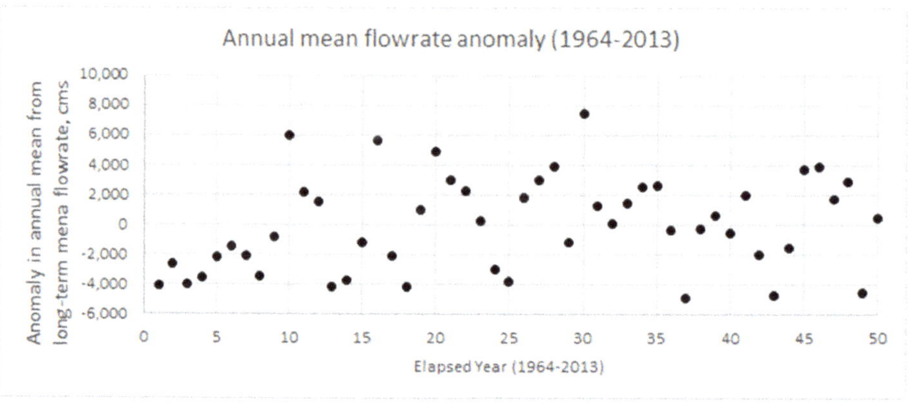

Figure 3. Annual mean flow anomaly from 1964 to 2013.

All of the analyses presented here (with the exception of the current conditions run, which used observed boundary conditions) were conducted using the medium scenario for the 2017 Coastal Master Plan [35]. The medium scenario uses 0.63 m of eustatic sea-level rise (ESLR) from 2015 through the end of 2064. The acceleration term increased over time, and the rate corresponded to 1.5 m of ESLR by 2100 compared to the 1992 sea level. The relative sea-level rise rates were assigned based upon these ESLR rates and subsidence values, which varied spatially across the model domain [42]. For the area included in this study, the subsidence ranges from 2.6 mm/year in the Upper Pontchartrain basin to 17 mm/year in the Birdsfoot Delta (refer to Figure S1 in Supplementary Materials: ICM Overview, for a map showing subsidence zones for coastal Louisiana used in this analysis).

A set of simulations was designed to explore a range of uses of Mississippi River water and sediment. The locations of existing and planned connections from the Mississippi River to the estuarine basins are shown in Figure 1 and their configurations are shown in Table 1.

The six simulations conducted for this analysis were:

1. Current Conditions. Before the ICM was used to assess future hydrologic and landscape conditions throughout coastal Louisiana, the ICM-Hydro subroutine was run from 2006 to 2013 using observed salinity and water level data collected across coastal Louisiana. The descriptions of the data used for this run and discussions on general model performance and limitations can be found in previously published literature [28,30–32]. This run is used to illustrate the current pattern of salinity fluctuations within the estuaries, against which simulations that include sea-level rise can be compared.

2. Simulations of Future Conditions with Varying Levels of Freshwater Inflow.

  a. Future Without Action Run (FWOA). The ICM was run for a future 50-year period without any potential future projects included. This simulation is termed the future without action (FWOA). In the FWOA, all sediment and freshwater diversions that were already built were implemented in the model. A full discussion of all boundary conditions assumed for this 50-year future condition is available in Brown et al., [28]. This run is used to demonstrate the effects of relative sea-level rise on the estuarine salinity gradient in the absence of additional freshwater inflow.

  b. Base Run with Both Mid-Basin Sediment Diversions (Base Run). This 50-year model run includes all diversions that are currently in some phase of the engineering and design process (e.g., Mid-Barataria sediment diversion, Mid-Breton sediment diversion, and Maurepas freshwater diversion) and represents a likely future condition against which the need for additional management of river-estuarine can be assessed. In the base run, these diversions were implemented in addition to the existing diversion included in FWOA (Table 1). This run is used to demonstrate the effects of the diversion projects which are expected to be built within the next 5 years in the context of future sea-level rise.

  c. All 2017 Master Plan Diversions (MP17). This simulation shows the potential effect of all 2017 Coastal Master Plan diversions working together over a 50-year period. This includes 5 diversions in addition to those included in the base run (Table 1).

3. Simulations examining connectivity between the river and estuary. In addition to the above multi-year simulations, several single-year simulations were also conducted to explore specific options for adjusting connectivity between the river and the estuary and the effects of diversion operation on sediment delivery, which is a key factor in using sediment diversions to combat the effects of future sea-level rise. One simulation explores the modified exchange between the Mississippi River and Breton Sound by comparing the effect of a single larger diversion point versus multiple smaller exchange points. The currently active exchange points with the Mississippi River (e.g., Caernarvon Freshwater Diversion, Mardi Gras Pass, Bohemia Spillway, Fort St. Philip, etc. see Table 1) were modeled as inactive with the flow being maintained in the main channel of the Mississippi River.

4. Simulations examining diversion operations to deliver fine sediment. Many of the planned diversions have been justified on the basis of delivering sediment to the estuarine basins [43]. In addition to the information developed on sediment delivery using the simulations described above, an additional ICM run was conducted to assess the use of a diversion operational regime focused on the delivery of fine sediment, i.e., less focused on maximizing water flow during peak river flows as for the Mid-Breton and Mid-Barataria diversion (Table 1). This simulation used the Central Wetlands diversion (Table 1, Figure 3) as an example and was conducted on three years which represent years with low, average, and high rates of Mississippi River flow.

To examine the effect of diversions over time as the relative sea-level rise progresses, the salinity gradients within the receiving basins plots were constructed showing the area of the basin (wetland and open water combined) subject to different salinity levels. Six bins were identified representing different salinity ranges. These salinity ranges were selected on the basis of salinity tolerance of seven wetland species which have previously been used to assess the effects of Mississippi River diversions [44]. The range in salinity values in which these species are optimally located (as defined by the LAVegMod model rules [25]) are shown in Figure 4. The daily salinity values for the growing season for the Louisiana coastal wetland vegetation were used to calculate the area of the basin, experiencing the salinity range for each bin in each year of the simulation. The model output for each ICM-Hydro compartment, which included both open water and marsh areas, was summarized based on three ecoregions shown in Figure 1.

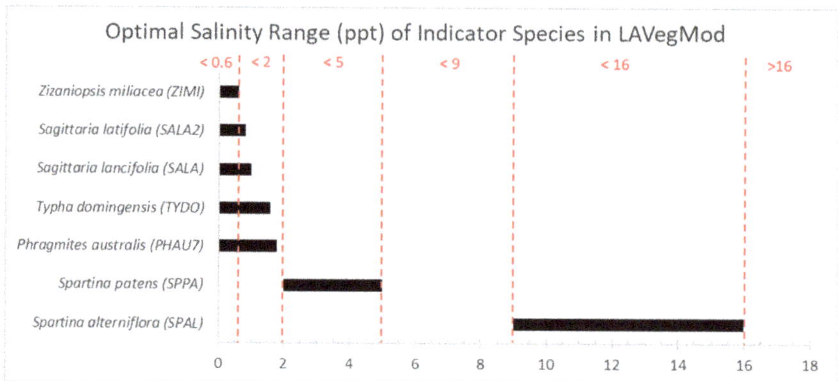

Figure 4. Salinity ranges considered optimal for seven indicator wetland vegetation species.

3. Results

3.1. Current Conditions

A comparison across the basins (Figure 5) shows the differences in the penetration of salinity into the basins. Both Pontchartrain and Barataria show extensive areas with salinity 16–35 ppt in all years, while for four of the years this salinity range does not occur in Breton. This is due to the proximity of Breton to the Mississippi River and the many small connections (see further analysis of the effects of these connections below).

Breton is also different at the lower end of the salinity range. Both Pontchartrain and Barataria extend inland some distance, are larger in size (increasing freshwater inputs through precipitation) and receive some municipal stormwater runoff. Pontchartrain also received freshwater inflow from tributary streams. Both Barataria and Pontchartrain include areas <0.6 ppt in all years in this run. However, the conditions in 2006, 2007 and 2012 eliminate this zone in Breton.

The period shown in Figure 5 included both a high flood year (2011) and a low river flow year (2012) as shown in Figure 3. However, there is a substantial change in the salinity distribution in only one of the three basins. The high flow in 2011 does not increase the area <0.6 ppt in any basin indicating the limited connections between the Mississippi River and estuarine basins in the current condition. However, Breton which has several small connections, does show a substantial decrease in the area with salinities <2 ppt in 2012 during the low flow year.

3.2. Simulations of Future Conditions with Varying Levels of Freshwater Inflow

Three of the simulations progressively increase the exchange between the Mississippi River and estuarine basins (Table 1). Table 2 summarizes the differences in the distribution of salinity for decadal intervals during the 50-year model runs, showing the relative area of the basins for three different salinity ranges (based on the six bins described above) for ease of comparison.

Table 2. Relative area of each basin subject to three salinity ranges by decade for future without action (FWOA), the base run and MP17 simulations.

	FWOA								
	Pontchartrain			Breton			Barataria		
	<2 ppt	2–16 ppt	>16 ppt	<2 ppt	2–16 ppt	>16 ppt	<2 ppt	2–16 ppt	>16 ppt
Year 10	53%	47%	0%	54%	46%	0%	60%	27%	13%
Year 20	53%	47%	0%	48%	52%	0%	54%	31%	15%
Year 30	54%	46%	0%	54%	46%	0%	49%	35%	16%
Year 40	50%	50%	0%	27%	73%	0%	25%	45%	29%
Year 50	31%	66%	3%	17%	83%	0%	4%	52%	44%

Table 2. *Cont.*

Base Run—Mid-Basin Diversions								
Pontchartrain			Breton			Barataria		
<2 ppt	2–16 ppt	>16 ppt	<2 ppt	2–16 ppt	>16 ppt	<2 ppt	2–16 ppt	>16 ppt
Year 10 58%	42%	0%	55%	45%	0%	65%	25%	10%
Year 20 54%	46%	0%	55%	45%	0%	66%	24%	10%
Year 30 58%	42%	0%	70%	30%	0%	72%	18%	10%
Year 40 50%	50%	0%	49%	51%	0%	60%	28%	12%
Year 50 31%	63%	6%	29%	71%	0%	56%	29%	16%

All 2017 Master Plan Diversions								
Pontchartrain			Breton			Barataria		
<2 ppt	2–16 ppt	>16 ppt	<2 ppt	2–16 ppt	>16 ppt	<2 ppt	2–16 ppt	>16 ppt
Year 10 61%	39%	0%	58%	42%	0%	67%	23%	10%
Year 20 58%	42%	0%	55%	45%	0%	67%	23%	10%
Year 30 61%	39%	0%	70%	30%	0%	76%	14%	10%
Year 40 50%	50%	0%	48%	52%	0%	63%	27%	10%
Year 50 36%	58%	6%	29%	71%	0%	60%	30%	10%

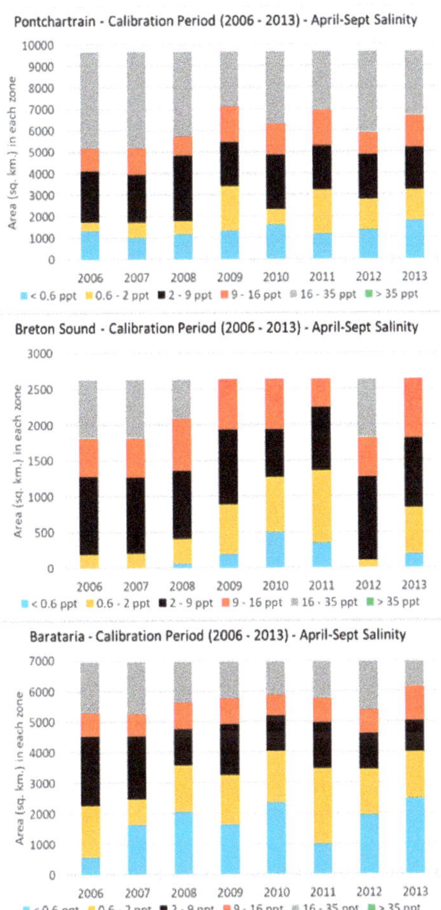

Figure 5. April to September salinity region variation in Pontchartrain (**top**), Breton (**middle**) and Barataria (**lower**) for the current conditions simulation.

The FWOA run maintains the same connections between the basins and Mississippi River as the current conditions simulation. However, the 50-year simulation includes the effects of the relative sea-level rise and the changing wetland landscape within the basins [33]. Under FWOA, all basins show an increase over time in the proportion of the basin in the 2–16 ppt range, and both Pontchartrain and Breton show an increase in the area >16 ppt by year 50. The rate of eustatic sea-level rise accelerates over time in the simulation and is assumed to be the lead cause of that significant increase in salinities in the last decade.

The base run shows the same increase in salinity levels in Pontchartrain, Breton, and Barataria for April–September as found in FWOA with some individual years, fresher or more saline. The simulation including all the 2017 Master Plan diversions does not result in a substantially different salinity regime compared to the base run, mostly because both simulations include several diversion projects that divert large freshwater volumes to the basins. The operation plans are different between the simulations but not sufficiently to create major differences in the salinity regime. Table 2 shows the least amount of area >16 ppt at year 50 in this run compared to the other runs in Barataria, with the other basins being similar.

The annual snapshots shown in Table 2 mask the extensive interannual variation, and the differences in the response among the basins to events included in the 50-year boundary conditions. Of note, here is a severe drought (low rainfall) in the simulation in years 42–43. Figure 6 shows the effect of this drought in Pontchartrain in years 42–43 for the three simulations. For each simulation, these years show the greatest extent >16 ppt and the least extent <0.6 ppt.

Figure 6 shows a few major trends over time in Pontchartrain with substantial interannual variability, as might be expected where there are connections to riverine inflows. The last 15 years of the simulations show a generally smaller area of <0.6 ppt. In Pontchartrain, there appears to be an increase in the frequency of years with large areas >16 ppt in the later years which may be associated with the progressive influence of sea-level rise. This occurs in all three simulations and may be more related to the changes in freshwater from tributary inflows to the basin from inland streams, which remains the same for all simulations, rather than the river diversion inflows.

The changes in the Breton basin are shown in Figure 7. This basin has no inputs of freshwater apart from the precipitation and multiple relatively small controlled or uncontrolled connections to the Mississippi River (Table 1). The last 15 years of the simulations show a generally smaller area of <0.6 ppt in all simulations shown in Figure 7, and in FWOA there is also a decrease in the area <2 ppt. The FWOA also shows some incursion of water >16 ppt in isolated years in the last decade in contrast to earlier years of the simulation.

With the mid-basin diversion, the area in Breton <0.6 ppt increases compared to FWOA, showing that the large diversion has a greater impact on salinity than the existing connections (see next section). However, there are more years in the base and MP17 runs compared to FWOA, especially early in the simulations, when Breton includes areas >16 ppt. This seems counterintuitive when more freshwater is entering the basin through the diversion. However, as diversions into Barataria are also operating in these simulations, the flow toward the mouth of the Mississippi is diminished, reducing the discharge through distributaries in the Birdsfoot Delta, and increasing salinity in the lower parts of the Breton basin (Figure 3).

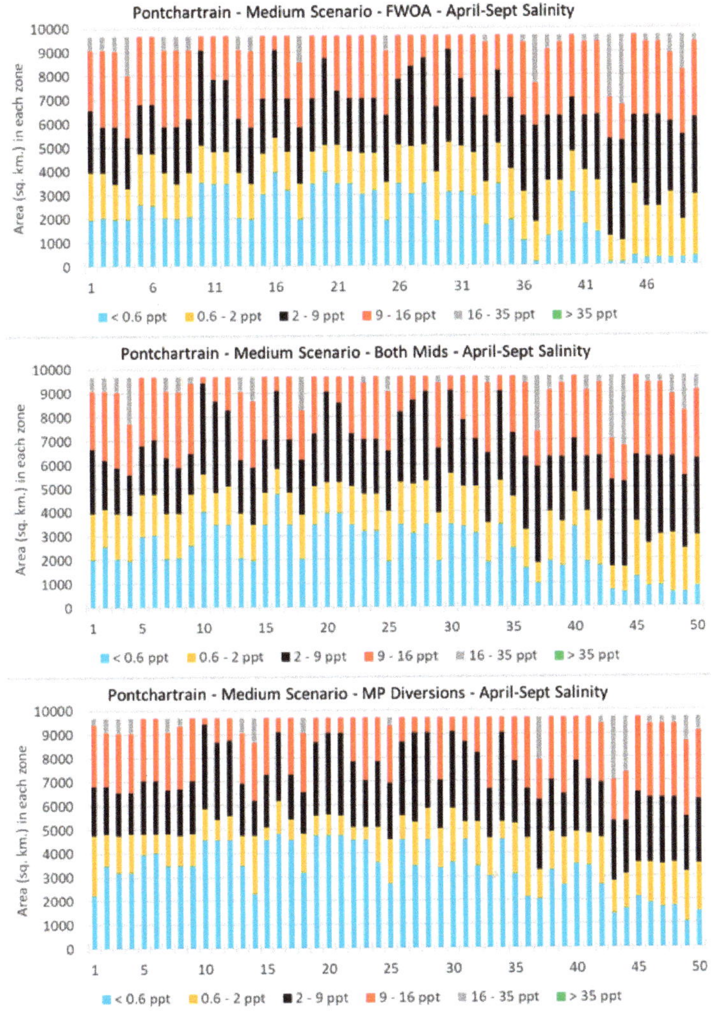

Figure 6. April to September salinity region variation in Pontchartrain for runs with varying levels of freshwater inflow

Barataria shows a very different pattern from the other basins, especially under FWOA (Figure 8). The FWOA run maintains the same connections between the basins and Mississippi River as current conditions. The 50-year simulation includes the effects of the relative sea-level rise and the changing wetland landscape and these effects are most dramatically shown in FWOA. There is a major increase in the extent of the basin with 16–35ppt salinity after year 30 (Figure 8). This period also shows a major decrease in the area <0.6 ppt. In the last few years of the simulation, only a small area of the basin experiences salinities <2 ppt.

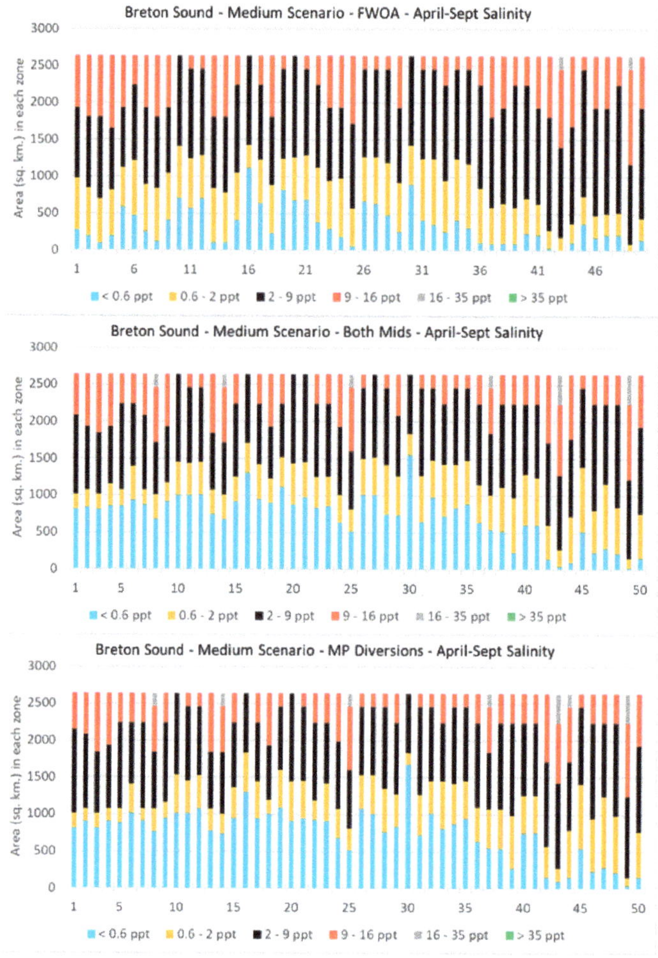

Figure 7. April to September salinity region variations in Breton for runs with varying levels of freshwater inflow.

With the diversions operating, Barataria does not experience the dramatic increase in the areas >16 ppt found in FWOA. Figure 8 shows remarkably little change over time. There is an increase for most of the 50-year simulation in the extent <0.6 ppt in Barataria in the base run but during the drought period, a large area <2 ppt remains. The MP17 simulation includes an additional diversion into the Barataria basin at Ama (Table 1, Figure 1) which appears to have little effect on the salinity distribution for most of the simulation, but reduces the effect of the drought. Even with this diversion, there are still conditions in specific years later in the simulation, that dramatically reduce the area of the basin <0.6 ppt.

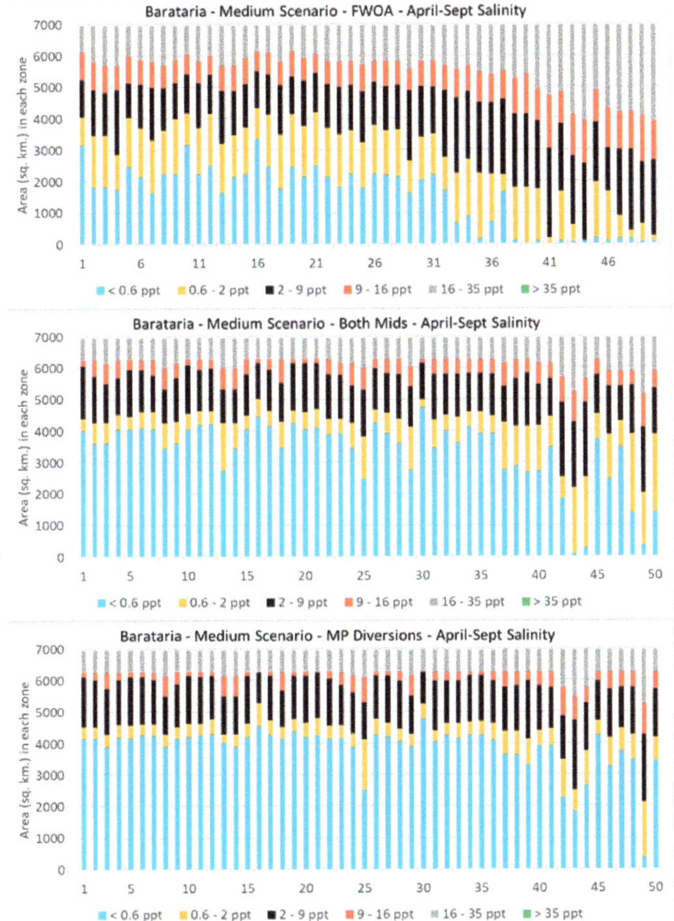

Figure 8. April to September salinity region variations in Barataria for runs with varying levels of freshwater inflow.

3.3. Modified Exchange between the Mississippi River and Breton Sound

The currently active exchange points (distributaries) between the Mississippi River and Breton Sound (e.g., Caernarvon Freshwater Diversion, Mardi Gras Pass, Bohemia Spillway, Fort St. Philip, etc.) were modeled as inactive with the flow being maintained in the main channel of the Mississippi River and eventually reaching the Birdsfoot Delta. All other simulation conditions (including diversion operations) and environmental assumptions were as represented in the 2017 Coastal Master Plan FWOA and the base run (Table 1).

Figure 9 shows the salinity difference maps for the run with modified exchange compared to the base run at year 25 and 50, respectively. The simulation results show a salinity increase that extends east into the Lower Pontchartrain basin due to the reduced freshwater into Breton. The April-September salinity distribution plots in Figure 10 show the salinity gradient in Breton due to the disconnection from the Mississippi River, which shows a greater extent of saline bins compared to the base run (Figure 7) while little change is observed in Barataria or Pontchartrain.

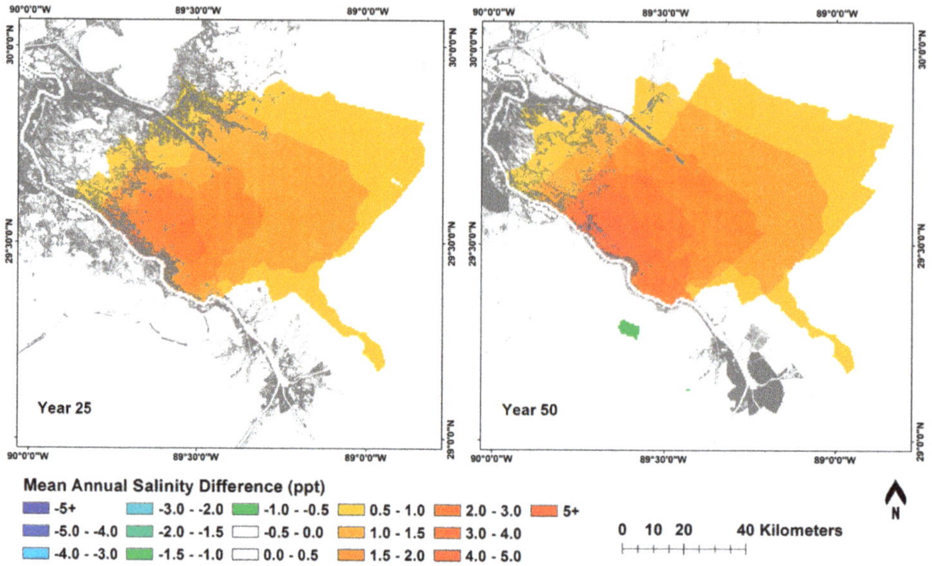

Figure 9. The salinity difference when exchanges with Breton Sound through distributary channels downstream of the Mid-Basin diversions are reduced (modified exchange run compared to the base run with both Mid-Basin diversions) at year 25 (**left**) and year 50 (**right**). Eustatic sea level rise by year 25 was 0.33 m and 0.63 m by year 50.

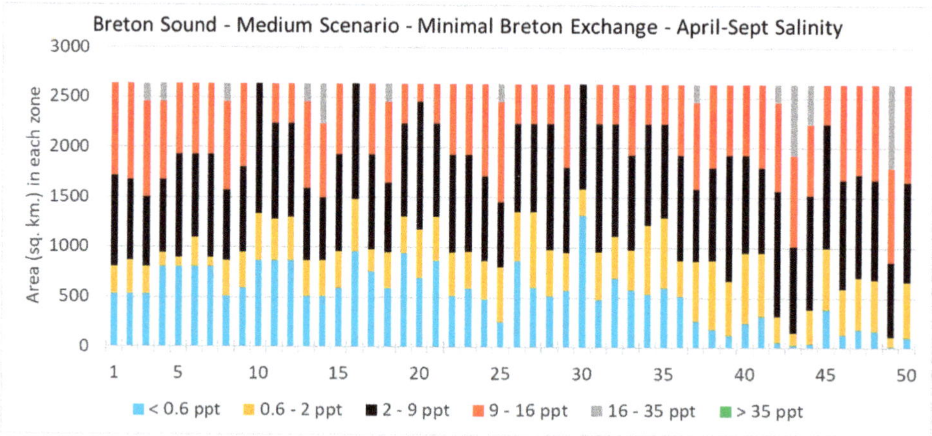

Figure 10. April to September salinity region variation in Breton Sound eco-region with minimal exchange between the Mississippi River and Breton Sound.

The vegetation coverage for the base run and minimal exchange for year 25 and 50 are shown in Figure 11. The vegetation in lower Breton Sound and along the east bank side of the Mississippi River converts towards a more saline type due to the freshwater reduction with the modified exchange as the direct freshwater input through Mardi Gras Pass and Fort St. Philip (Figure 3) are reduced. While less vegetated land is shown in year 50 as the effects of the sea-level rise and subsidence have resulted in land loss, a similar effect is apparent. At year 50, with 0.63 m of additional sea-level rise, the diversion alone (Modified Exchange in Figure 11) maintains fresh marsh in the upper part of the Breton basin but is insufficient to prevent saline marsh along the east bank of the river.

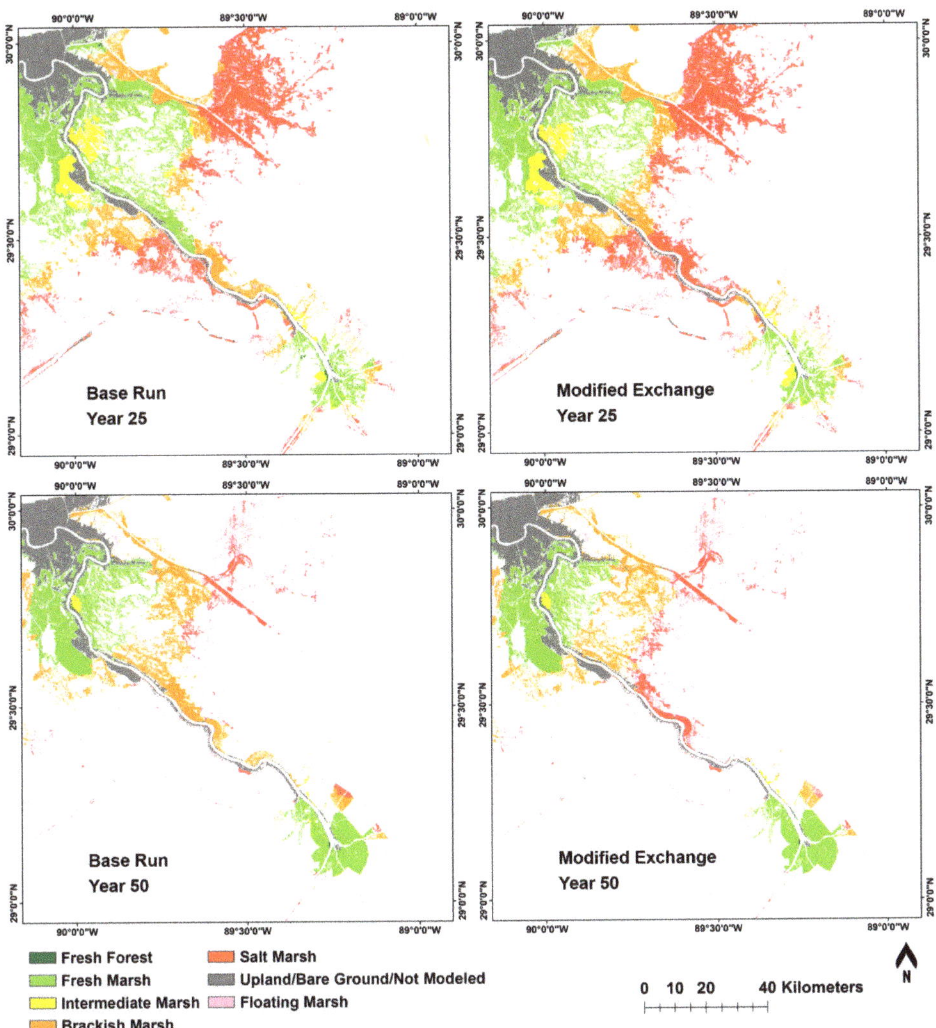

Figure 11. Vegetation coverage for base run with both of the Mid-Basin diversions (**left**) and the modified exchange run with both diversions active but other connections to Breton Sound reduced in flow capacity (**right**). Eustatic sea level rise by year 25 (**top**) was 0.33 m and 0.63 m by year 50 (**bottom**).

There is also a marked difference in the extent of the brackish marsh zone between year 25 and year 50. At year 25, this is a very narrow band in both simulations. The freshwater from the diversion and saline water from the Gulf squeeze out the brackish zone. However, as the sea-level rise progresses, saltwater penetrates further, mixing with the fresh conditions maintained by the diversion to produce a more extensive zone of brackish marsh. It should be noted that in addition to the vegetation shifts over time and under different freshwater exchange scenarios, there is a marked decrease in the amount of wetland area throughout lower Barataria and lower Breton basins under this assumed relative sea level rise scenario. This conversion of wetland area to open water is a consequence of persistent inundation and salinity penetration due to the combined effects of eustatic sea level rise and subsidence [33].

3.4. Operational Regime and Sediment Delivery

The purpose of most planned river diversions in Louisiana is to provide sediment for rebuilding of coastal wetlands and their maintenance under a future sea-level rise [45]. Their location and design are often optimized based on the potential for delivery of sand from the river through the diversion structure [46,47]. Table 3 presents the cumulative suspended sand load (not including bedload or fines in suspension) diverted for the simulations described above. These observations also point to the need for careful management of the operation of proposed restoration strategies (with dynamic components).

Table 3. Suspended sand load diverted from the Mississippi River for each of the simulations (total over 50 years).

Sand Load Diverted from the Mississippi River (Million Tonnes)	Breton	Barataria
Base Run (two Mid-Basin diversions)	136.7	47.7
All 2017 Master Plan Diversions	146.8	44.8
Modified Exchange between the Mississippi River and Breton Sound	84.5	N/A

These simulations reflect the combined influence of different combinations of diversions with and without various existing connections between the river and estuary, over 50 years. Table 3 shows greater potential for sand delivery into Breton, which may be the result of the Mid-Breton diversion being upstream of the Mid-Barataria diversion, reducing the flow available for distribution when they are both operating simultaneously.

Varying the operational regimes shown in Table 1 would modify these sand loads. However, as fine sediments are important for coastal marsh maintenance in the face of sea-level rise [45], an additional simulation was conducted to assess the effects of an operational regime and the river flow on fine sediment delivery. This simulation used the Central Wetlands diversion as an example (see Figure 1 for location). The single-year runs were conducted to examine the different configurations and operational rules. All other simulation conditions (including diversion operations—see Table 1) and environmental assumptions were as represented in the 2017 Coastal Master Plan FWOA and the base run.

Three Central Wetlands diversion configurations were evaluated:

1. The operation of 141 cms assumed in the 2017 Coastal Master Plan was used.
2. The diversion was operated similar to the Mid-Barataria diversion in the base run (see Table 1)
3. The diversion opened when river flow was above 12,742 cms with a discharge of 0.83% of the Mississippi River flow. This diversion to river ratio was determined as the sum of the total diversion discharge if it operated with a constant 141 cms flowrate in 50-years and divided by the total river flow when the diversion is opened (river flow above 12,742 cms) in 50 years. The idea is to operate the diversion only during the high river condition while retaining the same total annual discharge volume as with a constant flowrate (configuration 1).

The diversion discharge for these three different operational configurations in the selected low, medium, and high river years (see Methods) are shown in Figure S1 in Supplementary Materials: ICM Simulation Settings. The three diversion operations were evaluated in these pre-selected river flow conditions and the sand concentration in the diverted water was set to zero in order to evaluate sediment accumulation from fine particles.

The fine sediment accumulation in the receiving basin for the Central Wetlands diversion for different diversion configurations in different river years are shown in Figure 12.

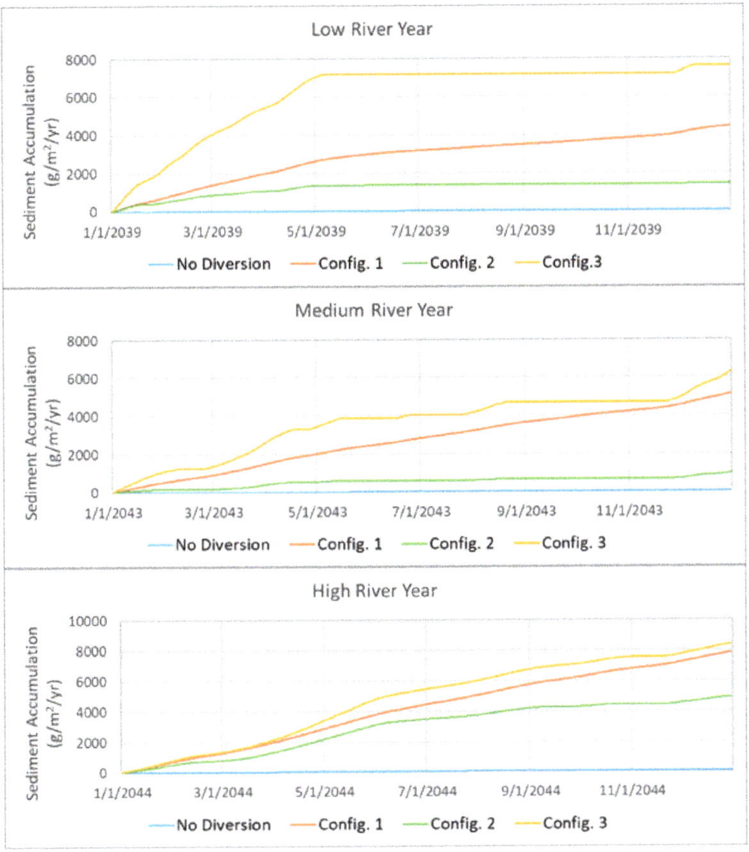

Figure 12. Sediment accumulation in the receiving compartment for the three Central Wetlands diversion configurations: A constant year-round flowrate of 141 cms as modeled for the 2017 Master Plan (**1**), a flowrate activated when the river flow was above a given threshold with a maximum diverted rate of 141 cms (**2**), and a diversion such that the annual volume diverted was equal to the annual volume diverted in configuration 1, but only during periods of peak flow in the river (**3**).

The receiving compartment is a 17 km² tract of marsh (16% of which is open water) on the east bank of the Mississippi River. The Central Wetlands tract is bounded on all sides by hydrologic barriers (flood walls, levees, roadways) and currently receives freshwater inflows from several pump station discharge lines which drain the eastern portions of New Orleans and neighboring municipalities. The sediment accumulation for diversion configuration 3 is higher compared to diversion configuration 1 and 2 in all simulated river conditions. Diversion configuration 3 performs better in low and medium river years when compared to the high river year.

The fifty-year simulations with diversion operational configuration 1 and 3 were conducted. Salinity in the immediate receiving area is highly influenced by the different operational configurations. As illustrated in Figure 13, operational configuration 1 resulted in a sustained low salinity in the diversion receiving area, while in operational configuration 3, salinity is only slightly different from the no diversion condition.

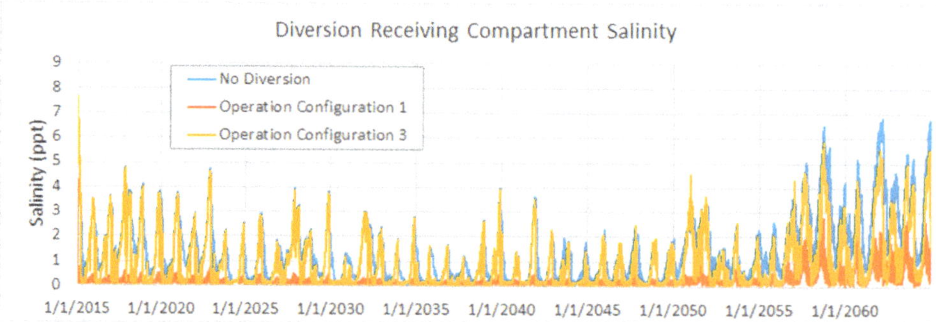

Figure 13. Salinity of the diversion receiving compartment for different operational configurations.

4. Discussion

4.1. Future Effects of Relative Sea-Level Rise

The effect of 50 years of sea-level rise on estuarine environments is often focused on the fate of the coastal wetlands and the impact of their loss of habitat for fish and wildlife [48–50]. The simulations described here show that the effects of the sea-level rise on these estuarine basins can be largely offset by more extensive use of river resources. Barataria basin (Figure 8) shows the most variation in salinity during the FWOA with a marked increase in later years of the simulation. It has a few existing connections to the Mississippi River, other than the Davis Pond freshwater diversion (Table 1), and no other freshwater inputs other than precipitation, stormwater runoff from local communities. Under FWOA, the fresher parts of the estuarine gradient, i.e., <2 ppt, are greatly reduced and almost eliminated during drought years with low rainfall (Figure 8, Table 2).

Barataria also has a relatively intact chain of barrier islands, separated by tidal inlets, at the Gulfward margin (Figure 3). Barrier islands have been shown to modulate salinity in Louisiana estuaries [51]. The simulations described here do not discuss barrier island dynamics over time. Further, how the barrier shoreline is maintained in the future, as it also faces impacts from the sea-level rise, will be important for future salinities and the relative influence of diversion in the Barataria basin.

As this modeling has shown, the interaction of the future sea-level rise with other climatic factors is more complex. The inclusion of the drought years in year 42–43 of the simulations showed that even in Pontchartrain where sea-level rise did not cause a major progressive increase in salinity > 16 ppt, the effect of a higher sea-level and reduced precipitation caused a temporary increase (Figure 6). The potential for adaptive operations for diversions to mitigate such events is discussed below. However, one other key influence on the issues discussed here is the effect of future climate change on Mississippi River discharge. The simulations included here are all based on the same 50-year record of the Mississippi River flow in order that the effects of diversions on the change in the estuaries due to the sea-level rise can be specifically explored. Studies have shown that trends in flooding on the Mississippi River are influenced by climate variability including the El Niño–Southern Oscillation and the Atlantic Multidecadal Oscillation [52], although predicting future river flow is challenging in large river systems [53].

4.2. Effects of River Diversions on Estuarine Conditions

This study has shown the influence of different magnitudes of river inflow on salinity gradients within the estuaries. The comparison of the future with only minor connections between the river and estuarine basin (FWOA) with two major diversions (base run) and then with additional diversions (MP17) for salinity distribution shows differences among the basins (Figures 5–7). For Pontchartrain (Figure 6) which does not directly receive additional inflows under the base run, the difference is the greatest in the last decade in the extent of the basin <0.6 ppt. The incremental addition of diversions increases the area <0.6 ppt. The area <0.6 ppt, however, is smaller in extent than experienced in earlier decades, reflecting the progressive influence of the sea-level rise on the system. Even though the diversions are not directly influencing Pontchartrain basin, the extent of the freshwater influence from large diversions into Breton basin extends into lower Pontchartrain (Figure 9) perhaps ameliorating the penetration of higher salinities from the Gulf. There is much less difference in Pontchartrain among the higher salinity levels.

In contrast, in FWOA Breton (Figure 7) has fluctuating areas of <0.6 ppt, likely due to fluctuations in the river flow (Figure 2) and the many small existing connections between the river and the basin (Table 1). Breton Sound basin both experienced fewer periods of the freshest salinity range, with three of the years having no portion of the basin with a mean salinity <0.6 ppt. While in four other years, the basin had no portion with the mean salinities greater than 16 ppt. These freshening events occur during the wetter-than-average years of the flow in the Mississippi River. This was the case in five of the last six years in this simulation, 2008 through 2013 with the exception of 2012 (Figure 2). All five of these years were years with higher-than-average flowrates in the Mississippi River. The sixth year, 2012, was well below the average for the river flow. The sensitivity to the Mississippi River flow is due to the lack of any other pluvial source into Breton Sound. Additionally, the most Gulfward portions of the Breton Sound Basin are somewhat buffered from the open waters of the Gulf by both the distal barrier Chandeleur Islands enhancing the influence of Mississippi River fluctuations on salinities in the lower Breton basin [54]. This results in a basin that has less overall range to salinity in most years. However, the current conditions simulations ended in 2013, and did not include the flowrates for Mardi Gras Pass which initially formed during the 2011 flood. While one flow rating curve was used in this modeling (Table 1), the observed flows leaving this diversion and other connections to Breton had a noticeable impact on salinities in Breton, which was also seen in future simulations with these connections turned off (Figure 9).

Thus, the addition of the Mid-Breton diversion in the base run increases the freshwater area <0.6 ppt. However, Figure 7 also shows little change in the salinity zones between 2 and 16 ppt. This is shown in the vegetation patterns for year 25 (Figure 11) and the greater brackish area shown in Figure 11 at year 50 comes at the expense of the freshwater areas, the latter of which is limited in all three simulations in the last decade (Figure 7). Even the addition of the Lower Breton diversion (Table 1, Figure 1) in the MP17 simulation does not substantially change the basin scale conditions in the last decade, although there is likely localized freshening.

4.3. Effect on Estuarine Fisheries

Estuaries are an important habitat for the Gulf of Mexico's commercial and recreational fisheries with 97% (by weight) depending on estuaries for some part of their life cycle [55]. The importance of the freshwater inflow to estuarine ecosystems is widely recognized [56–58]. Often the focus of concern is the upstream impoundment of streams for water supply or flood management [59,60] but in the estuarine basins discussed here, which have been essentially isolated from riverine inflows for more than a century due to river management in the delta plain, the key issue is where, when, and how much inflow is required to meet certain goals and avoid undesirable outcomes for other parts of the ecosystem [61].

The existing diversions have changed the aquatic ecology of their receiving area. Studies of Davis Pond and Caervarvon have examined the effects of diversion influxes of freshwater on the ecology of the current system [62,63], including the effects on important commercial and recreational species, such as spotted seatrout *Cynoscion nebulosus*, red drum *Sciaenops ocellatus*, Gulf menhaden *Brevoortia patronus*, blue crab *Callinectes sapidus*, and penaid shrimp. These effects vary with discharge and are often localized [64]. Modeling has also been used as a tool in such evaluations. For example, the simulations of fish response to the operation of the Caernarvon freshwater diversion showed movement down the estuary of Bay Anchovy *Anchoa mitchilli*, which agrees with field observations [63]. The long-term simulations of fish communities in Barataria, Breton, and Pontchartrain basins in response to the sea-level rise and the inflow from four large river diversions (Mid-Barataria, Mid-Breton, Lower Breton in Table 1 and an additional diversion into Lower Barataria) showed local shifts in the distribution of six species of ecological and economic importance including the Eastern oyster *Crassostrea virginica* [65]. Some changes were the direct effects of decreased salinity while other effects were related to reduced Chl *a* due to increased turbidity, a trend also identified in the field and literature-based studies [66]. The simulations also showed that local biomass reductions were mostly the result of redistribution, as there was minimal impact on the total biomass compared to a future without action.

The ecological effect of the larger scale riverine inflow is likely complex. An examination of the salinity sensitivity of 44 estuarine fish species, including many of economic importance, in the Gulf of Mexico estuaries found many species showing low or moderate sensitivity to the salinity change [67]. However, freshwater fishes currently in the extensive areas of the basin >0.6 ppt, may be impacted by increased salinity through the impacts on physiological costs (e.g., osmoregulation) which may be offset through increased productivity and greater food availability [68]. Further, experimental studies using coastal marsh fish which are found across the salinity gradient showed that descendants of individuals from brackish populations had better survival when exposed to salinity than fish with no historical exposure to salinity, suggesting genetic adaptation [69].

4.4. Lessons for Future Management

It is not a new concept that large-scale river diversions into the Mississippi River estuary could utilize an operational regime which mimics the natural fluctuations of riverine flood events [70]. Crevasses and small uncontrolled diversions show such a flow regime [71,72] in contrast to the near continuous flow of a delta distributary channel. This analysis examined three distinct operational regimes to explore how a diversion could be operated to mimic a seasonal high flow and sediment pulse that is typically seen in the springtime Mississippi River flows [10,45,70]. Capturing an equivalent amount of diverted water, the flood-wave specific operations are able to capture and deliver a greater load of sediment from the river and distribute it to the estuary (Figure 12). This regime has the largest impact during river years with below-average flowrates when the total annual load must be scheduled during shorter periods (Figure S1 in Supplementary Material). This operation regime, while simple enough to implement in a model with assigned boundary conditions, the real-world implementation would be challenging, given the inability for accurate seasonal forecasts of the flow in the Mississippi River. In practice, such a regime would likely operate as soon as the river is high enough (typically in the spring) and continue until the annual sediment/freshwater load mass/volume were met.

An additional benefit from flood-mimicking operational regimes is that the concentrations of suspended sediments within the Mississippi River waters (both fines and sands) vary in relation to flood events and annual cycles [73]. The current flow-to-sediment rating curves for the lowermost Mississippi River show a hysteresis effect where the rising limb of the annual river flood has a flushing effect and carries a greater suspended sediment load than an equivalent flowrate on the falling limb of the flood hydrograph [10]. The scheduling diversion operations for the arrival of the floodwave allow for the most efficient delivery of suspended sediments from the Mississippi River into the estuary (Figure 12).

The simulations presented in this report provided insights on how the system responds to a variety of operation plans. Some of the responses are not intuitive, emphasizing the value of using such holistic landscape models that capture (to the extent possible) feedbacks and interactions among various processes. The FWOA simulation shows a dramatic change in salinity in parts of the system due to the sea-level rise. The rate of the eustatic sea-level rise accelerates over time in the simulation and is assumed to be the lead cause of that significant increase in salinities in the last decade. These changes have been shown to have dramatic effects on vegetation [32] and, in combination with the changes in temperature, can dramatically influence the distribution of aquatic species [34].

One important aspect of using the resources of the Mississippi River to manage the coastal ecosystem in the face of the sea-level rise is that it allows the management of conditions to achieve a variety of goals, e.g., land building or positioning suitable salinity conditions for fishery benefits, as well as the ability to adjust to changing conditions. The simulations here all apply a fixed set of rules over time. In reality, the diversions are likely to be managed in response to estuarine (and river) conditions. The effects of the drought event in the last decade of the simulation were in some way ameliorated by the various diversion configurations (Figures 5–7) even with operational rules which were not designed to specifically address such situations. As the implementation of diversions proceeds, near-term forecasting models of estuarine conditions and adaptable operational rules for diversion flows may be useful to further protect against massive interannual fluctuations on estuarine conditions. However, this study has shown that planning how many diversions are needed and at which locations within the estuary must focus on the long-term in order that the effects of the sea-level rise can be integrated into decision making.

5. Conclusions

The simulations described here show, for the estuarine basins of the Mississippi Delta Plain, that the effects of the sea-level rise decades into the future are potentially dramatic, but the effects on the estuarine salinity gradient varies according to the landscape setting. The analysis demonstrates that the ecological character of the future system is dependent on both the sea-level rise and the management of river resources. The pronounced incursion of higher salinity water into Barataria in later years of the simulation associated with the relative sea-level rise was largely offset by the freshwater inputs from the planned Mid-Barataria diversion. While sediment delivery is frequently the focus of future river diversions in the Mississippi River Delta, this analysis has shown how large-scale river diversions also affect spatiotemporal patterns in the salinity gradient throughout the estuarine basins, combatting the effects of a future sea-level rise by maintaining present-day salinity variations throughout the basin. The spatial extent of these salinity gradients varies annually due to the variability in riverine freshwater flows reaching the delta and are also susceptible to the number (and operational regime) of engineered diversions and natural distributary channels into the estuary.

Supplementary Materials: The following are available online at http://www.mdpi.com/2073-4441/11/10/2028/s1: *ICM Overview* and *ICM Simulation Settings*. Please contact the corresponding author for access to model source code repositories on GitHub.

Author Contributions: The co-authors contributed to the research and preparation of this manuscript in the following ways: conceptualization, E.M., D.R., N.P.S., A.R., and E.D.W.; methodology, E.D.W., Y.W., E.M., and D.R.; software, E.D.W., Y.W., E.M., and others listed in acknowledgements; investigation, E.D.W., Y.W., D.R., and E.M.; writing—original draft preparation, E.D.W., Y.W., D.R., E.M. and A.R.; writing—review and editing, D.R., A.R., E.D.W., N.P.S. and E.M.; visualization, E.D.W., Y.W. and A.R.; supervision, D.R., E.M., A.R. and N.P.S.; project administration, N.P.S. and A.R.; funding acquisition, N.P.S.

Funding: This analysis was funded by Environmental Defense Fund (EDF) (grant number 1048-000005-10650; #7557790/#7930492) through support from the Gilder Foundation.

Acknowledgments: The Integrated Compartment Model development and calibration discussed in this work was initially funded by the Louisiana Coastal Protection and Restoration Authority; the authors are indebted to the many researchers and engineers who participated in the 2017 Coastal Master Plan analyses and assisted in those efforts. We would like to acknowledge fellow model developers for the various subroutines of the ICM including (among many others): Alex McCorquodale, Jenneke Visser, Scott Duke-Sylvester, Brady Couvillion and Zhifei Dong.

Conflicts of Interest: The authors declare no conflicts of interest.

References

1. Nicholls, R.J.; Wong, P.P.; Burkett, V.; Codignotto, J.; Hay, J.; McLean, R.; Ragoonaden, S.; Woodroffe, C.D.; Abuodha, P.A.O.; Arblaster, J.; et al. Coastal systems and low-lying areas. In *Climate Change 2007: Impacts, Adaptation and Vulnerability*; Parry, M.L., Canziani, O.F., Palutikof, J.P., van der Linden, P.J., Hanson, C.E., Eds.; Contribution of Working Group II to the Fourth Assessment Report of the Intergovernmental Panel on Climate Change; Cambridge University Press: Cambridge, UK, 2007; pp. 315–356.

2. Wong, P.P.; Losada, I.J.; Gattuso, J.-P.; Hinkel, J.; Khattabi, A.; McInnes, K.L.; Saito, Y.; Sallenger, A. Coastal Systems and Low-Lying Areas. In *Climate Change 2014: Impacts, Adaptation, and Vulnerability*; Part A: Global and Sectoral Aspects. Contribution of Working Group II to the Fifth Assessment Report of the Intergovernmental Panel on Climate Change; Cambridge University Press: Cambridge, UK, 2014; p. 49.

3. Syvitski, J.P.M.; Kettner, A.J.; Overeem, I.; Hutton, E.W.H.; Hannon, M.T.; Brakenridge, G.R.; Day, J.; Vörösmarty, C.; Saito, Y.; Giosan, L.; et al. Sinking deltas due to human activities. *Nat. Geosci.* **2009**, *2*, 681–686. [CrossRef]

4. Coleman, J.M.; Roberts, H.H.; Stone, G.W. Coleman. Mississippi River Delta: An Overview. *J. Coast. Res.* **1998**, *14*, 689–716.

5. Couvillion, B.R.; Beck, H.; Schoolmaster, D.; Fischer, M. *Land Area Change in Coastal Louisiana (1932 to 2016)*; USGS Numbered Series 3381; Geological Survey: Reston, VA, USA, 2017. Available online: http://pubs.er.usgs.gov/publication/sim3381 (accessed on 11 December 2018).

6. Yuill, B.; Lavoie, D.; Reed, D.J. Understanding Subsidence Processes in Coastal Louisiana. *J. Coast. Res.* **2009**, 23–36. [CrossRef]

7. Boesch, D.F.; Josselyn, M.N.; Mehta, A.J.; Morris, J.T.; Nuttle, W.K.; Simenstad, C.A.; Swift, D.J.P. Scientific Assessment of Coastal Wetland Loss, Restoration and Management in Louisiana. *J. Coast. Res.* **1994**, *10*, 1–103.

8. Barry, J.M. *Rising Tide: The Great Mississippi Flood of 1927 and How It Changed America*; Simon and Schuster: New York, NY, USA, 2007.

9. Corthell, E. The delta of the Mississippi River. *Natl. Geogr.* **1897**, *12*, 351–354.

10. Allison, M.A.; Demas, C.R.; Ebersole, B.A.; Kleiss, B.A.; Little, C.D.; Meselhe, E.A.; Powell, N.J.; Pratt, T.C.; Vosburg, B.M. A water and sediment budget for the lower Mississippi—Atchafalaya River in flood years 2008–2010: Implications for sediment discharge to the oceans and coastal restoration in Louisiana. *J. Hydrol.* **2012**, *432*, 84–97. [CrossRef]

11. Nittrouer, J.A.; Best, J.L.; Brantley, C.; Cash, R.W.; Czapiga, M.; Kumar, P.; Parker, G. Mitigating land loss in coastal Louisiana by controlled diversion of Mississippi River sand. *Nat. Geosci.* **2012**, *5*, 534–537. [CrossRef]

12. Kim, W.; Mohrig, D.; Twilley, R.; Paola, C.; Parker, G. Is It Feasible to Build New Land in the Mississippi River Delta? *EOS* **2009**, *90*, 373–374. [CrossRef]

13. Paola, C.; Twilley, R.R.; Edmonds, D.A.; Kim, W.; Mohrig, D.; Parker, G.; Viparelli, E.; Voller, V.R. Natural Processes in Delta Restoration: Application to the Mississippi Delta. *Annu. Rev. Mar. Sci.* **2011**, *3*, 67–91. [CrossRef]

14. Elsey-Quirk, T.; Graham, S.; Mendelssohn, I.; Snedden, G.; Day, J.; Twilley, R.; Shaffer, G.; Sharp, L.; Pahl, J.; Lane, R. Mississippi river sediment diversions and coastal wetland sustainability: Synthesis of responses to freshwater, sediment, and nutrient inputs. *Estuar. Coast. Shelf Sci.* **2019**, *221*, 170–183. [CrossRef]

15. Reed, D.J. Planning for the Future of the Pontchartrain Coast. *J. Coast. Res.* **2009**, *SI54*, 198–205. [CrossRef]

16. Soniat, T.M.; Conzelmann, C.P.; Byrd, J.D.; Roszell, D.P.; Bridevaux, J.L.; Suir, K.J.; Colley, S.B. Predicting the Effects of Proposed Mississippi River Diversions on Oyster Habitat Quality; Application of an Oyster Habitat Suitability Index Model. *J. Shellfish. Res.* **2013**, *32*, 629–638. [CrossRef]

17. Adamack, A.; Stow, C.; Mason, D.; Rozas, L.; Minello, T. Predicting the effects of freshwater diversions on juvenile brown shrimp growth and production: A Bayesian-based approach. *Mar. Ecol. Prog. Ser.* **2012**, *444*, 155–173. [CrossRef]

18. Rozas, L.P.; Minello, T.J. Variation in penaeid shrimp growth rates along an estuarine salinity gradient: Implications for managing river diversions. *J. Exp. Mar. Biol. Ecol.* **2011**, *397*, 196–207. [CrossRef]

19. De Mutsert, K.; Cowan, J.H. A Before–After–Control–Impact Analysis of the Effects of a Mississippi River Freshwater Diversion on Estuarine Nekton in Louisiana, USA. *Chesap. Sci.* **2012**, *35*, 1237–1248. [CrossRef]

20. Coastal Protection and Restoration Authority of Louisiana. *Integrated Ecosystem Restoration and Protection: Louisiana's Comprehensive Master Plan for a Sustainable Coast*; Coastal Protection and Restoration Authority of Louisiana: Baton Rouge, LA, USA, 2007.

21. Coastal Protection and Restoration Authority of Louisiana. *Louisiana's Comprehensive Master Plan for a Sustainable Coast—2012*; Coastal Protection and Restoration Authority of Louisiana: Baton Rouge, LA, USA, 2012.

22. Coastal Protection and Restoration Authority of Louisiana. *Lousiana's Comprehensive Master Plan for a Sustainable Coast—2017*; Coastal Protection and Restoration Authority of Louisiana: Baton Rouge, LA, USA, 2017.

23. White, E.D.; Meselhe, E.; McCorquodale, A.; Couvillion, B.; Dong, Z.; Duke-Sylvester, S.M.; Wang, Y. *2017 Coastal Master Plan: Attachment C3-22: Integrated Compartment Model (ICM) Development*; Coastal Protection and Restoration Authority: Baton Rouge, LA, USA, 2017.

24. Meselhe, E.; McCorquodale, J.A.; Shelden, J.; Dortch, M.S.; Brown, T.S.; Elkan, P.; Rodrigue, M.D.; Schindler, J.K.; Wang, Z. Ecohydrology Component of Louisiana's 2012 Coastal Master Plan: Mass-Balance Compartment Model. *J. Coast. Res.* **2013**, *67*, 16–28. [CrossRef]

25. Visser, J.M.; Duke-Sylvester, S.M. LaVegMod v2: Modeling Coastal Vegetation Dynamics in Response to Proposed Coastal Restoration and Protection Projects in Louisiana, USA. *Sustainability* **2017**, *9*, 1625. [CrossRef]

26. Couvillion, B.R.; Steyer, G.D.; Wang, H.; Beck, H.J.; Rybczyk, J.M. Forecasting the Effects of Coastal Protection and Restoration Projects on Wetland Morphology in Coastal Louisiana under Multiple Environmental Uncertainty Scenarios. *J. Coast. Res.* **2013**, *67*, 29–50. [CrossRef]

27. Folse, T.M.; Sharp, L.A.; West, J.L.; Hymel, M.K.; Troutman, J.P.; McGinnis, T.; Weifenbach, D.; Boshart, W.M.; Rodrigue, L.B.; Richardi, D.C.; et al. *A Standard Operating Procedures Manual for the Coast-wide Reference Monitoring System-Wetlands: Methods for Site Establishment, Data Collection, and Quality Assurance/Quality Control*; Louisiana Coastal Protection and Restoration Authority, Office of Coastal Protection and Restoration: Baton Rouge, LA, USA, 2018; p. 228.

28. Couvillion, B. *2017 Coastal Master Plan Modeling: Attachment C3-27: Landscape Data*; Coastal Protection and Restoration Authority: Baton Rouge, LA, USA, 2017.

29. Patankar, S. *Numerical Heat Transfer and Fluid Flow*; CRC Press: Boca Raton, FL, USA, 1980; p. 214.

30. White, E.D.; Messina, F.; Moss, L.; Meselhe, E. Salinity and Marine Mammal Dynamics in Barataria Basin: Historic Patterns and Modeled Diversion Scenarios. *Water* **2018**, *10*, 1015. [CrossRef]

31. White, E.D.; Reed, D.J.; Meselhe, E.A. Modeled Sediment Availability, Deposition, and Decadal Land Change in Coastal Louisiana Marshes under Future Relative Sea Level Rise Scenarios. *Wetlands* **2019**. [CrossRef]

32. Baustian, M.M.; Clark, F.R.; Jerabek, A.S.; Wang, Y.; Bienn, H.C.; White, E.D. Modeling current and future freshwater inflow needs of a subtropical estuary to manage and maintain forested wetland ecological conditions. *Ecol. Indic.* **2018**, *85*, 791–807. [CrossRef]

33. Reed, D.; Meselhe, E.; White, E.; Wang, Y. Wetland transitions and loss in coastal Louisiana under scenarios of future relative sea-level rise. *Geomorphology* **2019**. under review.

34. Hijuelos, A.C.; Sable, S.E.; O'Connell, A.M.; Geaghan, J.P.; Lindquist, D.C.; White, E.D. Application of Species Distribution Models to Identify Estuarine Hot Spots for Juvenile Nekton. *Estuaries Coasts* **2017**, *40*, 1183–1194. [CrossRef]

35. Meselhe, E.; White, E.D.; Reed, D.J. *2017 Coastal Master Plan: Appendix C: Modeling Chapter 2–Future Scenarios*; Coastal Protection and Restoration Authority: Baton Rouge, LA, USA, 2017.

36. Brown, S.; Couvillion, B.; Dong, Z.; Meselhe, E.; Visser, J.; Wang, Y.; White, E. *2017 Coastal Master Plan: Attachment C3-23: ICM Calibration, Validation, and Performance Assessment*; Coastal Protection and Restoration Authority: Baton Rouge, LA, USA, 2017.

37. McCorquodale, A.; Couvillion, B.; Dortch, M.S.; Freeman, A.; Meselhe, E.; Reed, D.; Roth, B.; Shelden, J.; Wang, H.; White, E. *2017 Coastal Master Plan: Appendix C: Attachment C3-1: Sediment Distribution*; Coastal Protection and Restoration Authority: Baton Rouge, LA, USA, 2017.

38. Allison, M.A.; Chen, Q.J.; Couvillion, B.; Leadon, M.; McCorquodale, A.; Meselhe, E.; Ramatchandirane, C.; Reed, D.J.; White, E.D. *2017 Coastal Master Plan: Model Improvement Plan, Attachment C3-2: Marsh Edge Erosion*; Coastal Protection and Restoration Authority: Baton Rouge, LA, USA, 2017.

39. Pahl, J. *2017 Coastal Master Plan: Attachment C-2: Eustatic Sea Level Rise*; Coastal Protection and Restoration Authority: Baton Rouge, LA, USA, 2017.

40. Brown, S. *2017 Coastal Master Plan Modeling: Attachment C3-26: Hydrology and Water Quality Boundary Conditions*; Version Final; Coastal Protection and Restoration Authority: Baton Rouge, LA, USA, 2017.

41. Meselhe, E.; White, E.; Wang, Y. *2017 Coastal Master Plan: Attachment C3-24: Integrated Compartment Model Uncertainty Analysis*; Coastal Protection and Restoration Authority: Baton Rouge, LA, USA, 2017.

42. Reed, D.; Yuill, B. *2017 Coastal Master Plan: Attachment C2-2: Subsidence*; Coastal Protection and Restoration Authority: Baton Rouge, LA, USA, 2017.

43. Allison, M.A.; Meselhe, E.A. The use of large water and sediment diversions in the lower Mississippi River (Louisiana) for coastal restoration. *J. Hydrol.* **2010**, *387*, 346–360. [CrossRef]

44. Baustian, M.M.; Meselhe, E.; Jung, H.; Sadid, K.; Duke-Sylvester, S.M.; Visser, J.M.; Allison, M.A.; Moss, L.C.; Ramatchandirane, C.; Van Maren, D.S.; et al. Development of an Integrated Biophysical Model to represent morphological and ecological processes in a changing deltaic and coastal ecosystem. *Environ. Model. Softw.* **2018**, *109*, 402–419. [CrossRef]

45. Xu, K.; Bentley, S.J.; Day, J.W.; Freeman, A.M. A review of sediment diversion in the Mississippi River Deltaic Plain. *Estuar. Coast. Shelf Sci.* **2019**, *225*, 106241. [CrossRef]

46. Gaweesh, A.; Meselhe, E. Evaluation of Sediment Diversion Design Attributes and Their Impact on the Capture Efficiency. *J. Hydraul. Eng.* **2016**, *142*, 04016002. [CrossRef]

47. Allison, M.A.; Ramirez, M.T.; Meselhe, E.A. Diversion of Mississippi River Water Downstream of New Orleans, Louisiana, USA to Maximize Sediment Capture and Ameliorate Coastal Land Loss. *Water Resour. Manag.* **2014**, *28*, 4113–4126. [CrossRef]

48. McKee, K.L.; Cahoon, D.R.; Feller, I.C. Caribbean mangroves adjust to rising sea level through biotic controls on change in soil elevation. *Glob. Ecol. Biogeogr.* **2007**, *16*, 545–556. [CrossRef]

49. Morris, J.T.; Sundareshwar, P.V.; Nietch, C.T.; Kjerfve, B.; Cahoon, D.R. Responses of coastal wetlands to rising sea level. *Ecology* **2002**, *83*, 2869–2877. [CrossRef]

50. Craft, C.; Clough, J.; Ehman, J.; Joye, S.; Park, R.; Pennings, S.; Guo, H.; Machmuller, M. Forecasting the effects of accelerated sea-level rise on tidal marsh ecosystem services. *Front. Ecol. Environ.* **2009**, *7*, 73–78. [CrossRef]

51. Reyes, E.; Georgiou, I.; Reed, D.; McCorquodale, A. Using Models to Evaluate the Effects of Barrier Islands on Estuarine Hydrodynamics and Habitats: A Numerical Experiment. *J. Coast. Res.* **2005**, *SI44*, 176–185.

52. Munoz, S.E.; Giosan, L.; Therrell, M.D.; Remo, J.W.F.; Shen, Z.; Sullivan, R.M.; Wiman, C.; O'Donnell, M.; Donnelly, J.P. Climatic control of Mississippi River flood hazard amplified by river engineering. *Nature* **2018**, *556*, 95–98. [CrossRef] [PubMed]

53. Krysanova, V.; Vetter, T.; Eisner, S.; Huang, S.; Pechlivanidis, I.; Strauch, M.; Gelfan, A.; Kumar, R.; Aich, V.; Arheimer, B.; et al. Intercomparison of regional-scale hydrological models and climate change impacts projected for 12 large river basins worldwide—A synthesis. *Environ. Res. Lett.* **2017**, *12*, 105002. [CrossRef]

54. La Peyre, M.K.; Geaghan, J.; Decossas, G.; La Peyre, J.F.; Peyre, L. Analysis of Environmental Factors Influencing Salinity Patterns, Oyster Growth, and Mortality in Lower Breton Sound Estuary, Louisiana, Using 20 Years of Data. *J. Coast. Res.* **2016**, *319*, 519–530. [CrossRef]

55. Lellis-Dibble, K.A.; McGlynn, K.E.; Bigford, T.E. *Estuarine Fish and Shellfish Species in U.S. Commercial and Recreational Fisheries: Economic Value as an Incentive to Protect and Restore Estuarine Habitat*; NOAA Tech, Memo, NMFSF/SPO-90; U.S. Dep Commerce.: Silver Spring, MD, USA, 2008; pp. 1–94.

56. Chen, W.; Chen, K.; Kuang, C.; Zhu, D.Z.; He, L.; Mao, X.; Liang, H.; Song, H. Influence of sea level rise on saline water intrusion in the Yangtze River Estuary, China. *Appl. Ocean Res.* **2016**, *54*, 12–25. [CrossRef]

57. Cloern, J.E.; Abreu, P.C.; Carstensen, J.; Chauvaud, L.; Elmgren, R.; Grall, J.; Greening, H.S.; Johansson, J.O.R.; Kahru, M.; Sherwood, E.T.; et al. Human activities and climate variability drive fast-paced change across the world's estuarine–coastal ecosystems. *Glob. Chang. Biol.* **2016**, *22*, 513–529. [CrossRef]

58. Yang, Z.; Wang, T.; Voisin, N.; Copping, A. Estuarine response to river flow and sea-level rise under future climate change and human development. *Estuar. Coast. Shelf Sci.* **2015**, *156*, 19–30. [CrossRef]

59. Cloern, J.E.; Kimmerer, W.; Moyle, P.B.; Mueller–Solger, A.; Moyle, P.B.; Kay, J. Water Wasted to the Sea? *SFEWS* **2017**, *15*. [CrossRef]

60. Montagna, P.A.; Alber, M.; Doering, P.; Connor, M.S. Freshwater inflow: Science, policy, management. *Estuaries* **2002**, *25*, 1243–1245. [CrossRef]

61. Peyronnin, N.S.; Caffey, R.H.; Cowan, J.H.; Justic, D.; Kolker, A.S.; Laska, S.B.; McCorquodale, A.; Melancon, E.; Nyman, J.A.; Twilley, R.R.; et al. Optimizing Sediment Diversion Operations: Working Group Recommendations for Integrating Complex Ecological and Social Landscape Interactions. *Water* **2017**, *9*, 368. [CrossRef]

62. Mutsert, K.D.; Cowan, J.H., Jr.; Walters, C.J. Using Ecopath with Ecosim to Explore Nekton Community Response to Freshwater Diversion into a Louisiana Estuary. *Mar. Coast. Fish.* **2012**, *4*, 104–116. [CrossRef]

63. Rose, K.A.; Huang, H.; Justic, D.; De Mutsert, K. Simulating Fish Movement Responses to and Potential Salinity Stress from Large-Scale River Diversions. *Mar. Coast. Fish.* **2014**, *6*, 43–61. [CrossRef]

64. Piazza, B.P.; La Peyre, M.K. Nekton community response to a large-scale Mississippi River discharge: Examining spatial and temporal response to river management. *Estuar. Coast. Shelf Sci.* **2011**, *91*, 379–387. [CrossRef]

65. De Mutsert, K.; Lewis, K.; Milroy, S.; Buszowski, J.; Steenbeek, J. Using ecosystem modeling to evaluate trade-offs in coastal management: Effects of large-scale river diversions on fish and fisheries. *Ecol. Model.* **2017**, *360*, 14–26. [CrossRef]

66. Bargu, S.; Justic, D.; White, J.R.; Lane, R.; Day, J.; Paerl, H.; Raynie, R. Mississippi River diversions and phytoplankton dynamics in deltaic Gulf of Mexico estuaries: A review. *Estuar. Coast. Shelf Sci.* **2019**, *221*, 39–52. [CrossRef]

67. Christensen, J.D.; Monaco, M.E.; Lowery, T.A. An Index to Assess the Sensitivity of Gulf of Mexico Species to Changes in Estuarine Salinity Regimes. *Gulf Res. Rep.* **1997**, *9*, 219–229. [CrossRef]

68. Peterson, M.S.; Meador, M.R. Effects of salinity on freshwater fishes in coastal plain drainages in the southeastern U.S. *Rev. Fish. Sci.* **1994**, *2*, 95–121. [CrossRef]

69. Purcell, K.M.; Hitch, A.T.; Klerks, P.L.; Leberg, P.L. Adaptation as a potential response to sea-level rise: A genetic basis for salinity tolerance in populations of a coastal marsh fish. *Evol. Appl.* **2008**, *1*, 155–160. [CrossRef]

70. Day, J.W.; Lane, R.R.; D'Elia, C.F.; Wiegman, A.R.; Rutherford, J.S.; Shaffer, G.P.; Brantley, C.G.; Kemp, G.P.; D'Elia, C.F. Large infrequently operated river diversions for Mississippi delta restoration. *Estuar. Coast. Shelf Sci.* **2016**, *183*, 292–303. [CrossRef]

71. Yuill, B.T.; Khadka, A.K.; Pereira, J.; Allison, M.A.; Meselhe, E.A. Morphodynamics of the erosional phase of crevasse-splay evolution and implications for river sediment diversion function. *Geomorphology* **2016**, *259*, 12–29. [CrossRef]

72. Cahoon, D.R.; White, D.A.; Lynch, J.C. Sediment infilling and wetland formation dynamics in an active crevasse splay of the Mississippi River delta. *Geomorphology* **2011**, *131*, 57–68. [CrossRef]

73. Mossa, J. Sediment dynamics in the lowermost Mississippi River. *Eng. Geol.* **1996**, *45*, 457–479. [CrossRef]

Article

Natural Variability and Vertical Land Motion Contributions in the Mediterranean Sea-Level Records over the Last Two Centuries and Projections for 2100

Antonio Vecchio [1,2], **Marco Anzidei [3,*]**, **Enrico Serpelloni [4]** and **Fabio Florindo [3]**

[1]   Radboud Radio Lab, Department of Astrophysics/IMAPP-Radboud University, P.O. Box 9010, 6500GL Nijmegen, The Netherlands

[2]   LESIA, Observatoire de Paris, Université PSL, CNRS, Sorbonne Université, Université de Paris, 5 place Jules Janssen, 92195 Meudon, France

[3]   Istituto Nazionale di Geofisica e Vulcanologia, 00143 Rome, Italy

[4]   Istituto Nazionale di Geofisica e Vulcanologia, 40100 Bologna, Italy

*   Correspondence: marco.anzidei@ingv.it; Tel.: +39-06-5186-0214

Received: 14 June 2019; Accepted: 5 July 2019; Published: 16 July 2019

Abstract: We analyzed a set of geodetic data to investigate the contribution of local factors, namely the sea level natural variability (SLNV) and the vertical land motion (VLM), to the sea-level trend. The SLNV is analyzed through the Empirical Mode Decomposition (EMD) on tidal data (>60 years of recordings) and results are used to evaluate its effects on sea levels. The VLM is measured at a set of continuous GPS (cGPS) stations (>5 years of recordings), located nearby the tide gauges. By combining VLM and SLNV with IPCC-AR5 regional projections of climatic data (Representative Concentration Pathways (RCP) 2.6 and 8.5), we provide relative sea-level rise projections by 2100. Results show that the combined effects of SLNV and VLM are not negligible, contributing between 15% and 65% to the sea-level variability. Expected sea levels for 2100 in the RCP8.5 scenario are between 475 ± 203 (Bakar) and 818 ± 250 mm (Venice). In the Venice Lagoon, the mean land subsidence at 3.3 ± 0.85 mm a^{-1} (locally up to 8.45 ± 1.69 mm a^{-1}) is driving the local sea-level rise acceleration.

Keywords: sea-level rise; Mediterranean Sea; tide gauges; natural variability; vertical land motion

---

1. Introduction

In situ and remote sensing data show that the rate of global sea-level rise over the past two centuries has increased at faster rates than in the last two or three millennia [1–4]. Satellite radar altimetry and tide gauge data indicate a rise of the global mean sea-level rate at 1.7 mm a^{-1} and 3.2 mm a^{-1} for the 20th century and the last two decades, respectively [5–7]. The Mediterranean Sea has been rising at the mean rate of ~1.8 mm a^{-1} in the last two-three centuries [8–12]. At global scales, the sea-level change is described as the sum of eustatic, glacio-hydro-isostatic, and land-hydrology components [13], which are independently modeled to provide future sea-level predictions. Local sea-level change can significantly differ from mean global sea-level rise because local factors become relevant and their contribution adds up to the global components mentioned above [14]. Oceanographic and climatic effects, glacial isostatic adjustment (GIA), and continuous vertical land motion, especially coastal subsidence due to tectonics (volcanic included) or anthropic origins, are some of the factors that mainly affect sea-level changes at local scales [15]. Therefore, depending on the geographical location they cause spatial variability in the sea-level change pattern. While local effects have little influence at global scale, since they are smoothed out by averaging operations, they become critical in adaptation planning and risk management, where localized assessments are needed.

At time scales of years or decades, oceanographic and climatic factors drive non-stationary sea-level fluctuations, the so-called sea-level natural variability, whose assessment is important for future sea-level projections since their contribution is not negligible [16–26], (http://www.ipcc.ch). For example, in the Indian Ocean the effects of the natural variability, mainly driven by the decadal El Niño–Southern Oscillation (ENSO), Indian Ocean Dipole (IOD) and monsoon, dominate the external forcing contributing up 85% of the observed trend [27]. In the Mediterranean basin, the glacio-isostatic contribution, acting from the Last Glacial Maximum, has a predictable pattern that has been estimated and compared with direct observational data in deforming zones [4,9,15]. The vertical land motion due to natural (tectonics, isostasy and ground compaction) [4,9,13], and/or anthropogenic factors (fluid extraction) [28,29], produces changes in the shape of the basins and changes in the height of the land with respect to the sea (see e.g., [9,13] and references therein). Although this effect does not directly change the quantity of water and its physical properties it results in a net increase (in case of subsidence) or decrease (in case of uplift) of the sea level, when evaluated in restricted geographical areas (as for example the punctual measurements by tide gauges), since the land is moving. If the vertical land motion operates over time it locally results in a net sea-level rate that adds up to the rate induced by GIA, thus affecting the local sea-level change. Tide gauge records are strongly sensitive to the vertical land motion since these instruments are tied to land. Although the role and rates of the main contributors to the sea-level rise are still debated [30], it is a matter of fact that most of the coasts of the world are undergoing the effects of the continuing sea-level rise. The vertical motion due to tectonic effects is strongly variable in the Mediterranean region (as shown in [9] and references therein) and in some areas it can strongly enhance the effect of the sea-level rise. In particular, along subsiding coasts, the rising sea is expected to flood in advance, with respect to stable zones, over the coming decades [1,15,31].

In this framework, the Northern Mediterranean region, being characterized by coastal plains with low elevation (even placed below sea level as some areas of the Adriatic Sea) is therefore prone to marine flooding [9,32–34], (www.savemedcoasts.eu), and it represents an interesting case study. Several historical maritime densely-inhabited cities (i.e., Venice and Trieste) and valuable environmental and industrial areas (i.e., Genoa and Marseille) are located in this area. Therefore, detailed studies of local sea-level variations and future projections are relevant to support marine flooding scenarios for a cognizant coastal management.

This study aims to characterize the effect of local factors, such as the sea level natural variability (SLNV) and vertical land motion (VLM), on the sea-level changes at nine locations in the Northern Mediterranean Sea (Figure 1). The SLNV was inferred from tide gauge data through the Empirical Mode Decomposition (EMD) while VLM was measured by means of continuous GPS (cGPS) data. Our approach to estimate SLNV is comprised of two steps: (i) the EMD is applied to tide gauge data to characterize and isolate the natural variability; (ii) a simple theoretical model, based on EMD results, is developed to reproduce the observed SLNV. By combining the SLNV and VLM contributions with the local time-dependent trend based on regional IPCC- fifth Assessment Report (AR5) estimations, the expected sea levels for 2050 and 2100 at each tide gauge location were estimated.

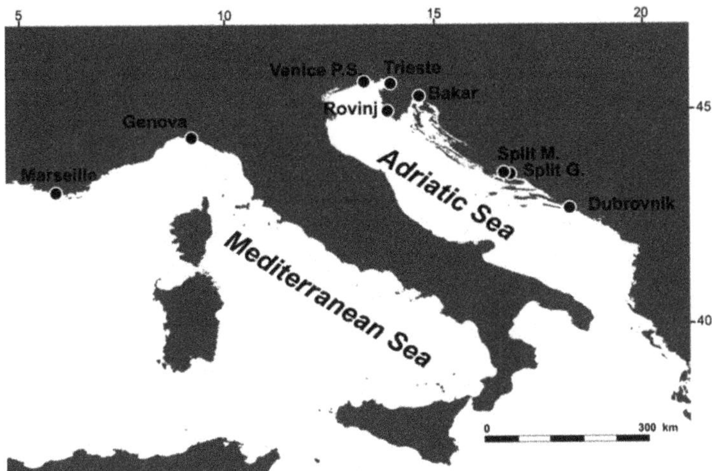

Figure 1. Location of the nine analyzed tidal stations: Marseille (5.35°–43.32°, 1885–2012), Trieste (13.76°–45.57°, 1875–2012), Genova (9.92°–44.41°, 1884–1997), Venice Punta Salute (12.32°–45.39°, 1909–2000), Bakar (14.5°–45.30°, 1930–2011), Dubrovnik (19.13°–42.65°, 1956–2009), Rovinj (13.64°–45.06°, 1955–2011), Split RT Marjana (16.39°–43.51°, 1952–2011), Split Gradska Luka (16.44°–43.51°, 1954–2011). Coordinates of the station and time span of recordings are indicated in parentheses.

2. Materials and Methods: Data Set and Natural Variability

In this study we analyzed the longest available monthly tide gauge records (>60 years of Revised Local Reference (RLR) data collected in the 1888–2008 time span) at nine tide gauge stations located in the Mediterranean Sea, belonging to the Permanent Service for Mean Sea Level (www.psmsl.org). The location of the stations, the time span of the records and the analyzed time series are shown in Figures 1 and 2. The choice of data sets with length exceeding 60 years aims to avoid the effect of inter-decadal variations affecting tidal recordings shorter than ~50 years [35,36]. In addition, to avoid the introduction of artificial signals in the records, interpolations on data gaps were not applied.

The identification of sea-level trends and large timescales variability from sea-level records is a debated argument and the best methods to discriminate between acceleration and intrinsic climate variability remains an open issue. Indeed, since in tide gauge data the long-term trend is embedded in fluctuations at many different time scales, from daily tides to the seasonal cycle as well as interannual and decadal variations, the filtering method used to remove these variations may impact the accuracy of the calculated trend. Different approaches to this problem have been proposed (for a detailed review see [37]). Among these approaches, the EMD has been used to characterize the intrinsic fluctuations and the long-term trend in sea-level data [38–45]. This technique has been developed to analyze non-stationary and non-linear data, such as tide gauge records, by decomposing the signal in modes that provide a description of the intrinsic timescales present in the time series. Being "adaptive,," the modes are empirically obtained from the original time series according to its local characteristics. The EMD reveals the proper components of the phenomenon under study better than other methods, such as Fourier analysis, whose modes with predefined functional form are often far from being eigenfunctions of the phenomenon at hand. This makes the EMD very powerful for identifying and isolating, for example, deformed wave profiles, associated with nonlinearities in the data, preserving their local properties. For the applications of this paper, two main reasons guided us in using the EMD. Firstly, the sea level natural variability is described by a few numbers of modes with a quite simple functional form (oscillating functions). This will be exploited in the following for building up a simple low dimensional mode for sea-level estimations. Secondly, the EMD does not require a predetermined timescale to carry out the filtering operation (as it happens e.g., in moving averages commonly used as low-pass filter to recover the large timescale variability in sea-level records): this has an important rational basis, since in a nonstationary process the local timescale is unknown a priori.

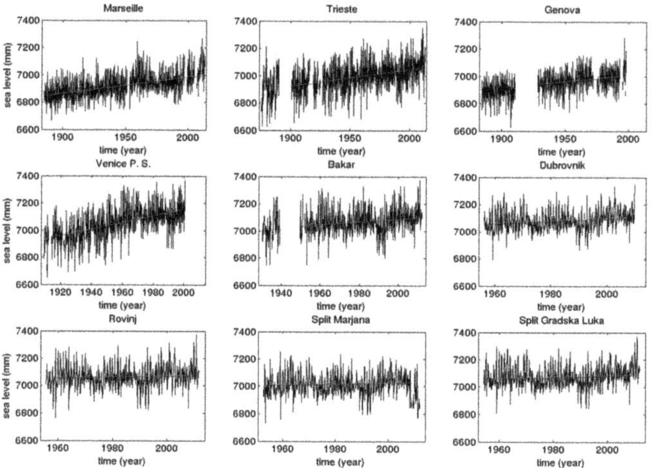

Figure 2. Time series of the monthly mean tidal records from stations in Figure 1 (data from www.psmsl.org). Red lines show linear regressions on the data (see the text for details).

In the EMD framework the measured sea level $L(t)$ is decomposed into a finite number n of band-limited oscillating components (called intrinsic mode functions, IMFs) $L_j(t)$, whose functional shape is not fixed a priori, but obtained from the data, and a residue $r_n(t)$ providing the nonlinear trend [40]:

$$L\,(t) = \sum\nolimits_{j=0}^{n} L_j(t) + r_n\,(t). \tag{1}$$

The procedure of IMF extraction consists of three steps: (i) local maxima and minima of $L(t)$ are identified; (ii) these are connected through a cubic spline identifying the lower and upper envelope; (iii) the signal $h(t)$, the difference between $L(t)$ and the mean of lower and upper envelopes, are calculated. The signal $h(t)$ represents the first IMF, $L_j(t)$, if it has exactly one zero crossing between two consecutive local extrema and its mean is zero. The first condition ensures a global narrowband requirement, namely periods which are too different are not mixed together into an IMF; the second requirement ensures getting oscillations about the zero level. Both these conditions ensure that $L_j(t)$ is an oscillating function with time-dependent amplitude and phase, namely it can be expressed as $L_j(t) = B_j(t)\cos[\phi_j(t)]$. If none of these conditions are fulfilled, then step (iii) is repeated by using $h(t)$ as input data. This procedure is iterated until the resulting signal fulfills the IMF properties and it is repeated to find the next IMF by using $L(t) - L_j(t)$ as new input data. When no more modes can be extracted, the difference between the original $L(t)$ and the sum of all IMFs results in the residue $r_n(t)$. Technical details on the procedure for IMF extraction are reported in [41]. The EMD allows calculating a meaningful instantaneous phase for each IMF by using the Hilbert transform

$$L_j^*\,(t) = \pi^{-1} \times P \times \int_{-\infty}^{\infty} \frac{L_j\,(t')}{t - t'} \times dt' \tag{2}$$

where P denotes the Cauchy principal value and L_j^* represents the complex conjugate of $L(t)$ as $\phi_j\,(t) = $ arctan (L_j^*/L_j). Finally, the instantaneous frequency, $\omega_j(t)$, is given by $d\phi_j(t)/dt$. For each $L_j(t)$ we can define a characteristic timescale τ_j, as the time average $<2\pi/\omega_j(t)>$, that represents an average period of the IMF and provides an estimate of the timescale characterizing the EMD mode. We remark that τ_j is not to be intended as the Fourier one. It just gives an indication of the timescale characterizing the EMD mode for which it is computed, although many modes with different average periods may contribute to the variability of the actual signal at a particular timescale. The IMF statistical significance with respect to white noise can be evaluated through a test based on the comparison between IMF amplitude from the real signal and from a white noise series [46] at different confidence levels. Since IMFs are a local, complete and orthogonal set, the EMD is useful to filter raw signal through partial sums in Equation (1). Moreover, since a low number of modes are produced, this property can be exploited to build up low order theoretical models to describe the dynamics of the investigated systems.

3. Results: The EMD Analysis and a Generalized Model for Long-Term Sea-Level Variations

When the EMD is used to analyze the nine tide gauge time series, the number of IMFs, n, varies between 10 and 15, depending on the station. This is mainly related to the different lengths of the time series which vary between 137 and 53 years: for the longest time series more components of the natural variability are recovered.

Let S be the set of IMF with timescale >15 years (IMFs timescales τ_j are shown in Table 1) and significant with respect to a white noise at the 90th significance level. The long-term variations of the measured sea level, L_{LT}, can be defined as:

$$L_{LT}(t) = \sum_{j \in S} L_j(t) + r_n(t), \tag{3}$$

where IMFs in S and the residue are added together.

Table 1. Characteristic periods (years) of the significant intrinsic mode functions (IMFs) with a characteristic timescale >15 years, obtained by the Empirical Mode Decomposition (EMD) for the analyzed data set. Errors are estimated as the standard deviation.

Tide Gauge Station	n	$j = 7$	$j = 8$	$j = 9$	$j = 10$	$j = 11$	$j = 12$	$j = 13$
Marseille	15				29.8 ± 14.5	36.5 ± 7.4	43.1 ± 15.6	63.1 ± 21.2
Trieste	15			19.8 ± 4.4	27.0 ± 3.6	37.6 ± 8.8	52.2 ± 12.7	76.3 ± 45.9
Genova	10	15.3 ± 2.5	32.2 ± 2.1					
Venice P.S.	12		16.0 ± 2.5	21.0 ± 2.1	44.3 ± 5.0			
Bakar	12		16.9 ± 2.8	29.4 ± 7.7	58.0 ± 4.9			
Dubrovnik	11		17.0 ± 2.5	49.4 ± 19.2				
Rovinj	11		16.2 ± 5.6	36.6 ± 8.3				
Split M.	13			18.6 ± 3.4	21.7 ± 4.5	33.7 ± 25.9		
Split G.	11	17.8 ± 3.7	49.6 ± 16.4					

The choice of a 15-year threshold follows the approach used in other studies to remove the short-term variability from sea-level data [1,18]. We remark that, as pointed out by [45], the results of single IMFs from tide gauge data need of a careful interpretation. However, a discussion about the physical meaning of the single IMF is out of the scope of this paper since we use the EMD as a low-pass filter by adding up several IMFs. In this case, the EMD has shown to be a useful tool to reduce the impact of high-frequency variability and noise [45].

The contribution L_{LT}, for the analyzed tide gauges, is shown in Figure 3. All records show a long-term variability in a well-defined range of timescales of about 20–30 and, for the longer records, >45 years, in the typical ranges where internal modes of natural variability are detected [47]. Observational datasets and simulations of the typical scales of Atlantic internal modes of natural variability show variations at decadal timescales that are linked to the ocean-atmosphere coupling [47–50]. In particular, the 20–30 year variability has been related to the Atlantic Meridional Overturning Circulation (MOC), while the >45 year fluctuations are plausibly due to salinity and matter exchange processes between the Atlantic and Arctic Ocean causing the Atlantic Multidecadal Oscillation (AMO) pattern. Both patterns are quasi-persistent and present up to 8000 years ago [51,52]. The similarity of timescales we found in the tide gauge data with those of the Atlantic modes of variability suggests a connection between climate variability and sea-level variations in the Mediterranean Sea as also indicated by [53,54]. The combined analysis of global oceanic circulation, climate models and observations, indeed, indicates that the main physical mechanism driving the connection between sea level in the Mediterranean Sea and internal modes of natural variability is the mass exchange through the Strait of Gibraltar [53]. This mechanism generates a coherent and uniform signal in the whole Mediterranean at decadal time scales [54,55]. The detected sea-level variations at 20–30 years can be thus related to the combined effect of the North Atlantic Oscillations (NAO) and AMO [26]. An additional contribution could come from the Inter-decadal Pacific Oscillation (IPO), causing variations of the global surface temperature at these timescales, which affects also the climate dynamics in the Atlantic Ocean [56,57]. Likewise, major volcanic eruptions, affecting the global temperature by acting on the radiative forcing [57,58], could also contribute to the SLNV. Fluctuations at longer timescales can be due to both atmospheric forcing [59] and exchange processes [47].

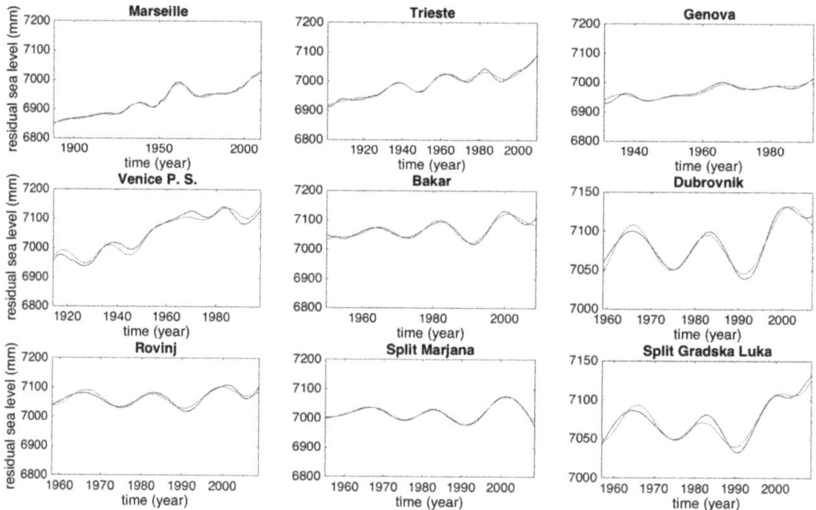

Figure 3. Function $L_{LT}(t)$ (black line), obtained by summing up IMFs at timescales >15 years and the residue, and the best-fit function $L_M(t)$ (red line) described by Equation (4).

A first check of the EMD results can be performed by comparing the SLNV obtained for the four closest stations in the North Adriatic sea, namely Venice, Trieste, Bakar, and Rovinij that are separated by a minimum and maximum distance of ~65 and ~150 km, respectively. Figure 4 shows a comparison of the EMD reconstructions, through IMF with timescale >15 years (the residue $r_n(t)$ has not been added up), for these records. For the four stations the variability is very similar with maxima and minima almost in phase. Small differences can be due to border effects for shorter series (e.g., Rovinij) and/or presence of data gaps affecting the IMF calculation or natural reasons such as small local differences in the signal variability. We note that the Venice station shows a different behavior with a flatter signal between 1945 and 1980 due to temporary anthropogenic subsidence induced by groundwater [28].

Results show that, for all the stations, the SLNV is described by a small number of EMD modes (from 3 to 6). Since IMFs are oscillating functions, we modelled the long-term behavior of the tide gauge signal as the superposition of a linear trend and the sum of oscillating functions of given amplitude, frequency and phase:

$$L_M(t) = (rt + c) + \sum_{i=j}^{k} A_j \times \cos[\Omega_j t + \Phi_j]. \tag{4}$$

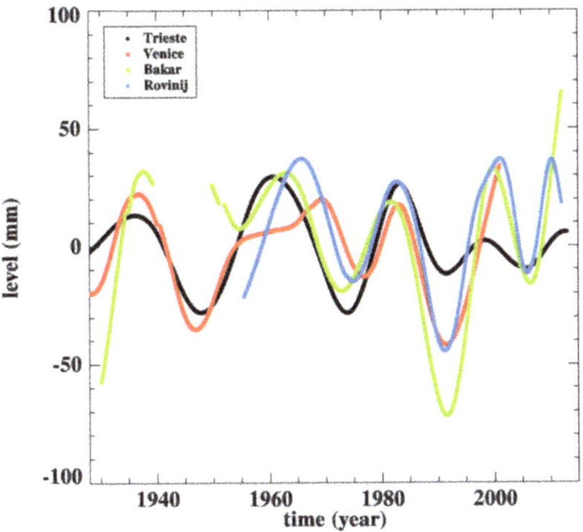

Figure 4. EMD reconstruction by using IMFs with timescale >15 years (the residue $r_n(t)$ has not been added up) for the four closest stations in the North Adriatic Sea.

In Equation (4) the trend is linear, and the long-term fluctuations are accounted by oscillating functions whose frequency Ω_j is the time average of the instantaneous frequency of IMFs in S. This represents the most elementary approximation to reproduce the long-term variability inferred by the EMD through modes with time-dependent amplitude and phase. The free parameters in Equation (4) (rate r, intercept c, A_j and Φ_j) are calculated by fitting $L_M(t)$ to $L_{LT}(t)$. To avoid possible end effects due to the procedure of IMF extraction and the computation of derivative at borders, we cut three years of data at both boundaries of $L_{LT}(t)$ before performing the fit. In addition, since data from Trieste, Genoa, Venice, and Bakar show gaps in the initial part of their records, for these stations the fit starts from year 1901, 1931, 1914, and 1949, respectively, when recordings begin to be continuous for many decades. Two sources of uncertainty were taken into account. Firstly, the uncertainty due to the use of constant frequencies, instead of time varying ones, is evaluated by building 1000 $L_M(t)$ realizations whose Ω_j are randomly extracted from a Gaussian distribution with mean $< \omega_j(t) >$ and standard deviation σ $[\omega_j(t)]$. Secondly, effects of possible undetected modes of climate variability, with periods of the order and/or longer than the record lengths, are taken into account as an increase of the estimated uncertainty. Indeed, since these very long period modes are not included in the model, the accuracy for estimating the sea level is reduced. This additional uncertainty is accounted for by adding to each realization $L_M(t)$ a normally distributed zero average white noise. Since we expect that modes with periods of the order and/or longer than the record lengths are settled on the EMD nonlinear residue, the standard deviation of the white noise signal was taken as the standard deviation of the EMD residue from which we subtracted a linear trend.

Figure 3 shows the best solutions obtained by fitting $L_M(t)$ to $L_{LT}(t)$. Fit parameters and adjusted R-square statistics are shown in Table 2. R-square values close to 1 indicate that a good fit is achieved for the analyzed records. Rates r obtained by a fit through Equation (4) are compatible or slightly higher than rates estimated by a linear fit to the raw data for the same time intervals shown in Figure 2 and varying between 0.35 ± 0.38 mm a^{-1} (Split M.) and 2.43 ± 0.23 mm a^{-1} (Venice P.S.). Only for Split Gradska Luka, the linear rate from Equation (4) is higher than the linear fit. This is possibly due to a steeper sea-level rise observed in L_{LT} at this station after 2006. In the raw data, this feature is hidden by short-term fluctuations, thus resulting in a lower rate when a linear fit is computed on the original time series. The origin of this steepening has been proposed to be the coincidence of AMO-NAO phase opposition and warm AMO phase [60]. The same feature is also present in L_{LT} for the Trieste and Rovinj data. However, for Trieste the linear rate from Equation (4) does not show significant differences with respect to the rate of the linear fit likely due to the longer duration of recordings for which the fitting procedure is very robust and not affected by variations in the last part of the data.

Table 2. Fit parameters and adjusted R-square values for the analyzed data set. Uncertainties define the 90% confidence limits.

Tide Gauge Station	Linear Fit (mm a^{-1})	r (mm a^{-1})	c (mm)	A (mm)	Φ	R^2
Marseille	1.22 ± 0.10	1.28 ± 0.01	4436 ± 12	10.1 ± 0.3 11.1 ± 0.3 16.1 ± 0.5 18.7 ± 0.5	4.03 ± 0.30 -0.84 ± 0.03 1.84 ± 0.04 -0.42 ± 1.54	0.9948
Trieste	1.22 ± 0.15	1.59 ± 0.05	3874 ± 86	10.8 ± 0.4 13.2 ± 0.8 10.3 ± 1.7 19.1 ± 2.9 21.9 ± 2.0	2.54 ± 2.11 2.34 ± 0.05 4.87 ± 0.73 -3.04 ± 0.13 2.52 ± 0.05	0.9844
Genova	1.07 ± 0.24	1.20 ± 0.03	4618 ± 54	7.8 ± 0.7 12.1 ± 0.6	-1.93 ± 0.08 2.58 ± 0.24	0.9254
Venice P.S.	2.43 ± 0.23	2.78 ± 0.04	1615 ± 74	13.1 ± 1.1 18.8 ± 1.3 23.8 ± 1.3	-0.61 ± 0.09 -2.03 ± 0.07 2.10 ± 0.05	0.9579
Bakar	0.9 ± 0.37	0.88 ± 0.15	5323 ± 306	21.1 ± 1.3 21.9 ± 2.2 12.1 ± 0.7	3.36 ± 0.89 -1.63 ± 0.11 -0.66 ± 2.36	0.9515
Dubrovnik	0.90 ± 0.44	1.24 ± 0.05	4520 ± 104	27.4 ± 0.6 20.6 ± 0.8	2.41 ± 0.02 -1.74 ± 0.04	0.9662
Rovinj	0.38 ± 0.43	0.97 ± 0.10	5144 ± 190	26.5 ± 0.9 20.3 ± 1.2	-0.45 ± 0.03 1.76 ± 0.09	0.9014
Split M.	0.35 ± 0.38	0.17 ± 0.02	6671 ± 29	18.8 ± 0.3 26.8 ± 0.4 26.4 ± 0.4	2.62 ± 0.02 -1.84 ± 0.01 1.35 ± 0.01	0.9888
Split G. L.	0.51 ± 0.40	1.87 ± 0.07	3360 ± 137	17.1 ± 0.7 35.1 ± 0.1	0.68 ± 0.04 1.48 ± 0.03	0.9434

4. Discussion: The Vertical Land Motion and Relative Sea Levels by 2050 and 2100

The VLM trends were derived from the processing of raw data collected by regional continuous GPS networks operating in the Euro-Mediterranean region for more than 7 years [61] and located near the tide gauges. The geodetic rates of VLM were obtained from the analysis of position-times series, following the procedures described in [62], according to the IGS recommendations (http://acc.igs.org/reprocess2), to produce solutions with the most updated standards and models. GPS velocities were estimated and represented in the absolute geocentric reference frame (the IGb08 realization of the global ITRF08 [63]. Measured VLM and general information about GPS stations are shown in Figure 5 and Table 3. Due to the poor quality of the data, the VLM at ROGS (Bakar) and RIGS (Rovinj) were not used in this study.

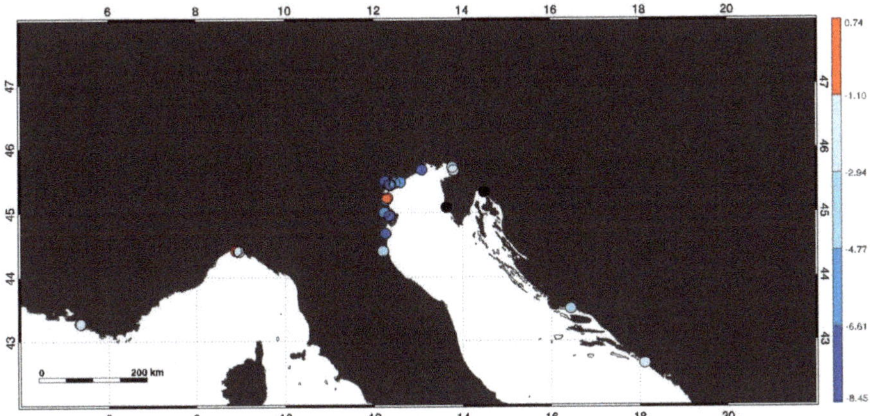

Figure 5. Map of the absolute (i.e., IGb08 frame) vertical velocities for the continuous GPS (cGPS) stations used in this work. Black dots refer to the stations ROGS (Bakar) and RIGS (Rovinj) for which, due to the limited available data, calculated vertical land motion (VLM) are not reliable.

Table 3. VLM measured at the cGPS station closest to the tide gauges used in this work. The cGPS station name, position, rate of VLM, time interval of the time series, mean VLM, and the closest tide gauges to the cGPS stations are indicated. Negative sign in the VLM corresponds to land subsidence.

GNSS Station	Lon	Lat	VLM (mm a^{-1})	Time Interval	Average VLM (mm a^{-1})	Tide Gauge Location
PRIE	5.3727	43.2768	−0.65 ± 0.55	2007.6397–2018.0890	−0.61 ± 0.43	Marseille
MARS	5.3538	43.2788	−0.58 ± 0.31	1998.5465–2018.0917		
TRIE	13.7635	45.7098	−0.20 ± 0.36	2003.1054–2018.0917	−0.08 ± 0.49	Trieste
TRI1	13.7878	45.6606	0.04 ± 0.61	2007.7821–2016.8948		
GENV	8.8809	44.4152	0.74 ± 0.90	2008.4631–2014.7602	−0.12 ± 1.54	Genova
GENO	8.9211	44.4194	−0.25 ± 0.30	1998.5575–2018.0917		
GENA	8.9482	44.3976	−0.51 ± 4.30	2016.6270–2017.9273		
GENU	8.9593	44.4027	−0.50 ± 0.68	2009.9465–2017.9986		
DUB2	18.1103	42.6502	−1.13 ± 0.96	2011.9739–2018.0917	−1.56 ± 0.72	Dubrovnik
DUBR	18.1104	42.6500	−1.99 ± 0.48	2000.7226–2012.7363		
SPLT	16.4385	43.5066	0.56 ± 0.77	2005.0013–2012.2527	0.56 ± 0.77	Split
CGIA	12.2655	45.2065	−2.91 ± 0.83	2010.9630–2018.0917	−3.30 ± 0.85	Venice
SFEL	12.2913	45.2300	−4.57 ± 0.67	2001.5465–2011.1657		
VEAR	12.3578	45.4379	−2.17 ± 1.35	2006.1602–2010.7164		
VEN1	12.3541	45.4306	−1.46 ± 0.65	2009.8068–2018.0917		
VENE	12.3320	45.4370	−0.89 ± 0.82	2001.0863–2007.5630		
VE01	12.3339	45.4375	−2.22 ± 1.56	2007.8506–2011.1821		
MSTR	12.2386	45.4904	−2.50 ± 0.79	2007.8890–2014.6315		
TREP	12.4547	45.4677	−8.45 ± 1.69	2004.1871–2008.0724		
CAVA	12.5827	45.4794	−2.71 ± 0.58	2001.5438–2011.1657		
TGPO	12.2283	45.0031	−4.89 ± 0.53	2007.3164–2018.0917		
PTO1	12.3341	44.9515	−4.26 ± 0.59	2008.4631–2017.9301		
GARI	12.2494	44.6769	−3.25 ± 0.66	2009.3958–2018.0917		
RAVE	12.1919	44.4053	−4.23 ± 0.58	2008.0259–2017.9986		
BEVA	13.0694	45.6719	−1.69 ± 0.57	2008.0368–2018.0917		

The pattern of vertical GPS rates is variable: the VLM (the average value is considered when more than one cGPS station is available) is almost null in the Gulf of Trieste; it is ≤1 mm a^{-1} at Marseille, Genova, and Split and it is about 1.5 mm a^{-1} at Dubrovnik. Sudden temporary tectonic episodes, which can cause permanent signatures during the tidal recordings at these stations, have not occurred (www.psmsl.org).

On the other hand, vertical GPS velocities in the North Adriatic coast around the Venice lagoon, display significant subsidence rates, with values in agreement with previous studies [32,64]. The average VLM in the region, calculated as the average over 14 GPS stations, is -3.30 ± 0.85 mm a^{-1}. Part of the detected subsidence is due to the continuing global glacio-hydroisostatic signal, estimated in this region to be between -0.12 and -0.21 mm a^{-1} [15], and the remaining part can be related to natural and anthropogenic effects [64,65]. We can reasonably assume that the VLM rate will remain unchanged until 2100 AD, in the absence of additional episodes of land movement that may result from eventual earthquakes with significant magnitude in the area [66] (http://www.isc.ac.uk/). The remarkable subsidence in the Venice lagoon results in a steeper sea-level rate with respect to those provided by IPCC estimations in the area [29,67].

Equation (4) can be used to evaluate sea levels at future times by including both the sea-level trend and VLM, included in r, and the effects of the natural long-term variability, reproduced by the oscillating contributions. By keeping the linear rate derived from a linear fit on the raw data, we implicitly assume a constant trend scenario for which the sea-level trend remains unchanged for the next decades. This represents, of course, a very unlikely scenario since there is high confidence that the rate of sea-level rise is increasing (IPCC – AR5). Therefore, to evaluate more realistic sea levels, a time-dependent rate, obtained from the local IPCC projections and including the VLM contribution is substituted to the constant rate r in Equation (4). Concerning the modeling of the SLNV, since the Atlantic climatic variability is quasi-persistent on secular times [51,52] and since it is connected to the sea-level variations in the Mediterranean sea (see Section 3), we can reasonably assume that the modes of oscillation associated with the natural variability observed in the past, can be considered as reliable for the next ~90 years. The contribution of the natural variability is also significant in a system dominated by a forced CO_2 emission, as the Earth nowadays and possibly in future years. Studies on the effects of the long-term variability on regional sea levels from different climatic models indicate that the internal variability contribution is already relevant when 100 year integrations are considered [68–70]. Moreover, the estimated time of emergence, namely the time at which the CO_2-forced sea level signal starts to dominate the internal variability, is close to or longer than a century in most of the Oceans. Other studies [71,72] indicate that the interplay of natural and forced variability in sea-level changes is expected to persist in time throughout the 21st century. In addition, possible variations of the modes of natural variability, between 2016 and 2100, were taken into account in our analysis by using modes with several frequencies for the uncertainty calculation.

To check the robustness of our approach in estimating future sea levels and the accuracy of the SLNV evaluation, we carried out a simple test using data from Marseille, the station with the longest and almost complete time series. Firstly, we applied the EMD analysis on sea-level data collected in the interval 1888–1957 (70 years). Two IMFs with $\tau_j > 15$ years ($\tau_9 = 33.4 \pm 5.4$ years and $\tau_{10} = 58.1 \pm 7.6$ years) were found. Then, sea levels for the next 50 years (1957–2007) were evaluated through the model, by using the linear trend from the fit to Equation (4) and were compared to the real complete data set.

Figure 6 shows real data, both reduced at the 1 year of resolution (red dashed line) and smoothed by a 15-years sliding window (red full line), and the model curve (black line) with the 90% confidence interval (enclosed by black dashed lines). The model reproduces well variations in both curves derived from the full time series. For a quantitative comparison:

1. The normalized mean squared error (NMSE) between the 15 years smoothed and the model curves, after 1957, is 4.4×10^{-6} (if a model has a very low NMSE then it reproduces the data well);
2. The maximum and minimum discrepancy between the 15 years smoothed data and the model-derived sea levels are found in 1965 and 1997, respectively. Sea-level values are 6976 and 6977 mm for smoothed data, 6974 (6956, 6992) and 7000 (6968, 7032) mm for model data (values in parentheses represents the upper/lower 90% confidence interval).

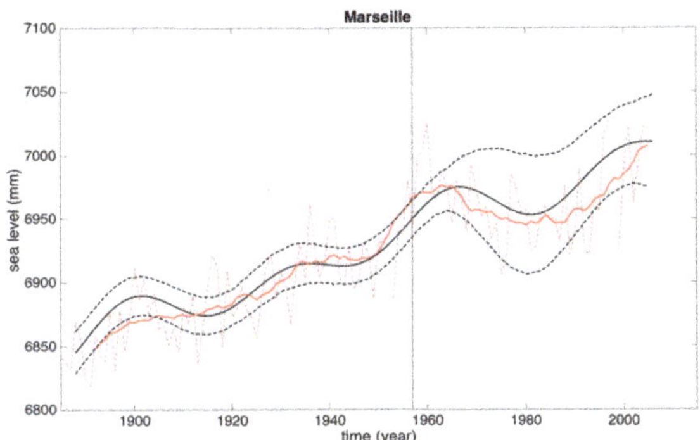

Figure 6. Sea levels in the period 1888–2007 at Marseille as estimated by deriving the model from the first 70 years (1888–1957) of tidal data. Fit parameters are: $r = 1.57 \pm 0.04$; $c = 3882 \pm 75$; $A = 13.8 \pm 1.0$, 9.8 ± 1.01; $\Phi = 1.22 \pm 0.06$, 1.89 ± 0.09, $R^2 = 0.91$. The model (black line) with 90% confidence interval (enclosed by black dashed lines), sea-level data at 1 year of resolution (red dashed line) and smoothed by a 15-year sliding window (red line), are shown. The vertical bar marks the boundary between sea level data used for the model (left) and estimated sea levels (right).

The test shows that the modeling of the last 50 years of sea levels, based on the previous 70 years of data, fits the real sea levels within a 90% confidence interval well, thus indicating that the approach used is reliable.

To get sea-level projections at future times, time dependent rates from regional IPCC AR5 sea-level projections (spatial resolution: $1° \times 1°$), discussed in the Fifth Assessment Report of the IPCC-AR5 [73,74], and available from the Integrated Climate data Center-ICDC of the University of Hamburg (http://icdc.cen.uni-hamburg.de/1/daten/ocean/ar5-slr.html), were considered. Rates obtained under the Representative Concentration Pathways RCP2.6 and RCP8.5, providing the least and most amounts of future sea level rise respectively, were used. These data consist of mean values and upper/lower 90% confidence bounds obtained by summing up the contributions of several geophysical sources [73,74]: the thermosteric/dynamic contribution, from 21 CMIP5 coupled atmosphere-ocean general circulation models (AOGCMs); the surface mass balance and dynamic ice sheet contributions from Greenland and Antarctica; the glacier and land water storage contributions; the GIA and the inverse barometer effect. For each tide gauge record the IPCC regional sea-level rate was evaluated at the grid point closest to the location of the corresponding station. To take into account the VLM, we substituted the modeled GIA contribution to the IPCC rate with the measured cGPS rate that includes both real GIA and tectonics/anthropic vertical motion. When more than one cGPS station was present, the average VLM value was used (see Table 3). The IPCC rate, corrected for the measured VLM, represents the function r in Equation (4). Uncertainties for the sea-level estimations were then calculated by combining lower and upper sea level bounds from IPCC projection, errors from GPS measurements, and uncertainties from the model. Note that for the Bakar and Rovinj sea levels we used the modeled GIA component calculated from the mean of the ICE-5G model [75] and the Australian National University's (ANU) ice sheet model [76] and subsequent improvements, since the measured VLM is not reliable.

Computed sea levels for the two considered IPCC scenarios are shown in Figure 7 and values at 2050 and 2100 (with respect to 2005) are indicated in Table 4. Let us first discuss the results without considering Venice P.S. A mean sea-level rise of 198 ± 102 mm by 2050 and 570 ± 255 mm by 2100 in the most severe RCP8.5 scenario, and 174 ± 97 mm and 342 ± 211 mm by 2050 and 2100, respectively, for the RCP2.6 scenario results from our modelling. In the most favorable scenario (RCP2.6) a sea-level rise between 142 ± 82 (Trieste) and 191 ± 106 (Split M.) mm by 2050, and between 259 ± 165 (Bakar) and 445 ± 200 (Dubrovnik) mm by 2100 is expected.

Table 4. Sea levels for 2050 and 2100, with respect to 2005, at the tide gauge for the two scenarios RCP2.6 and RCP8.5 including both the VLM and sea level natural variability (SLNV) contribution. Uncertainties define the 90% confidence limits. Time span used for the modeling and duration of raw records (in parenthesis) are also indicated. Since cGPS data are not reliable, the contribution of VLM is not included for Bakar and Rovinj.

Tide Gauge Station (Duration)	Sea Level (mm) RCP2.6		Sea Level (mm) RCP8.5	
	2050	2100	2050	2100
Marseille 1888–2009 (128 years)	182 ± 79	364 ± 167	208 ± 79	602 ± 240
Trieste 1901–2009 (138 years)	142 ± 82	336 ± 197	150 ± 86	523 ± 237
Genova 1931–1992 (92 years)	163 ± 150	337 ± 306	193 ± 156	581 ± 347
Venice P.S. 1914–1997 (92 years)	283 ± 103	603 ± 217	311 ± 114	818 ± 258
Bakar 1949–2008 (82 years)	166 ± 69	259 ± 165	182 ± 70	475 ± 203
Dubrovnik 1959–2006 (54 years)	225 ± 91	445 ± 200	246 ± 95	681 ± 246
Rovinj 1958–2008 (57 years)	149 ± 65	295 ± 164	177 ± 80	510 ± 216
Split M. 1955–2008 (60 years)	191 ± 106	322 ± 213	220 ± 112	567 ± 249
Split G.L. 1957–2008 (58 years)	174 ± 106	376 ± 240	204 ± 112	621 ± 273

The most severe scenario projections indicate a sea-level rise between 150 ± 86 (Trieste) and 246 ± 95 (Dubrovnik) mm by 2050 and between 475 ± 203 (Bakar) and 681 ± 246 (Dubrovnik) mm by 2100. The combined effect of the VLM and SLNV produces an average increase of the sea level, with respect to pure IPCC projections, of 23/58 mm in RCP2.6 scenario and 37/65 mm in RCP8.5 scenario for 2050/2100 corresponding to a variation of about the 20% (2050) and 13% (2100). The maximum sea-level variation, with respect to pure IPCC projections, of 146/205 mm in RCP2.6 and 150/222 mm in RCP2.6 for 2050/2100 is found at Dubrovnik where the highest VLM rate is measured. The natural variability produces an average variation on the estimated sea level of about 20 (2050) and 10 (2100) mm with maximum values reaching about 65 mm with respect to the IPCC projections. Its average contribution, for the analyzed stations, is about the 9% in 2050 and 2.5% in 2100 (note that the SLNV contribution, in percentage, decreases when the effects of global changes and VLM increase). We remark that, although the contribution of the natural variability in the analyzed stations is of the order of few centimeters, its contribution is not negligible especially for coastlines that are more prone to marine flooding. In these areas, indeed, changes of some centimeters in projected sea levels can change the flooding scenarios [33].

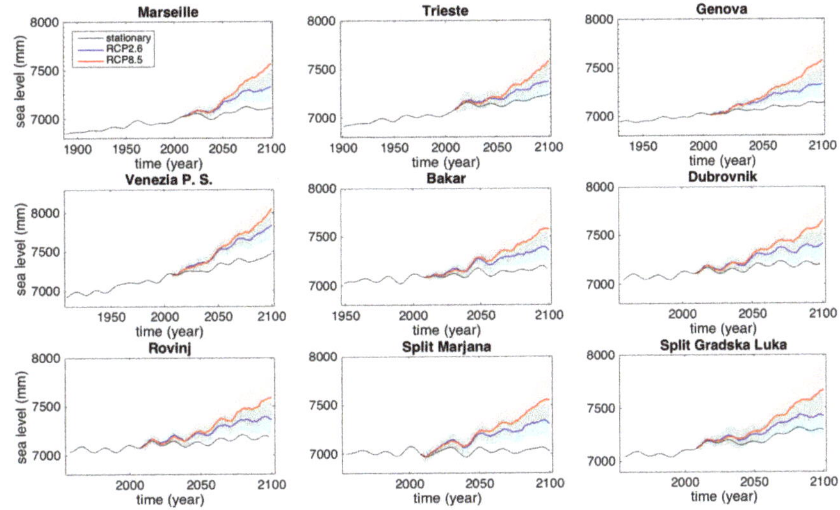

Figure 7. Sea-level height up to 2100 for the analyzed stations, as obtained by the model of Equation (4) that includes VLM and AR5 RCP2.6 (blue line) and RCP8.5 (red line) rate. Color bands represent the 90% confidence interval, obtained by including uncertainties from the model, VLM, and AR5 trend. The black line represents sea levels when the constant trend obtained from fits is used. Since cGPS data are not reliable, the contribution of VLM is not included for Bakar and Rovinj.

The sea level at Venice deserves a separate discussion. At Venice P.S., future sea levels are strongly driven by the land subsidence that accounts alone up to 68% (2050) and 60% (2100) of the sea-level rise. VLM, together with the natural variability, accounts for up to 78% (2050) and 64% (2100) of the level rise. In the RCP2.6/RCP8.5 scenarios, the expected sea levels are 283 ± 103 mm/311 ± 114 mm (2050) and 603 ± 217 mm/818 ± 258 mm (2100) well above the values found for the other analyzed records. These values exceed, by about 220 (2050) and 385 mm (2100), the values obtained for Venice when pure IPCC scenarios are considered. These results have considerable implications for the area of the Venice lagoon, characterized by high-density of population, valuable historical setting, residential and industrial infrastructures, besides areas for agricultural use and national parks, for which sea-level rise can become a serious hazard.

The sea-level projections obtained in this study were finally compared with previous investigations in the same geographical areas. References [15,32] estimated the sea-level rise in the Venice area by making use of IPCC predictions, GIA modeling, and including the VLM contribution from geological and archeological data. By using the global IPCC 2007, in [15] the authors found a sea-level rise in 2100 (relative to the sea level average 1986–2005) of about 315 mm; an average sea level between 580 (minimum) and 996 mm (maximum) was estimated by [32], using the global IPCC AR5 RCP8.5 scenario. By modeling and combining the contributions of terrestrial ice melt, GIA and steric sea-level components, in [26] the authors elaborated a minimum and maximum sea-level rise scenario for sea level in 2050: sea levels at Marseille, Venice P.S., Trieste, Bakar were 60, 113, 113, 113 mm and 221, 227, 227, 229 mm respectively. Reference [14] presented a worldwide set of local sea-level projections up to 2100 and 2200 by modeling the contribution to the sea level through three ice sheet components, glacier and ice cap, oceanographic processes, and land water storage. A long-term, local, non-climatic sea-level change calculated through a Gaussian process model, based on historical tide gauge data, was also included to estimate the non-climatic contribution. A quite different approach consists of using a dynamical system model with spatial analysis capability [77–79] to assess the spatial variability of both temperature and sea-level changes over the 21st century. In this framework, temperature and sea level are considered as coupled variables whose behavior is described by two ordinary differential equations. After calibration through historical data, the model is able to provide reliable predictions of the two variables for the next century. For the Mediterranean Sea region, in particular, an average sea-level rise below 600 mm, relative to 1990, is expected.

The sea-level value from [15] (315 mm) is significantly lower than our RCP2.6 result for 2100 (603 ± 217 mm). This can be ascribed to the use of a global not upgraded IPCC model and an estimation of the VLM from geological and archeological data instead of GPS measurements. On the other hand, our RPC 8.5 results for 2100 are in agreement, within the error limits, with the results by [32]. Additionally, in this case small differences can arise from the use of a global IPCC model (instead of a local one) and the VLM estimation from geological data. Concerning the analysis proposed by [26], sea levels from their maximum scenario are in agreement with our predictions for both RCP2.6 and 8.5. Differences (mainly with their minimum scenario) possibly arise from differences in modeling ice melt, GIA, and steric components with respect to IPCC projections and from the lack of inclusion of the VLM contribution in their study.

Our results are in agreement with the mean sea-level values of [14], within the uncertainty limit, for both 2050 and 2100 at the nine analyzed locations. Differences between the two results are smaller (at most 10%) for the records spanning a longer period (Marseille, Trieste, and Genova) which indicates that both analyses seem to converge when longer time series are considered. We recall that both the approaches use observationally based rates to account for local effects in the sea-level predictions and both the EMD as well as the statistical approach from [14] are more accurate when longer data points are available. Small differences can also be due to differences in modeling ice sheet components, glacier and ice cap, oceanographic processes, and land water storage with respect to IPCC projections. Regarding the results from the Venice station, differences reaching 27% were found. Due to the strong subsidence affecting only the Venice area, and not the entire North Adriatic Sea, we believe that an approach that takes into account the punctual vertical land motion measured closer to the tide gauge station is more suitable for modeling this effect on the sea-level. A comparison of the sea level in 2100 in the Mediterranean Sea area from [77] shows very good agreement with our results when the RCP85 scenario is considered. Indeed, when averaged over all stations (excluding Venice) and reported to the same time basis (year 1990) as [77] we found an average sea-level of 626 ± 255 mm that fits well, within the uncertainty limits, the value found by [77].

Finally, the modeling of the effect of the natural variability on future sea-level projections, present in this study, can also introduce, to a lesser extent, differences with sea-level estimations from the works mentioned above.

5. Conclusions

Sea levels for 2050 and 2100 at nine locations in the Central-Northern Mediterranean coasts were estimated from tide gauge data by including the contribution of both the vertical land motion and the natural variability. The VLM was inferred at a set of locations nearby tide gauges through continuous GPS measurements. The SLNV was estimated by the EMD analysis on the tide gauge data. The EMD approach revealed modes at characteristic time scales in the ranges 20–30 and >45 years, associated with the long-term natural variability, resulting from the combined effect of teleconnection patterns. Expected sea levels in 2050 and 2100 were then estimated by using IPCC projection, for the time dependent sea-level rate, the measured VLM rate, and a superposition of periodic functions, whose properties were obtained from the EMD results, to model the contribution of the long-term variability. Results (excluding the Venice P.S. tide gauge station) provided a mean sea-level rise of 198 ± 102 mm by 2050 and 570 ± 255 mm by 2100 in the most severe RCP8.5 scenario, while a lower bound is estimated to be 174 ± 97 and 342 ± 211 mm by 2050 and 2100, respectively, for the RCP2.6 scenario. The contribution of natural variability to local sea-level variations is significant and can reach, on average, up to the 9% with respect to the sea-level changes induced only by global climate change. VLM and SLNV together account, on average, for about the 15% of the sea-level variation.

In the Venice lagoon, the land subsidence is increasing the local sea-level rise. By taking into account VLM rates measured by the cGPS stations in this area, relative sea levels are expected to reach 603 ± 217 and 818 ± 258 mm in 2100 for the RCP2.6 and RCP8.5 scenarios, respectively. In particular, at the Venice P.S. tidal station, VLM and SLNV contribute more than the 60% to the expected sea-level rise.

Finally, our analysis shows that the contribution of local effects, such as the natural variability and the vertical land motion, play a key role in local sea-level rise projections. Effects are even more relevant for low elevated coastal areas where marine ingression may represent a potential hazard for the environment and human activities. Even in the more optimistic scenario, the estimated sea-level rise will have an important impact along the coasts, causing diffuse erosion of the shorelines, hence determining serious impacts on many coastal areas, increasing the hazard related to flooding events, storm surges, and possibly tsunamis. In subsiding areas, like the Venice Lagoon, severe environmental impact with subsequent loss of economic value of shores and coastal infrastructures can be reasonably expected before 2100. In this regard, land planners and decision makers should take into account the rising sea levels reported in this study for cognizant coastal management.

Author Contributions: Conceptualization, A.V. and M.A.; Formal analysis, A.V. and E.S.; Funding acquisition, M.A.; Methodology, A.V.; Project administration, M.A.; Supervision, M.A. and F.F.; Validation, A.V.; Writing – original draft, A.V.; Writing—review & editing, M.A., E.S. and F.F.

Funding: This work was carried out under the umbrella of the SAVEMEDCOASTS project "Sea level rise scenarios along the Mediterranean coasts", funded by the EU (www.savemedcoasts.eu; agreement number ECHO/SUB/2016/742473/PREV16).

Acknowledgments: We acknowledge two anonymous reviewers for the useful comments.

Conflicts of Interest: The authors declare no conflict of interest.

References

1. Vermeer, M.; Rahmstorf, S. Global sea level linked to global temperature. *Proc. Natl. Acad. Sci. USA* **2009**, *106*, 21527–21532. [CrossRef] [PubMed]

2. Church, J.A.; White, N.J. Sea-Level Rise from the Late 19th to the Early 21st Century. *Surv. Geophys.* **2011**, *32*, 585–602. [CrossRef]

3. Kemp, A.C.; Horton, B.P.; Donnelly, J.P.; Mann, M.E.; Vermeer, M.; Rahmstorf, S. Climate related sea-level variations over the past two millennia. *Proc. Natl. Acad. Sci. USA* **2011**, *108*, 11017–11022. [CrossRef] [PubMed]

4. Lambeck, K.; Woodroffe, C.D.; Antonioli, F.; Anzidei, M.; Gehrels, W.R.; Laborel, J.; Wright, A.J. Paleoenvironmental Records, Geophysical Modeling, and Reconstruction of Sea-Level Trends and Variability on Centennial and Longer Timescales. In *Understanding Sea-Level Rise and Variability*; Blackwell: Hoboken, NJ, USA, 2010; pp. 61–121.

5. Meyssignac, B.; Cazenave, A. Sea level: A review of present-day and recent-past changes and variability. *J. Geodyn.* **2012**, *58*, 96–109. [CrossRef]

6. Mitchum, G.T.; Nerem, R.S.; Merrifield, M.A.; Gehrels, W.R. Modern sea level changes estimates. In *Understanding Sea Level Rise and Variability*; Church, J.A., Woodworth, P.L., Aarup, T., Wilson, W.S., Eds.; Wiley-Blackwell Publishing: London, UK, 2010.

7. Jevrejeva, S.; Moore, J.; Grinsted, A. Sea level projections to AD2500 with a new generation of climate change scenarios. *Glob. Planet. Chang.* **2012**, *80*, 14–20. [CrossRef]

8. Wöppelmann, G.; Marcos, M. Coastal sea level rise in southern Europe and the nonclimate contribution of vertical land motion. *J. Geophys. Res. Space Phys.* **2012**, *117*. [CrossRef]

9. Anzidei, M.; Lambeck, K.; Antonioli, F.; Furlani, S.; Mastronuzzi, G.; Serpelloni, E.; Vannucci, G. Coastal structure, sea-level changes and vertical motion of the land in the Mediterranean. *Geol. Soc. Lond. Spéc. Publ.* **2014**, *388*, 453–479. [CrossRef]

10. Tsimplis, M.N.; Calafat, F.M.; Marcos, M.; Jorda, G.; Gomis, D.; Fenoglio-Marc, L.; Struglia, M.V.; Josey, S.A.; Chambers, D.; Fenoglio-Marc, L. The effect of the NAO on sea level and on mass changes in the Mediterranean Sea. *J. Geophys. Res. Oceans* **2013**, *118*, 944–952. [CrossRef]

11. Fenoglio-Marc, L.; Rietbroek, R.; Grayek, S.; Becker, M.; Kusche, J.; Stanev, E. Water mass variation in the Mediterranean and Black Seas. *J. Geodyn.* **2012**, *59*, 168–182. [CrossRef]

12. Fenoglio-Marc, L.; Mariotti, A.; Sannino, G.; Meyssignac, B.; Carillo, A.; Struglia, M.; Rixen, M. Decadal variability of net water flux at the Mediterranean Sea Gibraltar Strait. *Glob. Planet. Chang.* **2013**, *100*, 1–10. [CrossRef]

13. Lambeck, K.; Purcell, A. Sea-level change in the Mediterranean Sea since the LGM: Model predictions for tectonically stable areas. *Quat. Sci. Rev.* **2005**, *24*, 1969–1988. [CrossRef]

14. Kopp, R.E.; Horton, R.M.; Little, C.M.; Mitrovica, J.X.; Oppenheimer, M.; Rasmussen, D.J.; Strauss, B.H.; Tebaldi, C. Probabilistic 21st and 22nd century sea-level projections at a global network of tide-gauge sites. *Earth's Future* **2014**, *2*, 383–406. [CrossRef]

15. Lambeck, K.; Antonioli, F.; Anzidei, M.; Ferranti, L.; Leoni, G.; Scicchitano, G.; Silenzi, S. Sea level change along the Italian coast during the Holocene and projections for the future. *Quat. Int.* **2011**, *232*, 250–257. [CrossRef]

16. Douglas, B.C. Global sea rise: A redetermination. *Surv. Geophys.* **1997**, *18*, 270–292. [CrossRef]

17. Knutti, R.; Stocker, T.F. Influence of the Thermohaline Circulation on Projected Sea Level Rise. *J. Clim.* **2000**, *13*, 1997–2001. [CrossRef]

18. Rahmstorf, S. A Semi-Empirical Approach to Projecting Future Sea-Level Rise. *Science* **2007**, *315*, 368–370. [CrossRef] [PubMed]

19. Rahmstorf, S.; Cazenave, A.; Church, J.A.; Hansen, J.E.; Keeling, R.F.; Parker, D.E.; Somerville, R.C.J. Recent Climate Observations Compared to Projections. *Science* **2007**, *316*, 709. [CrossRef] [PubMed]

20. Milne, G.A.; Gehrels, W.R.; Hughes, C.W.; Tamisiea, M.E. Identifying the causes of sea-level change. *Nat. Geosci.* **2009**, *2*, 471–478. [CrossRef]

21. Tsimplis, M.; Marcos, M.; Colin, J.; Somot, S.; Pascual, A.; Shaw, A. Sea level variability in the Mediterranean Sea during the 1990s on the basis of two 2d and one 3d model. *J. Mar. Syst.* **2009**, *78*, 109–123. [CrossRef]

22. Tsimplis, M.; Spada, G.; Marcos, M.; Flemming, N. Multi-decadal sea level trends and land movements in the Mediterranean Sea with estimates of factors perturbing tide gauge data and cumulative uncertainties. *Glob. Planet. Chang.* **2011**, *76*, 63–76. [CrossRef]

23. Church, J.A.; Aarup, T.; Woodworth, P.L.; Wilson, W.S.; Nicholls, R.J.; Rayner, R.; Lambeck, K.; Mitchum, G.T.; Steffen, K.; Cazenave, A.; et al. Sea-Level Rise and Variability: Synthesis and Outlook for the Future. In *Understanding Sea-Level Rise and Variability*, 1st ed.; Church, A.J., Woodworth, P.L., Aarup, T., Wilson, S., Eds.; Blackwell Publishing Ltd.: Hoboken, NJ, USA, 2010.

24. Cazenave, A.; Remy, F. Sea level and climate: Measurements and causes of changes. *Wiley Interdiscip. Rev. Clim. Chang.* **2011**, *2*, 647–662. [CrossRef]

25. Zhang, X.; Church, J.A. Sea level trends, interannual and decadal variability in the Pacific Ocean. *Geophys. Res. Lett.* **2012**, *39*, 21701. [CrossRef]

26. Galassi, G.; Spada, G. Sea-level rise in the Mediterranean Sea by 2050: Roles of terrestrial ice melt, steric effects and glacial isostatic adjustment. *Glob. Planet. Chang.* **2014**, *123*, 55–66. [CrossRef]

27. Han, W.; Stammer, D.; Meehl, G.A.; Hu, A.; Sienz, F.; Zhang, L. Multi-Decadal Trend and Decadal Variability of the Regional Sea Level over the Indian Ocean since the 1960s: Roles of Climate Modes and External Forcing. *Climate* **2018**, *6*, 51. [CrossRef]

28. Brambati, A.; Carbognin, L.; Quaia, T.; Teatini, P.; Tosi, L. The Lagoon of Venice: Geological setting, evolution and land subsidence. *Episodes* **2003**, *26*, 264–268.

29. Carbognin, L.; Teatini, P.; Tosi, L. Eustacy and land subsidence in the Venice Lagoon at the beginning of the new millennium. *J. Mar. Syst.* **2004**, *51*, 345–353. [CrossRef]

30. De Conto, R.M.; Pollard, D. Contribution of Antarctica to past and future sea-level rise. *Nature* **2016**, *531*, 591–597. [CrossRef]

31. Syvitski, J.P.M.; Kettner, A.J.; Overeem, I.; Hutton, E.W.H.; Hannon, M.T.; Brakenridge, G.R.; Day, J.; Vörösmarty, C.; Saito, Y.; Giosan, L.; et al. Sinking deltas due to human activities. *Nat. Geosci.* **2009**, *2*, 681–686. [CrossRef]

32. Antonioli, F.; Anzidei, M.; Amorosi, A.; Presti, V.L.; Mastronuzzi, G.; Deiana, G.; De Falco, G.; Fontana, A.; Fontolan, G.; Lisco, S.; et al. Sea-level rise and potential drowning of the Italian coastal plains: Flooding risk scenarios for 2100. *Quat. Sci. Rev.* **2017**, *158*, 29–43. [CrossRef]

33. Bosman, A.; Carluccio, R.; Casalbore, D.; Caracciolo, F.D.; Esposito, A.; Nicolosi, I.; Pietrantonio, G.; Vecchio, A.; Carmisciano, C.; Chiappini, M.; et al. Flooding scenarios due to land subsidence and sea-level rise: A case study for Lipari Island (Italy). *Terra Nova* **2017**, *29*, 44–51.

34. Marsico, A.; Lisco, S.; Presti, V.L.; Antonioli, F.; Amorosi, A.; Anzidei, M.; Deiana, G.; De Falco, G.; Fontana, A.; Fontolan, G.; et al. Flooding scenario for four Italian coastal plains using three relative sea level rise models. *J. Maps* **2017**, *13*, 961–967. [CrossRef]

35. Douglas, B.C. Global sea level rise. *J. Geophys. Res.* **1991**, *96*, 6981–6992. [CrossRef]

36. Douglas, B.C. Global sea level rise acceleration. *J. Geophys. Res.* **1992**, *97*, 12699–12796. [CrossRef]

37. Visser, H.; Dangendorf, S.; Petersen, A.C. A review of trend models applied to sea level data with reference to the "acceleration-deceleration debate". *J. Geophys. Res. Oceans* **2015**, *120*, 3873–3895. [CrossRef]

38. Huang, N.E.; Shen, Z.; Long, S.R.; Wu, M.C.; Shih, E.H.; Zheng, Q.; Tung, C.C.; Liu, H.H. The empirical mode decomposition and the Hilbert spectrum for non-stationary time series analysis. *Proc. R. Soc. Lond.* **1998**, *454*, 903–995. [CrossRef]

39. Vecchio, A.; Capparelli, V.; Carbone, V. The complex dynamics of the seasonal component of USA's surface temperature. *Atmos. Chem. Phys.* **2010**, *10*, 9657–9665. [CrossRef]

40. Vecchio, A.; Anzidei, M.; Capparelli, V.; Carbone, V.; Guerra, I. Has the Mediterranean Sea felt the March 11th, 2011, Mw 9.0 Tohoku-Oki earthquake? *Europhys. Lett.* **2012**, *98*, 59001. [CrossRef]

41. Vecchio, A.; Anzidei, M.; Carbone, V. New insights on the tsunami recording of the May, 21, 2003, Mw 6.9 Boumerdès earthquake from tidal data analysis. *J. Geodyn.* **2014**, *79*, 39–49. [CrossRef]

42. Breaker, L.C.; Ruzmaikin, A. The 154-year record of sea level at San Francisco: Extracting the long-term trend, recent changes, and other tidbits. *Clim. Dyn.* **2011**, *36*, 545–559. [CrossRef]

43. Ezer, T.; Corlett, W.B. Is sea level rise accelerating in the Chesapeake Bay? A demonstration of a novel new approach for analyzing sea level data. *Geophys. Res. Lett.* **2012**, *39*, L19605. [CrossRef]

44. Ezer, T.; Atkinson, L.P.; Corlett, W.B.; Blanco, J.L. Gulf Stream's induced sea level rise and variability along the U.S. mid-Atlantic coast. *J. Geophys. Res. Ocean* **2013**, *118*, 685–697. [CrossRef]

45. Chambers, D.P. Evaluation of empirical mode decomposition for quantifying multi-decadal variations and acceleration in sea level records. *Nonlinear Proc. Geophys.* **2015**, *22*, 157–166. [CrossRef]

46. Wu, Z.; Huang, N.E. A study of the characteristics of white noise using the empirical mode decomposition method. *Proc. R. Soc. Lond. A* **2004**, *460*, 1597–1611. [CrossRef]

47. Frankcombe, L.M.; Von der Heydt, A.; Dijkstra, H.A. North Atlantic multidecadal climate variability: An Investigation of dominant time scales and processes. *J. Climatol.* **2010**, *23*, 3626–3638. [CrossRef]

48. Marshall, J.; Johnson, H.; Goodman, J. A study of the interaction of the North Atlantic Oscillation with the ocean circulation. *J. Clim.* **2001**, *14*, 1399–1421. [CrossRef]

49. Bellucci, A.; Gualdi, S.; Scoccimarro, E.; Navarra, A. NAO-ocean circulation interactions in a coupled general circulation model. *Clim. Dyn.* **2008**, *31*, 759–777. [CrossRef]

50. Fan, M.; Schneider, E.K. Observed decadal North Atlantic tripole SST variability. Part I: Weather noise forcing and coupled response. *J. Atmos. Sci.* **2012**, *69*, 35–50.

51. Knudsen, M.F.; Seidenkrantz, M.-S.; Jacobsen, B.H.; Kuijpers, A. Tracking the Atlantic Multidecadal Oscillation through the last 8,000 years. *Nat. Commun.* **2011**, *2*, 178. [CrossRef]

52. Hubeny, J.B.; King, J.W.; Santos, A. Subdecadal to multidecadal cycles of Late Holocene North Atlantic climate variability preserved by estuarine fossil pigments. *Geology* **2006**, *34*, 569–572. [CrossRef]

53. Calafat, F.M.; Jordà, G.; Marcos, M.; Gomis, D. Comparison of Mediterranean sea level variability as given by three baroclinic models. *J. Geophys. Res.* **2012**, *117*, c02009. [CrossRef]

54. Calafat, F.M.; Chambers, D.P.; Tsimplis, M.N. Mechanisms of decadal sea level variability in the eastern North Atlantic and the Mediterranean Sea. *J. Geophys. Res.* **2012**, *117*, c09022. [CrossRef]

55. Tsimplis, M.N.; Shaw, A.G.P. The forcing of mean sea level variability around Europe. *Glob. Planet. Chang.* **2008**, *63*, 196–202. [CrossRef]

56. Kosaka, Y.; Xie, S.-P. Recent global-warming hiatus tied to equatorial Pacific surface cooling. *Nature* **2013**, *501*, 403–407. [CrossRef] [PubMed]

57. Zhang, L. The roles of external forcing and natural variability in global warming hiatuses. *Clim. Dyn.* **2016**, *47*, 3157–3169. [CrossRef]

58. Kosaka, Y.; Xie, S.-P. The tropical Pacific as a key pacemaker of the variable rates of global warming. *Nat. Geosci.* **2016**, *9*, 669–673. [CrossRef]

59. Marullo, S.; Artale, V.; Santoleri, R. The SST Multidecadal Variability in the Atlantic–Mediterranean Region and Its Relation to AMO. *J. Clim.* **2011**, *24*, 4385–4401. [CrossRef]

60. Vigo, M.; Sánchez-Reales, J.; Trottini, M.; Chao, B. Mediterranean Sea level variations: Analysis of the satellite altimetric data, 1992–2008. *J. Geodyn.* **2011**, *52*, 271–278. [CrossRef]

61. Devoti, R.; D'Agostino, N.; Serpelloni, E.; Pietrantonio, G.; Riguzzi, F.; Avallone, A.; Cavaliere, A.; Cheloni, D.; Cecere, G.; D'Ambrosio, C.; et al. The mediterranean crustal motion map compiled at INGV. *Ann. Geophys.* **2017**, *60*. [CrossRef]

62. Serpelloni, E.; Faccenna, C.; Spada, G.; Dong, D.; Williams, S.D.P. Vertical GPS ground motion rates in the Euro-Mediterranean region: New evidence of velocity gradients at different spatial scales along the Nubia-Eurasia plate boundary. *J. Geophys. Res. Solid Earth* **2013**, *118*, 6003–6024. [CrossRef]

63. Altamimi, Z.; Collilieux, X.; Métivier, L. ITRF2008: An improved solution of the international terrestrial reference frame. *J. Geod.* **2011**, *85*, 457–473. [CrossRef]

64. Antonioli, F.; Anzidei, M.; Lambeck, K.; Auriemma, R.; Gaddi, D.; Furlani, S.; Orrù, P.; Solinas, E.; Gaspari, A.; Karinja, S.; et al. Sea level change during Holocene from Sardinia and northeastern Adriatic (Central Mediterranean sea) from archaeological and geomorphological data. *Quat. Sci. Rev.* **2007**, *26*, 2463–2524. [CrossRef]

65. Carbognin, L.; Gatto, P.; Mozzi, G.; Gambolati, G.; Ricceri, G. New trend in the subsidence of Venice. In Proceedings of the Anaheim Symposium, Anaheim, CA, USA, 13–17 December 1976.

66. Guidoboni, E.; Comastri, A. *Catalogue of Earthquakes and Tsunami in the Mediterranean Area from the 11th to the 15th Century*; INGV-SGA: Rome, Italy, 2007.

67. Carbognin, L.; Teatini, P.; Tomasin, A.; Tosi, L. Global change and relative sea level rise at Venice: What impact in term of flooding. *Clim. Dyn.* **2010**, *35*, 1039–1047. [CrossRef]

68. Lyu, K.; Zhang, X.; Church, J.A.; Slangen, A.B.A.; Hu, J. Time of emergence for regional sea-level change. *Nat. Clim. Chang.* **2014**, *4*, 1006–1010. [CrossRef]

69. Bordbar, M.H.; Martin, T.; Latif, M.; Park, W. Effects of long-term variability on projections of twenty-first century dynamic sea level. *Nat. Clim. Chang.* **2015**, *5*, 343–347. [CrossRef]

70. Carson, M.; Köhl, A.; Stammer, D. The Impact of Regional Multidecadal and Century-Scale Internal Climate Variability on Sea Level Trends in CMIP5 Models. *J. Clim.* **2015**, *28*, 853–861. [CrossRef]

71. Kopp, R.E.; Hay, C.C.; Little, C.M.; Mitrovica, J.X. Geographic Variability of Sea-Level Change. *Curr. Clim. Chang. Rep.* **2015**, *1*, 192–204. [CrossRef]

72. Little, C.M.; Horton, R.M.; Kopp, R.; Oppenheimer, M.; Yip, S. Uncertainty in Twenty-First-Century CMIP5 Sea Level Projections. *J. Clim.* **2015**, *28*, 838–852. [CrossRef]

73. Church, J.A.; Clark, P.U.; Cazenave, A.; Gregory, J.; Jevrejeva, S.; Levermann, A.; Merrifield, M.; Milne, G.; Nerem, R.S.; Nunn, P.; et al. Sea level change. In *Climate Change 2013: The Physical Science Basis*; Stocker, T.F., Qin, D.G.-K., Plattner, M., Tignor, S.K., Allen, J., Boschung, A., Nauels, Y., Bex, X.V., Midgley, P.M., Eds.; Cambridge University Press: Cambridge, UK; New York, NY, USA, 2013.

74. Church, J.A.; Clark, P.U.; Cazenave, A.; Gregory, J.; Jevrejeva, S.; Levermann, A.; Merrifield, M.; Milne, G.; Nerem, R.S.; Nunn, P.; et al. Sea Level Change Supplementary Material. In *Climate Change 2013: The Physical Science Basis*; Contribution of Working Group I to the Fifth Assessment Report of the Intergovernmental Panel on Climate Change; Stocker, T.F., Qin, D.G.-K., Plattner, M., Tignor, S.K., Allen, J., Boschung, A., Nauels, Y., Bex, X.V., Midgley, P.M., Eds.; Cambridge University Press: Cambridge, UK; New York, NY, USA, 2013.

75. Peltier, W.R. Global glacial isostasy and the surface of the ice-age earth: The ICE-5G (VM2) model and GRACE. *Annu. Rev. Earth Planet. Sci.* **2004**, *32*, 111–149. [CrossRef]

76. Lambeck, K.; Smither, C.; Johnston, P. Sea-level change, glacial rebound and mantle viscosity for northern Europe. *Geophys. J. Int.* **1998**, *134*, 102–144. [CrossRef]

77. Aral, M.M.; Chang, B. Spatial Variation of Sea Level Rise at Atlantic and Mediterranean Coastline of Europe. *Water* **2017**, *9*, 522. [CrossRef]

78. Chang, B.; Guan, J.; Aral, M.M. A Scientific Discourse: Climate Change and Sea-Level Rise. *ASCE J. Hydrol. Eng.* **2015**, *20*, A4014003. [CrossRef]
79. Chang, B.; Guan, J.; Aral, M.M. Modeling Spatial Variations of Sea Level Rise and Corresponding Inundation Impacts: A Case Study for Florida, USA. *Water Qual. Expo. Health* **2013**, *6*, 39–51. [CrossRef]

Article

Beach-Foredune Sediment Budget Response to Sea Level Fluctuation. Curonian Spit, Lithuania

Darius Jarmalavičius *, Donatas Pupienis, Gintautas Žilinskas, Rasa Janušaitė and Viktoras Karaliūnas

Laboratory of Geoenvironmental Research, Nature Research Centre, Akademijos 2, LT-08412 Vilnius, Lithuania; donatas.pupienis@gamtc.lt (D.P.); gintautas.zilinskas@gamtc.lt (G.Ž.); rasa.janusaite@gamtc.lt (R.J.); viktoras.karaliunas@gamtc.lt (V.K.)
* Correspondence: darius.jarmalavicius@gamtc.lt

Received: 14 January 2020; Accepted: 17 February 2020; Published: 20 February 2020

Abstract: Beach-foredune sediment exchange maintains a coastal system's stability. Sea level fluctuation is one of the most important factors that modifies the beach and foredune sediment budget. This study aims to assess beach and foredune sand budget changes depending on sea level fluctuations. On the basis of annual measurements of cross-shore profiles on the Curonian Spit in Lithuania, the sediment volumes on the beach and foredune and their changes between 2002 and 2019 were calculated. The sea level fluctuations were examined in parallel. The obtained data revealed that in the case of a sand surplus, a relatively low sea level rise does not have a significant impact on the development of a foredune (and a minimal impact on a beach) on a decadal time-scale. Short-term sea level fluctuations are reflected in year-to-year variability in a beach sediment budget. However, no significant relationship between year-to-year variability in sea level fluctuation and the foredune sediment budget has yet been identified, nor is there a reliable year-to-year variability relationship between the foredune and beach sediment budget. The foredune sediment budget remained positive both through an increase and a reduction in the sediment volume on the beach.

Keywords: sea level; coastal erosion; coastal morphometry; Baltic Sea

1. Introduction

The global mean sea level has been rising during the past century [1]. Based on satellite data, the rate of sea level rising has peaked at 3.36 mm/yr since 1993 [2]. The sea level trend in the Baltic Sea is similar to the global mean sea level trend [3].

Global climate change and sea level rise are often related to increasing coastal erosion and associated socio-economic transformations in coastal regions [4,5]. One of the most important problems in predicting future coastal evolution is determining the impact of a potential rise in sea level on coastal systems. However, there are other significant factors that also influence coastal dynamics including the geologic framework [6,7] and sand availability [8–13]. It is worth noting that coastal changes on short-term and decadal time scales are relevant to coastal managers and planners [14]. Within a decadal time period, there are other factors besides sea level fluctuation and geologic framework: extreme events [15–19], vegetation cover [20–23], human activity [24–26], and river discharge [27]. Since there is a complex interrelationship between sea level change and other factors on a short-term time scale, it is difficult to separate the direct impact of sea level on coastal systems from other factors. Depending on the interaction between various factors, accumulation processes may predominate on the coast despite the rising sea level [9,13,17,28,29]. Due to different driving forces, there can be high uncertainties related to the prediction of sea level rise impact on coastal systems [30]. The sand spits with predominant alongshore sediment transport are good examples for illustrating erosion and accretion processes along the shore [31–33]. Similar shoreline dynamic processes can be observed at the Curonian Spit where at the base of the spit erosion dominates, whereas on the distal end of the spit accretion dominates [34,35]. Due to the reduction of sediment supply in the southern part of the spit and sea level rise, there is intensification of erosion [36]. Alongshore sediment transport directed northward creates favorable conditions for accumulation near the distal part of the spit. Thus, a huge sand mass is involved in the formation of coastal landforms [37]. The development of foredunes is closely linked to beach evolution and the formation of a beach-foredune system [38–40]. Despite close interaction between beach and foredune, these two components of the system do not necessarily evolve in one direction. For example, the foredune sand budget may be positive while the beach sand budget is in equilibrium or slightly negative [29,39,41–43].

Since 2002, annual changes in the beach and foredune sand volume have been monitored along the Lithuanian part of the Curonian Spit Baltic Sea coast. Between 2002 and 2019, the coastal dynamics in coastal systems with different geomorphologies were determined based on the obtained data. The aim of this paper is to evaluate the interrelationship between the beach and foredune sand budget, and the influence of annual mean sea level fluctuation on the Baltic Sea coast in the Lithuanian part of the Curonian Spit on a decadal time scale.

2. Study Area

Of the Lithuanian part of the Curonian Spit (51 km long), the foredune ridge occupies an area of about 262.3 ha (Figure 1). A continuous foredune ridge along the entire Curonian Spit was formed in the second half of the 19th century. The formation of the foredune was started along the shore in order to protect the coastal settlements from sand being blown and the coast from wave action. Semi-permeable sand fences were constructed along the shore between 1810–1892 [41].

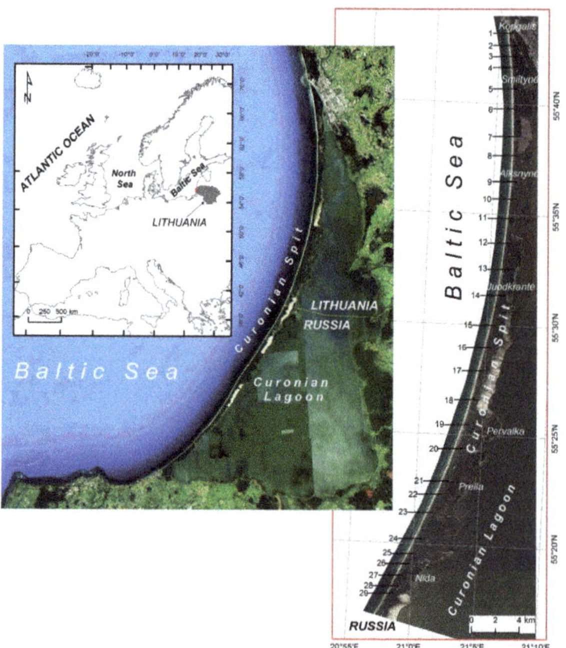

Figure 1. Location map. The black lines and numbers represent cross-shore profiles (Google Earth, earth.google.com/web/.).

When the ridge of accumulated sand reached 3 m in height it was planted with marram grass. Later on, the ridge developed without human intervention, except for being reinforced with tree branches in some places. Therefore, human activity has no crucial effect on foredune development. The height of the foredune ranges from 6 m at Juodkrantė to 16 m at Alksnynė (Figure 2). Striking differences have also occurred in the foredune volume. The largest foredune is at Smiltynė (up to 2200 m^3/m) and the smallest is at Juodkrantė (110 m^3/m). The foredunes are densely covered with marram grass. The widths of the beaches vary from up to 65 m at Smiltynė to 30 m at Juodkrantė. The volume of beach sediment ranges between 124 m^3/m at the Smiltynė and 42 m^3/m at the Juodkrantė. The Curonian Spit is formed exclusively of Quaternary deposits. The upper part of the Quaternary deposits in the Curonian Spit is composed of sediments that have been formed in the basins of the various stages of the Baltic Sea's development—starting from the Baltic Ice Lake and ending with recent marine sediments [34]. The beach of the Curonian Spit is composed of fine and medium sand. Mean sand grain size ranged from 0.2 mm at Smiltynė to 0.5 mm at Juodkrantė.

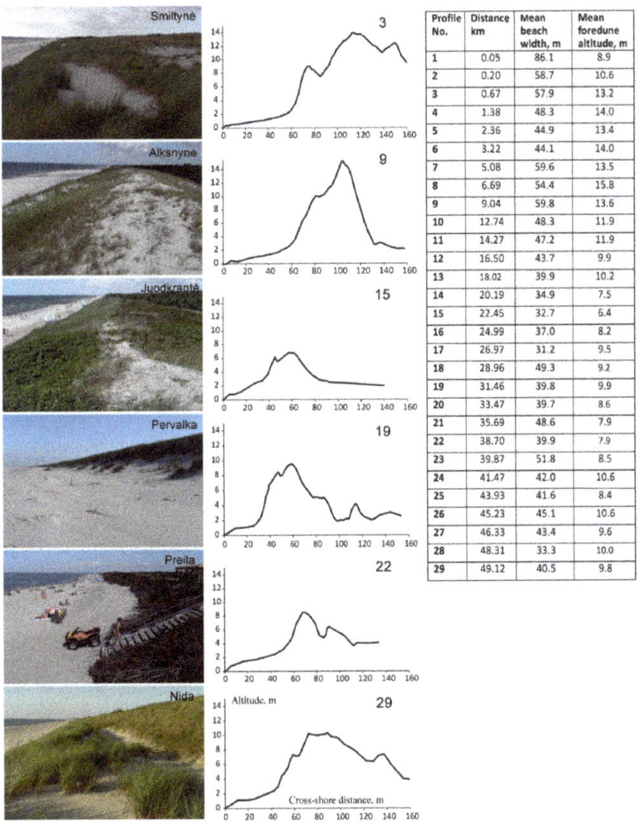

Profile No.	Distance km	Mean beach width, m	Mean foredune altitude, m
1	0.05	86.1	8.9
2	0.20	58.7	10.6
3	0.67	57.9	13.2
4	1.38	48.3	14.0
5	2.36	44.9	13.4
6	3.22	44.1	14.0
7	5.08	59.6	13.5
8	6.69	54.4	15.8
9	9.04	59.8	13.6
10	12.74	48.3	11.9
11	14.27	47.2	11.9
12	16.50	43.7	9.9
13	18.02	39.9	10.2
14	20.19	34.9	7.5
15	22.45	32.7	6.4
16	24.99	37.0	8.2
17	26.97	31.2	9.5
18	28.96	49.3	9.2
19	31.46	39.8	9.9
20	33.47	39.7	8.6
21	35.69	48.6	7.9
22	38.70	39.9	7.9
23	39.87	51.8	8.5
24	41.47	42.0	10.6
25	43.93	41.6	8.4
26	45.23	45.1	10.6
27	46.33	43.4	9.6
28	48.31	33.3	10.0
29	49.12	40.5	9.8

Figure 2. Typical cross-shore profiles and mean values of beach width and foredune height.

As the tidal range of the South-Eastern Baltic coasts is not significant and barely reaches 3.5–4.0 cm [44,45], the wind-generated waves [46], the prevailing alongshore currents, and accordingly the sand transport from south to north [35,47–49] and the aeolian processes [42] are the main beach-forming factors on the Lithuanian Baltic Sea coast. The most significant short-term sea level fluctuations near the Lithuanian coast occur due to storm surges. The most extreme storm surges are observed between November and February. In extreme cases, the sea level can rise up to 185 cm above the mean sea level [50] and wave height can reach up to between 4 and 6 m [46]. The most frequent recurrences of the storm wind directions are south-westerly (35.6%) and westerly winds (24.3%) [51].

3. Materials and Methods

The sediment volume of the Curonian Spit beach and foredune (m^3/m—cubic meters per meter of beach length) was assessed based on cross-shore profiles. The cross-shore profile was measured at 29 sites in May of each year using Global Navigation Satellite System (GNSS) Topcon HiPer SR (Livermore, CA, USA) with a horizontal accuracy of ±1.0 cm and vertical accuracy of ±1.5 cm. The leveling of all 29 cross-profiles was done in calm weather over 16 hours, therefore the morphometric changes are insignificant. Cross-shore leveling has been conducted at fixed positions from benchmarks installed some distance inland from foredune to waterline. The placement of each cross-profile has been chosen to correspond with the best representation of the general geomorphological characteristics. Cross-shore profiles are denser where there is relatively high variability in morphometrical characteristics. The sediment budget (m^3/m) was established by comparing cross-shore profiles in two successive years. Beach and foredune volumes were calculated separately. Cross-shore beach volume is defined as the sand volume enclosed by beach surface and the horizontal line from the waterline to the foredune toe. Foredune volume is defined as the volume of enclosed sand vertically from the foredune toe to some distance inland where the vertical variability is negligible. The cross-shore profile area was calculated using ArcMap software.

Spacing between individual profiles is on average 1.5 km. The observation periods were from 2002 to 2019. Changes in the sediment volume were calculated over time from the first observation in 2002. The trend of beach and foredune volume (m^3/m per year) was determined from the linear regression of each individual profile. A 95% confidence interval was chosen (p-value < 0.05). The height of the foredune in different years was also determined. The sea level and frequency of the westerly wind (SW-NW direction) data from the Klaipėda station between 2002 and 2019 were collected from the Department of Marine Research of the Environmental Protection Agency. Annual mean values of sea level and frequency of the westerly wind were calculated from daily data. This way, seasonal variation was eliminated.

4. Results

During the research period (2002–2019), the recorded sea level rise was 0.12 cm/yr (Figure 3). The obtained trend is not statistically significant ($p > 0.05$). A reason for this is the large interannual variability. However, it was determined that the sea level has been rising at a similar rate (0.16 ± 0.02 cm/yr) since the beginning of the 20th century.

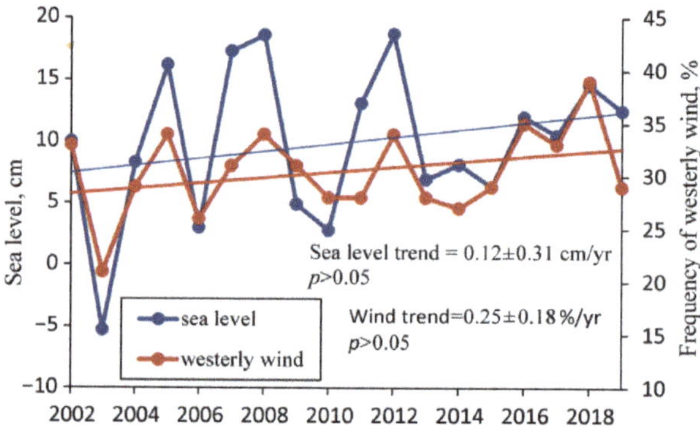

Figure 3. Sea level changes between 2002 and 2019 and their linear trends.

The marked sea level fluctuations in different years could have been the result of meteorological factors, in particular, cyclonic activity. This is proven by the frequency of westerly winds in 2002–2019, as shown in Figure 3. The relation between these characteristics (r = 0.77) is statistically significant ($p < 0.05$).

During the research period (2002–2019), no extreme storms occurred. The strongest storm, Ervin, was recorded on 9 January 2005, when gusts of W-SW winds reached 28 m/s and the storm surge rose up to 153 cm [50]; on 5 December 2013 the storm surge resulting from southwesterly winds (up to 21 m/s) generated by storm Xaver reached up to 120 cm, and on 11 January, 2015 (storm Felix) SW winds of 28 m/s elevated storm surge by up to 137 cm. However, due to a rapid beach recovery, the erosion had no determinant influence on the long-term trend of the shoreline displacement and beach and foredune sand volume variations.

Despite sea level rise, a high alongshore variation of shoreline displacement was determined (Figure 3). In the northern part, Klaipėda port jetties caused a seaward shoreline displacement up to 1.5 m per year, whereas in the central and southern parts the shoreline displacement rate ranges from −1.0 m/yr to 2.0 m/yr (Figure 4). It was noticed that shoreline fluctuations vary over a wide range in individual years. Annual changes in individual cases may reach 20 m or more. This is also reflected in the *p*-value (red line in Figure 4). Of 29 cases, nine trends are reliable ($p < 0.05$). This indicates that changes in shoreline position did not have a clear trend.

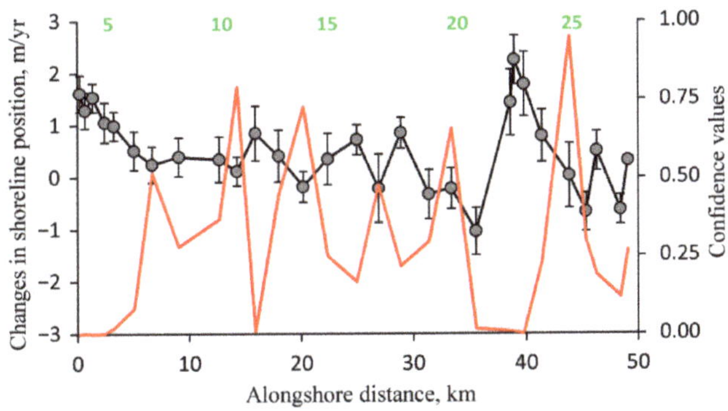

Figure 4. Linear trends (m/yr) in the shoreline position between 2002 and 2019. Red line—significance level (*p*-value). Green labels indicate cross-shore profile numbers.

Similarly, high variations of beach volume were determined in different coastal sectors (Figure 5A). In the northern part, the amount of beach sand has tended to increase (on average by 1–5 m^3/m per year), whereas in the southern part it has been decreasing (Figure 5A). It should be pointed out that these changes in beach sand volume in different coastal sectors are asynchronous. For example, until 2007, sand accretion had been the dominant trend in the southern part of the investigated sector. In 2008, this trend changed to erosion. Meanwhile, in the northern part of the Curonian Spit, this pattern was not observed. In general, permanent sand accretion on the beach in 2002–2019 only took place in the northern part of the spit (the Smiltynė–Alksnynė coastal sector), predetermined by a blockage of the alongshore sediment transport by the Klaipėda port jetties [35,49]. In the southern part (Pervalka–Nida), the variations in beach sand volume followed no definite pattern.

During the investigation period, the volume of foredune sand increased along the entire spit coast except for short stretches near Pervalka and Preila (Figure 5B). The foredune sand volume increased from 0.3 m^3/m per year in the southern part to 3.9 m^3/m/yr in the northern part. It should be noted that in the long term, no considerable variations in foredune sand volume were observed. Similar to the beach sand volume budget, the foredune sand volume variations in different coastal sectors were also asynchronous. Until 2007, sand volume variations in the northern part had been negligible, whereas after 2007 the accretion rates have been increasing. In the southern part, this pattern has not been observed.

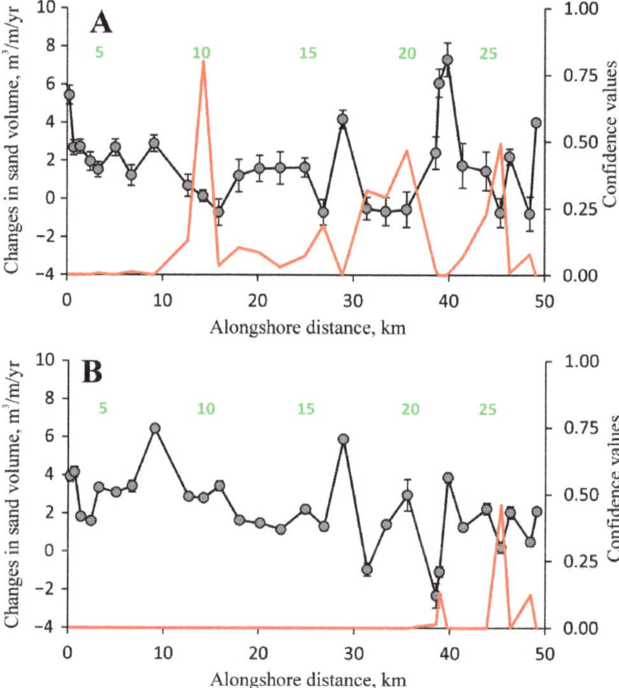

Figure 5. Linear trends (m³/m per year) in the (**A**) beach and (**B**) foredune volume alongshore. Distance from Klaipėda port jetties. Red line—significance level (*p*-value). Green labels indicate cross-shore profile numbers.

It is worth noting that the foredune in the Lithuanian part of the Curonian Spit has grown since the end of the 19th century [52]. Between 1859 and 1910 (from the beginning of foredune formation) its height increased from 2.3 m (Juodkrantė) to 8.9 m (Smiltynė) (Figure 6). During the investigation period (2002–2019) the rate of foredune growth slowed but remained positive. Averaged foredune height increases at a rate of 1.4 cm/yr. It was observed that despite the high variability of beach sand volume, almost all (26 of 27 cases (Figure 5B)) trends are statistically significant ($p < 0.05$). The reason for the slowing down of foredune growth is the high elevation reached in the middle of the 20th century. Due to high elevation and dense coverage with marram grass on the western slope, sand particles hardly reach the top of the foredune. Therefore, the most accretion takes place at the foredune toe and western slope. Often, especially in the northern part, incipient dunes are formed. In recreational zones, due to high pedestrian pressure, foredunes are reinforced with tree branches, which is the second reason for accretion at the foredune toe and western slope. The height of the foredune has increased more in the southern part where the foredune is lower, while in the northern part where the foredune height reaches up to 16 m, the height of the foredune has remained practically unchanged for the last 20 years. Thus, the foredune grew rapidly upwards in the early 20th century, and currently, with similar accumulation rates, it has been widening. It should also be noted that between 2002 and 2019 there was not an appreciable shoreline and foredune migration.

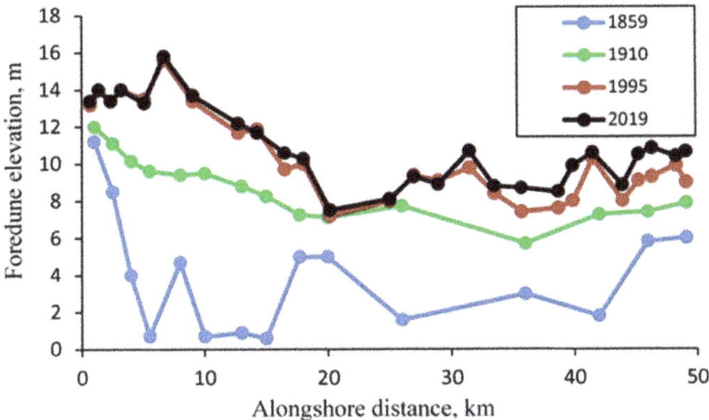

Figure 6. Changes in foredune elevation since 1859. Distance from Klaipėda port jetties.

5. Discussion

It can be seen in Figure 4 that over the last 17 years both the beach and foredune accretion processes have prevailed despite a rise in sea level of up to 0.12 cm/yr. The alongshore sediment transport from the southern part of the Curonian Spit (Kaliningrad district) is the main cause of this sand accretion. Due to the sand transport being blocked by the Klaipėda port jetties, the most intense accretion takes place in the northern part of the spit [35,49]. The observed rate of sea level rise does not stop the dominant trend of sand accumulation and coastal accretion [43]. Even during storms, the sand washed away from the coast is returned during the recovery phase and is not reflected in the long-term sand budget [41,53].

The influence of the sea level rise taking place during decadal time span on coastal sand volume changes is inconspicuous. Due to erosion at the Curonian Spit base (Kaliningrad district, Russia) where coastal regression occurred already in the 19th century [52] large amounts of sand transported northward helped to maintain a stable shoreline position in the central part of the spit (between Nida and Juodkrantė), while in the northern part (Alksnynė–Smiltynė) shoreline displacement seaward took place [52]. The same process takes place in the early 21st century. A comparison has shown that the shoreline displacement coincides with sand volume variations on the beach (Figure 7) (the correlation is significant at $p < 0.05$). Short erosion periods do not reverse the general accretion trend over a long-term period.

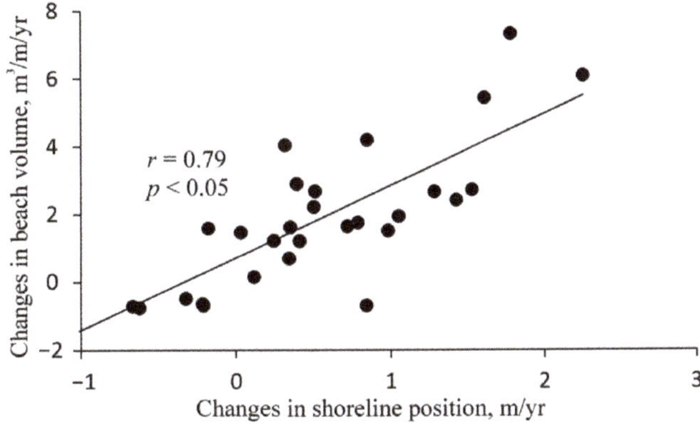

Figure 7. Correlation between mean changes in the shoreline position and beach sand volume.

An analysis of the interdependence of changes in shoreline position and foredune sand volume variations has shown no significant correlation ($p > 0.05$) (Figure 8). Even during years with predominated storm activity, the foredune sand budget has remained positive. The dominant sand accretion in the foredune takes place despite sea level oscillations or the beach sand budget.

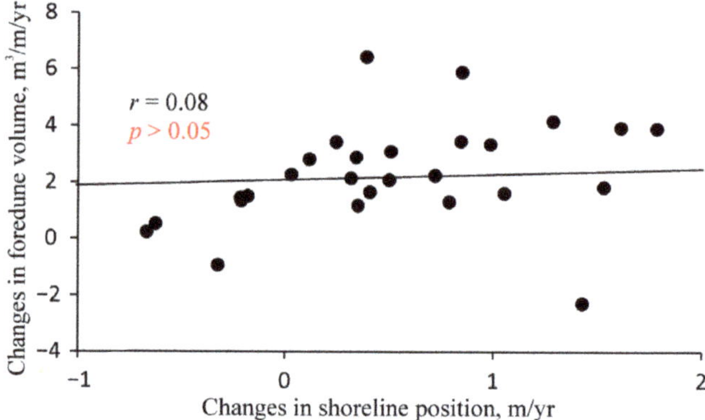

Figure 8. Correlation between changes in the shoreline position and foredune sand volume.

Despite a certain link between the beach and foredune sand budgets [39–41,54,55], as shown in Figure 9, no significant correlations between the changes in the beach and foredune sand budgets have been determined. Yet, the weak correlation between the beach and foredune sand budgets does not imply an absence of interdependence. The interdependence is presumably complicated by a few circumstances. One is the morphological properties of the beach [56]. During storms, narrow beaches can be replenished by foredune sand, in this way retaining their shape, whereas wider beaches can protect the foredunes at the expense of their sand budget. In the first case, when the foredune sand budget is negative, the beach sand volume will change insignificantly. In the second case, the beach sand budget will be negative, whereas in the foredune it will remain unchanged or even become positive [57]. It is also important that the beach and foredune budget are not synchronized in time [58]. Sand accretion on a beach takes place during calm meteorological conditions when a sandbar welds to the beach. Meanwhile, foredune accretion occurs during a storm season [58,59]. This process may also be influenced by interrelations with adjacent coastal sectors due to alongshore sediment transport [60]. The obtained results showed that if there is a sufficient amount of sand, even a slow sea level rise has a positive effect on the foredune sand budget. This is related not only to the alongshore transport of huge amounts of sediments from the southern part of the spit [35,47–49], but also to small nearshore inclines, which create favorable conditions for sandbar welding to the beach during fair weather conditions [61]. In these circumstances, foredunes that are built up at distal ends of some spits can build upwards despite a sea level rise [62]. Accretion was dominant in foredunes, even in coastal sectors where a low trend for sand decrease was observed on the beach. Such a situation is observed at 16, 27, 35, and 48 km from Klaipėda port jetties (Figure 5). This has also been supported by investigations carried out on other coasts [29,39,56,60].

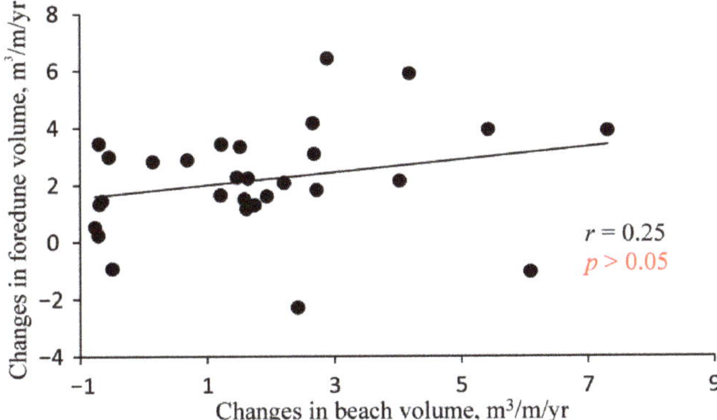

Figure 9. Correlation between changes in the beach and foredune sand volume.

It is worth noting that prevailing accretion in the Lithuanian part of the Curonian Spit has taken place since the end of the 19th century [52]. During the 20th century, there was not an appreciable shoreline or foredune migration. Due to alongshore sediment transport directed northward, rising sea level may increase coastal erosion in the southern part of the spit (Kaliningrad district) and the amount of sand transported, which may enhance sediment accumulation in the northern part of the spit) and herewith raise the transported sand amount, which may lead to the accumulation of sediments in the northern part of the spit.

It is assumed that coastal erosion processes start only when the sea level rise reaches a threshold value higher than the present rate on the Lithuanian coast of the Baltic Sea. According to Pye and Blott [63], the effects of these changes might not be significant for at least 30–50 years. However, it is worth noting that it is impossible to strictly predict threshold value determining erosional processes to be started because the aforementioned changes happen very slowly, so the coastal system can adapt to these alterations. It is also unclear what variations may happen to the sand supply, etc.

6. Conclusions

The present research showed that due to alongshore sand transport, the analyzed coastal sector is distinguished by a sand surplus. Therefore, a slow sea level rise (up to 0.12 cm/yr) has no negative effect on the foredune and only a minimal effect on beach development on a decadal time scale. No significant correlation was determined between the changes in foredune sand budget and sea level fluctuations. Also, no significant correlation between the changes in foredune and beach sand budgets was determined. Despite the rise in sea level and changes in beach sand volume, the foredune sand budget remained positive almost in the whole area of investigation. However, the absence of an interrelation between the beach and foredune budgets does not mean an absence of interdependence. The interrelation may have been complicated by a few circumstances: a time lag in different links (the beach in close contact with the sea tends to more readily adjust to changing sea levels, whereas in the foredune these processes lag behind). Moreover, the foredune sand washed away during more powerful storms may compensate for beach sand losses, thus preserving the unchanged beach sand budget.

Author Contributions: Conceptualization, D.J.; methodology, D.J. and D.P.; formal analysis, D.J., D.P., and G.Ž.; investigation, D.J., D.P., and G.Ž.; data curation, D.P. and R.J.; writing—original draft preparation, D.J.; writing—review and editing, D.P. and G.Ž.; visualization, R.J. and V.K. All authors have read and agree to the published version of the manuscript.

Funding: This research received no external funding.

Acknowledgments: The authors wish to thank Laurynas Jukna (Vilnius University, Lithuania) for technical support.

Conflicts of Interest: The authors declare no conflicts of interest.

References

1. Church, J.A.; White, N.J. Sea-level rise from late 19th to the early 21st century. *Surv. Geophys.* **2011**, *32*, 585–602. [CrossRef]
2. Beckley, B.D.; Lemoine, F.G.; Luthcke, S.B.; Ray, R.D.; Zelensky, N.P. A reassessment of global and regional mean sea level trends from TOPEX and Jason-1 altimetry based on revised reference frame and orbits. *Geophys. Res. Lett.* **2007**, *34*, L14608. [CrossRef]
3. Stramska, M.; Chudziak, N. Recent multiyear trends in the Baltic Sea level. *Oceanologia* **2013**, *55*, 319–337. [CrossRef]
4. Nichols, M.M. Sediment accumulation rates and relative sea-level rise in lagoons. *Mar. Geol.* **1989**, *88*, 201–219. [CrossRef]
5. Torresan, S.; Crito, A.; Valle, M.D.; Harvey, N.; Marcomini, A. Assessing coastal vulnerability to climate change comparing segmentation at global and regional scales. *Sustain. Sci.* **2008**, *3*, 45–65. [CrossRef]
6. Riggs, S.R.; Cleary, W.J.; Snyder, S.W. Influence of inherited geologic framework on barrier shore face morphology and dynamics. *Mar. Geol.* **1995**, *126*, 213–234. [CrossRef]
7. Honeycutt, M.R.; Krantz, D. Influence of the geologic framework on spatial variability in long-term shoreline change, Cape Henlopen to Rehoboth Beach, Delaware. *J. Coast. Res.* **2003**, *SI 38*, 147–167.
8. Thom, B.G. Transgresive and regressive stratigraphies of coastal sand barriers in Southern Australia. *Mar. Geol.* **1984**, *56*, 137–158. [CrossRef]
9. Carter, R.W.G.; Johnston, T.W.; McKenna, J.; Orford, J.D. Sea level, sediment supply and coastal changes: Examples from the coast of Ireland. *Prog. Oceanogr.* **1987**, *18*, 79–101. [CrossRef]
10. Jackson, D.W.T.; Cooper, A.G. Beach fetch distance and Aeolian sediment transport. *Sedimentology* **1999**, *46*, 517–522. [CrossRef]
11. Anthony, E.J.; Vanhee, S.; Ruz, M.-H. Short-term beach-dune sand budgets on the north sea coast of France: Sand supply from shoreface to dunes, and the role of wind and fetch. *Geomorphology* **2006**, *81*, 316–329. [CrossRef]
12. Healy, T. Sea level rise and impact on nearshore sedimentation: An overview. *Geol. Rundsch.* **1996**, *85*, 546–553. [CrossRef]
13. Storms, J.E.A.; Weltje, G.J.; van Dijke, J.J.; Geel, C.R.; Kroonenberg, S.B. Process-response modelling of wave-dominated coastal systems: Simulating evolution and stratigraphy on geological time scales. *J. Sediment. Res.* **2002**, *72*, 226–239. [CrossRef]
14. Karnauskaitė, D.; Schernewski, G.; Schumacher, J.; Grunert, R.; Povilanskas, R. Assessing coastal management case studies around Europe using an indicator based tool. *J. Coast. Conserv.* **2018**, *22*, 549–570. [CrossRef]
15. Claudino-Sales, V.; Wang, P.; Horwitz, M.H. Factors controlling the survival of coastal dunes during multiple hurricane impacts in 2004 and 2005: Santa Rosa barrier island, Florida. *Geomorphology* **2008**, *95*, 295–315. [CrossRef]
16. Houser, C.; Hapke, C.; Hamilton, S. Controls on coastal dune morphology, shoreline erosion and barrier island response to extreme storms. *Geomorphology* **2008**, *100*, 223–240. [CrossRef]
17. Pye, K.; Blott, S.J. Decadal-scale variation in dune erosion and accretion rates: An investigation of the significance of changing storm tide frequency and magnitude on the Sefton coast, UK. *Geomorphology* **2008**, *102*, 652–666. [CrossRef]
18. Suanez, S.; Cariolet, J.-M.; Cancouët, R.; Ardhuin, F.; Delacourt, C. Dune recovery after storm erosion on a high-energy beach: Vougot Beach, Brittany (France). *Geomorphology* **2012**, *139–140*, 16–33. [CrossRef]
19. Roelvink, D.; Reniers, A.; Van Dongeren, A.; Van Thiel de Vries, J.; McCall, R.; Lescinski, J. Modelling storm inpacts on beach, dunes and barrier islands. *Coast. Eng.* **2009**, *56*, 1133–1152. [CrossRef]
20. Hesp, P. Morphology, dynamics and internal stratification of some established foredunes in southeast Australia. *Sediment. Geol.* **1988**, *55*, 17–41. [CrossRef]
21. Lancaster, N.; Baas, A. Influence of vegetation cover on sand transport by wind: Field studies at Owens Lake, California. *Earth Surf. Proc. Land.* **1998**, *23*, 69–82. [CrossRef]

22. Nield, J.M.; Baas, A.C.W. The influence of different environmental and climatic conditions on vegetated aeolian dune landscape development and response. *Glob. Planet. Chang.* **2008**, *64*, 76–92. [CrossRef]

23. Zarnetske, P.L.; Ruggiero, P.; Seabloom, E.W.; Hacker, S.D. Coastal foredune evolution: The relative influence of vegetation and sand supply in the US Pacific Northwest. *J. Roy. Soc. Interface* **2015**, *12*, 20150017. [CrossRef] [PubMed]

24. Corbau, C.; Simeoni, U.; Melchiorre, M.; Rodella, I.; Utizi, K. Regional variability of coastal dunes observed along the Emilia-Romagna littoral, Italy. *Aeolian Res.* **2015**, *18*, 169–183. [CrossRef]

25. Nordstrom, K.F. Beaches and dunes of human-alerted coasts. *Prog. Phys. Geog.* **1994**, *18*, 497–516. [CrossRef]

26. Nordstrom, K.F. Aeolian sediment transport on a human-alerted foredune. *Earth Surf. Proc. Land.* **2007**, *32*, 102–115. [CrossRef]

27. De Vincenzo, A.; Covelli, C.; Molino, A.J.; Pannone, M.; Ciccaglione, M.; Molino, B. Long-term management policies of reservoirs: Possible re-use of dredged sediments for coastal nourishment. *Water* **2019**, *11*, 15. [CrossRef]

28. Nicholls, R.J.; Leatherman, S.P.; Dennis, K.C.; Volonte, C.R. Impact and responses to sea-level rise: Qualitative and quantitative assessments. *J. Coast. Res.* **1995**, *SI 14*, 26–43.

29. Battiau-Queney, Y.; Billet, J.F.; Chaverot, S.; Lanoy-Ratel, P. Recent shoreline mobility and geomorphologic evolution of macrotidal sandy beaches in the north of France. *Mar. Geol.* **2003**, *194*, 31–45. [CrossRef]

30. FitzGerald, D.M.; Fenster, M.S.; Argow, B.A.; Buynevich, V. Coastal impacts due to sea-level rise. *Annu. Rev. Earth Planet. Sci.* **2008**, *36*, 601–647. [CrossRef]

31. Thomas, T.; Lynch, S.K.; Phillips, M.R.; Williams, A.T. Long-term evolution of sand spit, physical forcing and links to coastal flooding. *Appl. Geogr.* **2014**, *53*, 187–201. [CrossRef]

32. Miot da Silva, G.; Hesp, P. Coastline orientation, aeolian sediment transport and foredune dunefield dynamics of Moçambique Beach, Southern Brasil. *Geomorphology* **2010**, *120*, 258–278. [CrossRef]

33. Chubarenko, B.; Babakov, A. Sediment transport near the Vistula Spit (Baltic Sea). In *Managing Risk to Coastal Regions and Communities in a Changing World, Proceedings of the International Conference EMECS'11-Sea Coast XXVI, Saint-Petersburg, Russia, 22–27 August 2016*; RSHU: Saint-Petersburg, Russia, 2016; pp. 174–185.

34. Badyukova, E.N.; Zhindarev, L.A.; Lukyanova, S.A.; Solovieva, G.D. Geology and evolution history of the Curonian Spit (SE Baltic Sea). *Oceanology* **2007**, *47*, 554–563. [CrossRef]

35. Pupienis, D.; Buynevich, I.; Ryabchuk, D.; Jarmalavičius, D.; Žilinskas, G.; Fedorovič, J.; Kovaleva, O.; Sergeev, A.; Cichon-Pupienis, A. Spatial patterns in heavy-mineral concentrations along the Curonian Spit coast, southeastern Baltic Sea. *Estuar. Coast. Shelf S.* **2017**, *195*, 41–50. [CrossRef]

36. Kharin, G.S.; Zhukovskaya, I.P. Types of sediments and sections in the upper quarternary cover and geological stability of the Curonian Spit (Baltic Sea). *Lithol. Miner. Resour.* **2013**, *48*, 198–2015. [CrossRef]

37. Badyukova, E.N.; Solovieva, G.D. Coastal eolian landforms and sea level fluctuations. *Oceanology* **2015**, *55*, 124–130. [CrossRef]

38. Short, A.; Hesp, P. Wave, beach and dune interactions in Southeastern Australia. *Mar. Geol.* **1982**, *48*, 259–284. [CrossRef]

39. Sherman, D.J.; Bauer, B.O. Dynamics of beach-dune systems. *Prog. Phys. Geog.* **1993**, *17*, 413–447. [CrossRef]

40. Houser, C.; Ellis, J. Beach and dune interaction. In *Treatise on Geomorphology*; Shroder, J.F., Ed.; Academic Press: San Diego, CA, USA, 2013; Volume 10, pp. 267–288.

41. Cooper, J.A.G.; Jackson, D.W.T. Geomorphological and dynamic constraints on mesoscale coastal response to storms, Western Ireland. In *Coastal Sediments' 03, Proceedings of the 6th International Symposium on Coastal Engineering and Science of Coastal Sediment Processes, Sheraton Sand Key Resort, Clearwater Beach, FL, USA, 18–23 May 2003*; Davis, A., Howd, P.A., Kraus, N.C., Eds.; World Scientific Pub Co. Inc.: Hackensack, NJ, USA, 2003; pp. 1–13.

42. Jarmalavičius, D.; Satkūnas, J.; Žilinskas, G.; Pupienis, D. The influence of coastal morphology on wind dynamics. *Est. J. Earth Sci.* **2015**, *61*, 120–130. [CrossRef]

43. Jarmalavičius, D.; Žilinskas, G.; Pupienis, D.; Kriaučiūnienė, J. Subaerial beach volume change on a decadal time scale: The Lithuanian Baltic Sea coast. *Z. Geomorphol.* **2017**, *61*, 149–158. [CrossRef]

44. Hupfer, P. *Die Ostsee–Kleines Meer mit Grossen Problemen*; Teubner Verlagsgesellschaft: Leipzig, Germany, 1979.

45. Medvedev, I.P.; Rabinovich, A.B.; Kulikov, E.A. Tidal oscillations in the Baltic Sea. *Oceanology* **2013**, *53*, 596–609. [CrossRef]

46. Jakimavičius, D.; Kriaučiūnienė, J.; Šarauskienė, D. Assessment of wave climate and energy resources in the Baltic Sea nearshore (Lithuanian territorial water). *Oceanologia* **2018**, *60*, 207–218. [CrossRef]

47. Ostrowski, R.; Pruszak, Z.; Babakov, A. Cond it ion of south-easte rn Bal tic Sea shores and meth ods of pro tect ing them. *Arch. Hydro Eng. Environ. Mech.* **2014**, *61*, 17–37. [CrossRef]

48. Krek, A.; Stont, Z.; Ulyanova, M. Along shore bed load trans port in the south east ern part of the Bal tic Sea un der chang ing hydrom eteorol ogic al cond it ions. *Reg. Stud. Mar. Sci.* **2016**, *7*, 81–87. [CrossRef]

49. Žilinskas, G.; Jarmalavičius, D.; Pupienis, D. The influence of natural and anthropogenic factors on grain size distribution along the southeaster Baltic spits. *Geol. Q.* **2018**, *62*, 375–384.

50. Jarmalavičius, D.; Šmatas, V.; Stankūnavičius, G.; Pupienis, D.; Žilinskas, G. Factors controlling coastal erosion during storm events. *J. Coast. Res.* **2016**, *SI75*, 1112–1116. [CrossRef]

51. Kriaučiūnienė, J.; Gailiušis, B.; Kovalenkovienė, M. Peculiarities of sea wave propagation in the Klaipėda Strait, Lithuanica. *Baltica* **2006**, *19*, 20–29.

52. Musset, M. Untersuchungen über die erfolge der dünenarbeiten auf der Kurische Nehrung. *Z. Bauwes.* **1916**, *66*, 253–260.

53. Vespremeanu-Stroe, A.; Preoteasa, L. Beach-dune interactions on the dry-temperate Danube delta coast. *Geomorphology* **2007**, *86*, 267–286. [CrossRef]

54. Psuty, N.P. Sediment budget and dune/beach interaction. *J. Coast. Res.* **1988**, *SI 3*, 1–4.

55. Psuty, N.P. The coastal foredune: A morphological basis for regional coastal dune development. In *Coastal Dunes. Ecology and Conservation*; Martinez, M.L., Psuty, N.P., Eds.; Springer: Berlin/Heidelberg, Germany, 2007; pp. 11–27.

56. Sabatier, F.; Anthony, E.J.; Héquette, A.; Suanez, S.; Musereau, J.; Ruz, M.; Regnaud, H. Morphodynamics of beach/dune systems: Examples from the coast of France. *Géomorphologie* **2009**, *15*, 3–22. [CrossRef]

57. Jarmalavičius, D.; Satkūnas, J.; Žilinskas, G.; Pupienis, D. Dynamics of beaches of the Lithuanian coast (the Baltic Sea) for period 1993-2008 based on morphometric indicators. *Environ. Earth Sci.* **2012**, *65*, 1727–1736. [CrossRef]

58. Cohn, N.; Ruggiero, P.; de Vries, S.; Kaminsky, G.M. New insights on coastal foredune growth: The relative contributions of marine and aeolian processes. *Geophys. Res. Lett.* **2018**, *45*, 4965–4973. [CrossRef]

59. Houser, C. Synchronization of transport and supply in beach-dune interaction. *Prog. Phys. Geog.* **2009**, *33*, 733–746. [CrossRef]

60. Saye, S.E.; van der Wal, D.; Pye, K.; Blott, S.J. Beach-dune morphological relationships and erosion/accretion: An investigation at five sites in England and Wales using LIDAR data. *Geomorphology* **2005**, *72*, 128–155. [CrossRef]

61. Aagaard, T.; Sørensen, P. Coastal profile response to sea level rise: A process-based approach. *Earth Surf. Proc. Land.* **2012**, *37*, 354–362. [CrossRef]

62. Hesp, P.A.; Walker, I.J.; Chapman, C.; Davidson-Arnott, R.; Bauer, B.O. Aeolian dynamics over a coastal foredune, Prince Edward Island, Canada. *Earth Surf. Proc. Land.* **2013**, *38*, 1566–1575. [CrossRef]

63. Pye, K.; Blott, S.J. Coastal processes and morphological change in the Dunwich-Sizewell area, Suffolk, UK. *J. Coast. Res.* **2006**, *22*, 453–473. [CrossRef]

MDPI

St. Alban-Anlage 66

4052 Basel

Switzerland

Tel. +41 61 683 77 34

Fax +41 61 302 89 18

www.mdpi.com

Water Editorial Office

E-mail: water@mdpi.com

www.mdpi.com/journal/water